Lecture Notes in Control and Information Sciences 407

Editors: M. Thoma, F. Allgöwer, M. Morari

AF902499

Jean Lévine and Philippe Müllhaupt (Eds.)

Advances in the Theory of Control, Signals and Systems with Physical Modeling

Springer

Series Advisory Board

P. Fleming, P. Kokotovic,
A.B. Kurzhanski, H. Kwakernaak,
A. Rantzer, J.N. Tsitsiklis

Editors

Jean Lévine
CAS, Unité Mathématiques et Systèmes
MINES-ParisTech
35, rue Saint-Honoré
77300 Fontainebleau
France
E-mail: jean.levine@mines-paristech.fr

Philippe Müllhaupt
Laboratoire d'Automatique
Faculté des Sciences
de l'Ingénieur
Station 9
CH-1015 Lausanne
Switzerland
E-mail: philippe.muellhaupt@epfl.ch

ISBN 978-3-642-16134-6 e-ISBN 978-3-642-16135-3

DOI 10.1007/978-3-642-16135-3

Lecture Notes in Control and Information Sciences ISSN 0170-8643

Library of Congress Control Number: 2010936509

© 2010 Springer-Verlag Berlin Heidelberg

This work is subject to copyright. All rights are reserved, whether the whole or part of the material is concerned, specifically the rights of translation, reprinting, reuse of illustrations, recitation, broadcasting, reproduction on microfilm or in any other way, and storage in data banks. Duplication of this publication or parts thereof is permitted only under the provisions of the German Copyright Law of September 9, 1965, in its current version, and permission for use must always be obtained from Springer. Violations are liable for prosecution under the German Copyright Law.

The use of general descriptive names, registered names, trademarks, etc. in this publication does not imply, even in the absence of a specific statement, that such names are exempt from the relevant protective laws and regulations and therefore free for general use.

Typeset & Cover Design: Scientific Publishing Services Pvt. Ltd., Chennai, India.

Printed on acid-free paper

5 4 3 2 1 0

springer.com

Advances in the Theory of Control, Signals, and Systems, with Physical Modeling

Jean Lévine[1] and Philippe Müllhaupt[2]

1 Introduction

This book gathers articles that have been invited for presentation in the framework of a *Bernoulli Programme*, held at the Bernoulli Center in Lausanne (Switzerland) from January to June 2009.

This Programme mainly consisted of three workshops aiming at reviewing the advances in the theory of control, signals, and systems, with a particular emphasis on their relationship to physical modeling.

More precisely, the aim of this series of three workshops was to

- bring together knowledge and know-how from the communities of control, signals and systems,
- focus on the theoretical advances in these areas and examine the possibilities of new convergences between them,
- contribute to the enhancement of the dialogue between theoretical laboratories and more practically oriented units and industries.

In the 60's, control, signals and systems had a common linear algebraic background and, according to their evolution, their respective backgrounds have now dramatically differed. Recovering such a common background, especially in the nonlinear context, is currently a fully open question.

In most contributions, emphasis has been put on physical modeling, which serves as an Ariadne's thread between the diverse fields of interest. This idea is not new, however. As an example, mechanical system modeling, which heavily relies on analytical mechanics and in particular its conservation laws, has greatly inspired control theory. As another example, control of chemical processes also gained in the use of

[1] Centre Automatique et Systèmes, Unité Mathématiques et Systèmes, Mines-ParisTech, E-mail: jean.levine@mines-paristech.fr

[2] Laboratoire d'Automatique, Ecole Polytechnique Fédérale de Lausanne E-mail: philippe.muellhaupt@epfl.ch

sophisticated modeling software tools based on theories of mass balance conservation and entropy laws. Hence one purpose of this program was to force the interaction of probably uncorrelated disciplines thanks to these theoretical modeling aspects.

Another important aspect of the conferences was to present and develop new applications of the above approaches, and contribute to the enhancement of the dialogue between theoretical laboratories and more practically oriented research units and industries, in both classical areas and emerging fields of research.

The first workshop, entitled *Electrical and Mechatronical Systems Workshop* looked at various applications stemming from Mechatronics, Electrical and Mechanical Engineering, such as MEMS, eletrical machines, robots and car suspension. From the modeling and methodological side, finite dimensional systems (described by ordinary differential equations or difference equations) and infinite dimensional systems (delayed systems, distributed systems, PDEs', non-integer derivations) were approached for control and signal processing, as well as model-free techniques Indeed, the influence of physical modeling contributed to outline some convergences. In particular, a unifying Lagrangian formalism has been sketched so as to integrate electrical, electronical, magnetic and mechanical aspects of systems, potentially leading to significant simplifications in the analysis of control systems. Both finite dimensional and infinite dimensional models are shown to ease some estimation, adaptative control and observation problems. New applications in emerging fields of mechatronic systems, such as MEMS, or new suspension technologies, have been presented, showing that Mechanics, Mechatronics and Electronics remain a major source of inspiration for control and system theorists.

The aim of the second workshop, entitled *Mathematical Tools Workshop*, was to serve as a think tank for mathematical paradigms in the fields of Control, Signals and Systems. Again, both finite dimensional and infinite dimensional models have been explored. Various approaches, in the framework of differential geometry and algebra have been examined. Group theory and Riemannian Geometry appeared in many presentations with, in particular, robotics, mechanical systems or quantum control as background applications. Recent advances, in the fields of hamiltonian, lagrangian, quantum, energy-based and flat or non flat control systems have also been presented.

Finally, the third and last workshop, entitled *Chemical and Life Science Workshop*, concerned new approaches in the analysis of biomedical, biomechanical and reaction systems, possibly coupled with fluid dynamics, with many challenging applications such as cancer treatment and diagnosis. Important results concerning unifying approaches to deal with complex chemical and biochemical reactions have been presented taking into account the network structure of the reactions while ensuring robustness with respect to various unknown parameters and perturbations. The influence of noisy data in the biological and chemical reaction systems has also been approached. Time-scales, transients and bifurcations in ecological systems, population dynamics and biological systems have also received a great attention and their control theoretical perspectives have been envisaged.

The reader will find in the present volume key contributions and surveys, giving a precise account of the above topics. The book is organized in three parts, according to the three aforementioned workshops. In each part, the articles follow the alphabetic order of the first author. This order has been prefered to a more sophisticated, but often artificial, clustering by sub-themes. We hope that these readings will be most inspiring and informative to PhD students and researchers in Mathematics, Electrical, Mechanical, Chemical or Bio Engineering, and more generally to every people of both the academic and industrial spheres curious of the recent developments in control, signals and systems.

We are very grateful both to the Swiss National Science Foundation for funding such an endeavor and to the Centre Bernoulli for providing the required infrastructure. In particular, we thank Mrs. Christiane De Paola, Talya Van Woerden, Sabrina Martone, and Rana Gherzeddine for the important administrative and organisational work and Mr. Marc Perraudin for maintaining the internet server. Last, but not least, we are deeply indebted to Prof. Tudor Ratiu for his constant encouragements to organize the above program.

Jean Lévine and Philippe Müllhaupt
Mines-ParisTech and EPFL

Contents

Part I: Electrical and Mechatronical Systems

Modeling and Control of Multi-Body Mechanical Systems: Part I A Riemannian Geometry Approach 3
Suguru Arimoto

Modeling and Control of Multi-Body Mechanical Systems: Part II Grasping under Rolling Contacts between Arbitrary Shapes 17
Suguru Arimoto

Sliding Mode Control for a High-Speed Linear Axis Driven by Pneumatic Muscles 31
Harald Aschemann, Dominik Schindele

Using Hamiltonians to Model Saturation in Space Vector Representations of AC Electrical Machines 41
Duro Basic, Al Kassem Jebai, François Malrait, Philippe Martin, Pierre Rouchon

Iterative Learning Control Using Stochastic Approximation Theory with Application to a Mechatronic System 49
Mark Butcher, Alireza Karimi

Elimination Theory for Nonlinear Parameter Estimation 65
John Chiasson, Ahmed Oteafy

Controlling Underactuated Mechanical Systems: A Review and Open Problems 77
Zhong-Ping Jiang

Time Scaling in Motion Planning and Control of Tree-Like Pendulum Structures 89
Matthias Krause, Joachim Rudolph, Frank Woittennek

Mechanical Version of the CRONE Suspension 99
Alain Oustaloup, Xavier Moreau

Electrostatic MEMS: Modelling, Control, and Applications 113
Guchuan Zhu

Part II: Mathematical Tools

Flatness Characterization: Two Approaches 127
Felix Antritter, Jean Lévine

Nonholonomic Mechanics, Dissipation and Quantization 141
Anthony M. Bloch

Controlled Lagrangians 153
Dong Eui Chang

Compensation of Input Delay for Linear, Nonlinear, Adaptive, and PDE Systems 161
Miroslav Krstic

Boundary Value Problems and Convolutional Systems over Rings of Ultradistributions 179
Hugues Mounier, Joachim Rudolph, Frank Woittennek

Wei-Norman Technique for Control Design of Bilinear ODE Systems with Application to Quantum Control 189
Markku Nihtilä

Interval Methods for Verification and Implementation of Robust Controllers 201
Andreas Rauh, Harald Aschemann

Rational Interpolation of Rigid-Body Motions 213
J.M. Selig

Contact Geometry and Its Application to Control 225
Peter J. Vassiliou

Part III: Chemical Processes and Life Sciences

Piecewise Affine Models of Regulatory Genetic Networks: Review and Probabilistic Interpretation 241
Madalena Chaves, Jean-Luc Gouzé

A Control Engineering Model for Resolving the TGF-β Paradox in Cancer 255
Seung-Wook Chung, Carlton R. Cooper, Mary C. Farach-Carson, Babatunde A. Ogunnaike

A Mathematical Model of Air-Flow Induced Regional Over-Distention during Mechanical Ventilation: Comparing Pressure-Controlled and Volume-Controlled Modes 269
P.S. Crooke, A.M. Kaynar, J.R. Hotchkiss

Positive Feedbacks Contribute to the Robustness of the Cell Cycle with Respect to Molecular Noise 283
Didier Gonze, Marc Hafner

Guaranteed and Randomized Methods for Stability Analysis of Uncertain Metabolic Networks 297
Heinz Koeppl, Stefano Andreozzi, Ralf Steuer

Coexistence of Three Predators Competing for a Single Biotic Resource 309
Claude Lobry, Tewfik Sari, Karim Yadi

Control Problems for One-Dimensional Fluids and Reactive Fluids with Moving Interfaces 323
Nicolas Petit

A Port-Hamiltonian Formulation of Open Chemical Reaction Networks 339
Arjan van der Schaft, Bernhard Maschke

Bifurcations of Dynamical Systems, Logistic and Gompertz Growth Laws in Processes of Aggregation 349
Alex Shoshitaishvili, Andrei Raibekas

Global Uncertainty Analysis for a Model of TNF-Induced NF-κB Signalling 365
Steffen Waldherr, Jan Hasenauer, Malgorzata Doszczak, Peter Scheurich, Frank Allgöwer

Author Index 379

Part I
Electrical and Mechatronical Systems

Modeling and Control of Multi-Body Mechanical Systems: Part I A Riemannian Geometry Approach

Suguru Arimoto

Abstract. Control problems of motion of multi-body mechanical systems under constraints and/or with redundancy in system's degrees-of-freedom (DOF) are treated from the standpoint of Riemannian geometry. A multi-joint reaching problem with excess DOF is tackled and it is shown that a task space PD feedback with damping shaping in joints maneuvers the end-point of the robot arm to reach a given target in the sense of exponentially asymptotic convergence. An artificial potential inducing the position feed-back in task space can be regarded as a Morse-Bott function introduced in Riemannian geometry, from which the Lagrange stability theorem can be directly extended to this redundant case. The speed of convergence of both the orbit of the endpoint in task space and the trajectory of joint vector in joint space can be adjusted by damping shaping and adequately choosing a single stiffness parameter. In the case that the endpoint is constrained on a hypersurface in $\boldsymbol{E}^3$, the original Lagrange dynamics expressed in an implicit form by introducing a Lagrange multiplier is decomposed into two partial dynamics with the aid of decomposition of the tangent space into the image of the endpoint Jacobian matrix and the kernel orthogonally complemented to the image. The stability problem of point-to-point endpoint movement on the constraint surface is reduced to the former case without constraint.

1 Introduction

Motion of a multi-joint robot arm that is a mechanical system of multi-bodies serially connected through rotational joints is characterized by the Lagrange

Suguru Arimoto
Research Organization of Science and Engineering, Ritsumeikan University
e-mail: arimoto@fc.ritsumei.ac.jp
and
RIKEN-TRI Collaboration Center, RIKEN

J. Lévine & P. Müllhaupt (Eds.): Adv. in Theory of Control, Signals, LNCIS 407, pp. 3–16.
springerlink.com © Springer-Verlag Berlin Heidelberg 2010

equation of motion that is nonlinear and has strong couplings between joints. Notwithstanding the complexity of its dynamics, it is shown [1] that position control is feasible by designing a PD (Position and its Derivative) feedback with damping shaping, provided that the gravity term is adequately compensated. This control methodology is extended to the cases that 1) the target position is given and described in task space [1] and 2) the robot endpoint is constrained on a surface in $\boldsymbol{E}^3$ (Eucleadian Space) [2]. However, in both the cases, it is implicitly assumed that system's degrees-of-freedom (DOF) is non-redundant.

On the other hand, Nicholai A. Bernstein pointed out more than a half century ago the importance and difficulty of the kinematic DOF problem in relation to human movement [3], since the inverse kinematics becomes illposed generally in case of systems with redundant DOFs. Although the book [3] presents Bernstein's major ideas on the development and control of voluntary movement in general and the notion of dexterity, even nowadays Bernstein's problem is controversial not only in developmental psychology, neurophysiology, and kinesiology [4, 5], but also in robotics, as discussed in the author's previous paper [6]. From the viewpoint of robot control for a class of redundant multi-joint arms, it is shown in the paper [7] that a multi-joint point-to-point reaching movement can be established by using a task-space PD feedback with damping shaping in joint space. Nevertheless, the proof of asymptotic convergence was rather sophisticated, lacking the mathematical rigor. Therefore, there still remains unsolved a lot of important problems for control of nonlinear mechanical systems that are both constrained geometrically and redundant in DOF. One practical example of such systems is related to a stablization control for grasping a rigid object under rolling contact constraints by using multiple robot fingers with multi-joints. Even in this case, only a rough and rather intuitive sketch for proving the asymptotic convergence of motion of the overall fingers/object system is presented so far [8, 9]. Another interesting example is found in control of a hand-writing robot [10].

This paper tackles such a difficult problem of control for a class of multi-body robotic systems that are constrained geometrically or/and subject to redundancy in system's DOF from the standpoint of Riemannian geometry. It is widely known among roboticsists that kinematics and planning of multi-joint robots are tretaed in the configuration space regarded as an n-dimensional numerical space $\boldsymbol{R}^n$ [11, 12]. On the other hand, Arnold [13] pointed out the importance of Riemannian geometry in the analysis of mechanical systems and shown that the dynamics of motion of a double pendulum can be described by an orbit on a two-dimensional torus T^2 that is regarded as $T^2 = S^1 \times S^1$, where S^1 denotes a unit circle. In line with this notion, an n-DOF robot arm can be treated on an n-dimensional Rimeannian manifold like an n-dimensional torus T^n, and the stability problems of PD feedback with damping shaping [1] were retreated in a Riemannian-geometric manner [14, 15]. More recently, the author and his group showed that, given a robot arm, the set of all possible postures can be

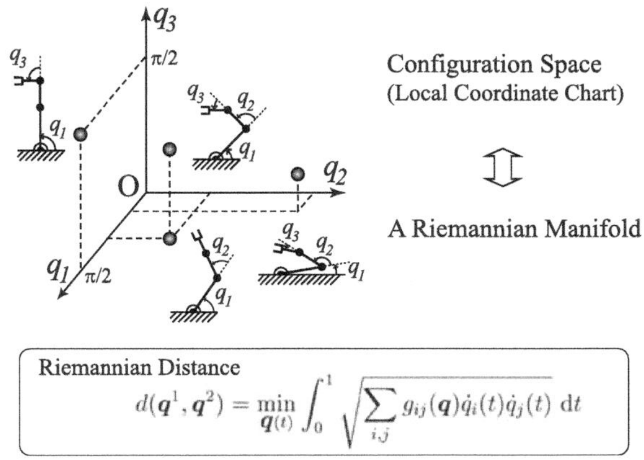

Fig. 1. A set of all possible postures of a planar robot with three joints regarded as a Riemannian manifold with Riemannian metric $g_{ij}(q)$ of the inertia matrix

regarded as a Riemannian manifold with the Riemannian metric that constitutes the inertia matrix [16] (see Fig. 1). Thus, an orbit of motion as a geodesic to the Euler equation can be regarded as an inertia-induced motion without affection of damping and gravity forces [17].

2 Euler-Lagrange Equations and Geodesics

It is well known that motion of a robot manipulator as a serially connected rigid-body system is governed by the Euler-Lagrange equation shown (see [18])

$$G(q)\ddot{q} + \left\{\frac{1}{2}\dot{G}(q) + S(q,\dot{q})\right\}\dot{q} + g(q) = u \tag{1}$$

where $q = (q_1, \cdots, q_n)^{\mathrm{T}}$ denotes the vector of joint angles, $G(q) = (g_{ij})$ does the $n \times n$ inertia matrix, u a control toque vector, $g(q) = \partial P(q)/\partial q$ with a scalar function $P(q)$ called the gravity potential, and $S(q,\dot{q})$ a skew-symmetric matrix $S = (S_{ij})$ defined as

$$S_{ij} = \frac{1}{2}\left\{\frac{\partial}{\partial q_j}\left(\sum_{k=1}^{n}\dot{q}_k g_{ik}\right) - \frac{\partial}{\partial q_i}\left(\sum_{k=1}^{n}\dot{q}_k g_{jk}\right)\right\} \tag{2}$$

If we consider a control torque that can exactly compensate the gravity term, that is, $u = g(q)$, then it reduces (1) to

$$G(q)\ddot{q} + \left\{\frac{1}{2}\dot{G}(q) + S(q,\dot{q})\right\}\dot{q} = 0 \tag{3}$$

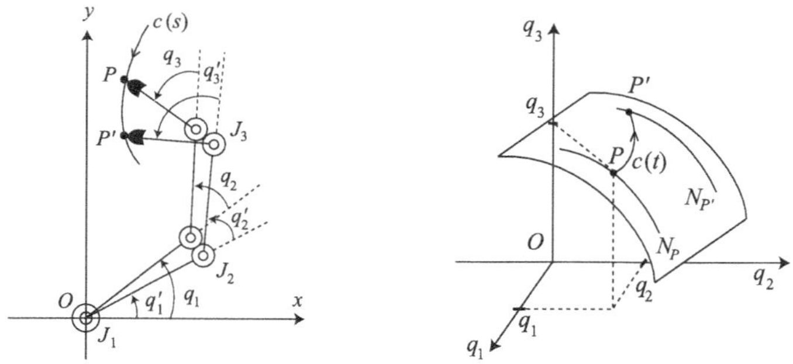

(a) A planar arm with three joints

(b) Submanifolds N_P and $N_{P'}$ correspond to positions P and P'

Fig. 2. A homeomorphic map of "postures" of an arm given in $\boldsymbol{E}^2$ to "points" in $\boldsymbol{R}^3$ (the configuration space)

which is considered an ideal equation of motion without affection of the gravity and joint damping like a robot arm on an artificial satellite in space. It is pointed out in [8] that (3) is equivalently written in the form

$$g_{ik}(q)\ddot{q}_i + \Gamma_{ikj}(q)\dot{q}_j\dot{q}_i = 0, \quad k = 1, \cdots, n \tag{4}$$

or in another equivalent form called the Euler equation

$$\ddot{q}_k + \Gamma^k_{ij}\dot{q}_i\dot{q}_j = 0, \quad k = 1, \cdots, n \tag{5}$$

where Γ_{ikj} and Γ^k_{ij} are defined as

$$\Gamma_{ikj} = \frac{1}{2}\left(\frac{\partial g_{jk}}{\partial q_i} + \frac{\partial g_{ik}}{\partial q_j} - \frac{\partial g_{ij}}{\partial q_k}\right) \tag{6}$$

$$\Gamma^k_{ij} = \frac{1}{2}\sum_{l=1}^{n} g^{lk}\left(\frac{\partial g_{jl}}{\partial q_i} + \frac{\partial g_{il}}{\partial q_j} - \frac{\partial g_{ij}}{\partial q_l}\right) = \frac{1}{2}\sum_{l=1}^{n} g^{lk}\Gamma_{ilj} \tag{7}$$

and (g^{lk}) denotes the inverse of G $(= (g_{ij}))$. The equivalence of (3) to (4) is shown in Appendix A.

Let us now consider reaching movements of a redundant planar robot arm shown in Fig. 2(a), when the target endpoint position is specified by point P or equivalently the position vector $\boldsymbol{x}_d$ $(= (x_d, y_d)^{\mathrm{T}})$ in $\boldsymbol{E}^2$. Since the endpoint position is a function of joint vector $q = (q_1, q_2, q_3)^{\mathrm{T}}$ of C^∞-class denoted by $\boldsymbol{x}(q)$, a set of possible postures of the robot, $N_P = \{q|\boldsymbol{x}(q) = P\}$, constitutes a Riemannian submanifold of 1-dimension, and similarly $N_{P'} = \{q|\boldsymbol{x}(q) = P' \ (= \boldsymbol{x}_d)\}$. If a starting posture $q(0)$ with $\boldsymbol{x}(q(0)) = P$ is given and fixed, there arises an infinite number of trajectories (or curves) in joint space that start from $q(0)$ on N_P and reach some point on $N_{P'}$ satisfying

$\boldsymbol{x}(q(T)) = P' \ (= \boldsymbol{x}_d)$. In this paper, we suppose an appropriate chart F in the base manifold $\{M, g_{ij}\}$ so that its homeomorphic map $\phi(F)$ is an open set $U \ (= \phi(F))$ in joint space and denote the single-dimensional submanifold $N_P \cap U$ and $N_{P'} \cap U$ (restrictions to $U = \phi(F)$) by N_P and $N_{P'}$ renewedly. In this local sense, it is reasonable to suppose that there exists an optimal orbit $c(t)$ that minimizes the Riemannian distance from P to $N_{P'}$ such that

$$\begin{aligned} d(P, N_{P'}) &= \inf_{q(t)} \int_0^{\mathrm{T}} \sqrt{\sum g_{ij}(q(t))\dot{q}_i(t)\dot{q}_j(t)}\ \mathrm{d}t \\ &= \int_0^{\mathrm{T}} \sqrt{\sum g_{ij}(c(t))\dot{c}^i(t)\dot{c}^j(t)}\ \mathrm{d}t \end{aligned} \tag{8}$$

where the infimum is taken over all of the orbits lying in F (or equivalently in $U \ (= \phi(F))$ in the configuration space) and connecting $q(0)$ on N_P and some point on $N_{P'}$. Since the chart F is local (compact) and connected, the optimal orbit $c(t) = (c^1(t), \cdots, c^n(t))$ must satisfy the Euler equation (see [19])

$$\ddot{c}^k + \Gamma^k_{ij}\dot{c}^i\dot{c}^j = 0, \qquad k = 1, \cdots, n \tag{9}$$

Once the geodesic curve $c(t)$ of C^∞-class is specified, the distance from P to P' can be calculated by any other parameter $s(= \alpha t + s_0)$ for $\alpha > 0$ and $s_1 = \alpha T + s_0$ as follows:

$$d(P, N_{P'}) = \int_{s_0}^{s_1} \sqrt{\sum g_{ij}\left(c(s)\right)\left(\frac{dc^i}{ds}\right)\left(\frac{dc^j}{ds}\right)} ds \tag{10}$$

3 Morse-Bott Function and an Extension of the Lagrange Stability

As for the robot dynamics of (1), consider a PD feedback in task space with damping shaping in joint space, that is expressed in the following form

$$u = -C\dot{q} - J_q^{\mathrm{T}}(\boldsymbol{x})k(\boldsymbol{x} - \boldsymbol{x}_d) \tag{11}$$

where k denotes a positive constant called "stiffness", C a positive definite constant matrix called "damping matrix" in joint space, and $J_q(\boldsymbol{x}) = \partial\boldsymbol{x}/\partial q^{\mathrm{T}}$ the 2×3 Jacobian matrix of $\boldsymbol{x}$ in q. Substitution of (11) into (1) yields

$$G(q)\ddot{q} + \left\{\frac{1}{2}\dot{G}(q) + S(q, \dot{q})\right\}\dot{q} + C\dot{q} + J_q^{\mathrm{T}}(q)k\Delta\boldsymbol{x} = 0 \tag{12}$$

where $\Delta\boldsymbol{x} = \boldsymbol{x} - \boldsymbol{x}_d$ and the term $g(q)$ is ignored in this case since motion of the robot is confined to the horizontal plane. Then, the inner product of (12) and $\dot{q}$ reduces to

$$\frac{\mathrm{d}}{\mathrm{d}t}\left\{\frac{1}{2}\dot{q}^{\mathrm{T}}G(q)\dot{q}+\frac{k}{2}\|\Delta\boldsymbol{x}\|^2\right\}+\dot{q}^{\mathrm{T}}C\dot{q}=0 \tag{13}$$

In what follows, we use the symbols

$$E(q,\dot{q})=\frac{1}{2}\dot{q}^{\mathrm{T}}G(q)\dot{q}+U(q), \quad U(q)=\frac{k}{2}\|\Delta\boldsymbol{x}\|^2 \tag{14}$$

The quantity $E(q,\dot{q})$ is positive definite in $\dot{q}$, but it is not positive definite in q. Apparently, $U(q)$ defined over the chart F in $\{M, g_{ij}\}$ is a non-negative function that attains its minimum on $q \in N_{P'}$. In other words, there aises an infinite number of critical points that satisfy $\partial U(q)/\partial q = 0$ (i.e., $\boldsymbol{x}(q) = \boldsymbol{x}_d = P'$). The set of critical points in $\phi(F)$ constitutes a submanifold of 1-dimension.

In order to analyze the behavior of motion of the robot that is characterized by a solution trajectory $q(t)$ to the closed-loop dynamics of (12), it is necessary to see an important relation between $N_{P'}$ at some point $q = q_d$ with $\boldsymbol{x}(q_d) = \boldsymbol{x}_d$ and the Hessian of $U(q)$ at that point. Since the gradient of $U(q)$ is given by

$$\frac{\partial U}{\partial q}=kJ_q(\boldsymbol{x}(q))(\boldsymbol{x}(q)-\boldsymbol{x}_d) \tag{15}$$

the Hessian at $\boldsymbol{x}(q_d) = \boldsymbol{x}_d$ is formulated as follows:

$$H_{\boldsymbol{x}_d}=\left.\frac{\partial^2 U}{\partial q \partial q^{\mathrm{T}}}\right|_{q=q_d}=kJ_q^{\mathrm{T}}(\boldsymbol{x}_d)J_q(\boldsymbol{x}_d) \tag{16}$$

We call such a q $(= q_d)$ a critical (or equilibrium) point in the configuration space that $\partial U/\partial q|_{q=q_d} = 0$. According to (15), any point q_d in $N_{P'}$ is a critical point. Now, let us introduce the following definition (see [20, 21]):

Definition (Nondegenerate critical manifold of U): If a Riemannian submanifold $N_{P'}$ satisfies the following two conditions, it is called a nondegenerate critical manifold of the function U:

1) $N_{P'}$ is a smooth connected submanifold of M (or $\phi(F)$) and every point of $N_{P'}$ is a critical point of $U(q)$.

2) For any $q \in N_{P'}$, the nullspace of the Hessian of $U(q)$ at $\boldsymbol{x}(q) = \boldsymbol{x}_d$ is coincident with the tangent space of $N_{P'}$, that is, null $(H_{\boldsymbol{x}_d}) = T_q N_{P'}$ on $q \in N_{P'}$.

From this definition and the property of $N_{P'}$, $N_{P'}$ becomes a nondegenerate critical manifold of $U(q)$. Hence, we call $U(q)$ the Morse-Bott function.

In order to consider the stability of motion as a solution to the closed-loop dynamics of (12) it is necessary to introduce a Riemannian ball that contains a part of $N_{P'}$ in a neighborhood of a given starting posture $q(0)$ in F or $\phi(F)$ that satisfies $\|\boldsymbol{x}(q(0)) - \boldsymbol{x}_d\| \le \delta$ for some specified number $\delta > 0$. Since $N_{P'}$ is not a single point, it is necessary for us to suppose that there is a desired reference posture q^* belonging to $N_{P'}$ but actually the initial posture $q(0)$

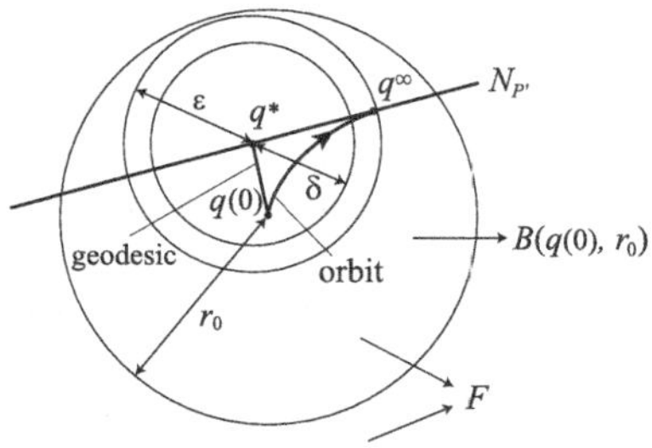

Fig. 3. Definition of the Riemannian ball $B(q^*, \delta)$ included in a chart F. Any orbit starting from $B(q^*, \delta)$ remains in $B(q^*, \varepsilon)$

differes from q^* leaving the critical manifold $N_{P'}$. Thus, we introduce the concept of neighborhoods of q^* in the chart F by defining a Riemannian ball based upon the Riemannian distance (see Fig. 3) such that

$$B(q^*, r_0) = \{q | d(q, q^*) < r_0, q \in F\} \tag{17}$$

Following the definition of stability on a manifold previously discussed (see [10], in which we call $N_{P'}$ an equilibrium-point manifold and denote it by EM_1), we define:

Definition 1 If for any $\varepsilon > 0$ there exists $\delta(\varepsilon) > 0$ and $r_1 > 0$ (that is independent of ε but may be less than r_0) such that a solution $q(t)$ of (12) starting from $q(0)$ with $\dot{q}(0) = 0$ in $B(q^*, r_1)$ remains inside $B(q^*, r_0)$ and its endpoint satisfies $\|\boldsymbol{x}(q(t)) - \boldsymbol{x}_d\| < \varepsilon$, then the reference critical point q^* is said to be stable on a manifold.

It should be remarked that the quantities ε and $\delta(\varepsilon)$ are taken on the basis of physical unit [m] in $\boldsymbol{E}^2$ but r_0 and r_1 are based on the unit of the Riemannian metric originally introduced for measuring the distance $d(q, \bar{q})$ connecting two postures q and $\bar{q}$.

This definition of the stability of motion around a critical point may be a natural extension of Lyapunov's stability. However, the Lyapunov-like relation (13) does not directly conclude the stability of motion in the sense of Definition 1, because the existence of excess DOF of the system may incur "self-motion" as discussed in [22]. Therefore, a more severe notion of stability is necessary when the system is subject to redundancy in system's DOF.

Definition 2 If for any $\varepsilon > 0$ there exists a number $\delta(\varepsilon) > 0$ such that any trajectory of (12) starting from an arbitrary initial posture $q(0)$ inside $B(q^*, \delta(\varepsilon))$ with $\dot{q}(0) = 0$ remains inside $B(q^*, \varepsilon)$ for any $t > 0$ and further approaches asymptotically to some posture q^∞ on $N_{P'}$ together with convergence of $\dot{q}(t)$ to zero, then the posture q^* on $N_{P'}$ is said to be asymptotically stable on a manifold (see Fig. 3).

In this definition the quantities ε and $\delta(\varepsilon)$ are based on the unit of the Riemannian distance. At this point, we assume that the concerned robot has a regular physical size like a human arm, hand, or fingers and numerical values of each entry of the inertia matrix $G(q)$ or the damping matrix C is given on the basis of each physical unit kgm^2 or kgm^2/s. The stiffness parameter k is given on the basis of kg/s^2. Then, if we set $C = \sqrt{k}C_0$ and choose a constant positive definite matrix C_0 adequately (see Appendix B), it is possible to prove:

Theorem (Extended Lagrange Stability)

As to the closed-loop dynamics of (12), the reference posture q^* on $N_{P'}$ is asymptotically stable on a manifold. Further, the speed of convergence to the manifold $N_{P'}$ is exponential with an exponent $-\sigma\sqrt{k}$ with some constant $\sigma > 0$.

The proof is given in Appendix B.

4 Extension of Dirichlet-Lagrange Stability

It is well-known that motion of a hand-writing robot shown in Fig. 4 under the endpoint constraint $\varphi(\boldsymbol{x}(q)) = 0$ is governed by the Euler-Lagrange equation

$$G(q)\ddot{q} + \left\{\frac{1}{2}\dot{G}(q) + S(q,\dot{q})\right\}\dot{q} + g(q) = -\lambda J_q^{\mathrm{T}}(q)\frac{\partial\varphi}{\partial\boldsymbol{x}^{\mathrm{T}}} + u \tag{18}$$

no matter how the arm is redundant in DOF and $\varphi(\boldsymbol{x}(q)) = 0$ expresses an arbitrary hypersurface. In (18) $J_q(q)$ denotes the Jacobian matrix of $\boldsymbol{x}(q)$ with respect to q and hence it follows that

$$\frac{\partial\varphi}{\partial q^{\mathrm{T}}} = \frac{\partial\varphi}{\partial\boldsymbol{x}^{\mathrm{T}}}\left(\frac{\partial\boldsymbol{x}}{\partial q^{\mathrm{T}}}\right) = \frac{\partial\varphi}{\partial\boldsymbol{x}^{\mathrm{T}}}J_q(q) \tag{19}$$

The scalar λ in (18) signifies a Lagrange multiplier corresponding to the constraint $\varphi(\boldsymbol{x}(q)) = 0$. From the physical meaning of the constraint, the 3-dimensional vector $\partial\varphi/\partial\boldsymbol{x}^{\mathrm{T}}$ stands for the normal vector at the contacting point P $(= (x,y,z)^{\mathrm{T}})$ in $\boldsymbol{E}^3$. Therefore, it is reasonable to define the unit normal along the locus of the contact point as $\boldsymbol{n}(s) = (\partial\varphi/\partial\boldsymbol{x}^{\mathrm{T}})\|\partial\varphi/\partial\boldsymbol{x}^{\mathrm{T}}\|^{-1}$, where s denotes the length parameter along the locus of the contact point on the hypersurface. To simplify the notation, we denote the position of the contact at $s(t)$ on the locus by $\boldsymbol{x}(s(t))$ $(= P(s(t)))$ or $\boldsymbol{x}(s)$ $(= P(s))$. Further, if we redefine the Lagrange multiplier in (18) as $f = \lambda\|\partial\varphi/\partial\boldsymbol{x}^{\mathrm{T}}\|$, then (18) can be rewritten into

$$G(q)\ddot{q} + \left\{\frac{1}{2}\dot{G}(q) + S(q,\dot{q})\right\}\dot{q} + g(q) = -fJ_q^{\mathrm{T}}(\boldsymbol{x}(s))\boldsymbol{n}(s) + u \tag{20}$$

To keep the constraint condition in (20), the sign of the constraint force f should not change during motion in accordance with the physical meaning of

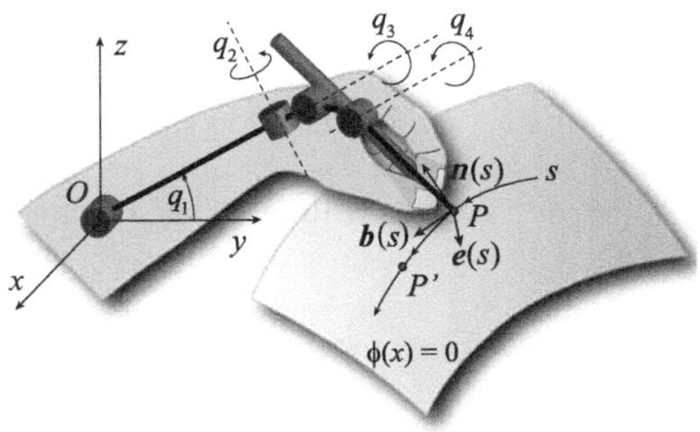

Fig. 4. A hand-writing robot with four DOFs

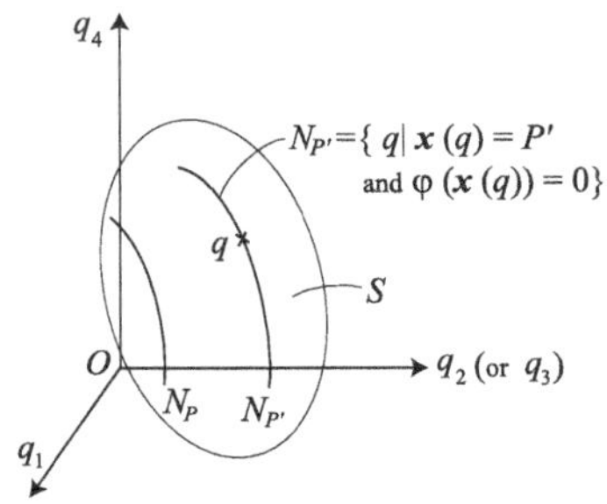

Fig. 5. N_P and $N_{P'}$ express a 1-dim. submanifold and S a 3-dim. submanifold in $\boldsymbol{R}^4$

the contact constraint. This condition is ensured by lifting (or pressing) the dynamics by introducing a force control signal

$$u_1 = f_d J_q^{\mathrm{T}}(\boldsymbol{x}(s))\boldsymbol{n}(s) \tag{21}$$

provided that the vector $J_q^{\mathrm{T}}(\boldsymbol{x}(s))\boldsymbol{n}(s)$ or equivalently $J_q^{\mathrm{T}}(\boldsymbol{x})\partial\varphi/\partial\boldsymbol{x}$ can be computed in real time. For a given target endpoint position $\boldsymbol{x}_d$ on the constraint surface, a position control signal can be designed as follows:

$$u_2 = g(q) - C\dot{q} - J_q^{\mathrm{T}}(\boldsymbol{x})\left\{\zeta\dot{\boldsymbol{x}} + k(\boldsymbol{x} - \boldsymbol{x}_d)\right\} \tag{22}$$

where $g(q)$ in the right hand side means the direct compensation of the gravity term. Thus, substitution of the control signal $u = u_1 + u_2$ into (18) leads to the closed-loop dynamics

$$G(q)\ddot{q} + \left\{\frac{1}{2}\dot{G}(q) + S(q,\dot{q})\right\}\dot{q} + J_q^{\mathrm{T}}(\boldsymbol{x})\left\{\zeta\dot{\boldsymbol{x}} + k\Delta\boldsymbol{x} + \Delta f\boldsymbol{n}(s)\right\} = 0 \tag{23}$$

where $\Delta\boldsymbol{x} = \boldsymbol{x} - \boldsymbol{x}_d$ and $\Delta f = f - f_d$. Since the endpoint velocity $\mathrm{d}\boldsymbol{x}(q)/\mathrm{d}t$ at $\boldsymbol{x}$ on the surface is defined by $\dot{\boldsymbol{x}} = J_q(\boldsymbol{x})\dot{q}$ and it must be orghogonal to the normal $\boldsymbol{n}(\boldsymbol{x}(s))$, the inner product of (22) and $\dot{q}$ leads to

$$\frac{\mathrm{d}}{\mathrm{d}t}\left\{\frac{1}{2}\dot{q}^{\mathrm{T}}G(q)\dot{q} + \frac{k}{2}\|\Delta\boldsymbol{x}\|^2\right\} = -\dot{q}C\dot{q} - \zeta\|\dot{\boldsymbol{x}}\|^2 \tag{24}$$

where $\|\Delta\boldsymbol{x}\|$ and $\|\dot{\boldsymbol{x}}\|$ denote the Euclidean norm in $\boldsymbol{E}^3$.

Now, consider a set of all postures of the hand-writing robot whose endpoint is constrained to the surface $\varphi(\boldsymbol{x}(q)) = 0$, that is denoted by

$$S = \left\{q|\, \varphi(\boldsymbol{x}(q)) = 0 \text{ in } \boldsymbol{E}^3 \text{ and } q \in F\right\} \tag{25}$$

Apparently, S is a three-dimensional submanifold of $\{M, g_{ij}\}$. Given a target posture $P' = \boldsymbol{x}_d$ on the surface, the set $N_{P'} = S \cap \{q|\boldsymbol{x}(q) = \boldsymbol{x}_d\}$ constitutes a one-dimensional manifold as shown in Fig. 5. The hybrid position/force control problem is now to prove that, given a starting posture $q(0)$ of the robot lying on N_P $(= S \cap \{q|\boldsymbol{x}(q) = P\})$, the orbit of motion as a solution to the closed-loop dynamics (23) converges asymptotically to some point belonging to $N_{P'}$ as t tends to infinity.

In order to prove this stability problem, it is convenient to split the dynamics of (23) into the two dynamics, the one is a quotient dynamics expressed in the kernel space of $w = J_q^{\mathrm{T}}(\partial\varphi/\partial\boldsymbol{x}^{\mathrm{T}})$ and the other is expressed in the image space of w. To do this, let us introduce the orthogonal transformation in the tangent space T_qF defined by

$$\dot{q} = (P, w/\|w\|)\begin{pmatrix}\dot{\boldsymbol{\eta}}\\ \dot{\varphi}\end{pmatrix} = Q\dot{\bar{q}} \tag{26}$$

where $Q = (P, w/\|w\|)$, $\dot{\bar{q}} = (\dot{\boldsymbol{\eta}}^{\mathrm{T}}, \dot{\varphi})^{\mathrm{T}}$, and $\|w\|$ denotes the Euclidean norm of w, P is a 4×3 matrix whose column vectors are orthogonal to each other and to w and have the unit norm. Then, Q becomes an orthogonal matrix satisfying $Q^{-1} = Q^{\mathrm{T}}$. Hence, if $\dot{q} \in \ker(w)$ (the kernel space of vector w), then $\dot{\varphi} = w^{\mathrm{T}}\dot{q} = 0$. Restriction of (23) to the kernel space of w can be attained by multiplying (23) by P^{T} from the left in the following way

$$\begin{aligned} &P^{\mathrm{T}}G(q)\frac{\mathrm{d}}{\mathrm{d}t}(P\dot{\boldsymbol{\eta}}) + P^{\mathrm{T}}\left\{\frac{1}{2}\dot{G} + S\right\}P\dot{\boldsymbol{\eta}} \\ &\quad + P^{\mathrm{T}}CP\dot{\boldsymbol{\eta}} + P^{\mathrm{T}}J_q^{\mathrm{T}} + \{\zeta\dot{\boldsymbol{x}} + kP_{\boldsymbol{x}}\Delta\boldsymbol{x}\} = 0 \end{aligned} \tag{27}$$

where $P_{\boldsymbol{x}}$ signifies the orthogonal projection of $\Delta\boldsymbol{x}$ to the tangent plane of the constraint surface at $\boldsymbol{x}$ in $\boldsymbol{E}^3$. This equation can be rewritten into the form

$$\bar{G}(q)\ddot{\boldsymbol{\eta}} + \left\{\frac{1}{2}\dot{\bar{G}} + \bar{S}\right\}\dot{\boldsymbol{\eta}} + \bar{C}\dot{\boldsymbol{\eta}} + (J_qJ_q^{\mathrm{T}})^{1/2}\{\zeta\dot{\boldsymbol{x}} + kP_{\boldsymbol{x}}\Delta\boldsymbol{x}\} = 0 \tag{28}$$

since P is given as $P = J_q^{\mathrm{T}}(J_qJ_q^{\mathrm{T}})^{-1/2}$ and $\bar{G}(q) = P^{\mathrm{T}}G(q)P$, $\bar{C} = P^{\mathrm{T}}CP$, and $\bar{S} = P^{\mathrm{T}}SP - \frac{1}{2}\dot{P}^{\mathrm{T}}GP + \frac{1}{2}P^{\mathrm{T}}G\dot{P}$. Note that $\bar{S}$ is skew symmetric, too. It should be remarked that equation (28) is similar to (12) by the reason that $\dot{\boldsymbol{x}}$ in (28) stands for the velocity on the 2-dimensional tangent plane of the constraint surface, but $\dot{\boldsymbol{\eta}}$ in (28) signifies motion of the joint orthogonally complemented to w. Since $\dot{\boldsymbol{\eta}}$ is of 3-dimension, it is still redundant. Nevertheless, by using a similar argument given in the previous section, the asymptotic stability of position control can be established by choosing appropriate gains C, ζ, and k provided that the normal curvature in any direction of the constraint surface is small in comparison with the reciprocal of a representative length of rigid links constituting the robot arm. It should be noted that in this case the Hessian matrix of the scalar function $U = (k/2)\|\Delta\boldsymbol{x}\|^2$ is given by the same form as (16) with $q_d \in N_{P'}$. Apparently, the null space of H_{q_d} at $q = q_d \in N_{P'}$ is coincident with $T_{q_d}N_{P'}$. Even if the Hessian is taken over the 3-dimensional submanifold S (see Fig. 5), it coincides with the Hessian of (16) when q reaches some $q_d \in N_{P'}$.

Finally it can be concluded easily from (23) that the convergences $\dot{q}(t) \to 0$ and $\boldsymbol{x}(t) \to \boldsymbol{x}_d$ as $t \to \infty$ imply the convergence of $f(t)$ to f_d as $t \to \infty$.

The stability notion of Definition 2 can be extended so as to permit joint motions to start with non-zero velocities $\dot{q}(0) \neq 0$ at $t = 0$ (see the details [8]).

References

1. Takegaki, M., Arimoto, S.: A new feedback method for dynamic control of manipulators. Trans. ASME J. of Dynamic Systems Measurement, and Control 103, 119–125 (1981)
2. Wang, D., McClamroch, N.H.: Position and force control for constrained manipulator motion: Lyapunov's direct method. IEEE Trans. on Robotics and Automation 9(3), 308–313 (1993)
3. Bernstein, N.A.: On dexterity and its development. In: Latash, M.L., Turvey, M.T. (eds.) Dexterity and Its Development, pp. 3–275. Lawrence Erlbaum Associates, Mahway (1996)
4. Latash, M.L.: The Bernstein problem: How does the central nervous system make its choices? In: Latash, M.L., Turvey, M.T. (eds.): ibid, pp. 277–303
5. Latash, M.L.: Neurophysiological Basis of Movement. Human Kinetics, New York (1998)
6. Arimoto, S., Hashiguchi, H., Sekimoto, M., Ozawa, R.: Generation of natural motions for redundant multi-joint systems: A differential-geometric approach based upon the principle of least actions. Journal of Robotic Systems 22(11), 583–605 (2005)
7. Arimoto, S., Sekimoto, M., Hashiguchi, H., Ozawa, R.: Natural resolution of ill-posedness of inverse kinematics for redundant robots: A challenge to Bernstein's degrees-of-freedom problem. Advanced Robotics 19(4), 401–434 (2005)
8. Arimoto, S., Yoshida, M., Sekimoto, M., Tahara, K.: A Riemannian-geometry approach for control of robotic systems under constraints. SICE J. of Control, Measurement, and System Integration 2(2), 107–116 (2009)
9. Arimoto, S., Yoshida, M., Sekimoto, M., Tahara, K.: Modeling and control of 2-D grasping of an object with arbitrary shape under rolling contact. SICE J. of Control, Measurement, and System Integration 2(6), 379–386 (2009)
10. Arimoto, S., Hashiguchi, H., Ozawa, R.: A simple control method coping with a kinematically ill-posed inverse problem of redundant robots: Analysis in case of a handwriting robot. Asian Journal of Control 7(2), 112–123 (2005)
11. Sicilliano, B., Khatib, O. (eds.): Springer Handbook of Robotics. Springer, New York (2008)
12. Latombe, J.C.: Robot Motion Planning. Kluwer Academic Publishers, Norwell (1991)
13. Arnold, V.I., (Trs.K. Vogtmann and A. Weinstein): Methematical Methods of Classical Mechanics, 2nd edn. Springer, New York (1989)

14. Bullo, F., Lewis, A.D.: Geometric Control of Mechanical Systems: Modeling, Analysis, and Design for Simple Mechanical Control Systems. Springer, New York (2000)
15. Oliva, W.M.: Geometric Mechanics. In: Kopacek, P., Moreno-Díaz, R., Pichler, F. (eds.) EUROCAST 1999. LNCS, vol. 1798. Springer, Heidelberg (2000)
16. Arimoto, S., Yoshida, M., Sekimoto, M., Tahara, K.: A Riemannian-geometry approach for dynamics and control of object manipulation under constraints. In: Proc. of the 2009 IEEE Int. Conf. on Robotics and Automation, Kobe, Japan, May 12-17, pp. 1683–1690 (2009)
17. Sekimoto, M., Arimoto, S., Kawamura, S., Bae, J.-H.: Skilled-motion plannings of multi-body systems based upon Riemannian distance. In: Pasadena, C.A. (ed.) Proc. of the 2008 IEEE Int. Conf. on Robotics and Automation, Pasadena, CA, USA, May 19-23, pp. 1233–1238 (2008)
18. Arimoto, S.: Control Theory of Nonlinear Mechanical Systems: A Passivity-based and Circuit-theoretic Approach. Oxford Univ. Press, Oxford (1996)
19. Jost, J.: Riemannian Geometry and Geometric Analysis. Springer, Berlin (2002)
20. Bott, R.: The stable homotopy of the classical groups. Annals of Mathematics 70(2), 313–337 (1959)
21. Milnor, J.: Morse Theory. Princeton Univ. Press, Princeton (1963)
22. Seraji, H.: Configuration control of redundant manipulators: Theory and implementation. IEEE Trans. on Robotics and Automation 5(4), 472–490 (1989)

Appendix A

First note that $\dot{G} = \sum_i \{\partial G/\partial q_i\}\dot{q}_i$ the skew-symmetric matrix $S(q,\dot{q})$ in (3) is expressed as (2). Evidently, the second term in bracket ( ) of (6) corresponds to the first term in { } of (2) and the third term in ( ) of (6) does to the second term in { } of (2). Hence, it follows from (2)

$$\sum_{j=1}^{n} S_{kj}\dot{q}_j = \sum_{j=1}^{n} \frac{1}{2}\left[\left\{\frac{\partial}{\partial q_j}\left(\sum_{i=1}^{n}\dot{q}_i g_{ki}\right)\right\}\dot{q}_i - \left\{\frac{\partial}{\partial q_k}\left(\sum_{i=1}^{n}\dot{q}_i g_{ij}\right)\right\}\dot{q}_j\right]$$
$$= \sum_{j=1}^{n}\sum_{i=1}^{n}\frac{1}{2}\left\{\left(\frac{\partial g_{ik}}{\partial q_j} - \frac{\partial g_{ij}}{\partial q_k}\right)\right\}\dot{q}_i\dot{q}_j \tag{A-1}$$

Substituting this into (4) by comparing the last two terms of (6) with the last bracket { } of (A-1) results in the equivalence of (3) to (4).

Appendix B (Proof of the Theorem)

First we remark that the Riemannian distance $d(q,\bar{q})$ connecting q and $\bar{q}$ in $\{M, g_{ij}\}$ is numerically given on the basis of physical unit $[\sqrt{\text{kg}}\ \text{m}]$ and $\|\boldsymbol{x}\|$ of vector $\boldsymbol{x}$ in $\boldsymbol{E}$ expresses the Euclidean norm. Then, numerical values of G are given based upon $[\text{kgm}^2]$, C on $[\text{kgm}^2/\text{s}]$, stiffness k on $[\text{kg/s}^2]$, and hence C_0 on $[\sqrt{\text{kg}}\ \text{m}^2]$, since $C = \sqrt{k}C_0$. Next we remark that, according to the physical size of the concerned robot, the maximum eigenvalue of $G(q)$ is at most 0.5 [kgm^2] and the maximum length of robot links is less than 0.5 [m]. Then, it is possible to find a positive definite matrix C_0 such that it satisfies

$$C_0 \geq 2\alpha G(q) + \frac{1}{4\alpha} J^{\mathrm{T}}(q) J(q), \quad J(q) C_0^{-1} J^{\mathrm{T}}(q) \leq \alpha I_3 \tag{B-1}$$

with $\alpha = 1.0$ [$1/\sqrt{\mathrm{kg}}$] for all $q \in B(q_d^*, r_0)$. Since $J(q)$ is nondegenerate in $B(q_d^*, r_0)$, there exists a constant σ_0 [$1/\sqrt{\mathrm{kg}}$] such that

$$J(q) C_0^{-1} J^{\mathrm{T}}(q) \geq \sigma_0 I_3 \tag{B-2}$$

for all $q \in B(q_d^*, r_0)$. In other words, it is desirable to choose C_0 so as to make σ_0 in (B-2) as large as possible with the conditions of (B-1). Now, let

$$V = \sqrt{k} \Delta \boldsymbol{x}^{\mathrm{T}} J C_0^{-1} G \dot{q} + \frac{k}{2} \|\Delta \boldsymbol{x}\|^2, \quad W = \frac{1}{4} \dot{q}^{\mathrm{T}} G \dot{q} + \frac{k}{2} \|\Delta \boldsymbol{x}\|^2 \tag{B-3}$$

Then, it follows from multiplying (B-1) by JC_0^{-1} from the left and by $C_0^{-1}J^{\mathrm{T}}$ from the right that $JC_0^{-1}GC_0^{-1}J^{\mathrm{T}} \leq (1/2)I_3$, which makes it possible to show

$$0 \leq W \leq E + V \leq 3W \tag{B-4}$$

Next, we differentiate $E + V$ in t, that results in

$$\begin{aligned}
\dot{E} + \dot{V} &= -\sqrt{k} \dot{q}^{\mathrm{T}} C_0 \dot{q} + \dot{V} \\
&= -\sqrt{k} \left\{ \dot{q}^{\mathrm{T}} C_0 \dot{q} + k \Delta \boldsymbol{x}^{\mathrm{T}} J C_0^{-1} J^{\mathrm{T}} \Delta \boldsymbol{x} \right\} \\
&\quad + \sqrt{k} \dot{\boldsymbol{x}}^{\mathrm{T}} J C_0^{-1} G \dot{q} + \sqrt{k} \Delta \boldsymbol{x}^{\mathrm{T}} H(\dot{q}) \dot{q}
\end{aligned} \tag{B-5}$$

where $H(\dot{q})$ is a 3×4-matrix described as

$$H(\dot{q}) = \dot{J} C_0^{-1} G + J C_0^{-1} \left\{ \frac{1}{2} \dot{G} - S \right\} \tag{B-6}$$

By noting again the inequality

$$\begin{aligned}
\dot{\boldsymbol{x}}^{\mathrm{T}} J C_0^{-1} G \dot{q} &\leq \frac{\alpha}{2} \dot{q}^{\mathrm{T}} G \dot{q} + \frac{1}{2\alpha} \dot{\boldsymbol{x}}^{\mathrm{T}} J C_0^{-1} G C_0^{-1} J^{\mathrm{T}} \boldsymbol{x} \\
&\leq \frac{1}{2} \dot{q}^{\mathrm{T}} \left\{ \alpha G + \frac{1}{2\alpha} J^{\mathrm{T}} J \right\} \dot{q}
\end{aligned} \tag{B-7}$$

and by substituting (B-7) into (B-5) with setting $\alpha = 1.0$ [$1/\sqrt{\mathrm{kg}}$], we have

$$\begin{aligned}
\dot{E} + \dot{V} \leq & -\sqrt{k} \dot{q}^{\mathrm{T}} \left(C_0 - \frac{1}{2} G - \frac{1}{4} J^{\mathrm{T}} J \right) \dot{q} \\
& -\sqrt{k} k \Delta \boldsymbol{x}^{\mathrm{T}} J C_0^{-1} J^{\mathrm{T}} \Delta \boldsymbol{x} + \sqrt{k} \Delta \boldsymbol{x}^{\mathrm{T}} H(\dot{q}) \dot{q}
\end{aligned} \tag{B-8}$$

On account of (B-1) with $\alpha = 1.0$, this reduces to

$$\begin{aligned}\dot{E}+\dot{V} &\le -\sqrt{k}\left\{\frac{1}{2}\dot{q}^{\mathrm{T}}G\dot{q}+\sigma_0 k\|\Delta\boldsymbol{x}\|^2\right\}-\sqrt{k}\left\{\dot{q}^{\mathrm{T}}G\dot{q}-\Delta\boldsymbol{x}^{\mathrm{T}}H(\dot{q})\dot{q}\right\}\\ &\le -2\sigma_0\sqrt{k}W-\sqrt{k}\left\{\dot{q}^{\mathrm{T}}G\dot{q}-\Delta\boldsymbol{x}^{\mathrm{T}}H(\dot{q})\dot{q}\right\}\end{aligned} \tag{B-9}$$

At this stage, we note that (13) implies $\|\Delta\boldsymbol{x}(t)\| \le \|\Delta\boldsymbol{x}(0)\|$ and, since $H(\dot{q})\dot{q}$ is quadratic in $\dot{q}$, there exists a positive constant $\gamma_0 > 0$ with dimension [m] such that $\|\Delta\boldsymbol{x}(t)\| < \gamma_0$ implies

$$\dot{q}G(q)\dot{q} \ge \Delta\boldsymbol{x}^{\mathrm{T}}H(\dot{q})\dot{q} \tag{B-10}$$

On account of (B-4), this makes (B-9) reduce to

$$\dot{E}+\dot{V} \le -\frac{2}{3}\sigma_0\sqrt{k}(E+V) \tag{B-11}$$

provided that $\|\Delta\boldsymbol{x}(0)\| < \gamma_0$. Thus, from (B-11) it follows that

$$\begin{aligned}\sqrt{E(t)+V(t)} &\le \sqrt{E(0)+V(0)}e^{-(\sigma_0\sqrt{k}/3)t}\\ &= \sqrt{k/2}\|\Delta\boldsymbol{x}(0)\|e^{-(\sigma\sqrt{k})t}\end{aligned} \tag{B-12}$$

where we set $\sigma = \sigma_0/3$ $[1/\sqrt{\mathrm{kg}}\,]$. Hence

$$\begin{aligned}d(q(0),q(t)) &\le \int_0^t \sqrt{g_{ij}\dot{q}_i(\tau)\dot{q}_j(\tau)}\,\mathrm{d}\tau \le \int_0^t \sqrt{2\{E(\tau)+V(\tau)\}}\,\mathrm{d}\tau\\ &\le \sqrt{k}\|\Delta\boldsymbol{x}(0)\|\int_0^t e^{-\sigma\sqrt{k}\tau}\mathrm{d}\tau \le \|\Delta\boldsymbol{x}(0)\|/\sigma\end{aligned} \tag{B-13}$$

Since $\boldsymbol{x}(q)$ is of C^∞-class, for a given $\varepsilon > 0$ there exists $\delta_1(\varepsilon) > 0$ $[\sqrt{\mathrm{kg}}\ \mathrm{m}]$ such that any $q(0) \in B(q^*, \delta_1(\varepsilon))$ implies $\|\boldsymbol{x}(q(0)) - \boldsymbol{x}(q^*)\| \le \|\Delta\boldsymbol{x}(0)\| < \min\{\gamma_0, \sigma\varepsilon/2\}$. Then, we see from (B-13) and taking $\delta(\varepsilon) = \min\{\delta_1(\varepsilon), \varepsilon/2\}$ that any $q(0) \in B(q^*, \delta(\varepsilon))$ implies

$$\begin{aligned}d(q(t),q^*) &\le d(q(t),q(0)) + d(q(0),q^*)\\ &\le \|\Delta\boldsymbol{x}(t)\|/2 + \varepsilon/2 < \varepsilon/2+\varepsilon/2 = \varepsilon\end{aligned} \tag{B-14}$$

that proves the stability of q^* on a manifold. The asymptotic convergence of $q(t)$ to $N_{P'}$ is now apparent from (B-14) and the exponential decay of $\|\Delta\boldsymbol{x}(q(t))\|$ to zero shown in (B-12).

Modeling and Control of Multi-Body Mechanical Systems: Part II Grasping under Rolling Contacts between Arbitrary Shapes

Suguru Arimoto

Abstract. Modeling of 2-dimensional grasping and object manipulation under rolling contacts by a pair of multi-joint robot fingers with an arbitrary fingertip contour curve is discussed. Stabilization of grasping by using a control signal based on the fingers-thumb opposability is discussed from the analysis of a Morse-Bott function introduced as an artificial potential. An exentsion of modeling of 3-D grasping under rolling contact constraints is discussed under the circumstance of arbitrary shapes of the fingertips and object.

1 Modeling of 2-D Grasping under Arbitrary Geometry

A mathematical model of 2-dimensional grasping of a rigid object with an arbitrary shape by a pair of robot fingers with arbitrarily fingertip shapes (see Fig. 1) is derived on the basis of the following differential-geometric assumptions of rolling contacts [1]:

1) Two contact points on each contour curve must coincide at a single common point without mutual penetration, and

2) the two contours must have the same tangent at the common contact point.

The assumptions are equivalent to Nomizu's definition of a rolling contact [2], which is described by a mathematical form by using the common tangent

Suguru Arimoto
Research Organization of Science and Engineering, Ritsumeikan University
e-mail: arimoto@fc.ritsumei.ac.jp
and
RIKEN-TRI Collaboration Center, RIKEN

J. Lévine & P. Müllhaupt (Eds.): Adv. in Theory of Control, Signals, LNCIS 407, pp. 17–29.
springerlink.com © Springer-Verlag Berlin Heidelberg 2010

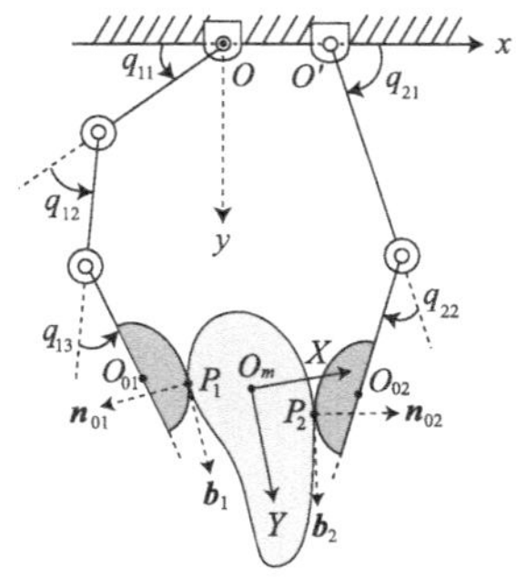

Fig. 1. 2-D pinching of a 2-D object with arbitrary shape by a pair of robot fingers with arbitrary fingertips

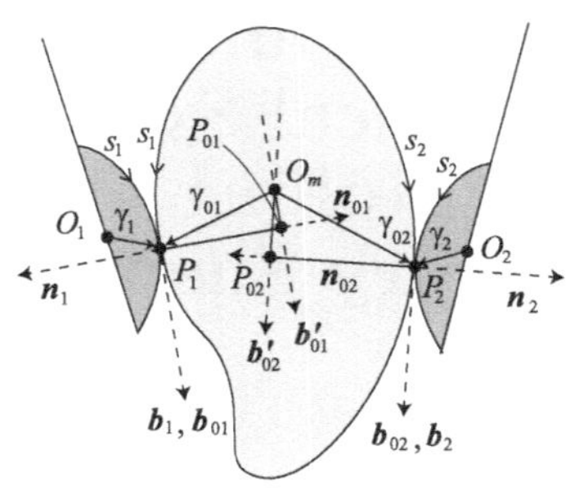

Fig. 2. Definitions of tangent vectors $\boldsymbol{b}_i$, $\boldsymbol{b}_{0i}$ and normals $\boldsymbol{n}_i$ and $\boldsymbol{n}_{0i}$ at contact points P_i for $i = 1, 2$

and two normals at the contact. In [1], a set of Euler-Lagrange's equations of motion of the overall fingers/object system is given in the form:

$$m\ddot{\boldsymbol{x}} - \sum_{i=1,2} (f_i \bar{\boldsymbol{n}}_{0i} + \lambda_i \bar{\boldsymbol{b}}_{0i}) = 0 \tag{1}$$

$$I\ddot{\theta} + \sum_{i=1,2} (-1)^i \left\{ f_i(\boldsymbol{b}_{0i}^{\mathrm{T}} \gamma_{0i}) - \lambda_i (\boldsymbol{n}_{0i}^{\mathrm{T}} \gamma_{0i}) \right\} = 0 \tag{2}$$

$$G_i(q_i)\ddot{q}_i + \left\{ \frac{1}{2}\dot{G}_i(q_i) + S_i(q_i, \dot{q}_i) \right\} \dot{q}_i + f_i \left\{ J_i^{\mathrm{T}}(q_i)\bar{\boldsymbol{n}}_{0i} - (-1)^i (\boldsymbol{b}_i^{\mathrm{T}} \gamma_i) \boldsymbol{e}_i \right\}$$
$$+ \lambda_i \left\{ J_i^{\mathrm{T}}(q_i)\bar{\boldsymbol{b}}_{0i} - (-1)^i (\boldsymbol{n}_i^{\mathrm{T}} \gamma_i) \boldsymbol{e}_i \right\} = u_i, \ i = 1, 2 \tag{3}$$

Here q_i stands for the joint vectors as $q_1 = (q_{11}, q_{12}, q_{13})^{\mathrm{T}}$ and $q_2 = (q_{21}, q_{22})^{\mathrm{T}}$, $\dot{\theta}$ denotes the angular velocity of rotation of the object around the object mass center O_m expressed by position vector $\boldsymbol{x} = (x, y)^{\mathrm{T}}$ in terms of the inertial frame coordinates O-xy. Equation (1) expresses the translational motion of the object with mass m and (2) its rotational motion with inertia moment I around the mass center O_m. At the contact point P_i, $\boldsymbol{b}_i$ denotes the unit tangent vector expressed in local coordinates of O_i-X_iY_i fixed to the fingertip of finger i ($i = 1, 2$) as shown in Fig. 1 and Fig. 2, and $\boldsymbol{n}_i$ denotes the unit normal to the tangent expressed in terms of O_i-X_iY_i. Similarly, $\boldsymbol{b}_{0i}$ and $\boldsymbol{n}_{0i}$ are the unit tangent and normal at P_i expressed in terms of local coordinates O_m-XY fixed to the object. All these unit vectors are determined uniquely from the assumptions 1) and 2) on the rolling contact constraints at each contact point P_i dependent on each corresponding value s_i for $i = 1, 2$ as shown in Fig. 2. Equation (3) denotes joint motions of finger i with the inertia matrix $G_i(q_i)$ for $i = 1, 2$, and $\boldsymbol{e}_1 = (1, 1, 1)^{\mathrm{T}}$ and $\boldsymbol{e}_2 = (1, 1)^{\mathrm{T}}$. All position vectors γ_i and γ_{0i} for $i = 1, 2$ are defined as in Fig. 2 and expressed in their corresponding local coordinates. The unit vectors $\bar{\boldsymbol{b}}_{0i}$ and $\bar{\boldsymbol{n}}_{0i}$ are expressed in the inertial frame coordinates as follows:

$$\bar{\boldsymbol{b}}_{0i} = \Pi_0 \boldsymbol{b}_{0i}, \quad \bar{\boldsymbol{n}}_{0i} = \Pi_0 \boldsymbol{n}_{0i}, \quad \Pi_0 = (\boldsymbol{r}_X, \boldsymbol{r}_Y) \tag{4}$$

where $\Pi_0 \in SO(2)$ and $\boldsymbol{r}_X$ and $\boldsymbol{r}_Y$ denote the unit vectors of X- and Y-axes of the object in terms of the frame coordinates O-xy. In (1) to (3), f_i and λ_i are Lagrange's multipliers that correspond to the following rolling contact constraints, respectively:

$$\begin{cases} Q_{bi} = (\boldsymbol{r}_i - \boldsymbol{r}_m)^{\mathrm{T}} \bar{\boldsymbol{b}}_{0i} + \boldsymbol{b}_i^{\mathrm{T}} \gamma_i - \boldsymbol{b}_{0i}^{\mathrm{T}} \gamma_{0i} = 0, \ i = 1, 2 \\ Q_{ni} = (\boldsymbol{r}_i - \boldsymbol{r}_m)^{\mathrm{T}} \bar{\boldsymbol{n}}_{0i} + \boldsymbol{n}_i^{\mathrm{T}} \gamma_i - \boldsymbol{n}_{0i}^{\mathrm{T}} \gamma_{0i} = 0, \ i = 1, 2 \end{cases} \tag{5}$$

where $\boldsymbol{r}_i$ denotes the position vector of the fingertip center O_i expressed in terms of the frame coordinates O-xy and $\boldsymbol{r}_m$ the position vector of O_m in terms of O-xy. In parallel with Euler-Lagrange equations (1) to (3), arclength parameters s_i $(i = 1, 2)$ should be governed by the first order differential equations

$$\{\kappa_{0i}(s_i) + \kappa_i(s_i)\} \frac{\mathrm{d}s_i}{\mathrm{d}t} = (-1)^i (\dot{\theta} - \dot{p}_i), \quad i = 1, 2 \tag{6}$$

where $\kappa_i(s_i)$ denotes the curvature of the fingertip contour for $i = 1, 2$, $\kappa_{0i}(s_i)$ the curvature of the object contour at contact point P_i for $i = 1, 2$, and $p_i = q_i^{\mathrm{T}} \boldsymbol{e}_i$ for $i = 1, 2$. Throughout the paper we use $(\dot{\ })$ for denoting the differentiation of the content of bracket ( ) in time t as $\dot{\theta} = \mathrm{d}\theta/\mathrm{d}t$ in (6) and $(\ ')$ for that of ( ) in length parameter s_i as illustrated by $\gamma_i'(s_i) = \mathrm{d}\gamma_i(s_i)/\mathrm{d}s_i$. As discussed in the paper [1], we have

$$\begin{cases} \boldsymbol{b}_i(s_i) = \gamma_i'(s_i) \left(= \dfrac{\mathrm{d}\gamma_i(s_i)}{\mathrm{d}s_i} \right), \ \boldsymbol{b}_{0i}(s_i) = \gamma_{0i}'(s_i), \ i = 1, 2 \\ \boldsymbol{n}_i(s_i) = \kappa_i(s_i) \boldsymbol{b}_i'(s_i), \quad \boldsymbol{n}_{0i}(s_i) = \boldsymbol{b}_{0i}'(s_i), \quad i = 1, 2 \\ \boldsymbol{b}_i(s_i) = -\kappa_i(s_i) \boldsymbol{n}_i'(s_i), \ \boldsymbol{b}_{0i}(s_i) = -\kappa_{0i}(s_i) \boldsymbol{n}_{0i}'(s_i), \ i = 1, 2 \end{cases} \tag{7}$$

It is well known in text books on differential geometry of curves and surfaces (for example, see [3]) that the last two equations of (7) constitute Frenet-Serret's formulae for the fingertip contour curves and object contours. Note that all equations of (1) to (3) are characterized by length parameters s_i for $i = 1, 2$ through unit vectors $\boldsymbol{n}_{0i}$, $\boldsymbol{b}_{0i}$, $\boldsymbol{b}_i$, and $\boldsymbol{n}_i$ and vectors γ_{0i} and γ_i expressed in each local coordinates, but quantities of the second fundamental form of contour curves, that is, $\kappa_i(s_i)$ and $\kappa_0(s_i)$ for $i = 1, 2$, do not enter into (1) to (3). It is shown that the Euler-Lagrange equations (1) to (3) can be derived by applying the variational principle to the Lagrangian of the system

$$L(X; s_1, s_2) = K(X, \dot{X}) - \sum_{i=1,2} (f_i Q_{ni} + \lambda_i Q_{bi}) \tag{8}$$

where X denotes the position state vector $X = (x, y, \theta, q_1^{\mathrm{T}}, q_2^{\mathrm{T}})^{\mathrm{T}}$ and

$$K(X, \dot{X}) = \frac{m}{2}(\dot{x}^2 + \dot{y}^2) + \frac{I}{2}\dot{\theta}^2 + \sum_{i=1,2} \frac{1}{2}\dot{q}_i G_i(q_i)\dot{q}_i \tag{9}$$

Note that $K(X, \dot{X})$ is independent of the shape parameters s_1 and s_2 but Q_{ni} and Q_{bi} defined in (5) are depnedent on s_i for $i = 1, 2$, respectively. The corresponding variational principle is written in the following form:

$$\int_{t_0}^{t_1} \left\{\delta L + u_1^{\mathrm{T}} \delta q_1 + u_2^{\mathrm{T}} \delta q_2\right\} \mathrm{d}t = 0 \tag{10}$$

When $u_1 = 0$ and $u_2 = 0$, (1) to (3) constitute the Euler equation for the base manifold $\{M, g_{ij}\}$ under the constraints of (5), where M stands for a set of all possible postures of the system specified by X and g_{ij} the Riemannian metric induced by the inertia matrix $G(X) = \mathrm{diag}(m, m, I, G_1, G_2)$.

2 Integrability of Rolling Contact Constraints

In order to derive the traditional rolling contact condition as discussed in the text book [4], first note that the assumption 1) implies

$$\boldsymbol{r}_i + \Pi_i \gamma_i = \boldsymbol{r}_m + \Pi_0 \gamma_{0i}, \quad i = 1, 2 \tag{11}$$

from which (5) follows through taking the inner product of (11) and $\bar{\boldsymbol{b}}_{0i}$ or $\bar{\boldsymbol{n}}_{0i}$ for $i = 1, 2$. On the other hand, differentiation of (11) concerning t yields

$$\dot{\boldsymbol{r}}_i + \dot{\Pi}_i \gamma_i + \Pi_i \gamma_i' \frac{\mathrm{d}s_i}{\mathrm{d}t} = \dot{\boldsymbol{r}}_m + \dot{\Pi}_0 \gamma_{0i} + \Pi_0 \gamma_{0i}' \frac{\mathrm{d}s_i}{\mathrm{d}t} \tag{12}$$

which is reduced to

$$\begin{cases} R_{bi} = (\dot{\boldsymbol{r}}_i - \dot{\boldsymbol{r}}_m)^{\mathrm{T}} \bar{\boldsymbol{b}}_{0i} - (-1)^i \left\{\dot{p}_i \boldsymbol{n}_i^{\mathrm{T}} \gamma_i + \dot{\theta} \boldsymbol{n}_{0i}^{\mathrm{T}} \gamma_{0i}\right\} = 0 \\ R_{ni} = (\dot{\boldsymbol{r}}_i - \dot{\boldsymbol{r}}_m)^{\mathrm{T}} \bar{\boldsymbol{n}}_{0i} - (-1)^i \left\{\dot{p}_i \boldsymbol{b}_i^{\mathrm{T}} \gamma_i - \dot{\theta} \boldsymbol{b}_{0i}^{\mathrm{T}} \gamma_{0i}\right\} = 0 \end{cases} \tag{13}$$

by taking the inner product of (12) and $\bar{\boldsymbol{b}}_{0i}$ or $\bar{\boldsymbol{n}}_{0i}$ for $i = 1, 2$. The first equation of (13) shows the traditional rolling constraint as the zero relative velocity of rolling of the contact at the fingerend to the object contour. It imports us to know that the Pfaffian form of (12) is integrable in t, that is,

$$\frac{\mathrm{d}}{\mathrm{d}t} Q_{ni} = R_{ni}, \qquad \frac{\mathrm{d}}{\mathrm{d}t} Q_{bi} = R_{bi}, \qquad i = 1, 2 \tag{14}$$

which can be derived through (6) that reflects the assumption 2) (see [1]). More precisely, the rolling contact constraints define the distribution over the tangent space $T_X M$ according to

$$\dot{Q}_{bi} = \Re_{bi}^{\mathrm{T}} \dot{X}, \qquad \dot{Q}_{ni} = \Re_{ni}^{\mathrm{T}} \dot{X}, \qquad i = 1, 2 \tag{15}$$

Here, the vectors are given as

$$\Re_{b1} = \begin{bmatrix} -\bar{\boldsymbol{b}}_{01} \\ \boldsymbol{n}_{01}^{\mathrm{T}} \gamma_{01} \\ J_1^{\mathrm{T}} \bar{\boldsymbol{b}}_{01} + (\boldsymbol{n}_1^{\mathrm{T}} \gamma_1) \boldsymbol{e}_1 \\ 0_2 \end{bmatrix}, \quad \Re_{n1} = \begin{bmatrix} -\bar{\boldsymbol{n}}_{01} \\ -\boldsymbol{b}_{01}^{\mathrm{T}} \gamma_{01} \\ J_1^{\mathrm{T}} \bar{\boldsymbol{n}}_{01} + (\boldsymbol{b}_1^{\mathrm{T}} \gamma_1) \boldsymbol{e}_1 \\ 0_2 \end{bmatrix} \tag{16}$$

and, similarly, $\Re_{b2}$ and $\Re_{n2}$.

3 Stabilization Based on a Morse-Bott Function

In this paper, we solve a stabilization problem of grasping in a special case that the object has a pair of parallel flat surfaces. Therefore it can be regarded as a rectangular object in a two dimensional horizontal plane. In this case, it is possible to consider a class of control signals defined by

$$u_i = -c_i \dot{q}_i + (-1)^i k J_i^{\mathrm{T}}(q_i)(\boldsymbol{r}_1 - \boldsymbol{r}_2) - k\alpha_i \hat{N}_i \boldsymbol{e}_i, \ i = 1, 2 \tag{17}$$

where k stands for a position feedback gain common for $i = 1, 2$ with the physical unit N/m, α_i in m^2 is also a positive constant for $i = 1, 2$. The variable $\hat{N}_i$ is defined as

$$\hat{N}_i = \boldsymbol{e}_i^{\mathrm{T}} \{q_i(t) - q_i(0)\} = p_i(t) - p_i(0), \quad i = 1, 2 \tag{18}$$

and c_i denotes a positive constant for joint damping for $i = 1, 2$. The first term of the right hand side of (17) stands for damping shaping, the second term plays a role of fingers-thumb opposition, and the last term adjusts possibly some abundant motion of rotation of the object through contacts. Note that the sum of inner products of u_i and $\dot{q}_i$ for $i = 1, 2$ is given by

$$\sum_{i=1,2} \dot{q}_i^{\mathrm{T}} u_i = -\frac{\mathrm{d}}{\mathrm{d}t} \left\{ \frac{k}{2} \|\boldsymbol{r}_1 - \boldsymbol{r}_2\|^2 + \sum_{i=1,2} \frac{\alpha_i}{2} \hat{N}_i^2 \right\} - \sum_{i=1,2} c_i \|\dot{q}_i\|^2 \tag{19}$$

Substitution of control signals of (17) into (3) yields

$$G_i \ddot{q}_i + \left\{ \frac{1}{2} \dot{G}_i + S_i \right\} \dot{q}_i + c_i \dot{q}_i - (-1)^i k J_i^{\mathrm{T}} (\boldsymbol{r}_1 - \boldsymbol{r}_2) + k\alpha_i \hat{N}_i \boldsymbol{e}_i$$
$$+ f_i \left\{ J_i^{\mathrm{T}} \bar{\boldsymbol{n}}_{0i} - (-1)^i (\boldsymbol{b}_i^{\mathrm{T}} \gamma_i) \boldsymbol{e}_i \right\} + \lambda_i \left\{ J_i^{\mathrm{T}} \bar{\boldsymbol{b}}_{0i} - (-1)^i (\boldsymbol{n}_i^{\mathrm{T}} \gamma_i) \boldsymbol{e}_i \right\} = 0, \ i = 1, 2 \tag{20}$$

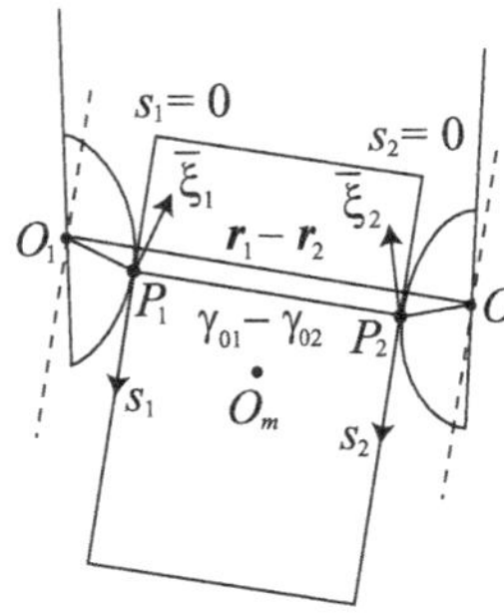

Fig. 3. Minimization of the squared norm $\|\boldsymbol{r}_1 - \boldsymbol{r}_2\|^2$ over rolling motions is attained when the straight line $\overline{P_1P_2}$ connecting the two contact points becomes parallel to the vector $(\boldsymbol{r}_1 - \boldsymbol{r}_2)$, that is, $\overline{O_1O_2}$ becomes parallel to $\overline{P_1P_2}$

Hence, the overall closed-loop dynamics is composed of the set of Euler-Lagrange equations of (1), (2), and (20) that are subject to four algebraic constraints of (5) and the pair of first-order differential equations of (6) that governs the update law of arclength parameters s_1 and s_2. It should be also remarked that, according to (19), the sum of inner products of (1) and $\dot{\boldsymbol{x}}$, (2) and $\dot{\theta}$, and (20) and $\dot{q}_i$ for $i = 1, 2$ yields the energy relation

$$\frac{\mathrm{d}}{\mathrm{d}t}\left\{E(X, \dot{X}) + P(X)\right\} = -\sum_{i=1,2} c_i \|\dot{q}_i\|^2 \tag{21}$$

with

$$P(X) = \frac{k}{2}\|\boldsymbol{r}_1 - \boldsymbol{r}_2\|^2 + \sum_{i=1,2} \frac{k\alpha_i}{2}\hat{N}_i^2 \tag{22}$$

Here, $P(X)$ represents the artificial potential energy that is a scalar function depending on only q_1 and q_2. It is importnat to note that the closed-loop dynamics of (1), (2), and (19) can be written in a general form

$$G(X)\ddot{X} + \left\{\frac{1}{2}\dot{G}(X) + S(X, \dot{X}) + C\right\}\dot{X} + \frac{\partial P(X)}{\partial X} + \sum_{i=1,2}(f_i \Re_{ni} + \lambda_i \Re_{bi}) = 0 \tag{23}$$

where $C = \mathrm{diag}(0_2, 0, c_1 I_3, c_2 I_2)$.

Now, $P(X)$ can be regarded as a Morse-Bott function of X on the base manifold, since both $\|\boldsymbol{r}_1 - \boldsymbol{r}_2\|^2$ and $\hat{N}_i^2$ are positive semi-definite functions of q_1 and q_2. Further, since in this case $\bar{\boldsymbol{b}}_{0i} \perp \bar{\boldsymbol{n}}_{0i}$ (see Fig. 3), the first term of $P(X)$ can be regarded as a function of length parameters such that

$$U(X) = \frac{k}{2}\|\boldsymbol{r}_1 - \boldsymbol{r}_2\|^2 = \frac{k}{2}\{d^2(s_1, s_2) + l^2(s_1, s_2)\} = U(s_1, s_2) \tag{24}$$

with the object width l_w and

$$\begin{cases} d(s_1, s_2) = s_1 - s_2 - \boldsymbol{b}_1^{\mathrm{T}}\gamma_1 + \boldsymbol{b}_2^{\mathrm{T}}\gamma_2 \\ l(s_1, s_2) = -l_w + (\boldsymbol{n}_1^{\mathrm{T}}\gamma_1 + \boldsymbol{n}_2^{\mathrm{T}}\gamma_2) \end{cases} \tag{25}$$

(see Figs. 2 and 3). From this geometric meaning of U, $U(s_1, s_2)$ is locally minimized when the line connecting the contact points P_1 and P_2 becomes parallel to the line $\overline{O_1O_2}$ shown in Fig. 3. Since s_i $(i = 1, 2)$ are also dependent on p_i $(i = 1, 2)$ and θ as shown in (5), it is still not trivial to find the condition for locally minimizing the potential function $P(X)$. Instead, we transform (1), (2), and (20) into

$$m\ddot{\boldsymbol{x}} - \sum_{i=1,2} (\Delta f_i \bar{\boldsymbol{n}}_{0i} + \Delta\lambda_i \bar{\boldsymbol{b}}_{0i}) = 0 \tag{26}$$

$$I\ddot{\theta} + \sum_{i=1,2} (-1)^i (\Delta f_i (\boldsymbol{b}_{0i}^{\mathrm{T}}\gamma_{0i}) - \Delta\lambda_i (\boldsymbol{n}_{0i}^{\mathrm{T}}\gamma_{0i})) + S_N = 0 \tag{27}$$

$$G_i\ddot{q}_i + \left\{\frac{1}{2}\dot{G}_i + S_i\right\}\dot{q}_i + c_i\dot{q}_i + \Delta N_i \boldsymbol{e}_i + \Delta f_i \left\{J_i^{\mathrm{T}}\bar{\boldsymbol{n}}_{0i} - (-1)^i(\boldsymbol{b}_i^{\mathrm{T}}\gamma_i)\boldsymbol{e}_i\right\}$$
$$+\Delta\lambda_i\left\{J_i^{\mathrm{T}}\bar{\boldsymbol{b}}_{0i} - (-1)^i(\boldsymbol{n}_i^{\mathrm{T}}\gamma_i)\boldsymbol{e}_i\right\} = 0, \quad i = 1, 2 \tag{28}$$

by using the lifting

$$\begin{cases} \Delta f_i = f_i + kl(s_1, s_2) \\ \Delta\lambda_i = \lambda_i - (-1)^i kd(s_1, s_2) \end{cases} \quad i = 1, 2 \tag{29}$$

Here, S_N and ΔN_i are given by

$$\begin{cases} S_N = k\{(s_1 - s_2)l + l_w d\} \\ \Delta N_i = kN_i + k\alpha_i\{p_i - p_i(0)\}, \quad i = 1, 2 \end{cases} \tag{30}$$

where

$$N_i = (-1)^i(\boldsymbol{b}_i^{\mathrm{T}}\gamma_i)l - (\boldsymbol{n}_i^{\mathrm{T}}\gamma_i)d, \quad i = 1, 2 \tag{31}$$

The details of the derivation of (26) to (31) are given in Appendix A. Then, it is possible to show (see Appendix A) that holds

$$\mathrm{d}P = S_N\,\mathrm{d}\theta + \sum_{i=1,2} \Delta N_i \boldsymbol{e}_i^{\mathrm{T}}\mathrm{d}q_i \tag{32}$$

or

$$\frac{\partial P}{\partial \theta} = S_N \quad \text{and} \quad \frac{\partial P}{\partial q_i} = \Delta N_i \boldsymbol{e}_i \quad (i = 1, 2) \tag{33}$$

That is, a local minimum of P as a function $P(X, s_1, s_2)$ of X, s_1 and s_2 is attained when $S_N = 0$ and $\Delta N_i = 0$ $(i = 1, 2)$. This condition is satisfied when $\overline{O_1 O_2}$ is parallel to $\overline{P_1 P_2}$ and $\Delta N_i = 0$. Finally, the closed-loop dynamics of (26) to (28) can be written in the form, correspondingly to (23),

$$G(X)\ddot{X} + \left(\frac{1}{2}\dot{G}(X) + S(X, \dot{X})\right)\dot{X} + C\dot{X} + \Phi\Lambda + \frac{\partial P}{\partial X} = 0 \tag{34}$$

where $\Lambda = (\Delta f_1, \Delta f_2, \Delta\lambda_1, \Delta\lambda_2)^{\mathrm{T}}$ and Φ is the 8×4-matrix defined by $\Phi = (\Re_{n1}, \Re_{n2}, \Re_{b1}, \Re_{b2})$. We note that, at a regular position of object pinching like Fig. 1, four 8-dimensional column vectors of Φ are independent to each other. That is, Φ is nondegenerate.

Now, we derive the Hessian matrix of $P(X, s_1, s_2)$ at $\partial P(X)/\partial X = 0$ (that is, $S_N = 0$ and $\Delta N_i = 0$ for $i = 1, 2$) by differentiating S_N and ΔN_i in t. Since $P(X, s_1, s_2)$ is regarded as a function of p_i and s_i as seen in (22) and (24), we obtain first the Hessian of P with respect to $\boldsymbol{z} = (\theta, p_1, p_2)^{\mathrm{T}}$ in the following form (see Appendix B):

$$\frac{\partial P}{\partial \boldsymbol{z}} = \left(\frac{\partial P}{\partial \theta}, \frac{\partial P}{\partial p_1}, \frac{\partial P}{\partial p_2}\right)^{\mathrm{T}} = (S_N, \Delta N_1, \Delta N_2)^{\mathrm{T}} \tag{35}$$

$$\left.\frac{\partial^2 P}{\partial \boldsymbol{z} \partial \boldsymbol{z}^{\mathrm{T}}}\right|_{\partial P/\partial \boldsymbol{z} = 0} = k \begin{pmatrix} v_1 + v_2 & -v_1 & -v_2 \\ -v_1 & \alpha_1 + v_{11} & -v_{12} \\ -v_2 & -v_{21} & \alpha_2 + v_{22} \end{pmatrix} = H_z \tag{36}$$

where the details of v_i and v_{ij} are given in Appendix B. Then, it is possible to verify that the Hessian of P with respect to $\boldsymbol{z}$ at $\partial P/\partial \boldsymbol{z} = 0$ is positive by choosing α_i to satisfy $\alpha_i > -l/\kappa_i$ (note that $l < 0$ according to (25)). Finally, the Hessian of P with respect to the position vector X is given by

$$H_X = \frac{\partial^2 P}{\partial X \partial X^{\mathrm{T}}} = \left(\frac{\partial \boldsymbol{z}^{\mathrm{T}}}{\partial X}\right) \frac{\partial^2 P}{\partial \boldsymbol{z} \partial \boldsymbol{z}^{\mathrm{T}}} \left(\frac{\partial \boldsymbol{z}}{\partial X^{\mathrm{T}}}\right) = D^{\mathrm{T}} H_z D \tag{37}$$

where D is a constant 3×8-matrix of the form

$$D = \begin{pmatrix} 1\,0\,0\,0\,0\,0\,0\,0 \\ 0\,0\,0\,1\,1\,1\,0\,0 \\ 0\,0\,0\,0\,0\,0\,1\,1 \end{pmatrix} \tag{38}$$

because $\boldsymbol{z}$ can be expressed by $\boldsymbol{z} = DX$. Thus, the Hessian matrix H_X is degenerate, but it is possible to see that the function $P(X)$ can be regarded as a Morse-Bott function as discussed in the next section.

4 Dirichlet-Lagrange Stability for 2-D Precision Prehension

The equilibrium manifold is determined by the set of all postures that have a form depicted in Fig. 3, in which $\overline{O_1O_2}$ becomes parallel to $\overline{P_1P_2}$. More rigorously, the equilibrium manifold denoted by $N_{P'}$ can be regarded as a set of all X that satisfy $S_N = 0$, $\Delta N_1 = 0$ and ΔN_2, that is, $\partial P/\partial \boldsymbol{z} = 0$. Since at any point X on $N_{P'}$ the tangent space $T_X N_{P'}$ must be the kernel space of the transformation matrix D that is a mapping $\boldsymbol{z} = DX$. In other words, the tangent space $T_X N_{P'}$ is coincident with the null space of the Hessian matrix of P with respect to X. Thus, $P(X)$ can be regarded as a Morse-Bott function whose minimum is attained on $N_{P'}$. It is then possible to prove the asymptotic stability of motion of the prehension by applying a similar method discussed in section 4 of Part I. In fact, by using the orthogonal transformation

$$\dot{X} = \left(R, \Phi(\Phi^{\mathrm{T}}\Phi)^{-1/2}\right)\begin{pmatrix}\dot{\eta} \\ \dot{\varphi}\end{pmatrix} \tag{39}$$

a quotient dynamics of (34) is obtained in such a way that holds

$$\bar{G}\ddot{\eta} + \left(\frac{1}{2}\dot{\bar{G}} + \bar{S}\right)\dot{\eta} + R^{\mathrm{T}}CR\dot{\eta} + R^{\mathrm{T}}D^{\mathrm{T}}\frac{\partial P}{\partial \boldsymbol{z}} = 0 \tag{40}$$

Here, R is composed of 4 column vectors with the unit norm orthogonal to each other satisfying $R^{\mathrm{T}}\Phi = 0$, and $\bar{G} = R^{\mathrm{T}}GR$, and

$$\bar{S} = R^{\mathrm{T}}SR - \frac{1}{2}\dot{R}^{\mathrm{T}}GR + \frac{1}{2}R^{\mathrm{T}}G\dot{R} \tag{41}$$

If, during maneuvering the system, R is always nondegenerate, it is possible to apply the method developed in section 4 of Part I to prove the exponentially asymptotic stability of motion on a manifold.

It should be remarked that, in order to confirm the stability on a manifold regarding the closed-loop dynamics of (34) by applying a similar method to Appendix A in Part I, the damping term $C\dot{X}$ in (34) is not fully dissipated, because there does not originally arise any damping term in object dynamics of (26) and (27). Notwithstanding this, it is important to note that the velocity constraints (13) imply

$$(\dot{\boldsymbol{r}}_1 - \dot{\boldsymbol{r}}_2)\bar{\boldsymbol{n}}_{01} + \dot{\theta}\sum_{i=1,2}\boldsymbol{n}_{0i}^{\mathrm{T}}\gamma_{0i} + \sum_{i=1,2}\dot{p}_i\boldsymbol{n}_i^{\mathrm{T}}\gamma_i = 0 \tag{42}$$

which results in

$$l_w^2\dot{\theta}^2 \leq c_{\theta 1}\|\dot{q}_1\|^2 + c_{\theta 2}\|\dot{q}_2\|^2 \tag{43}$$

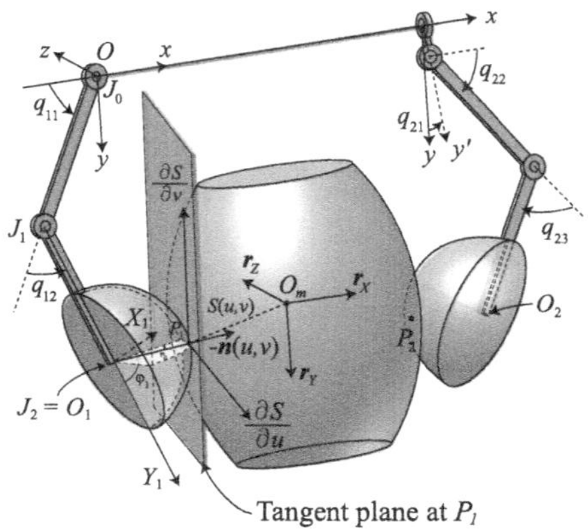

Fig. 4. A pair of robot fingers is grasping a rigid object with arbitrary smooth surfaces. The inertial frame is denoted by the coordinates O-xyz and the body coordinates are expressed by O_m-XYZ with body-fixed unit vectors $\boldsymbol{r}_X$, $\boldsymbol{r}_Y$, and $\boldsymbol{r}_Z$.

with some positive constants $c_{\theta 1}$ and $c_{\theta 2}$. Similarly to this argument, we obtain from (13)

$$\|\dot{\boldsymbol{x}}\|^2 \leq c_{01}\|\dot{q}_1\|^2 + c_{02}\|\dot{q}_2\|^2 + c_{03}\dot{\theta}^2 \tag{44}$$

These two inequalities imply that $R^{\mathrm{T}}CR$ in (40) is positive definite and hence the dynamics of (40) is fully dissipated.

A naive discussion of stability proof of the closed-loop dynamics of pinching with DOF-redundancy is found in [5] for robot fingers with spheric fingerends.

5 Modeling of 3-D Grasping

Dynamics of 3-D grasping (see Fig. 4) of a rigid object with arbitrary surfaces by a pair of robot fingers with smooth surfaces can be also derived under the same assumptions 1) and 2) given in section I. We show the Euler-Lagrange equation of motion of the fingers-object system under rolling contact constraints in the following:

$$M\ddot{\boldsymbol{x}} - f_1\bar{\boldsymbol{n}}_1 - f_2\bar{\boldsymbol{n}}_2 - \lambda_1\bar{\boldsymbol{b}}_1 - \lambda_2\bar{\boldsymbol{b}}_2 - \xi_1\bar{\boldsymbol{e}}_1 - \xi_2\bar{\boldsymbol{e}}_2 = 0 \tag{45}$$

$$H\dot{\boldsymbol{\omega}} + \boldsymbol{\omega}\times H\boldsymbol{\omega} - \sum_{i=1,2}\{f_i(\gamma_{0i}\times\boldsymbol{n}_{0i}) + \lambda_i(\gamma_{0i}\times\boldsymbol{b}_{0i}) + \xi_i(\gamma_{0i}\times\boldsymbol{e}_{0i})\} = 0 \tag{46}$$

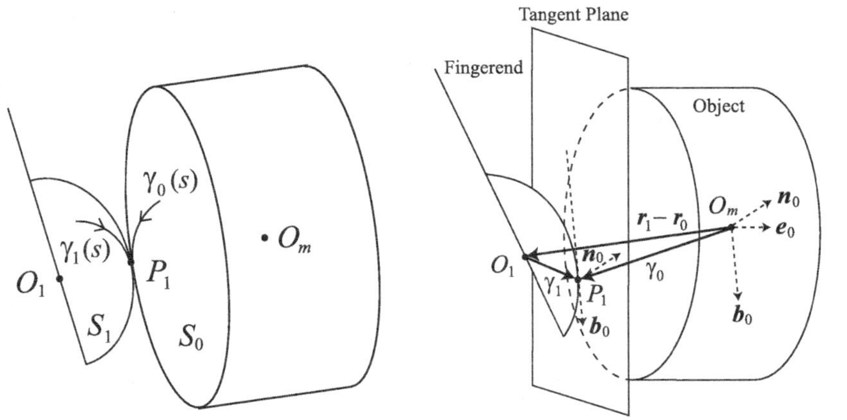

(a) A locus of contact points on the left fingerend surface S_1 is expressed by $\gamma_1(s)$ and that on the object surface S_0 by $\gamma_0(s)$ in each local coordinates.

(b) The contact condition at P_1 is expressed as $\boldsymbol{r}_1 - \boldsymbol{r}_0 = \Pi_0\gamma_0 - \Pi_1\gamma_1$, where $\Pi_0 = (\boldsymbol{r}_X, \boldsymbol{r}_Y, \boldsymbol{r}_Z)$ and $\Pi_1 = (\boldsymbol{r}_{X1}, \boldsymbol{r}_{Y1}, \boldsymbol{r}_{Z1})$ in the frame coordinates.

Fig. 5. Rolling contact constraints

$$G_i(q_i)\ddot{q}_i + \left\{\frac{1}{2}\dot{G}_i + S_i\right\}\dot{q}_i + J_i^{\mathrm{T}}(q_i)\left\{\bar{\boldsymbol{n}}_i f_i + \bar{\boldsymbol{b}}_i\lambda_i + \bar{\boldsymbol{e}}_i\xi_i\right\}$$
$$-W_i^{\mathrm{T}}\left\{\gamma_i \times (\boldsymbol{n}_i f_i + \boldsymbol{b}_i\lambda_i + \boldsymbol{e}_i\xi_i)\right\} = u_i, \quad i = 1, 2 \qquad (47)$$

Here, $q_1 = (q_{11}, q_{12})^{\mathrm{T}}$, $q_2 = (q_{21}, q_{22}, q_{23})^{\mathrm{T}}$, $\gamma_i(s_i)$ denote the locus of contact point P_i as shown in Fig. 5(a), $(\boldsymbol{b}_0, \boldsymbol{n}_0, \boldsymbol{e}_0)$ constitutes the orthogonal matrix expressed in the local coordinates O_m-XYZ (see Fig. 5(b)), $(\boldsymbol{b}_{0i}, \boldsymbol{n}_{0i}, \boldsymbol{e}_{0i})$ has a similar meaning to the 2-D case. Further, $J_i(q_i) = \partial\boldsymbol{r}_i/\partial q_i^{\mathrm{T}}$, $W_i = \partial\boldsymbol{\omega}_i/\partial\dot{q}_i^{\mathrm{T}}$, $\boldsymbol{\omega} = (\omega_X, \omega_Y, \omega_Z)^{\mathrm{T}}$, H denotes the inertia tensor of the object, and

$$\begin{cases} p_1 = q_{11} + q_{12}, & p_2 = q_{22} + q_{23} \\ \boldsymbol{\omega}_1 = (0, 0, \dot{p}_1)^{\mathrm{T}}, & \boldsymbol{\omega}_2 = (\dot{q}_{21}, \dot{p}_2\sin q_{21}, \dot{p}_2\cos q_{21})^{\mathrm{T}} \end{cases} \qquad (48)$$

and further $\Pi_0 = (\boldsymbol{r}_X, \boldsymbol{r}_Y, \boldsymbol{r}_Z)$ is subject to $\dot{\Pi}_0 = \Pi_0\boldsymbol{\omega}\times$. It should be noted that the length parameter s_i is updated by the first order differential equation

$$\{\kappa_{0i}(s_i) + \kappa_{in}(s_i)\}\frac{\mathrm{d}s_i}{\mathrm{d}t} = -\boldsymbol{e}_{0i}^{\mathrm{T}}\boldsymbol{\omega} - \boldsymbol{e}_i^{\mathrm{T}}\boldsymbol{\omega}_i, \quad i = 1, 2 \qquad (49)$$

where κ_{in} denotes the normal curvature of the fingerend of the finger i and κ_{0i} that of the object surface at the contact point P_i. We remark that (46) is expressed in local coordinates O_m-XYZ fixed to the object but (45) and (47) are expressed in frame coordinates O-xyz. Interestingly, (45) and (46) are described in a "wrench" vector form, that is, (45) and (46) as a whole express the Euler-Lagrange equation of the object in wrench space.

The rolling contact constraints are derived from the assumptions 1) and 2) that can be stated as

$$\begin{cases} Q_{bi} = (\boldsymbol{r}_i - \boldsymbol{r}_0)^{\mathrm{T}}\bar{\boldsymbol{b}}_{0i} + (\gamma_i^{\mathrm{T}}\boldsymbol{b}_i - \gamma_{0i}^{\mathrm{T}}\boldsymbol{b}_{0i}) = 0 \\ Q_{ni} = (\boldsymbol{r}_i - \boldsymbol{r}_0)^{\mathrm{T}}\bar{\boldsymbol{n}}_{0i} + (\gamma_i^{\mathrm{T}}\boldsymbol{n}_i - \gamma_{0i}^{\mathrm{T}}\boldsymbol{n}_{0i}) = 0 \\ Q_{ei} = (\boldsymbol{r}_i - \boldsymbol{r}_0)^{\mathrm{T}}\bar{\boldsymbol{e}}_{0i} + (\gamma_i^{\mathrm{T}}\boldsymbol{e}_i - \gamma_{0i}^{\mathrm{T}}\boldsymbol{e}_{0i}) = 0 \end{cases} \tag{50}$$

by taking inner products of the contact condition $\boldsymbol{r}_1 - \boldsymbol{r}_0 = \Pi_0\gamma_0 - \Pi_1\gamma_1$ and $\bar{\boldsymbol{b}}_1$, $\bar{\boldsymbol{n}}_1$, or $\bar{\boldsymbol{e}}_1$. It is possible to show (see [7]) that

$$0 = \frac{\mathrm{d}}{\mathrm{d}t}Q_{bi} = R_{bi} = (\dot{\boldsymbol{r}}_i - \dot{\boldsymbol{r}}_0)^{\mathrm{T}}\bar{\boldsymbol{b}}_{0i} + (\gamma_i \times \boldsymbol{b}_i)^{\mathrm{T}}\boldsymbol{\omega}_i - (\gamma_{0i} \times \boldsymbol{b}_{0i})^{\mathrm{T}}\boldsymbol{\omega} \tag{51}$$

holds and, similarly, $0 = \mathrm{d}Q_{ni}/\mathrm{d}t = R_{ni}$ and $0 = \mathrm{d}Q_{ei}/\mathrm{d}t = R_{ei}$.

6 Conclusions

Dirichlet-Lagrange stability of motion of mechanical systems under geometric constraints is discussed in an extensive way for a class of nonlinear mechanical systems with redundancy in the system's DOF. Specifically, the structual details of the Hessian matrix of an artificial potential related to the problem of stabilization of 2-D precision prehension is presented. It is shown that such an artificial potential function introduced to position control under constraints plays a crucial role as a Morse-Bott function in stability of motions subject to DOF-redundancy. Any discussions on stability of motions of 2-D grasping of a rigid object with arbitrary contours are not yet presented. In the case of 3-D grasping, none of control problems of grasping from the dynamical point of view has been tackled except a limited class of ball-plate problems [7].

References

1. Arimoto, S., Yoshida, M.: Modeling and control of 2-D grasping under rolling contact constraints between arbitrary shapes: A Riemannian-geometry approach. Journal of Robotics 2010, Article ID 926579, 13
2. Nomizu, K.: Kinematics and differential geometry of submanifolds — Rolling a ball with a prescribed locus of contact —, Tohoku Math. Journ. 30, 623–637 (1978)
3. Gray, A., Abbena, E., Salamon, S.: Modern Differential Geometry of Curves and Surfaces with Mathematica. Chapman & Hall/CRC, Boca Raton, Florida, USA (2006)
4. Murray, R.M., Li, Z., Sastry, S.S.: A Mathematical Introduction to Robotic Manipulation. CRC Press, Boca Raton (1994)
5. Arimoto, S.: Intelligent control of multi-fingered hands. Annual Review in Control 28(1), 75–85 (2004)

6. Arimoto, S.: Dynamics of grasping a rigid object with arbitrary smooth surfaces under rolling contacts. SICE J. of Control, Measurement, and System Integration 3 (2010) (to be published)
7. Arimoto, S.: Control Theory of Multi-fingered Hands. Springer, London (2008)

Appendix A

It is evident to see that substitution of (29) into (26) to (28) yields (1), (2), and (20). Next, from (22), (24), (6), and (29) to (31) it follows that

$$
\begin{aligned}
\frac{\mathrm{d}}{\mathrm{d}t}P &= \frac{\mathrm{d}}{\mathrm{d}t}U(s_1,s_2) + \sum_{i=1,2} k\alpha_i \hat{N}_i \frac{\mathrm{d}\hat{N}_i}{\mathrm{d}t} \\
&= \sum_{i=1,2} \left[(-1)^i k \left\{ (\boldsymbol{n}_i^{\mathrm{T}}\gamma_i)d - (-1)^i(\boldsymbol{b}_i^{\mathrm{T}}\gamma_i)l \right\} \kappa_i \frac{\mathrm{d}s_i}{\mathrm{d}t} + k\alpha_i \hat{N}_i \frac{\mathrm{d}P_i}{\mathrm{d}t} \right] \\
&= \sum_{i=1,2} \left[k \left\{ (\boldsymbol{n}_i^{\mathrm{T}}\gamma_i)d - (-1)^i(\boldsymbol{b}_i^{\mathrm{T}}\gamma_i)l \right\} (\dot{\theta} - \dot{p}_i) + k\alpha_i \hat{N}_i \dot{p}_i \right] \\
&= k \left\{ (\boldsymbol{n}_1^{\mathrm{T}}\gamma_1 + \boldsymbol{n}_2^{\mathrm{T}}\gamma_2)d + (\boldsymbol{b}_1^{\mathrm{T}}\gamma_1 - \boldsymbol{b}_2^{\mathrm{T}}\gamma_2)l \right\} \dot{\theta} + \sum_{i=1,2} \left(kN_i + k\alpha_i \hat{N}_i \right) \dot{p}_i \\
&= k \left\{ (l + l_w)d + (s_1 - s_2 - d)l \right\} \dot{\theta} + \sum_{i=1,2} \Delta N_i \dot{p}_i
\end{aligned}
\tag{A-1}
$$

In the light of (30), this equation yields (32).

Appendix B

Calculating directly the derivatives of S_N and ΔN_i for $i = 1, 2$ in t, we have

$$
\begin{cases}
\dfrac{\mathrm{d}S_N}{\mathrm{d}t} = (v_1 + v_2)\dfrac{\mathrm{d}\theta}{\mathrm{d}t} - v_1 \dfrac{\mathrm{d}p_1}{\mathrm{d}t} - v_2 \dfrac{\mathrm{d}p_2}{\mathrm{d}t} \\
\dfrac{\mathrm{d}\Delta N_i}{\mathrm{d}t} = -v_i \dfrac{\mathrm{d}\theta}{\mathrm{d}t} + (k\alpha_i + v_{ii})\dfrac{\mathrm{d}p_i}{\mathrm{d}t} - v_{ij}\dfrac{\mathrm{d}p_j}{\mathrm{d}t}
\end{cases}
\tag{B-1}
$$

for $i \neq j$, where

$$
\begin{cases}
v_i = k \left\{ -l/\kappa_i + l_w(\boldsymbol{n}_i^{\mathrm{T}}\gamma_i) + (-1)^i l_w d(\boldsymbol{b}_i^{\mathrm{T}}\gamma_i)/l \right\} \\
v_{ij} = -k \left\{ (\boldsymbol{n}_1^{\mathrm{T}}\gamma_1)(\boldsymbol{n}_2^{\mathrm{T}}\gamma_2) - (\boldsymbol{b}_1^{\mathrm{T}}\gamma_1)(\boldsymbol{b}_2^{\mathrm{T}}\gamma_2) \right\} \\
v_{ii} = k \left\{ -l/\kappa_i - l\boldsymbol{n}_i^{\mathrm{T}}\gamma_i + d\boldsymbol{b}_i^{\mathrm{T}}\gamma_i + (\boldsymbol{n}_i^{\mathrm{T}}\gamma_i)^2 + (\boldsymbol{b}_i^{\mathrm{T}}\gamma_i)^2 \right\}
\end{cases}
\tag{B-2}
$$

for $i = 1, 2$ and $j \neq i$. Note that $l < 0$, all v_1, v_2, v_{11}, and v_{22} are positive, and $v_{12} = v_{21}$ when $S_N = 0$ and $\Delta N_i = 0$ for $i = 1, 2$.

Sliding Mode Control for a High-Speed Linear Axis Driven by Pneumatic Muscles

Harald Aschemann and Dominik Schindele

Abstract. This paper presents a cascaded sliding mode control scheme for a new pneumatic linear axis. Its guided carriage is driven by a nonlinear mechanism consisting of a rocker with a pair of pneumatic muscle actuators arranged at both sides. Modelling leads to a system of four nonlinear differential equations including polynomial approximations of the volume characteristic as well as the force characteristic of the pneumatic muscles. The differential flatness of the system is exploited in combination with sliding mode techniques to stabilize the error dynamics. Furthermore, a proxy-based sliding mode controller was designed, which is a modified version of sliding mode control as well as an extension of PID control. It allows for accurate tracking during normal operation and smooth recovery from large position errors after unexpected incidents. The internal pressure of each pneumatic muscle is controlled by a fast underlying control loop, whereas in an outer control loop the carriage position and the mean internal pressure of the muscles are controlled. Remaining model uncertainties are compensated by a disturbance observer. Experimental results show an excellent control performance.

1 Introduction

Linear electrical direct drives provide for both high dynamics as well as small tracking errors but they are subject to a significant disadvantage: their thermal behaviour when high forces have to be applied for a longer period of time. As an alternative, current research focusses on a novel pneumatical linear drive that is actuated by pneumatic muscles in combination with a nonlinear mechanism: this low-cost solution does not show the mentioned thermal problem, provides large maximum forces, and allows for maximum velocities of approx. 1.3 m/s in a workspace of approx.

Harald Aschemann · Dominik Schindele
Chair of Mechatronics, University of Rostock, 18059 Rostock, Germany
e-mail: {harald.aschemann,dominik.schindele}@uni-rostock.de

J. Lévine & P. Müllhaupt (Eds.): Adv. in Theory of Control, Signals, LNCIS 407, pp. 31–40.
springerlink.com © Springer-Verlag Berlin Heidelberg 2010

1 m. The main advantages of pneumatic muscles as compared to classical cylinders are given by a larger maximum force, a significantly reduced weight, the absence of stick-slip effects, and the insensitivity to dirty working environment. Furthermore, pneumatic muscle actuators are significantly cheaper than pneumatic cylinders or electric direct drives with the same maximum force. Due to the nonlinear characteristics of a pneumatic muscle nonlinear control approaches are mandatory. A nonlinear drive mechanism is employed as depicted in Fig. 1. The carriage is driven by a rocker. A roller bearing unit at the tip of the rocker allows for both a rotational and translatory relative motion and transmits the drive force to the carriage. The rocker is actuated by a pair of pneumatic muscles in an antagonistic arrangement. The mounting points of the pneumatic muscles at the rocker have be defined so as to gain a reasonable tradeoff between increase in maximum velocity and reduction of the achievable drive force. The mass flow rate of compressed air into and out of each pneumatic muscle is controlled by means of a separate proportional valve, respectively. Pressure declines in the case of large mass flow rates are avoided by using an air accumulator for each muscle. The paper is structured as follows: first, the

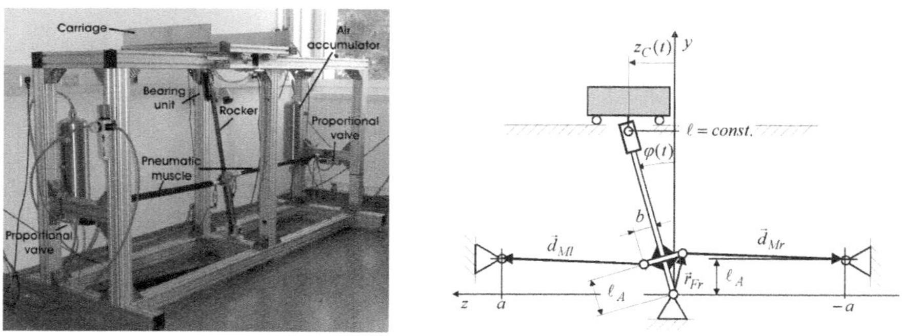

Fig. 1. Experimental setup (left), kinematical structure of the high-speed linear axis (right)

control-oriented modelling of the mechatronic system is addressed. For the nonlinear characteristics of the pneumatic muscle, i.e. the muscle volume and the muscle force, polynomial descriptions are used in terms of contraction length and internal muscle pressure. Second, by taking advantage of differential flatness, sliding mode control techniques are employed to design a nonlinear cascade control. The inner control loops involve a fast pressure control for each muscle, respectively. The outer control loop achieves a decoupling of rocker angle and mean muscle pressure as controlled variables and provides the reference pressures for the inner pressure control loops. In order to account for model uncertainties in the equation of motion, a disturbance torque is introduced and estimated by a nonlinear reduced-order disturbance observer. As shown by experimental results, desired trajectories for both rocker angle and mean pressure can be tracked independently with high accuracy.

2 Modelling of the Mechanical Structure

As for modelling, first the mechanical system part is regarded. The chosen mechanical model for the high-speed axis consists of three elements (Fig. 1): a rigid body for the rocker as actuated link (mass m_R), a single lumped mass m_A for the lateral connecting rods and a lumped mass m_C for the carriage. The inertial yz-coordinate system is chosen in the base joint of the rocker. The mounting points of the pneumatic muscles at the rocker are characterised by the distance l_A in longitudinal direction and the perpendicular distance b of the lateral connecting rods (right part of Fig. 1). The motion of the linear axis is completely described by the generalized coordinate $\varphi(t)$, which denotes the inclination of the rocker w.r.t. the plumb line. The carriage position is related to the rocker angle by the horizontal component $z_C(t) = \ell \cdot \tan\varphi(t)$, where ℓ denotes the constant distance in y-direction between the rotary joint at the carriage and the rocker joint. The nonlinear equation of motion directly follows from Lagrange's equations in form of a second order differential equation

$$J(\varphi)\ddot{\varphi} + k(\varphi, \dot{\varphi}) = \tau - \tau_U , \qquad (1)$$

with the resulting mass moment of inertia $J(\varphi)$ and the term $k(\varphi, \dot{\varphi})$, which takes into account the centrifugal as well as the gravity forces [10]. The drive torque τ resulting from the muscle forces F_{Mi}, $i = \{l, r\}$ can be stated as

$$\tau = \vec{e}_x \cdot (F_{Mr} \cdot \vec{r}_{Fr} \times \vec{e}_{Mr} + F_{Ml} \cdot \vec{r}_{Fl} \times \vec{e}_{Ml}) , \qquad (2)$$

with the unity vector $\vec{e}_x$ in x-direction and the unity vectors $\vec{e}_{Mi} = \vec{d}_{Mi}/d_{Mi}$ in direction of the pneumatic muscle forces. The position vectors $\vec{r}_{Fi}$ describe the connecting points, where the muscle forces act on the rocker. All remaining model uncertainties are taken into account by the disturbance torque τ_U. On the one hand, these uncertainties stem from approximation errors concerning the static muscle force characteristics. On the other hand, time-varying damping and friction acting on the carriage as well as on the rocker depend in a complex manner on lots of influence factors and cannot be accurately represented by a simple friction model.

3 Modelling of the Pneumatic System Part

In this section the modelling of the pneumatic system part is adressed. In this way the dynamics of the internal muscle pressures is regarded, in contrast to the model of [3], [8]. A mass flow $\dot{m}_{Mi}$ into the pneumatic muscle leads to an increase in internal pressure p_{Mi} and a contraction $\Delta\ell_{Mi}$ in longitudinal direction due to specially arranged fibers. This contraction effect can be exploited to generate muscle forces. The force F_{Mi} and the volume V_{Mi} of a pneumatic muscle depend in a nonlinear way on the according internal pressure p_{Mi} as well as the contraction length $\Delta\ell_{Mi}$. Given the length of the uncontracted muscle ℓ_M, the contraction length $\Delta\ell_{Mi} = \ell_M - d_{Mi}(\varphi)$

can be calculated with the distance d_{Mi} between both connecting points of each muscle [9]. The dynamics of the internal muscle pressure can be derived from a mass flow balance in combination with the pressure-density relationship. As the internal muscle pressure is limited by a maximum value of 7 bar, the ideal gas equation represents an accurate description of the thermodynamic behaviour. The thermodynamic process is modelled as a polytropic change of state with $n = 1.26$ as identified polytropic exponent [5]. The identified volume characteristic of the pneumatic muscle can be described by a polynomial function of both contraction length and muscle pressure [1]

$$V_{Mi}(\Delta\ell_{Mi}, p_{Mi}) = \sum_{j=0}^{3} a_j \cdot \Delta\ell_{Mi}^j \cdot \sum_{k=0}^{1} b_k \cdot p_{Mi}^k . \tag{3}$$

Finally, the resulting pressure dynamics for the muscle i is given by [9]

$$\dot{p}_{Mi} = \frac{n}{V_{Mi} + n \cdot \frac{\partial V_{Mi}}{\partial p_{Mi}} \cdot p_{Mi}} \left[R \cdot T_{Mi} \cdot \dot{m}_{Mi} - \frac{\partial V_{Mi}}{\partial \Delta\ell_{Mi}} \cdot \frac{\mathrm{d}\Delta\ell_{Mi}}{\mathrm{d}\varphi} \cdot p_{Mi} \cdot \dot{\varphi} \right], \tag{4}$$

where R_L denotes the gas constant of air. The temperature T_{Mi} in a pneumatic muscle is not measured but can be approximated with good accuracy by the temperature T_{amb} of the ambiance. The muscle force F_{Mi} depends on the internal pressure p_{Mi} as well as the contraction length $\Delta\ell_{Mi}$ and represents the connection of the mechanical and the pneumatic system part. Its nonlinear characteristic has been identified by measurements and, then, approximated by the following polynomial description [1]

$$F_{Mi} = \begin{cases} \bar{F}_{Mi}, & \bar{F}_{Mi} > 0 \\ 0 & , \text{ else} \end{cases} , \quad \bar{F}_{Mi} = \sum_{m=0}^{3} (a_m \cdot \Delta\ell_{Mi}^m) \, p_{Mi} - \sum_{n=0}^{4} (b_n \cdot \Delta\ell_{Mi}^n) . \tag{5}$$

4 Tracking Control Design

The sliding mode control design is performed by exploiting the differential flatness property of the system under consideration [11]. The robustness of the resulting control structure w.r.t. external disturbances as well as unmodeled dynamics is increased as compared to the usual flatness based control design [11].

4.1 Sliding Mode Control of Internal Muscle Pressure

With the internal muscle pressure as flat output candidate $y_i = p_{Mi}$, the nonlinear state equation(4) can be solved for the mass flow as input variable $u_{pi} = \dot{m}_{Mi}$

$$\dot{m}_{Mi} = \frac{1}{k_{ui}(\Delta\ell_{Mi}, p_{Mi})} \cdot \left[\dot{p}_{Mi} + k_{pi}(\Delta\ell_{Mi}, \Delta\dot{\ell}_{Mi}, p_{Mi}) \cdot p_{Mi} \right] . \tag{6}$$

The contraction length $\Delta\ell_{Mi}$ as well as its time derivative $\Delta\dot{\ell}_{Mi}$ can be considered as scheduling parameters in a gain-scheduled adaptation of k_{ui} and k_{pi}. With the internal pressure as flat output, its first time derivative $\dot{p}_{Mi} = \upsilon_{pi}$ is introduced as new control input (Fig. 3). According to the given first order pressure dynamics the following sliding surfaces s_{pi} are defined [12], [6]

$$s_{pi} = p_{Mid} - p_{Mi} = e_{pi}. \tag{7}$$

The convergence to the sliding surfaces in face of model uncertainty can be achieved by introducing the following dynamics involving discontinuous signum-functions

$$\dot{s}_{pi} = -W_{pi} \cdot sign(s_{pi}), \;\; W_{pi} > 0\,. \tag{8}$$

With properly chosen coefficients W_{pi} the sliding surfaces $s_{pi} = 0$ are attained in a finite amount of time depending on the initial conditions. Consequently, the corresponding control laws become

$$\upsilon_{pi} = \dot{p}_{Mid} + W_{pi} \cdot sign(s_{pi}). \tag{9}$$

4.2 *Sliding Mode Decoupling Control*

For the outer control loop, the following candidates can be chosen as flat outputs [1]: the rocker angle $y_1 = \varphi$ and the mean pressure $y_2 = p_M = (p_{Ml} + p_{Mr})/2$. The trajectory control of the mean pressure allows for increasing stiffness concerning disturbance forces acting on the carriage [2]. The input variables are represented by the muscle pressures p_{Ml} and p_{Mr} and can be calculated by the inverse dynamics [1]

$$\mathbf{u} = \begin{bmatrix} u_l \\ u_r \end{bmatrix} = \begin{bmatrix} p_{Ml}\,(y_1, \dot{y}_1, \upsilon_1, \upsilon_2, \tau_U) \\ p_{Mr}\,(y_1, \dot{y}_1, \upsilon_1, \upsilon_2, \tau_U) \end{bmatrix}, \tag{10}$$

with the control inputs $\upsilon_1 = \ddot{y}_1$ and $\upsilon_2 = y_2$. A sliding surface s_φ is defined for the outer control loop in the form

$$s_\varphi = (\dot{\varphi}_d - \dot{\varphi}) + \alpha_1 \cdot (\varphi_d - \varphi) = \dot{e}_\varphi + \alpha_1 \cdot e_\varphi, \tag{11}$$

At this, the coefficient α_1 must be chosen positive in order to obtain a Hurwitz-polynomial. The convergence to the sliding surfaces in face of model uncertainty can be achieved by specifying a discontinuous signum-function

$$\dot{s}_\varphi = -W_\varphi \cdot sign(s_\varphi), \;\; W_\varphi > 0\,. \tag{12}$$

With a properly chosen positive coefficient W_φ, the sliding surface $s_\varphi = 0$ is reached in finite time depending on the initial conditions. This leads to the stabilizing control law for the rocker angle

$$\upsilon_1 = \upsilon_\varphi = \ddot{\varphi}_d + \alpha_1 \cdot (\dot{\varphi}_d - \dot{\varphi}) + W_\varphi \cdot sign(s_\varphi). \tag{13}$$

For the second stabilizing control input υ_2 the desired trajectory for the mean pressure p_{Md} is directly utilised in a feedforward manner, i.e., $\upsilon_2 = p_{Md}$.

4.3 Reduction of Chattering

For implementation, an ideal actuator switching behaviour is not preferable in view of noise emission, wear and resulting reduced lifetime of the proportional valves. Hence, instead of the discontinuous switching function $sign(s_j)$ the continuous hyperbolic tangent function can be advantageously employed

$$W_j \cdot tanh\left(\frac{s_j}{\varepsilon}\right), \; j = \{pl, pr, \varphi\}. \tag{14}$$

Using these switching functions, high-frequency chattering can be reduced. This comes at the price of a non-ideal sliding mode within a resulting boundary layer determined by the parameter ε in the switching function.

4.4 Proxy-Based Sliding Mode Decoupling Control

Proxy-based sliding mode control is a modification of sliding mode control as well as an extension of PID-control [7], [13]. The basic idea is to introduce a virtual carriage, called proxy, which is controlled using sliding mode techniques, whereas the proxy is connected to the real carriage by a PID-type coupling force, see Fig. 2. The goal of proxy-based sliding mode is to achieve precise tracking during normal operation and smooth, overdamped recovery in case of large position errors, which leads to an inherent safety property. The sliding mode control law for the virtual carriage results from equation (13) with φ_s denoting the rocker angle of the proxy

$$\upsilon_a = \ddot{\varphi}_d + \alpha_1 \cdot (\dot{\varphi}_d - \dot{\varphi}_s) + W_\varphi \cdot tanh\left(\frac{\dot{\varphi}_d - \dot{\varphi}_s + \alpha_1\,(\varphi_d - \varphi_s)}{\varepsilon}\right). \tag{15}$$

The PID-type virtual coupling between the proxy and the real carriage is given by

$$\upsilon_c = K_I \int (\varphi_s - \varphi)\,\mathrm{dt} + K_P\,(\varphi_s - \varphi) + K_D\,(\dot{\varphi}_s - \dot{\varphi}). \tag{16}$$

Assuming a proxy with vanishing inertia, the condition $\upsilon_a = \upsilon_c$ holds. By introducing the new variable a as integrated difference between the real and the virtual rocker angle $a = \int (\varphi_s - \varphi)\,\mathrm{dt}$, the virtual coupling (16) and the stabilizing proxy-based sliding mode control law (15) result in [7]

$$\upsilon_c = K_I a + K_P \dot{a} + K_D \ddot{a}\,, \tag{17}$$

$$\upsilon_a = \ddot{\varphi}_d + \alpha_1 \dot{e}_\varphi - \alpha_1 \ddot{a} + W_\varphi tanh\left(\frac{\dot{e}_\varphi + \alpha_1 e_\varphi - \alpha_1 \dot{a} - \ddot{a}}{\varepsilon}\right). \tag{18}$$

The implementation of the control law is shown in the right part of Fig. 2.

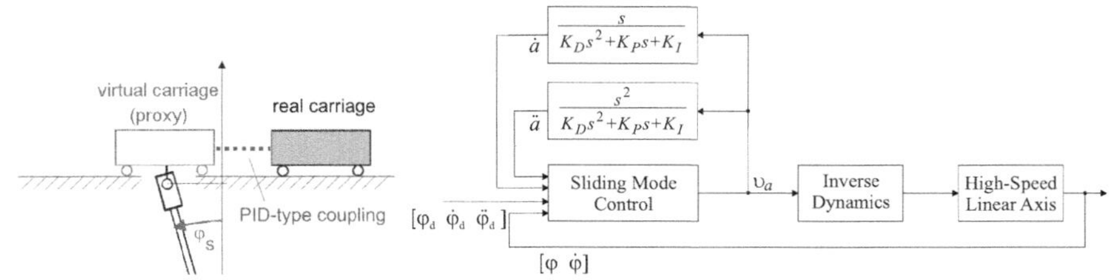

Fig. 2. Coupling between virtual and real carriage (left). Implementation of the proxy-based sliding mode control (right).

4.5 Nonlinear Reduced-Order Disturbance Observer

Disturbance behaviour and tracking accuracy in view of model uncertainties can be significantly improved by introducing a nonlinear reduced-order disturbance observer as described in [4]. The observer design is based on the equation of motion. The key idea is to extend the state equation with an integrator as disturbance model

$$\dot{\mathbf{y}} = \mathbf{f}(\mathbf{y}, \tau_U, \mathbf{u}) \ , \dot{\tau}_U = 0 \ , \tag{19}$$

where $\mathbf{y} = [\varphi, \dot{\varphi}]^T$ denotes the measurable state vector. The estimated disturbance torque and the state equation for z are given by

$$\hat{\tau}_U = \mathbf{h}^T (\mathbf{y}, \hat{\tau}_U, \mathbf{u}) \cdot \mathbf{y} + z \ , \tag{20}$$

$$\dot{z} = \Phi (\mathbf{y}, \hat{\tau}_U, \mathbf{u}) \ , \tag{21}$$

with the chosen observer gain $\mathbf{h}^T = \left[h_1 \ h_1 \right]$. The observer gain $\mathbf{h}$ and the nonlinear function Φ have to be chosen properly, so that the steady-state observer error $e = \tau_U - \hat{\tau}_U$ converges to zero. Thus, the function Φ can be determined as follows

$$\dot{e} = 0 = \dot{\tau}_U - \mathbf{h}^T (\mathbf{y}, \hat{\tau}_U, \mathbf{u}) \cdot \dot{\mathbf{y}} - \Phi (\mathbf{y}, \tau_U - 0, \mathbf{u}) \ . \tag{22}$$

In view of $\dot{\tau}_U = 0$, equation (22) yields

$$\Phi (\mathbf{y}, \tau_U - 0, \mathbf{u}) = -\mathbf{h}^T (\mathbf{y}, \hat{\tau}_U, \mathbf{u}) \cdot \dot{\mathbf{y}} \ . \tag{23}$$

The linearized error dynamics has to be made asymptotically stable. Accordingly all eigenvalues of the Jacobian $\mathbf{J}_e = \frac{\partial \Phi(\mathbf{y}, \hat{\tau}_U, \mathbf{u})}{\partial (\hat{\tau}_U)}$ must lie in the left complex half-plane. This can be achieved by proper choice of the observer gain h_1. The stability of the closed-loop control system has been investigated by thorough simulations.

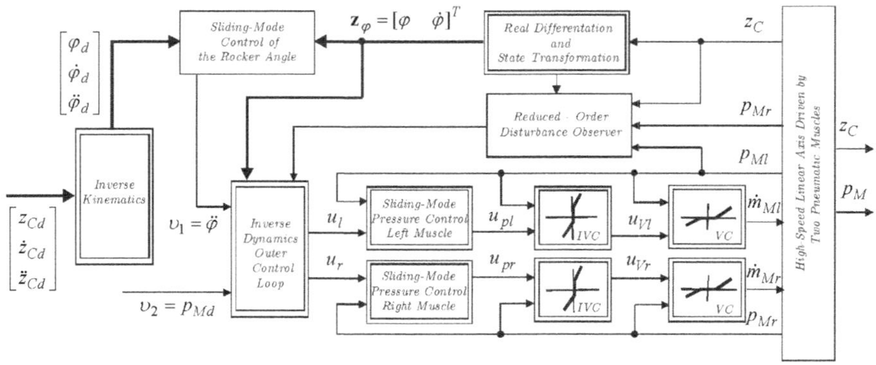

Fig. 3. Implementation of the sliding mode control

5 Experimental Results

For experiments at the test rig, the control structure depicted in Fig. 3 has been implemented. Within the inverse kinematics, the desired values for the rocker angle as well as the according first two time derivatives are calculated from the desired trajectory for the carriage position. The only imperfectly known inverse dynamics is corrected by an estimate of the resulting disturbance torque that acts on the rocker. Underlying fast sliding mode pressure control loops achieve an accurate tracking behaviour for the desired pressures provided by the outer control loop. The nonlinear valve characteristic (VC) has been identified by measurements [1] and is compensated by its approximated inverse valve characteristic (IVC) in each input channel. For investigation of both tracking performance and steady-state accuracy, the tracking of synchronized desired trajectories for the controlled variables as shown in Fig. 4 has been considered. Here, the desired trajectories for the carriage position, the corresponding desired velocity, the acceleration, and the mean pressure are depicted. These desired trajectories are subject to a maximum carriage velocity of 1.3 m/s and a maximum carriage acceleration of 5 m/s^2. The resulting tracking error for the carriage e_z is depicted in the middle lower part of Fig. 4. As for the carriage position, the maximum tracking error during the acceleration and deceleration intervals is approx. 3.5 mm. The steady-state error is negligible. The load stiffness of the carriage regarding external disturbances can be enlarged by specifying higher mean pressures within the admissible pressure interval. The specified reference trajectory for the mean pressure varies between 4 bar and 5 bar. Concerning the mean pressure, maximum control errors of approx. 0.3 bar occur during the movements, whereas the steady-state error is less than 0.03 bar, see the right lower part of Fig. 4. By introducing the estimated disturbance torque into the inverse dynamics, the tracking error is reduced significantly, as shown in the left part of Fig. 5. In this figure, also a comparison of sliding mode control and proxy-based sliding mode control is depicted. The robustness of the proposed solution is shown by a non-modelled additional mass of 25 kg, which represents almost the double of the

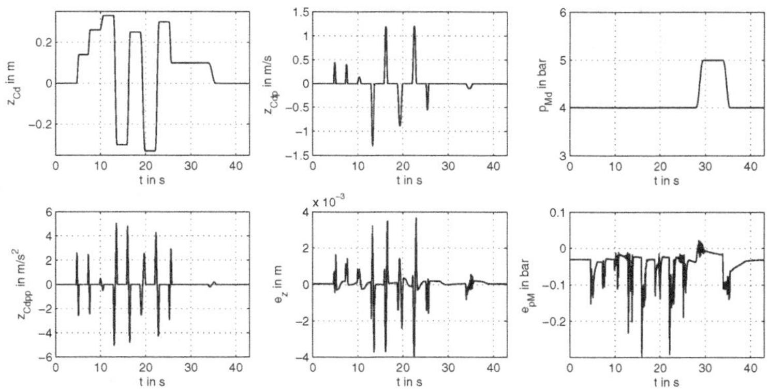

Fig. 4. Desired values for the carriage position, velocity, acceleration and mean pressure; corresponding control errors $e_z = z_{Cd} - z_C$ and $e_{pM} = p_{Md} - p_M$

nominal value. The corresponding tracking errors with and without additional mass are depicted in the right part of Fig. 5. Whereas the steady-state errors remain almost unchanged, the maximum tracking error increases up to approx. 8 mm due to the unmodelled inertia forces during the acceleration and deceleration phases. The closed-loop stability, however, is not affected by this parametric uncertainty.

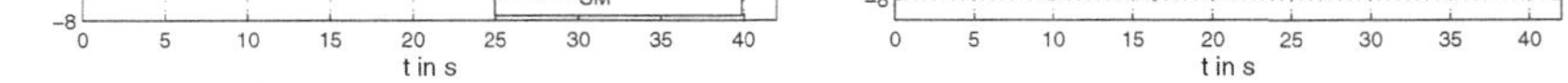

Fig. 5. Comparison of sliding mode control (SM) with and without disturbance compensation and proxy-based sliding mode control (PBSM) concerning the corresponding control error $e_z = z_{Cd} - z_C$ (left); Tracking error e_z with and without an additional mass of 25 *kg* (right)

6 Conclusion

In this paper, a nonlinear cascaded trajectory control is presented for a new linear axis driven by pneumatic muscles that offers a significant increase in both workspace and maximum velocity as compared to a directly actuated solution. The modelling of this mechatronic system leads to nonlinear system equations of fourth order containing identified polynomial descriptions of the nonlinear characteristics of the pneumatic muscle. The nonlinear valve characteristic is linearized using its approximated inverse characteristic. The inner control loops involve a sliding mode control of the internal muscle pressure with high bandwidth. The outer control loop achieves a sliding mode decoupling control of the rocker angle and the mean muscle pressure as controlled variables. As alternative to the standard sliding mode controller, a proxy-based sliding mode controller is introduced for the outer control loop, which aims at increasing the safety property during operation. Uncertainties in the muscle force characteristics as well as nonlinear friction are directly taken into account by a compensation scheme based on a nonlinear disturbance observer. Experimental results emphasize the excellent closed-loop performance with maximum position errors of approx. 3.5 mm during the movements, negligible steady-state position error and steady-state pressure error of less than 0.03 bar.

References

1. Aschemann, H., Schindele, D.: Sliding-mode control of a high-speed linear axis driven by pneumatic muscle actuators. IEEE Trans. Ind. Electronics 55(11), 3855–3864 (2008)
2. Bindel, R., Nitsche, R., Rothfuß, R., Zeitz, M.: Flatness based control of two valve hydraulic joint actuator of a large manipulator. In: Proc. of ECC 1999, Karlsruhe, Germany (1999)
3. Carbonell, P., Jiang, Z., Repperger, D.: Comperative study for three nonlinear control strategies for a pneumatic muscle actuator. In: Proc. of NOLCOS 2001, Saint-Petersburg, Russia, pp. 167–172 (2001)
4. Friedland, B.: Advanced Control System Design. Prentice-Hall, Englewood Cliffs (1996)
5. Hildebrandt, A., Sawodny, O., Neumann, R., Hartmann, A.: A flatness based design for tracking control of pneumatic muscle actuators. In: Proc. 7th Int. Conf. on Control, Automation, Robotics and Vision, Singapore, pp. 1156–1161 (2002)
6. Khalil, H.K.: Nonlinear systems, 2nd edn. Prentice Hall, Englewood Cliffs (1996)
7. Kikuuwe, R., Fujimoto, H.: Proxy-based sliding mode control for accurate and safe position control. IEEE Trans. on Industr. Electr. 53(5), 25–30 (2006)
8. Lilly, J., Yang, L.: Sliding mode control tracking for pneumatic muscle actuators in opposing pair configuration. IEEE Trans. on Contr. Syst. Techn. 13(4), 550–558 (2005)
9. Schindele, D., Aschemann, H.: Backstepping control of a high-speed linear axis driven by pneumatic muscles. In: Proc. of 17th IFAC World Congress, Seoul, Korea, pp. 7684–7689 (2008)
10. Schindele, D., Aschemann, H.: Disturbance compensation strategies for a high-speed linear axis driven by pneumatic muscles. In: Proc. of ECC 2009, Budapest, Hungary, pp. 436–441 (2009)
11. Sira-Ramirez, H., Llanes-Santiago, O.: Sliding mode control of nonlinear mechanical vibrations. J. of Dyn. Systems, Meas. and Control 122(12), 674–678 (2000)
12. Slotine, J.J.E., Li, W.: Applied Nonlinear Control. Prentice-Hall, Englewood Cliffs (1991)
13. Van-Damme, M., Vanderborght, R., Ham, R.V., Verrelst, B., Daerden, F., Lefeber, D.: Proxy-based sliding-mode control of a manipulator actuated by pleated pneumatic artificial muscles. In: Proc. IEEE Int. Conf. on Robotics and Automation, Rome, Italy, pp. 4355–4360 (2007)

Using Hamiltonians to Model Saturation in Space Vector Representations of AC Electrical Machines

Duro Basic, Al Kassem Jebai, François Malrait,
Philippe Martin, and Pierre Rouchon

Abstract. An Hamiltonian formulation with complex fluxes and currents is proposed. This formulation is derived from a recent Lagrangian formulation with complex electrical quantities. The complexification process avoids the usual separation into real and imaginary parts and notably simplifies modeling issues. Simple modifications of the magnetic energy underlying standard (α, β) models yield new (α, β) models describing machines with magnetic saturation and saliency. We prove that the usual expression of the electro-mechanical torque (wedge product of fluxes and currents) is related to a rotational invariance characterizing sinusoidal machines.

1 Introduction

In [1] a Lagrangian formulation with complex currents and fluxes is proposed. In this paper we develop the Hamiltonian counterpart only sketched in [1]. For three-phase electrical machines we recall the usual model linear in fluxes, currents and voltages, and give its Hamiltonian formulation based on magnetic energies depending quadratically on fluxes. We then propose a modification of the usual magnetic energies in order to take into account magnetic saturation. We prove that if these additional terms preserve the rotational invariance of the usual magnetic energies, then the resulting electro-magnetic torque always admits the usual form and is thus still proportional to the imaginary part of the product of complex conjugate of fluxes with stator currents.

Duro Basic · François Malrait
Schneider Electric, STIE, 33, rue André Blanchet, 27120 Pacy-sur Eure, France

Al Kassem Jebai · Philippe Martin · Pierre Rouchon
Mines ParisTech, Centre Automatique et Systèmes, Mathématiques et Systèmes, 60 Bd Saint-Michel, 75272 Paris cedex 06, France

J. Lévine & P. Müllhaupt (Eds.): Adv. in Theory of Control, Signals, LNCIS 407, pp. 41–48.
springerlink.com © Springer-Verlag Berlin Heidelberg 2010

Section 2 is devoted to permanent-magnet machines. In subsection 2.1 we present the Hamiltonian formulation of the usual model with saliency effects. In subsection 2.2 we introduce a class of saturation models and prove that, if we just replace in the usual model the constant inductances by inductances depending on the flux level, the resulting model does not admit in general a magnetic energy and thus is not correct from a physical ground. For sinusoidal machines where the magnetic energy is invariant with respect to the choice of angle origin, we prove in subsection 2.3 that the usual formula giving the electro-magnetic torque as a wedge product between the flux and current remains valid even in the presence of saliency and magnetic saturation. Section 3 is devoted to induction machines. In subsection 3.1 we present the Hamiltonian formulation of the usual model. In subsection 3.2 we introduce a class of saturation models. For machines with sinusoidally wound phases where the magnetic energy is invariant with respect to the choice of angle origin, we prove in subsection 3.3 that the usual formula giving the electro-magnetic torque as a wedge product between the flux and current remains valid even in the presence of magnetic saturation. In section 4 we suggest some further developments.

The authors acknowledge John Chiasson for interesting discussions and precious comments.

2 Permanent-Magnet Machines

2.1 Hamiltonian Modeling

In the (α, β) frame (total power invariant transformation), the usual dynamic equations read (see, e.g., [2, 4]):

$$\begin{cases} \frac{d}{dt}\left(J\dot{\theta}\right) = n_p \Im\left((\lambda \imath_s^* + \bar{\phi} e^{-\jmath n_p \theta} - \mu \imath_s e^{-2\jmath n_p \theta})\imath_s\right) - \tau_L \\ \frac{d}{dt}\left(\lambda \imath_s + \bar{\phi} e^{\jmath n_p \theta} - \mu \imath_s^* e^{2\jmath n_p \theta}\right) = u_s - R_s \imath_s \end{cases} \tag{1}$$

where

- * stands for complex-conjugation, $\Im$ means imaginary part, $\jmath = \sqrt{-1}$ and n_p is the number of pairs of poles.
- θ is the rotor mechanical angle, J and τ_L are the inertia and load torque, respectively.
- $\imath_s \in \mathbb{C}$ is the stator current, $u_s \in \mathbb{C}$ the stator voltage.
- $\lambda = (L_d + L_q)/2$ and $\mu = (L_q - L_d)/2$ (inductances $L_d > 0$ and $L_q > 0$, saliency when $L_d \neq L_q$).
- The constant $\bar{\phi} > 0$ represents the rotor flux due to the permanent magnets.

It is proved in [1] that (1) admits the following Hamiltonian formulation

$$\frac{d}{dt}\left(J\dot{\theta}\right) = -\frac{\partial \mathcal{H}_m}{\partial \theta} - \tau_L, \quad \frac{d}{dt}\phi_s = u_s - R_s \imath_s, \quad \imath_s = 2\frac{\partial \mathcal{H}_m}{\partial \phi_s^*} \tag{2}$$

where the magnetic energy $\mathcal{H}_m$

$$\begin{aligned}\mathcal{H}_m(\phi_s, \phi_s^*, \theta) &= \frac{1}{2L_d}\left(\Re(\phi_s e^{-\jmath n_p \theta}) - \bar{\phi}\right)^2 + \frac{1}{2L_q}\left(\Im(\phi_s e^{-\jmath n_p \theta})\right)^2 \\ &= \frac{1}{2L_d}\left(\frac{\phi_s e^{-\jmath n_p \theta} + \phi_s^* e^{\jmath n_p \theta}}{2} - \bar{\phi}\right)^2 + \frac{1}{2L_q}\left(\frac{\phi_s e^{-\jmath n_p \theta} - \phi_s^* e^{\jmath n_p \theta}}{2\jmath}\right)^2 \\ &= \frac{1}{8L_d}\left(\phi_s e^{-\jmath n_p \theta} + \phi_s^* e^{\jmath n_p \theta} - 2\bar{\phi}\right)^2 - \frac{1}{8L_q}\left(\phi_s e^{-\jmath n_p \theta} - \phi_s^* e^{\jmath n_p \theta}\right)^2\end{aligned} \tag{3}$$

where the rotor angle θ, the stator flux ϕ_s and its complex conjugate ϕ_s^* are considered independent variables when computing the partial derivatives of $\mathcal{H}_m$. In particular, $\imath_s = 2\frac{\partial \mathcal{H}_m}{\partial \phi_s^*}$ reads

$$\begin{aligned}\imath_s &= \tfrac{e^{\jmath n_p \theta}}{2L_d}\left(\phi_s e^{-\jmath n_p \theta} + \phi_s^* e^{\jmath n_p \theta} - 2\bar{\phi}\right) + \tfrac{e^{\jmath n_p \theta}}{2L_q}\left(\phi_s e^{-\jmath n_p \theta} - \phi_s^* e^{\jmath n_p \theta}\right) \\ &= \left(\tfrac{1}{2L_d} + \tfrac{1}{2L_q}\right)\phi_s - \tfrac{1}{L_d}\bar{\phi} e^{\jmath n_p \theta} + \left(\tfrac{1}{2L_d} - \tfrac{1}{2L_q}\right)\phi_s^* e^{2\jmath n_p \theta}.\end{aligned}$$

Inverting this relation we recover the usual relation between ϕ_s and the stator current

$$\phi_s = \lambda \imath_s + \bar{\phi} e^{\jmath n_p \theta} - \mu \imath_s^* e^{2\jmath n_p \theta}.$$

2.2 Magnetic Saturation

To take into account magnetic saturation, we keep the structure equations (2) and modify the magnetic energy $\mathcal{H}_m$ given in (3). For obvious physical reasons, L_d and L_q should be decreasing functions of $|\phi_s|^2$. The simplest magnetic saturation model will be given by setting

$$\mathcal{H}_m(\phi_s, \phi_s^*, \theta) = \frac{S_d(|\phi_s|^2)}{\bar{L}_d}\left(\Re(\phi_s e^{-\jmath n_p \theta}) - \bar{\phi}\right)^2 + \frac{S_q(|\phi_s|^2)}{\bar{L}_q}\left(\Im(\phi_s e^{-\jmath n_p \theta})\right)^2$$

where the saturation functions S_d and S_q are increasing function of $|\phi_s|^2$ with $S_d(0) = S_q(0) = 1$ and where $\bar{L}_d$ and $\bar{L}_q$ are the unsaturated values of L_d and L_q (low stator currents). The saturation model we propose is then given by (2) with this modified Hamiltonian.

Using this Hamiltonian formulation to define the relationships between $\imath_s$, ϕ_s and τ_{em} as in (2) automatically maintains energy conservation. This conservation results from the fact that mixed partial derivatives are independent of order,

$$\frac{\partial^2 \mathcal{H}_m}{\partial\theta\partial\phi_s^*} = \frac{\partial^2 \mathcal{H}_m}{\partial\phi_s^*\partial\theta}.$$

This implies

$$-2\frac{\partial\tau_{em}}{\partial\phi_s^*} = \frac{\partial\imath_s}{\partial\theta}$$

where τ_{em} and $\imath_s$ are considered as function of the independent variables ϕ_s, ϕ_s^* and θ.

On the other hand, an incorrect but seemingly "natural" way to include saturation in the usual (α, β) model

$$\begin{aligned}
\frac{d}{dt}\left(J\frac{d}{dt}\theta\right) &= n_p\Im\left(\phi_s^*\imath_s\right) - \tau_L \\
\frac{d}{dt}\phi_s &= u_s - R_s\imath_s \\
\imath_s &= \left(\frac{1}{2L_d} + \frac{1}{2L_q}\right)\phi_s + \left(\frac{1}{2L_d} - \frac{1}{2L_q}\right)\phi_s^* e^{2\jmath n_p\theta} - \frac{\bar{\phi}}{L_d}e^{\jmath n_p\theta}
\end{aligned}$$

consists in taking L_d and L_q as function of $\rho^2 = \phi_s\phi_s^*$, without changing the formula for the electro-magnetic torque. If we proceed like this we get

$$\tau_{em} = \frac{n_p}{2\jmath}\left(\left(\frac{1}{2L_d} - \frac{1}{2L_q}\right)\left((\phi_s^*)^2 e^{2\jmath n_p\theta} - (\phi_s)^2 e^{-2\jmath n_p\theta}\right) - \left(\frac{\bar{\phi}}{L_d}\right)\left(\phi_s^* e^{\jmath n_p\theta} - \phi_s e^{-\jmath n_p\theta}\right)\right)$$

where L_d and L_q depend on $\rho^2 = |\phi_s|^2$ Then some computations give

$$\begin{aligned}
-2\frac{\partial\tau_{em}}{\partial\phi_s^*} - \frac{\partial\imath_s}{\partial\theta} = \jmath n_p &\left(\frac{d\left(\frac{1}{2L_d} - \frac{1}{2L_q}\right)}{d\rho^2}\left((\phi_s^*)^2 e^{2\jmath n_p\theta} - (\phi_s)^2 e^{-2\jmath n_p\theta}\right)\right. \\
&\left. - \frac{d\left(\frac{\bar{\phi}}{L_d}\right)}{d\rho^2}\left(\phi_s^* e^{\jmath n_p\theta} - \phi_s e^{-\jmath n_p\theta}\right)\right)\phi_s.
\end{aligned}$$

Thus such modeling does not in general respect the commutation condition $-2\frac{\partial\tau_{em}}{\partial\phi_s^*} = \frac{\partial\imath_s}{\partial\theta}$: no magnetic energy exists for such non-physical models. The correct current relationships include additional terms with derivatives of the functions S_d and S_q:

$$\begin{aligned}
\imath_s = &\left(\frac{1}{2L_d} + \frac{1}{2L_q}\right)\phi_s + \left(\frac{1}{2L_d} - \frac{1}{2L_q}\right)\phi_s^* e^{2\jmath n_p\theta} - \frac{\bar{\phi}}{L_d}e^{\jmath n_p\theta} \\
&+ \frac{S_d'(|\phi_s|^2)}{2\bar{L}_d}\phi_s\left(\Re(\phi_s e^{-\jmath n_p\theta}) - \bar{\phi}\right)^2 + \frac{S_q'(|\phi_s|^2)}{2\bar{L}_q}\phi_s\left(\Im(\phi_s e^{-\jmath n_p\theta})\right)^2.
\end{aligned}$$

2.3 Sinusoidal Models

Assume that the magnetic energy $\mathcal{H}_m$ admits the following rotational invariance associated to sinusoidal back electro-magnetic force (bemf):

$$\forall \phi_s \in \mathbb{C},\ \forall \theta, \xi \in \mathbb{S}^1, \quad \mathcal{H}_m(e^{\jmath n_p \xi}\phi_s, e^{-\jmath n_p \xi}\phi_s^*, \xi + \theta) = \mathcal{H}_m(\phi_s, \phi_s^*, \theta).$$

Then with $\overline{\mathcal{H}}(\psi, \psi^*) = \mathcal{H}_m(\psi, \psi^*, 0)$, $\mathcal{H}_m$ admits the following form

$$\mathcal{H}_m(\phi_s, \phi_s^*, \theta) \equiv \overline{\mathcal{H}}(\phi_s e^{-\jmath n_p \theta}, \phi_s^* e^{\jmath n_p \theta}).$$

In this case

$$\tau_{em} = -\frac{\partial \mathcal{H}_m}{\partial \theta} = -\jmath n_p \left(\frac{\partial \overline{\mathcal{H}}}{\partial \psi^*} \phi_s^* e^{\jmath n_p \theta} - \frac{\partial \overline{\mathcal{H}}}{\partial \psi} \phi_s e^{-\jmath n_p \theta} \right)$$

$$\imath_s = 2 \frac{\partial \mathcal{H}_m}{\partial \phi_s^*} = 2 e^{\jmath n_p \theta} \frac{\partial \overline{\mathcal{H}}}{\partial \psi^*}.$$

Since $\overline{\mathcal{H}}$ is a real quantity $\imath_s^* = e^{-\jmath n_p \theta} \frac{\partial \overline{\mathcal{H}}}{\partial \psi}$. Thus we recover the usual formula relating the electro-magnetic torque to the flux ϕ_s and current $\imath_s$:

$$\tau_{em} = n_p \frac{\phi_s^* \imath_s - \phi_s \imath_s^*}{2\jmath} = n_p \Im \left(\phi_s^* \imath_s \right). \tag{4}$$

When $\mathcal{H}_m$ does not admit such rotational invariance, τ_{em} is different from $n_p \Im \left(\phi_s^* \imath_s \right)$. Thus (4) is a direct consequence of rotational invariance. The saturation models considered in the previous subsection admit this rotational invariance and yield electro-magnetic torques satisfying (4).

A simple example of a non sinusoidal model is a machine with a trapezoidal bemf $F(n_p\theta)$ (a sinusoidal model corresponds to $F(n_p\theta) = \cos(n_p\theta)$). In this case we change the Hamiltonian in (2) by

$$\mathcal{H}_m = \frac{1}{2L_d} \left(\Re(\phi_s(F(n_p\theta) + \jmath F(n_p\theta + \tfrac{\pi}{2}))) - \bar{\phi} \right)^2 + \frac{1}{2L_q} \left(\Im(\phi_s(F(n_p\theta) + \jmath F(n_p\theta + \tfrac{\pi}{2}))) \right)^2$$

where $e^{-\jmath n_p \theta}$ in (3) is replaced by $F(n_p\theta) + \jmath F(n_p\theta + \frac{\pi}{2})$. This Hamiltonian is not rotationally invariant.

3 Induction Machines

3.1 Hamiltonian Modeling

We will now proceed as for permanent-magnet machines. The standard T-model of an induction machine admit the following form:

$$\begin{cases} \frac{d}{dt}\left(J\dot{\theta}\right) = n_p \Im\left(L_m \imath_r^* e^{-\jmath n_p \theta} \imath_s\right) - \tau_L \\ \frac{d}{dt}\left(L_m\left(\imath_r + \imath_s e^{-\jmath n_p \theta}\right) + L_{fr}\imath_r\right) = -R_r \imath_r \\ \frac{d}{dt}\left(L_m\left(\imath_s + \imath_r e^{\jmath n_p \theta}\right) + L_{fs}\imath_s\right) = u_s - R_s \imath_s \end{cases} \tag{5}$$

where

- n_p is the number of pairs of poles, θ is the rotor mechanical angle, J and τ_L are the inertia and load torque, respectively.
- $\imath_r \in \mathbb{C}$ is the rotor current (in the rotor frame, different from the (d, q) frame) , $\imath_s \in \mathbb{C}$ the stator current (in the stator frame, i.e. the (α, β) frame) and $u_s \in \mathbb{C}$ the stator voltage (in the stator frame). The stator and rotor resistances are $R_s > 0$ and $R_r > 0$.
- The inductances L_m, L_{fr} and L_{fs} are positive parameters with $L_{fr}, L_{fs} \ll L_m$.
- The stator (resp. rotor) flux is $\phi_s = L_m\left(\imath_s + \imath_r e^{\jmath n_p \theta}\right) + L_{fs}\imath_s$ (resp. $\phi_r = L_m\left(\imath_r + \imath_s e^{-\jmath n_p \theta}\right) + L_{fr}\imath_r$).

The Hamiltonian formulation proposed in [1] reads:

$$\frac{d}{dt}\left(J\dot{\theta}\right) = -\frac{\partial \mathcal{H}_m}{\partial \theta} - \tau_L, \quad \frac{d}{dt}\phi_r = -2R_r \frac{\partial \mathcal{H}_m}{\partial \phi_r^*}, \quad \frac{d}{dt}\phi_s = u_s - 2R_s \frac{\partial \mathcal{H}_m}{\partial \phi_s^*} \tag{6}$$

where the magnetic energy $\mathcal{H}_m$ now depends on θ, the rotor flux ϕ_r and its complex conjugate ϕ_r^*, the stator flux ϕ_s and its complex conjugate ϕ_s^*. The rotor (resp. stator) current is then given by $2\frac{\partial \mathcal{H}_m}{\partial \phi_r^*}$ (resp. $2\frac{\partial \mathcal{H}_m}{\partial \phi_s^*}$). For the standard model (5), we have

$$\mathcal{H}_m = \tfrac{1}{2L_f}(\phi_s - e^{\jmath n_p \theta}\phi_r)(\phi_s^* - e^{-\jmath n_p \theta}\phi_r^*) + \tfrac{1}{2L_s}\phi_s\phi_s^* + \tfrac{1}{2L_r}\phi_r\phi_r^* \tag{7}$$

with $L_f = \frac{L_{fs}L_{fr}}{L_m} + L_{fs} + L_{fr}$, $L_s = L_{fs} + \frac{L_{fs}+L_{fr}}{L_{fr}}L_m$ and $L_r = L_{fr} + \frac{L_{fs}+L_{fr}}{L_{fs}}L_m$. Such Hamiltonian formulations based on fluxes are also named π-models whereas T-models based on currents correspond to Lagrangian formulations (see, e.g., [5]).

3.2 Magnetic Saturation

As in section 2.2, we will take into account magnetic saturation, we assume that in (7), L_s, L_r and L_f are decreasing function of $|\phi_s|^2$. Then the magnetic saturation model will be given by :

$$\frac{1}{L_s} = \frac{S_s(|\phi_s|^2)}{\bar{L}_s}, \quad \frac{1}{L_r} = \frac{S_r(|\phi_s|^2)}{\bar{L}_r}, \quad \frac{1}{L_f} = \frac{S_f(|\phi_s|^2)}{\bar{L}_f}$$

where the saturation functions S_s, S_r and S_f are increasing function of $|\phi_s|^2$ with $S_s(0) = S_r(0) = S_f(0) = 1$ and where $\bar{L}_s$, $\bar{L}_r$ and $\bar{L}_f$ are the unsaturated values of L_s , L_r and L_f. The saturated Hamiltonian is then

$$\mathcal{H}_m = \frac{S_f(|\phi_s|^2)}{2\bar{L}_f}|\phi_s - e^{\jmath n_p\theta}\phi_r|^2 + \frac{S_s(|\phi_s|^2)}{2\bar{L}_s}|\phi_s|^2 + \frac{S_r(|\phi_s|^2)}{2\bar{L}_r}|\phi_r|^2$$

With the dynamic equations then given by (6). This saturation model is the Hamiltonian counter-part of the saturation model proposed in [5].

3.3 Sinusoidal Models

The Hamiltonian $\mathcal{H}_m$ here above admits the following rotational invariance associated to a sinusoidal bemf:

$$\forall \phi_s \in \mathbb{C},\ \forall \theta, \xi \in \mathbb{S}^1,$$
$$\mathcal{H}_m(e^{\jmath n_p \xi}\phi_s, \phi_r, e^{-\jmath n_p \xi}\phi_s^*, \phi_r^*, \xi + \theta) = \mathcal{H}_m(\phi_s, \phi_r, \phi_s^*, \phi_r^*, \theta).$$

Then with $\overline{\mathcal{H}}(\psi_s, \psi_r, \psi_s^*, \psi_r^*) = \mathcal{H}_m(\psi_s, \psi_r, \psi_s^*, \psi_r^*, 0)$, $\mathcal{H}_m$ admits the following form

$$\mathcal{H}_m(\phi_s, \phi_r, \phi_s^*, \phi_r^*, \theta) \equiv \overline{\mathcal{H}}(e^{-\jmath n_p\theta}\phi_s, \phi_r, e^{\jmath n_p\theta}\phi_s^*, \phi_r^*).$$

In this case

$$\tau_{em} = -\frac{\partial \mathcal{H}_m}{\partial \theta} = -\jmath n_p \left(\frac{\partial \overline{\mathcal{H}}}{\partial \psi_s^*} \phi_s^* e^{\jmath n_p\theta} - \frac{\partial \overline{\mathcal{H}}}{\partial \psi_s} \phi_s e^{-\jmath n_p\theta} \right)$$

Since $\imath_s = 2\frac{\partial \mathcal{H}_m}{\partial \phi_s^*} = 2e^{\jmath n_p\theta}\frac{\partial \overline{\mathcal{H}}}{\partial \psi_s^*}$ and $\imath_s^* = 2\frac{\partial \mathcal{H}_m}{\partial \phi_s} = 2e^{-\jmath n_p\theta}\frac{\partial \overline{\mathcal{H}}}{\partial \psi_s}$ we recover the usual formula relating the electro-magnetic torque to stator flux and current:

$$\tau_{em} = -\jmath n_p \frac{\phi_s^* \imath_s - \phi_s \imath_s^*}{2} = n_p \Im\left(\phi_s^* \imath_s\right).$$

4 Concluding Remarks

It remains also to validate experimentally such magnetic-saturation models. Substantial modifications to such Hamiltonian formulation are needed to include, in parallel to magnetic-saturation, magnetic hysteresis and the associated energy losses [3].

References

1. Basic, D., Malrait, F., Rouchon, P.: Euler-Lagrange models with complex currents of three-phase electrical machines and observability issues. IEEE Trans. Automatic Control 55(1), 212–217 (2010)

2. Chiasson, J.: Modeling and High Performance Control of Electric Machines. IEEE Press Series on Power Engineering. Wiley-IEEE Press (2005)
3. Della Torre, E.: Magnetic Hysteresis. IEEE Press, Los Alamitos (1999)
4. Leonhard, W.: Control of Electrical Drives. Elsevier, Amsterdam (1985)
5. Sullivan, C.R., Sanders, S.R.: Models for induction machines with magnetic saturation of the main flux path. IEEE Trans. on Industry Applications 31(4), 907–914 (1995)

Iterative Learning Control Using Stochastic Approximation Theory with Application to a Mechatronic System

Mark Butcher and Alireza Karimi

Abstract. In this paper it is shown how Stochastic Approximation theory can be used to derive and analyse well-known Iterative Learning Control algorithms for linear systems. The Stochastic Approximation theory gives conditions that, when satisfied, ensure almost sure convergence of the algorithms to the optimal input in the presence of stochastic disturbances. The practical issues of monotonic convergence and robustness to model uncertainty are considered. Specific choices of the learning matrix are studied, as well as a model-free choice. Moreover, the model-free method is applied to a linear motor system, leading to greatly improved tracking.

1 Introduction

Iterative Learning Control (ILC) is a technique used to enhance the tracking performance of systems that perform repetitive operations. In this approach, information 'learnt' from previous repetitions is used to improve the performance of the system during the next repetition/iteration i.e. reduce the tracking error. ILC has been shown to be very effective for systems that are predominately affected by deterministic, repetitive disturbances, which are learnt from one iteration to the next. However, when the system is affected by stochastic disturbances the tracking performance is greatly diminished [9, 5]. It is, therefore, important to develop ILC algorithms that have reduced sensitivity to this type of disturbance.

Although the deterministic aspects of ILC have received more attention, certain researchers have already proposed algorithms that are robust to the presence of stochastic disturbances.

Mark Butcher · Alireza Karimi
Automatic Control Laboratory, Ecole Polytechnique Fédérale de Lausanne (EPFL), Switzerland
e-mail: alireza.karimi@epfl.ch

J. Lévine & P. Müllhaupt (Eds.): Adv. in Theory of Control, Signals, LNCIS 407, pp. 49–64.
springerlink.com © Springer-Verlag Berlin Heidelberg 2010

1) The use of a forgetting factor in ILC was first proposed in [12] for a D-type ILC law. It was then proposed in [2] for P-type ILC. It is shown that by introducing the forgetting factor the system's output converges to a neighbourhood of the desired one, despite the presence of norm-bounded initialisation errors, fluctuations of the dynamics and random disturbances. However, in [19] and [5], it is shown that the use of a forgetting factor can increase the expected value and variance of the error signal compared to standard ILC algorithms.

2) The filtering of the ILC command has been proposed in certain papers as a way of reducing the influence of noise on the error [15]. However, whilst it reduces the error variance, it causes a nonzero converged mean error.

3) Kalman filtering-type techniques have also been applied to ILC to estimate the controlled output, in the presence of disturbances [22, 20, 21, 14, 8, 1]. In the case of perfect knowledge of the disturbance covariance matrices and system parameters, convergence to the optimal input can be shown. However, perfect knowledge is unrealistic.

4) In [22] another ILC algorithm is proposed using a learning gain that decreases inversely proportionally to the iteration number and has the form of a Stochastic Approximation (SA) algorithm. No detailed analysis is, however, carried out. An algorithm with a similar iteration decreasing learning gain is also developed in [17] for repetitive disturbance rejection in the presence of measurement noise. This algorithm is derived in a similar way to recursive least squares identification algorithms, without mention to SA. The application of SA theory to ILC is most directly considered in [6] and [7] for the linear and nonlinear cases respectively. It is shown that the proposed ILC law converges almost surely to the optimal input and the output error is minimised in the mean square sense as the number of iterations tends to infinity. The algorithm requires only that the optimal input is realisable. Knowledge of neither the disturbance covariance matrix nor the system matrices is required because a simultaneous perturbation type algorithm is employed, which uses random perturbations to estimate the gradient. The disadvantage of this approach is slow convergence.

The main contribution of this paper is to show how ILC for linear systems affected by stochastic disturbances fits into the SA theory framework. Using SA theory it is possible to derive necessary conditions for well-known ILC algorithms to converge almost surely to the optimal input signal in the presence of stochastic disturbances. In addition, the important practical issues of monotonic convergence of the error signal and robustness to system uncertainty are addressed. Also two choices of learning matrix based on an uncertain model are studied, as well as a model-free choice. These choices are compared in a simulation example in [4].

In [6] the input is randomly perturbed and applied to the system in a second experiment at each iteration in order to estimate the gradient of the proposed cost function. In contrast, here either an uncertain system model or a second special experiment is considered. These choices will typically lead to faster convergence.

Steepest descent algorithms have been applied to ILC for the discrete-time case in [11]. Although certain similarities exist between the algorithms considered here and steepest descent algorithms, the major difference is the conditions SA sets on the step sizes between iterations. These conditions are necessary to ensure almost sure convergence to the optimal input in the presence of stochastic disturbances.

This paper is organised as follows. In Section 2 the notational framework is defined and the assumptions are stated. In Section 3 ILC is considered from an SA perspective. Then in Section 4 possible choices of the learning matrix are considered. In Section 5 experimental results obtained on a linear motor system are presented. Finally in Section 6 some conclusions are made.

2 Notation

We consider the linear time-invariant (LTI), discrete-time, stable SISO system $G(q)$, shown in Fig. 1, that carries out a finite-time, repetitive tracking task and whose controlled output $z_k(t)$, at time t and repetition k, is given by:

$$z_k(t) = G(q)u_k(t) + d_k(t), \tag{1}$$

where $u_k(t)$ is the input to the system, $d_k(t)$ is the load disturbance and q is the forward-shift time domain operator. The system's measured output, $y_k(t)$, is:

$$y_k(t) = z_k(t) + n_k(t), \tag{2}$$

where $n_k(t)$ is the measurement disturbance. It should be mentionned that if $G(q)$ represents a closed-loop transfer function then $d_k(t)$ and $n_k(t)$ will be the signals resulting from the filtering of external disturbances by the corresponding closed-loop transfer functions.

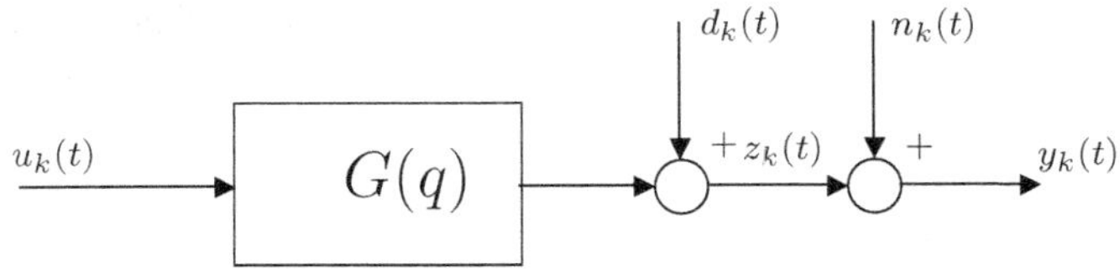

Fig. 1. System affected by stochastic disturbances

The controlled tracking error signal is defined as:

$$\epsilon_k(t) = y_d(t) - z_k(t), \tag{3}$$

where $y_d(t)$ is the bounded desired system output, which is defined over a finite repetition duration for $t = 0, \ldots, N-1$, and the measured error signal is given by:

$$e_k(t) = y_d(t) - y_k(t). \tag{4}$$

As the signals are defined over a finite duration, it is possible to express the system's input-output relationship by a matrix representation. Taking advantage of the non-causal filtering possibilities of ILC, the lifted-system representation is used. For a system with a relative degree m we define the vectors:

$$\mathbf{u}_k = [u_k(0), u_k(1), \ldots, u_k(N-m-1)]^T \tag{5}$$
$$\mathbf{z}_k = [z_k(m), z_k(m+1), \ldots, z_k(N-1)]^T. \tag{6}$$

The vectors $\mathbf{y}_k$, $\mathbf{d}_k$, $\mathbf{n}_k$ and $\mathbf{y_d}$ are defined similarly to $\mathbf{z}_k$. Using these vectors, the measured output of the system is:

$$\mathbf{y}_k = \mathbf{G}\mathbf{u}_k + \mathbf{d}_k + \mathbf{n}_k, \tag{7}$$

where $\mathbf{G}$ is:

$$\mathbf{G} = \begin{bmatrix} g_m & 0 & \ldots & 0 \\ g_{m+1} & g_m & \ldots & 0 \\ \vdots & \vdots & \ddots & \vdots \\ g_{N-1} & g_{N-2} & \cdots & g_m \end{bmatrix}, \tag{8}$$

g_i being the ith Markov parameter of $G(q)$. The controlled error vector is:

$$\boldsymbol{\epsilon}_k(\mathbf{u}_k) = \mathbf{y_d} - \mathbf{z}_k = \mathbf{y_d} - \mathbf{G}\mathbf{u}_k - \mathbf{d}_k \tag{9}$$

and the measured error vector:

$$\mathbf{e}_k(\mathbf{u}_k) = \mathbf{y_d} - \mathbf{y}_k = \boldsymbol{\epsilon}_k(\mathbf{u}_k) - \mathbf{n}_k, \tag{10}$$

where the errors' dependence on $\mathbf{u}_k$ is explicitly stated.

Furthermore, we have that the real system can be represented as:

$$G(q) = \hat{G}(q)[1 + \Delta(q)] \tag{11}$$

where $\hat{G}(q)$ is a model of the system and $\Delta(q)$ represents the multiplicative uncertainty. This representation is given in lifted-system form as:

$$\mathbf{G} = \hat{\mathbf{G}}[\mathbf{I} + \boldsymbol{\Delta}] \tag{12}$$

where $\mathbf{I}$ is the identity matrix, and $\hat{\mathbf{G}}$ and $\mathbf{I} + \boldsymbol{\Delta}$ are Toeplitz matrices formed similarly to (8) from the Markov parameters of $q^{\hat{m}}\hat{G}(q)$ and $q^{m-\hat{m}}[1 + \Delta(q)]$, respectively. $\hat{m}$ is the relative degree of $\hat{G}(q)$.

Definition: A real, square matrix $\mathbf{M}$ (not necessarily symmetric) is called positive definite $\mathbf{M} > 0$ if and only if all the eigenvalues of its symmetric part $(\mathbf{M} + \mathbf{M}^T)/2$ are positive.

2.1 Assumptions

(A1) The ideal input: $\mathbf{u}^* = \mathbf{G}^{-1}\mathbf{y_d}$ is realisable.
(A2) The system uncertainty satisfies: $\mathbf{I} + \boldsymbol{\Delta} > 0$.
(A3) The disturbances $\mathbf{d}_k$ and $\mathbf{n}_k$ are zero-mean, weakly stationary random vectors with unknown covariance matrices $\mathbf{R_d}$ and $\mathbf{R_n}$, respectively. Additionally, they have bounded, unknown cross-covariance matrices $\mathbf{R_{dn}}$ and $\mathbf{R_{nd}}$. Moreover, different realisations of $\mathbf{d}_k$ and $\mathbf{n}_k$ between iterations are mutually independent.
(A4) The mean input is bounded for all iterations: $E\{\mathbf{u}_k\} < \infty \quad \forall k$.

Remarks:

1) It is shown in [10] that a sufficient condition for Assumption **(A2)** is that the filter $q^{m-\hat{m}}[1+\Delta(q^{-1})]$ is strictly positive real (SPR). So when $m = \hat{m}$, Assumption **(A2)** is satisfied when $\|\Delta\|_\infty < 1$. This condition occurs frequently in the model uncertainty representation and so is a reasonable assumption.
2) The validity of Assumption **(A4)** will be discussed later in the chapter.

3 ILC from a SA Viewpoint

The ideal aim of tracking control is to achieve zero controlled error. When stochastic disturbances affect a system this objective is not possible. A reasonable aim is then to set the mean controlled error equal to zero. We can state a goal of the ILC algorithm, thus, as to iteratively calculate the optimal input signal $\mathbf{u}^*$ such that:

$$E\{\mathbf{L}\boldsymbol{\epsilon}_k(\mathbf{u}^*)\} = E\{\mathbf{L}\mathbf{e}_k(\mathbf{u}^*)\} = \mathbf{0}, \tag{13}$$

where $E\{\cdot\}$ denotes the mathematical expectation and $\mathbf{L}$ is a non-singular matrix.

It is straightforward to see that the solution to criterion (13) is $\mathbf{u}^*$ in Assumption **(A1)**. However, in order to calculate the ideal input $\mathbf{u}^*$ directly exact knowledge of $\mathbf{G}$ is needed, which is not available. Nevertheless, $\mathbf{u}^*$ can be found using an iterative stochastic approximation (SA) procedure, such as the Robbins-Monro algorithm [18], which does not require exact system knowledge. This algorithm calculates the input iteratively as:

$$\mathbf{u}_{k+1} = \mathbf{u}_k + \gamma_k \mathbf{L}\mathbf{e}_k(\mathbf{u}_k). \tag{14}$$

This algorithm clearly has the form of a standard P-type ILC law with an iteration varying learning gain γ_k. In the next subsection conditions will be given that, according to SA theory, ensure almost sure convergence of the algorithm to the ideal input.

3.1 Almost Sure Convergence

Theorem 1. *Under the Assumptions* **(A1)**, **(A3)** *and* **(A4)**, *the iterative update algorithm (14) converges almost surely to the solution* $\mathbf{u}^*$ *of (13) when* $k \to \infty$ *if:*

(C1) *The sequence* γ_k *of positive steps satisfies:*

$$\sum_{k=0}^{\infty} \gamma_k = \infty \quad \textit{and} \quad \sum_{k=0}^{\infty} \gamma_k^2 < \infty. \tag{15}$$

(C2) $E\{\mathbf{L}\mathbf{e}_k(\mathbf{u}_k)\}$ *is monotonically decreasing:*

$$\mathbf{Q}(\mathbf{u}_k) = \frac{d}{d\mathbf{u}_k} E\{\mathbf{L}\mathbf{e}_k(\mathbf{u}_k)\} < 0. \tag{16}$$

Proof. The proof is similar to that of the Robbins-Monro stochastic approximation algorithm.

Condition **(C1)** should be fulfilled by an appropriate choice of the sequence γ_k. $\mathbf{Q}(\mathbf{u}_k)$, in Condition **(C2)**, can be rewritten as:

$$\mathbf{Q}(\mathbf{u}_k) = \frac{d}{d\mathbf{u}_k} E\{\mathbf{L}\mathbf{e}_k(\mathbf{u}_k)\} = \frac{d}{d\mathbf{u}_k} E\{\mathbf{L}\mathbf{y}_\mathbf{d} - \mathbf{L}\mathbf{G}\mathbf{u}_k + \mathbf{L}\mathbf{d}_k + \mathbf{L}\mathbf{v}_k\}$$

$$= -\mathbf{L}\mathbf{G} = -\mathbf{L}\hat{\mathbf{G}}[\mathbf{I} + \boldsymbol{\Delta}] \tag{17}$$

and so Condition **(C2)** becomes:

$$\mathbf{L}\hat{\mathbf{G}}[\mathbf{I} + \boldsymbol{\Delta}] > 0. \tag{18}$$

Remark: By combining equations (9), (10), (14) and **A1** we can obtain the input error evolution as:

$$\mathbf{e}_{k+1}^u = \mathbf{u}^* - \mathbf{u}_{k+1} = (\mathbf{I} - \gamma_k \mathbf{L}\mathbf{G})\mathbf{e}_k^u + \gamma_k \mathbf{L}(\mathbf{y}_\mathbf{d} - \mathbf{d}_\mathbf{k} - \mathbf{n}_\mathbf{k}). \tag{19}$$

A necessary, but not sufficient, condition for asymptotic convergence of the input error, in the absence of disturbances, is:

$$|\lambda_i(\mathbf{I} - \gamma_k \mathbf{L}\mathbf{G})| < 1 \quad \forall k, \forall i \tag{20}$$

where $\lambda_i(\cdot)$ is the ith eigenvalue. If $\mathbf{L}$ represents a causal operator and is therefore a real, lower triangular matrix, a link between this condition and those given by SA theory can be made, as detailed below. Since $\mathbf{I} - \gamma_k \mathbf{L}\mathbf{G}$ will be a real, lower triangular matrix, its eigenvalues will be real. (20) therefore implies:

$$\overline{\lambda}(\mathbf{I} - \gamma_k \mathbf{L}\mathbf{G}) < 1 \iff 1 - \gamma_k \underline{\lambda}(\mathbf{L}\mathbf{G}) < 1 \iff \gamma_k \underline{\lambda}(\mathbf{L}\mathbf{G}) > 0 \tag{21}$$

and

$$\underline{\lambda}(\mathbf{I}-\gamma_k\mathbf{LG}) > -1 \iff 1-\overline{\lambda}(\gamma_k\mathbf{LG}) > -1 \iff \gamma_k\overline{\lambda}(\mathbf{LG}) < 2, \quad (22)$$

where $\underline{\lambda}(\cdot)$ and $\overline{\lambda}(\cdot)$ are the minimum and maximum eigenvalues, respectively. Moreover we have $\mathbf{LGx}_i = \lambda_i\mathbf{x}_i$, where $\mathbf{x}_i$ is the real eigenvector corresponding to λ_i. Taking the transpose of the both sides gives:

$$\mathbf{x}_i^T(\mathbf{LG})^T = \lambda_i\mathbf{x}_i^T. \quad (23)$$

Right multiplying (23) by $\mathbf{x}_i$ and adding it with its transpose gives:

$$\mathbf{x}_i^T\mathbf{LGx}_i + \mathbf{x}_i^T(\mathbf{LG})^T\mathbf{x}_i = 2\lambda_i\mathbf{x}_i^T\mathbf{x}_i \iff \mathbf{x}_i^T\left(\frac{\mathbf{LG}+(\mathbf{LG})^T}{2}\right)\mathbf{x}_i = \lambda_i\mathbf{x}_i^T\mathbf{x}_i. \quad (24)$$

So, if $\mathbf{LG}$ is positive definite, (21) is satisfied. (22) can be satisfied by an appropriate choice of γ_k.

3.2 Monotonic Convergence

Whilst almost sure convergence of the input sequence to the solution $\mathbf{u}^*$ when $k\to\infty$ is, obviously, of utmost importance, practically it is not the only type of convergence of interest. The monotonic convergence, from one iteration to the next, of a norm of the controlled error is also of concern.

To proceed, we will need the following lemma:

Lemma 1. *If a real, square matrix* $\mathbf{M}$ *(not necessarily symmetric) is positive definite, there exists an* $\alpha > 0$ *such that:*

$$\overline{\sigma}(\mathbf{I}-\alpha\mathbf{M}) < 1, \quad (25)$$

where $\overline{\sigma}(\cdot)$ *is the maximum singular value.*

Proof. Condition (25) is true iff:

$$\begin{aligned}
&\lambda_i\left(I-\alpha(\mathbf{M}^T+\mathbf{M})+\alpha^2\mathbf{M}^T\mathbf{M}\right) < 1 \quad \forall i\\
\iff &1-\lambda_i\left(\alpha(\mathbf{M}^T+\mathbf{M})-\alpha^2\mathbf{M}^T\mathbf{M}\right) < 1 \quad \forall i\\
\iff &\lambda_i\left(\mathbf{M}^T+\mathbf{M}-\alpha\mathbf{M}^T\mathbf{M}\right) > 0 \quad \forall i.
\end{aligned} \quad (26)$$

Furthermore the eigenvalues satisfy:

$$[\mathbf{M}^T+\mathbf{M}-\alpha\mathbf{M}^T\mathbf{M}]\mathbf{x}_i = \lambda_i\mathbf{x}_i. \quad (27)$$

Left multiplying (27) by $\mathbf{x}_i^T$ *we get:*

$$\mathbf{x}_i^T\left(\mathbf{M}^T+\mathbf{M}\right)\mathbf{x}_i - \alpha\mathbf{x}_i^T\mathbf{M}^T\mathbf{M}\mathbf{x}_i = \lambda_i\mathbf{x}_i^T\mathbf{x}_i. \quad (28)$$

So if $\mathbf{M} > 0$, *(26), and thus Condition (25), are satisfied when:*

$$0 < \alpha < \min_i \frac{\mathbf{x}_i^T \left(\mathbf{M}^T + \mathbf{M}\right) \mathbf{x}_i}{\mathbf{x}_i^T \mathbf{M}^T \mathbf{M} \mathbf{x}_i}. \tag{29}$$

Theorem 2. *If* $\hat{\mathbf{G}}[\mathbf{I} + \boldsymbol{\Delta}]\mathbf{L} > 0$, *there exists a sequence of positive step sizes* γ_k, *satisfying Condition* **(C1)**, *such that monotonic convergence of the 2-norm of the mean controlled error is achieved.*

Proof. *By combining equations (9), (10), (12) and (14) we can obtain the controlled error evolution equation as:*

$$\boldsymbol{\epsilon}_{k+1}(\mathbf{u}_{k+1}) = (\mathbf{I} - \gamma_k \hat{\mathbf{G}}[\mathbf{I} + \boldsymbol{\Delta}]\mathbf{L})\boldsymbol{\epsilon}_k(\mathbf{u}_k) + \mathbf{d}_k - \mathbf{d}_{k+1} + \gamma_k \hat{\mathbf{G}}[\mathbf{I} + \boldsymbol{\Delta}]\mathbf{L}\mathbf{n}_k. \tag{30}$$

The mean value of equation (30) is:

$$E\{\boldsymbol{\epsilon}_{k+1}(\mathbf{u}_{k+1})\} = (\mathbf{I} - \gamma_k \hat{\mathbf{G}}[\mathbf{I} + \boldsymbol{\Delta}]\mathbf{L})E\{\boldsymbol{\epsilon}_k(\mathbf{u}_k)\}. \tag{31}$$

Monotonic convergence of the 2-norm of the mean controlled error is obtained if the following condition is satisfied (see e.g. Theorem 2, [16]):

$$\overline{\sigma}(\mathbf{I} - \gamma_k \hat{\mathbf{G}}[\mathbf{I} + \boldsymbol{\Delta}]\mathbf{L}) < 1 \quad \forall k. \tag{32}$$

If a given sequence γ_k, *satisfying Condition* **(C1)**, *does not satisfy (32), a new, scaled sequence* $\gamma_k \triangleq \beta\gamma_k$, $\beta > 0$ *can always be defined that does, as follows from Lemma 1.*

Remarks:

1) Theorem 2's requirement that $\hat{\mathbf{G}}[\mathbf{I} + \boldsymbol{\Delta}]\mathbf{L}$ be positive definite is satisfied when $\mathbf{L}$ and $\hat{\mathbf{G}}[\mathbf{I} + \boldsymbol{\Delta}]$ commute, i.e. when $L(q)$ is causal, and condition (18) is satisfied.
2) Since the system $G(q)$ is assumed stable, its output and internal states will be bounded if its input is bounded. Combining equations (9), (10), (12) and (14) gives the input evolution equation as:

$$\mathbf{u}_{k+1} = (\mathbf{I} - \gamma_k \mathbf{L}\hat{\mathbf{G}}[\mathbf{I} + \boldsymbol{\Delta}])\mathbf{u}_k + \gamma_k \mathbf{L}(\mathbf{y_d} - \mathbf{d_k} - \mathbf{n_k}). \tag{33}$$

According to Theorem 5 of [16] the input will remain bounded from one iteration to the next if a) (33) is a uniformly exponentially stable iterative system, b) for a finite constant β, $\|\gamma_k \mathbf{L}\| < \beta \ \forall k$, and c) $\mathbf{y_d}$, $\mathbf{d_k}$ and $\mathbf{n_k}$ are bounded. As stated in Corollary 1 of [16], (33) is a uniformly exponentially stable iterative system if $\overline{\sigma}(\mathbf{I} - \gamma_k \mathbf{L}\hat{\mathbf{G}}[\mathbf{I} + \boldsymbol{\Delta}]) < 1 \quad \forall k$. This condition is considered in Lemma 1, implying that, when $\mathbf{L}\hat{\mathbf{G}}[\mathbf{I} + \boldsymbol{\Delta}] > 0$, a sequence γ_k exists that achieves uniform exponential stability. Furthermore, since $\|\gamma_k \mathbf{L}\| = |\gamma_k|\|\mathbf{L}\|$, there exists a sequence γ_k that satisfies the condition $\|\gamma_k \mathbf{L}\| < \beta \ \forall k$. So the boundedness of the system's signals requires the

disturbances to be bounded, which can usually be assumed to be the case in practice.
It should be noted that the mean input $E\{\mathbf{u}_k\}$ will be bounded if only the means of the disturbances are bounded, rather than the disturbances themselves.

3.3 Asymptotic Distribution of the Input Estimation Error

The asymptotic distribution of the input estimation error is given by the following theorem:

Theorem 3. *Assume that:*

i) Algorithm (14) converges almost surely to the solution $\mathbf{u}^*$ *as* $k \to \infty$.
ii) The sequence of step sizes is chosen as $\gamma_k = \frac{\alpha}{k+1}$.
iii) All the eigenvalues of the matrix $\mathbf{D} = \mathbf{I}/2 + \alpha\mathbf{Q}(\mathbf{u}^*)$ *have negative real parts.*

Then the sequence $\sqrt{k}(\mathbf{u}_k - \mathbf{u}^*) \in As\,\mathcal{N}(\mathbf{0}, \mathbf{V})$ *i.e it converges asymptotically in distribution to a zero-mean normal distribution with covariance:*

$$\mathbf{V} = \alpha^2 \int_0^\infty \exp(\mathbf{Dx})\mathbf{P}\exp(\mathbf{D}^T\mathbf{x})dx \tag{34}$$

where $\mathbf{P}$ *is the covariance matrix of* $\mathbf{Le}(\mathbf{u}^*)$*:*

$$\mathbf{P} = E\{\mathbf{Le}_k(\mathbf{u}^*)(\mathbf{Le}_k(\mathbf{u}^*))^T\}. \tag{35}$$

Proof. The proof can be found in [13] (Theorem 6.1 p.147).
Using Theorem 3 we have that:

$$\begin{aligned}\mathbf{P} &= E\{\mathbf{Le}_k(\mathbf{u}^*)(\mathbf{Le}_k(\mathbf{u}^*))^T\} = E\{(-\mathbf{L}(\mathbf{d}_k + \mathbf{n}_k))(-\mathbf{L}(\mathbf{d}_k + \mathbf{n}_k))^T\} \\ &= \mathbf{L}(\mathbf{R_d} + \mathbf{R_{dn}} + \mathbf{R_{nd}} + \mathbf{R_n})\mathbf{L}^T.\end{aligned} \tag{36}$$

Additionally, as $\mathbf{Q}(\mathbf{u}^*) = \frac{d}{d\mathbf{u}_k}E\{\mathbf{Le}_k(\mathbf{u}_k)\}|_{\mathbf{u}(k)=\mathbf{u}_0} = -\mathbf{LG}$, we have that:

$$\mathbf{D} = (\mathbf{I}/2 - \alpha\mathbf{LG}). \tag{37}$$

The covariance matrix $\mathbf{V}$ is then the unique symmetric solution of the following Lyapunov equation:

$$2\alpha^2\mathbf{L}(\mathbf{R_d} + \mathbf{R_{dn}} + \mathbf{R_{nd}} + \mathbf{R_n})\mathbf{L}^T + (\mathbf{I} - 2\alpha\mathbf{LG})\mathbf{V} + \mathbf{V}(\mathbf{I} - 2\alpha\mathbf{LG})^T = \mathbf{0}. \tag{38}$$

It is shown in [3] (Proposition 4, p.112) that if, instead of using a scalar learning gain α, we use a non-singular learning matrix $\mathbf{K}$, then the optimal matrix $\mathbf{K}^*$ to mimimise the trace of $\mathbf{V}$ is given by:

$$\mathbf{K}^* = -\mathbf{Q}(\mathbf{u}^*)^{-1} = (\mathbf{L}\mathbf{G})^{-1}. \tag{39}$$

Using this gain matrix results in the learning law:

$$\mathbf{u}_{k+1} = \mathbf{u}_k + \frac{\mathbf{G}^{-1}}{k+1}\mathbf{e}_k(\mathbf{u}_k), \tag{40}$$

and the optimal asymptotic covariance matrix:

$$\mathbf{V}^* = \mathbf{G}^{-1}(\mathbf{R_d} + \mathbf{R_{dn}} + \mathbf{R_{nd}} + \mathbf{R_n})\mathbf{G}^{-T}, \tag{41}$$

which means that the sequence $\sqrt{k}(\mathbf{u}_k - \mathbf{u}^*) \in \text{As}\,\mathcal{N}(\mathbf{0}, \mathbf{V}^*)$.

Moreover we have that $\boldsymbol{\epsilon}_k(\mathbf{u}_k) = -\mathbf{G}(\mathbf{u}_k - \mathbf{u}^*) - \mathbf{d}_k$ so the covariance matrix of $\boldsymbol{\epsilon}_k(\mathbf{u}_k)$ is then given by:

$$\text{cov}(\boldsymbol{\epsilon}_k(\mathbf{u}_k)) = E\{\boldsymbol{\epsilon}_k(\mathbf{u}_k)\boldsymbol{\epsilon}_k^T(\mathbf{u}_k)\} = \mathbf{G}E\{(\mathbf{u}_k - \mathbf{u}^*)(\mathbf{u}_k - \mathbf{u}^*)^T\}\mathbf{G}^T + \mathbf{R_d}. \tag{42}$$

Using the optimal gain matrix $\mathbf{K}^*$ means that the sequence $\boldsymbol{\epsilon}_k(\mathbf{u}_k)$ will have a converged covariance matrix given by

$$\text{cov}(\boldsymbol{\epsilon}_k(\mathbf{u}_k)) = \frac{1}{k}(\mathbf{R_d} + \mathbf{R_{dn}} + \mathbf{R_{nd}} + \mathbf{R_n}) + \mathbf{R_d}$$

and in the limit we have: $\lim_{k\to\infty} \text{cov}(\boldsymbol{\epsilon}_k(\mathbf{u}_k)) = \mathbf{R_d}$. However, $\mathbf{K}^*$ is not implementable because exact knowledge of $\mathbf{G}$ is not achievable. Nonetheless it gives an ideal law to aim for in the design of a stochastic ILC algorithm.

4 Specific Choices of L

In this section specific choices of the learning matrix $\mathbf{L}$ will be considered.

4.1 Use of the Uncertain System Inverse

We consider here the choice of $\mathbf{L} = \hat{\mathbf{G}}^{-1}$ i.e. the inverse of the uncertain system model. This choice is motivated by the fact that $\mathbf{L} = \hat{\mathbf{G}}^{-1}$ is an approximation of the optimal learning gain used in (40).

Theorem 4. *Under Assumption (**A2**) and when $\mathbf{L} = \hat{\mathbf{G}}^{-1}$, there exists a sequence of positive step sizes γ_k, satisfying Condition* (**C1**)*, that ensures that the ILC algorithm (14) converges almost surely to $\mathbf{u}^*$ and that the 2-norm of the mean controlled error convergences monotonically.*

*Proof. Condition (18) is automatically satisfied when $\mathbf{L} = \hat{\mathbf{G}}^{-1}$, under Assumption (**A2**). Therefore, when the sequence of positive step sizes γ_k satisfies Condition* (**C1**)*, the ILC algorithm (14) converges almost surely to $\mathbf{u}^*$, as stated by Theorem 1. Moreover, because $\mathbf{I} + \boldsymbol{\Delta}$ is a lower triangular Toeplitz matrix, $\mathbf{I}+\boldsymbol{\Delta}$ commutes with $\hat{\mathbf{G}}$ and, under Assumption (**A2**), $\hat{\mathbf{G}}[\mathbf{I}+\boldsymbol{\Delta}]\mathbf{L} > 0$. This result means Theorem 2 applies, implying the existence of a sequence, satisfying Condition* (**C1**)*, that ensures monotonic convergence.*

4.2 Use of the Uncertain System Transpose

Another choice is $\mathbf{L} = \hat{\mathbf{G}}^T$. This choice is motivated by the fact that it can be used when $\hat{\mathbf{G}}$ is ill conditioned, as may be the case when $\hat{G}(q)$ has unstable zeros. The previously considered choice of $\mathbf{L}$, on the other hand, may not be usable because the input signal generated by the ILC algorithm can grow unacceptably large before converging to the ideal input.

Theorem 5. *Under Assumption (**A2**) and when $\mathbf{L} = \hat{\mathbf{G}}^T$, there exists a sequence of positive step sizes γ_k, satisfying Condition (**C1**), that ensures that the ILC algorithm (14) converges almost surely to $\mathbf{u}^*$ and that the 2-norm of the mean controlled error convergences monotonically.*

Proof. *Since $\mathbf{I}+\boldsymbol{\Delta}$ is a lower triangular Toeplitz matrix, $\mathbf{I}+\boldsymbol{\Delta}$ commutes with $\hat{\mathbf{G}}$ and condition (18) can be written as $\hat{\mathbf{G}}^T[\mathbf{I}+\boldsymbol{\Delta}]\hat{\mathbf{G}} > 0$, when $\mathbf{L} = \hat{\mathbf{G}}^T$. This condition is fulfilled when $\hat{\mathbf{G}}$ is non-singular and $\mathbf{I} + \boldsymbol{\Delta} > 0$. The former is true because N is finite and the latter is Assumption (**A2**). Therefore, when the sequence of positive step sizes γ_k satisfies Condition (**C1**), the ILC algorithm (14) converges almost surely to $\mathbf{u}^*$, as stated by Theorem 1. Moreover, Theorem 2 applies, implying the existence of a sequence, satisfying Condition (**C1**), that ensures monotonic convergence.*

4.3 Use of an Experiment

So far the use of a model to give an $\mathbf{L}$ that can then be used in (14) to evaluate $\mathbf{L}\mathbf{e}_k(\mathbf{u}_k)$ has been considered. For the specific choice of $\mathbf{L} = \mathbf{G}^T$, it is, however, possible to use an extra experiment per iteration to evaluate $\mathbf{L}\mathbf{e}_k(\mathbf{u}_k)$. Condition (18) is automatically satisfied with this choice, and Theorem 2 also applies.

The fact that a special experiment can be used is seen by noting that $\mathbf{e2} = \mathbf{G}^T\mathbf{e}_k(\mathbf{u}_k)$ is equal to the following filtering operations:

$$e1(t) = G(q)e_k(N-t, u_k(t)) \tag{43}$$

$$e2(t) = e1(N-t). \tag{44}$$

We see that, in the disturbance free case, $\mathbf{e2}$ can be found using an experiment on the true system, where the time reversed error signal is fed into the system as its input, the system output is measured and then time reversed itself. In reality the special experiment will have its own disturbances $d2(t)$ and $v2(t)$ associated with it. Nonetheless, an unbiased estimate of $\mathbf{e2}$ can still be found since:

$$\begin{aligned} E\{\mathbf{e2}\} &= E\{\mathbf{G}^T\mathbf{e}_k(\mathbf{u}_k) + \mathbf{d2} + \mathbf{v2}\} = E\{\mathbf{G}^T\mathbf{e}_k(\mathbf{u}_k)\} + E\{\mathbf{d2}\} + E\{\mathbf{n2}\} \\ &= \mathbf{G}^T E\{\epsilon_k(\mathbf{u}_k)\} + \mathbf{0} + \mathbf{0}. \end{aligned} \tag{45}$$

This method of evaluating **e2** is attractive as it avoids the problems of model uncertainty. It does, however, require an additional, non-standard, experiment at each iteration, which, depending on the application, may not always be possible. One case where it may be useful is when ILC is used to tune the input to improve the system's performance before the system is used in its intended application.

Remarks:

1. So far the motivation of the ILC algorithms considered has been to find the input that solves the root-finding type criterion (13), which aims to set the mean controlled error to zero. The model-free algorithm can be motivated differently. Instead of criterion (13), a logical alternative objective is the minimisation of the trace of the controlled error covariance matrix i.e.:

$$\min_{\mathbf{u}_k} J_k(\mathbf{u}_k) = \min_{\mathbf{u}_k} \frac{1}{2}\mathrm{tr}\left(E\{\epsilon_k(\mathbf{u}_k)\epsilon_k(\mathbf{u}_k)^T\}\right). \tag{46}$$

The minimum of this criterion occurs when:

$$\left.\frac{dJ_k(\mathbf{u}_k)}{d\mathbf{u}_k}\right|_{\mathbf{u}_k=\mathbf{u}^*} = E\left\{\left(\left.\frac{\partial\epsilon_k(\mathbf{u}_k)}{\partial\mathbf{u}_k}\right|_{\mathbf{u}_k=\mathbf{u}^*}\right)^T \epsilon_k(\mathbf{u}^*)\right\} = -\mathbf{G}^T E\{\epsilon_k(\mathbf{u}^*)\} = \mathbf{0}. \tag{47}$$

$E\{\epsilon_k(\mathbf{u}_k)\}$ is not directly measurable. Nonetheless, because equation (47) can be written as:

$$\left.\frac{dJ_k(\mathbf{u}_k)}{d\mathbf{u}_k}\right|_{\mathbf{u}_k=\mathbf{u}^*} = -\mathbf{G}^T E\{\epsilon_k(\mathbf{u}^*)\} = -\mathbf{G}^T E\{\mathbf{e}_k(\mathbf{u}^*)\} = \mathbf{0} \tag{48}$$

it is possible to find the minimiser of the criterion, again, using the Robbins-Monro algorithm:

$$\mathbf{u}_{k+1} = \mathbf{u}_k + \gamma_k \mathbf{G}^T \mathbf{e}_k(\mathbf{u}_k), \tag{49}$$

i.e. (14) with $\mathbf{L} = \mathbf{G}^T$.

2. The model-free algorithm has similarities to that proposed in [23] where reversed time inputs are used to cancel the system phase and produce monotonic convergence. Stochastic aspects are not considered, however.
3. It also has similarities to [11], which uses the steepest descent method, and calls $\mathbf{G}^T$ the adjoint of $\mathbf{G}$. It shows that by using this 'adjoint' with an iteration-varying gain, monotonic convergence occurs. The gain sequence is calculated via an optimisation, which does not consider stochastic disturbances. The gain at iteration k is given by:

$$\gamma_k = \frac{\|\mathbf{G}^T\mathbf{e}_{k-1}\|^2}{w + \|\mathbf{G}\mathbf{G}^T\mathbf{e}_{k-1}\|^2}, \tag{50}$$

where w is a weight on γ_k in the cost function. Since the measured error signal is used to calculate the gain, it will be affected by stochastic disturbances. This means $\lim_{k\to\infty} \|\mathbf{G}^T\mathbf{e}_{k-1}\|^2 \neq 0$ and so $\lim_{k\to\infty} \gamma_k \neq 0$. This implies that the second series of condition **C1** cannot be satisfied. Therefore, whilst the algorithm developed can lead to fast deterministic convergence to the optimal input, this cannot be proved when stochastic disturbances are present.

5 Experimental Results

The model-free algorithm was applied to the tracking control of a linear, permanent magnet, synchronous motor (LPMSM), which forms the upper axis of an x-y positioning table. LPMSMs are very stiff and have no mechanical transmission components. They, therefore, do not suffer from backlash and so allow very high positioning accuracy to be achieved. Additionally they are capable of high velocities and accelerations. These properties make them a very appealing, and thus common, choice for use in industries where rapid, high precision movements are required.

A standard two-degree-of-freedom position controller is used to control the motor's position. It operates at a sampling frequency of 2kHz. An analog position encoder using sinusoidal signals with periods of 2μm, which are then interpolated with 8192 intervals/period to obtain a resolution of 0.24nm, is used to measure the motor's position. However, the accuracy of this type of encoders is limited to 20nm.

The input, $\mathbf{u}_k$, computed by the ILC algorithm, is used as the position reference signal of the closed-loop system.

The desired output position, $y_d(t)$, was a series of three low-pass filtered steps, each of amplitude 25mm in the positive direction, followed by a similar series of filtered steps in the negative direction. This movement represents a typical industrial positioning motion. It has $N = 8192$. This value corresponds to the maximum number of points in the look-up table into which the new reference signal is fed at each iteration.

The sequence $\gamma_k = \frac{\alpha}{k+1}$ is used with $\alpha = 0.85$, which was chosen to achieve monotonic convergence.

For the experiment $\mathbf{u}_0 = \mathbf{y}_d$ was used and 100 iterations were carried out. Figures 2 and 3 show the convergence of $\boldsymbol{\epsilon}_k^T(\mathbf{u}_k)\boldsymbol{\epsilon}_k(\mathbf{u}_k)$ and the initial and final tracking achieved, respectively.

As can be seen from the figures and the values $\boldsymbol{\epsilon}_0^T(\mathbf{u}_0)\boldsymbol{\epsilon}_0(\mathbf{u}_0) = 6.0143 \times 10^{-8}\text{m}^2$ and $\boldsymbol{\epsilon}_{100}^T(\mathbf{u}_{100})\boldsymbol{\epsilon}_{100}(\mathbf{u}_{100}) = 1.9913 \times 10^{-14}\text{m}^2$ the algorithm considerably improves the tracking.

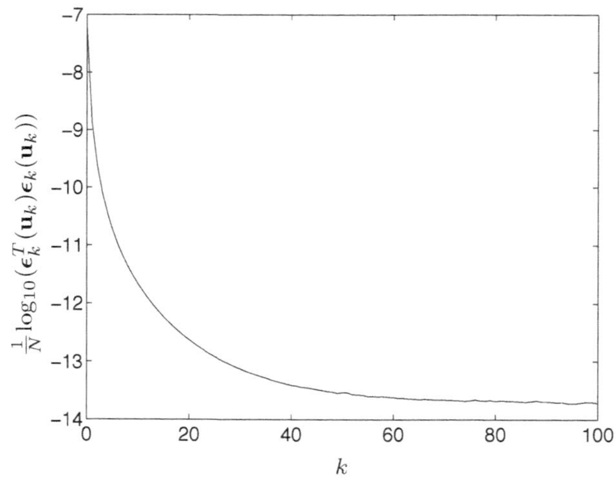

Fig. 2. $\epsilon_k^T(\mathbf{u}_k)\epsilon_k(\mathbf{u}_k)$ obtained using the model-free method

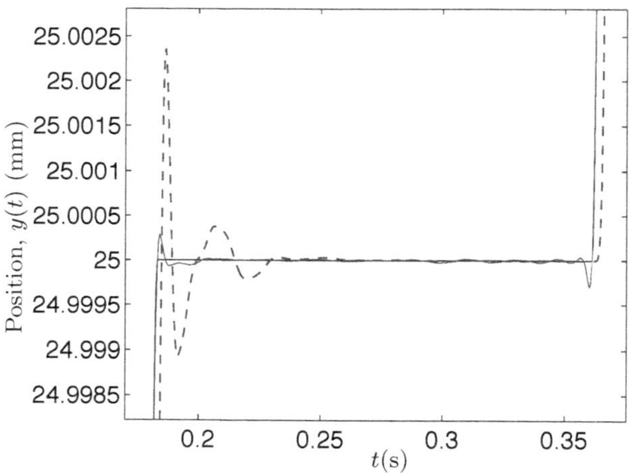

Fig. 3. Tracking at iteration $k = 0$ (green-dashed) and $k = 100$ (red) using the model-free method

6 Conclusions

The main contribution of this paper is to show how stochastic approximation theory can be used to derive and analyse Iterative Learning Control algorithms for linear time-invariant systems that are robust to non-repetitive disturbances. SA theory has provided general conditions that ensure almost sure convergence of the algorithm to the optimal input in the presence of stochastic disturbances.

ILC for LTI systems has been considered in this paper. The majority of the results apply, however, to linear time-varying (LTV) systems as well. In this case, however, the matrix $\mathbf{G}$ will not be lower triangular Toeplitz but a general lower triangular matrix instead. This implies that $\mathbf{L}$, $\hat{\mathbf{G}}$ and $\mathbf{I} + \boldsymbol{\Delta}$ will not, in general, commute.

The conditions imposed by SA require the learning gain to tend to zero as the iterations tend to infinity. This requirement is essential for stochastic learning algorithms. Practically it means that the learning ceases after a large number of iterations and if the desired output or repetitive disturbances change the algorithm will not react and the tracking will deteriorate. It is thus necessary to have a surveillance program that restarts the learning when the errors rise above a certain threshold.

References

1. Ahn, H.S., Moore, K.L., Chen, Y.: Kalman filter augmented iterative learning control on the iteration domain. In: IEEE American Control Conference, Minneapolis, U.S.A., pp. 250–255 (2006)
2. Arimoto, S.: Robustness of learning control for robotic manipulators. In: IEEE International Conference on Robotics and Automation, Cincinnati, Ohio USA, pp. 1528–1533 (May 1990)
3. Benveniste, A., Metivier, M., Priouret, P.: Adaptive Algorithms and Stochastic Approximation. Springer, Berlin (1990)
4. Butcher, M.: Data-driven methods for tracking improvement. PhD thesis, Ecole Polytechnique Fédérale de Lausane, Laboratoire d'Automatique (2009)
5. Butcher, M., Karimi, A., Longchamp, R.: A statistical analysis of certain iterative learning control algorithms. International Journal of Control 81(1), 156–166 (2008)
6. Chen, H.F.: Almost sure convergence of iterative learning control for stochastic systems. Sci. in China (Series F) 46(1), 1–13 (2003)
7. Chen, H.F., Fang, H.T.: Output tracking for nonlinear stochastic systems by iterative learning control. IEEE Transactions on Automatic Control 49(4), 583–588 (2004)
8. Dijkstra, B.G.: Iterative Learning Control with applications to a wafer stage. PhD thesis, Delft University of Technology, Delft, The Netherlands (2003)
9. Gunnarsson, S., Norrlöf, M.: On the disturbance properties of high order iterative learning control algorithms. Automatica 42(11), 2031–2034 (2006)
10. Hatonen, J.: Issues of algebra and optimality in Iterative Learning Control. PhD thesis, University of Oulu, Finland (2004)
11. Hatonen, J.J., Harte, T.J., Owens, D.H., Radcliffe, J.D., Lewin, P.L., Rogers, E.: A new robust iterative learning control algorithm for application on a gantry robot. In: IEEE Conference on Emerging Technologies in Factory Automation, Lisbon, Portugal, pp. 305–312 (2003)
12. Heinzinger, G., Fenwick, D., Paden, B., Miyaziki, F.: Robust learning control. In: 28th IEEE Conference on Decision and Control, Tampa, Florida USA, December 1989, pp. 436–440 (1989)
13. Nevelson, M.B., Hasminskii, R.Z.: Stochastic Approximation and Recursive Estimation. American Mathematical Society, Providence (1973)

14. Norrlöf, M.: An adaptive iterative learning control algorithm with experiments on an industrial robot. IEEE Transactions on Robotics and Automation 18(2), 245–251 (2002)
15. Norrlöf, M., Gunnarsson, S.: Disturbance aspects of iterative learning control. Engineering Applications of Artificial Intelligence 14(1), 87–94 (2001)
16. Norrlöf, M., Gunnarsson, S.: Time and frequency domain convergence properties in iterative learning control. International Journal of Control 75(14), 1114–1126 (2002)
17. Norrlöf, M., Gunnarsson, S.: Disturbance rejection using an ILC algorithm with iteration varying filter. Asian Journal of Control 6(3), 432–438 (2004)
18. Robbins, H., Monro, S.: A stochastic approximation method. Ann. Math. Stat. 22, 400–407 (1951)
19. Saab, S.S.: Optimal selection of the forgetting matrix into an iterative learning control algorithm. IEEE Transactions on Automatic Control 50(12), 2039–2043 (2005)
20. Saab, S.S.: A discrete-time stochastic learning control algorithm. IEEE Transactions on Automatic Control 46(6), 877–887 (2001)
21. Saab, S.S.: On a discrete-time stochastic learning control algorithm. IEEE Transactions on Automatic Control 46(8), 1333–1336 (2001)
22. Tao, M.K., Kosut, R.L., Gurcan, A.: Learning feedforward control. In: IEEE American Control Conference, Baltimore, Maryland USA, pp. 2575–2579 (June 1994)
23. Ye, Y., Wang, D.: Zero phase learning control using reversed time input runs. Journal of Dynamic Systems, Measurement and Control 127, 133–139 (2005)

Elimination Theory for Nonlinear Parameter Estimation

John Chiasson and Ahmed Oteafy

1 Introduction

The work presented here exploits elimination theory (solving systems of polynomial equations in several variables) [1][2] to perform nonlinear parameter identification. In particular show how this technique can be used to estimate the rotor time constant and the stator resistance values of an induction machine. Although the example here is restricted to an induction machine, parameter estimation is applicable to many practical engineering problems. In [3], L. Ljung has outlined many of the challenges of nonlinear system identification as well as its particular importance for biological systems. In these types of problems, the model developed for analysis is typically a nonlinear state space model with unknown parameter values. The typical situation is that only a few of the state variables are measurable requiring that the system be reformulated as a nonlinear input-output model. In turn, resulting the nonlinear input-output model is almost always nonlinear in the parameters. Towards that end, differential algebra tools for analysis of nonlinear systems have been developed by Michel Fliess [4][5] and Diop [6]. Moreover, Ollivier [7] as well as Ljung and Glad [8] have developed the use of the characteristic set of an ideal as a tool for identification problems. The use of these differential algebraic methods for system identification have also been considered in [9], [10]. The focus of their research has been the determination of *a priori* identifiability of a given system model. However, as stated in [10], the development of an efficient algorithm using these differential algebraic techniques is still unknown. Here, in contrast, a method for which one can actually numerically obtain the numerical value of the parameters is presented. We also point out that [11] has also done work applying elimination theory to systems problems.

John Chiasson · Ahmed Oteafy
ECE Department Boise State University Boise ID, USA
e-mail: `johnchiasson@boisestate.edu`, `ahmedoteafy@u.boisestate.edu`

J. Lévine & P. Müllhaupt (Eds.): Adv. in Theory of Control, Signals, LNCIS 407, pp. 65–75.
springerlink.com © Springer-Verlag Berlin Heidelberg 2010

Here, using the techniques of elimination theory, it is shown that a significant class of nonlinear identification problems can be formulated as a nonlinear least-squares problem whose solution is guaranteed to be found in a finite number of steps. The proposed methodology starts with obtaining an *over-parameterized* input-output model that is linear in the parameters. It is then assumed that the relationship between the actual parameters in the over-parameterized model are *rationally* related which is not atypical of many engineering systems. After making appropriate substitutions, the problem is transformed into a *nonlinear* least-squares problem which is- not overparameterized. It is then shown how the nonlinear least-squares problem can be solved in a finite number of steps using elimination theory.

2 Mathematical Model of an Induction Machine

An induction machine is now used as a realistic application to describe the methodology. Specifically, the identification of the rotor time constant and stator resistance are considered. As background, field-oriented control provides a means to obtain high-performance control of an induction machine for use in applications such as traction drives. This field-oriented control methodology requires knowledge of the rotor flux linkages, which are not usually measured [12][13]. To get around this problem, the rotor flux linkages are usually estimated using a state observer, and this observer requires the value of the rotor time constant T_R. However, $T_R = L_R/R_R$ varies due to ohmic heating and thus it is of considerable interest to estimate its value online in order to update the flux estimator with its current value.

A standard two-phase model of the induction machine is given by ([13])

$$\begin{aligned}
\frac{di_{Sa}}{dt} &= \frac{\beta}{T_R}\psi_{Ra} + \beta n_p \omega \psi_{Rb} - \gamma i_{Sa} + \frac{1}{\sigma L_S} u_{Sa} \\
\frac{di_{Sb}}{dt} &= \frac{\beta}{T_R}\psi_{Rb} - \beta n_p \omega \psi_{Ra} - \gamma i_{Sb} + \frac{1}{\sigma L_S} u_{Sb} \\
\frac{d\psi_{Ra}}{dt} &= -\frac{1}{T_R}\psi_{Ra} - n_p \omega \psi_{Rb} + \frac{M}{T_R} i_{Sa} \\
\frac{d\psi_{Rb}}{dt} &= -\frac{1}{T_R}\psi_{Rb} + n_p \omega \psi_{Ra} + \frac{M}{T_R} i_{Sb} \\
\frac{d\omega}{dt} &= \frac{M n_p}{J L_R}\left(i_{Sb}\psi_{Ra} - i_{Sa}\psi_{Rb}\right) - \frac{\tau_L}{J}
\end{aligned} \tag{1}$$

where the state variables are the rotor angular position θ, the rotor angular speed $\omega = d\theta/dt$, the (two-phase equivalent) stator currents i_{Sa}, i_{Sb}, and the (two-phase equivalent) rotor flux linkages ψ_{Ra}, ψ_{Rb}. The controllable inputs are the (two-phase equivalent) stator voltages u_{Sa}, u_{Sb} while the disturbance input is the load torque τ_L.

The parameters of the model are the stator and rotor resistances R_S and R_R, the mutual inductance M, the stator and rotor inductances L_S and L_R, the moment of inertia J and the number of pole-pairs n_p. The symbols

$$\begin{array}{ll} T_R = L_R/R_R & \sigma = 1 - M^2/\left(L_S L_R\right) \\ \beta = M/\left(\sigma L_S L_R\right) & \gamma = R_S/\left(\sigma L_S\right) + \beta M/T_R \end{array}$$

are used to simplify the expressions where σ is referred to as the total leakage factor.

This model is transformed into a coordinate system attached to the rotor as the signals in this new (x, y) rotor frame typically vary at the slower slip frequency rather than at the stator frequency in the (a, b) frame. The current variables are transformed according to

$$\begin{bmatrix} i_{Sx} \\ i_{Sy} \end{bmatrix} = \begin{bmatrix} \cos(n_p\theta) & \sin(n_p\theta) \\ -\sin(n_p\theta) & \cos(n_p\theta) \end{bmatrix} \begin{bmatrix} i_{Sa} \\ i_{Sb} \end{bmatrix}. \tag{2}$$

This transformation does not depend on any unknown parameter in contrast to the field-oriented (or dq) transformation which requires knowledge of the rotor fluxes. The stator voltages and the rotor fluxes are transformed in the same way as the currents resulting in the following model (see [14][15])

$$\frac{di_{Sx}}{dt} = \frac{u_{Sx}}{\sigma L_S} - \gamma i_{Sx} + \frac{\beta}{T_R}\psi_{Rx} + n_p\beta\omega\psi_{Ry} + n_p\omega i_{Sy} \tag{3}$$

$$\frac{di_{Sy}}{dt} = \frac{u_{Sy}}{\sigma L_S} - \gamma i_{Sy} + \frac{\beta}{T_R}\psi_{Ry} - n_p\beta\omega\psi_{Rx} - n_p\omega i_{Sx} \tag{4}$$

$$\frac{d\psi_{Rx}}{dt} = \frac{M}{T_R}i_{Sx} - \frac{1}{T_R}\psi_{Rx} \tag{5}$$

$$\frac{d\psi_{Ry}}{dt} = \frac{M}{T_R}i_{Sy} - \frac{1}{T_R}\psi_{Ry} \tag{6}$$

$$\frac{d\omega}{dt} = \frac{Mn_p}{JL_R}(i_{Sy}\psi_{Rx} - i_{Sx}\psi_{Ry}) - \frac{\tau_L}{J}. \tag{7}$$

As explained above, the interest here is in the online estimation of T_R as it changes due to ohmic heating so that an accurate value is available to the rotor flux estimator. However, the stator resistance value R_S will also vary due to ohmic heating, therefore its variation must also be taken into account in the estimation. The electrical parameters M, L_S, σ are assumed to be known and not varying. Measurements of the stator currents i_{Sa}, i_{Sb} and voltages u_{Sa}, u_{Sb} as well as the position θ of the rotor are assumed to be available; the velocity is then computed from the position measurements. The rotor flux linkages are not assumed to be measured.

3 Input-Output Model

Standard methods for parameter estimation are based on equalities where known signals depend *linearly* on unknown parameters. However, the induction motor model described above does not fit in this category unless the rotor flux linkages are measured. As this is not the case here, the fluxes ψ_{Rx}, ψ_{Ry} and their derivatives $d\psi_{Rx}/dt, d\psi_{Ry}/dt$ must be eliminated from the final identification model. The four equations (3), (4), (5), (6) are used to solve for the four unknowns $\psi_{Rx}, \psi_{Ry}, d\psi_{Rx}/dt, d\psi_{Ry}/dt$. Further, a new set of independent equations is found by differentiating equations (3) and (4) to obtain

$$\begin{aligned}\frac{1}{\sigma L_s}\frac{du_{Sx}}{dt} &= \frac{d^2 i_{Sx}}{dt^2} + \gamma\frac{di_{Sx}}{dt} - \frac{\beta}{T_R}\frac{d\psi_{Rx}}{dt} - n_p\beta\omega\frac{d\psi_{Ry}}{dt} - n_p\beta\psi_{Ry}\frac{d\omega}{dt}\\ &\quad - n_p\omega\frac{di_{Sy}}{dt} - n_p i_{Sy}\frac{d\omega}{dt}\end{aligned} \tag{8}$$

and

$$\begin{aligned}\frac{1}{\sigma L_s}\frac{du_{Sy}}{dt} &= \frac{d^2 i_{Sy}}{dt^2} + \gamma\frac{di_{Sy}}{dt} - \frac{\beta}{T_R}\frac{d\psi_{Ry}}{dt} + n_p\beta\omega\frac{d\psi_{Rx}}{dt} + n_p\beta\psi_{Rx}\frac{d\omega}{dt}\\ &\quad + n_p\omega\frac{di_{Sx}}{dt} + n_p i_{Sx}\frac{d\omega}{dt}.\end{aligned} \tag{9}$$

To simplify the presentation we now assume that the speed is held constant as in [16][17] (this is not necessary, see [18][19]). The expressions for $\psi_{Rx}, \psi_{Ry}, d\psi_{Rx}/dt, d\psi_{Ry}/dt$ found from solving equations (3), (4), (5), (6) are substituted into equations (8) and (9) with $d\omega/dt = 0$ to obtain

$$\begin{aligned}0 &= -\frac{d^2 i_{Sx}}{dt^2} + \frac{di_{Sy}}{dt}n_p\omega + \frac{1}{\sigma L_S}\frac{du_{Sx}}{dt} - (\gamma + \frac{1}{T_R})\frac{di_{Sx}}{dt}\\ &\quad - i_{Sx}(-\frac{\beta M}{T_R^2} + \frac{\gamma}{T_R}) + i_{Sy}n_p\omega(\frac{1}{T_R} + \frac{\beta M}{T_R}) + \frac{u_{Sx}}{\sigma L_S T_R}\end{aligned} \tag{10}$$

$$\begin{aligned}0 &= -\frac{d^2 i_{Sy}}{dt^2} - \frac{di_{Sx}}{dt}n_p\omega + \frac{1}{\sigma L_S}\frac{du_{Sy}}{dt} - (\gamma + \frac{1}{T_R})\frac{di_{Sy}}{dt}\\ &\quad - i_{Sy}(-\frac{\beta M}{T_R^2} + \frac{\gamma}{T_R}) - i_{Sx}n_p\omega(\frac{1}{T_R} + \frac{\beta M}{T_R}) + \frac{u_{Sy}}{\sigma L_S T_R}.\end{aligned} \tag{11}$$

As $\gamma = R_S/(\sigma L_S) + \beta M/T_R$, it follows that

$$\begin{aligned}-\beta M/T_R^2 + \gamma/T_R &= (R_S/T_R)/(\sigma L_S)\\ \gamma + 1/T_R &= R_S/(\sigma L_S) + (\beta M + 1)/T_R\end{aligned}$$

which is used to rewrite (10) and (11) as

$$0 = -\frac{d^2 i_{Sx}}{dt^2} + \frac{di_{Sy}}{dt} n_p\omega + \frac{1}{\sigma L_S}\frac{du_{Sx}}{dt} - \Big(R_S/\left(\sigma L_S\right) + \left(\beta M + 1\right)/T_R \Big) \frac{di_{Sx}}{dt}$$
$$- i_{Sx}(\frac{R_S}{T_R}\frac{1}{\sigma L_S}) + i_{Sy} n_p\omega((\beta M + 1)/T_R) + \frac{u_{Sx}}{\sigma L_S T_R} \tag{12}$$

$$0 = -\frac{d^2 i_{Sy}}{dt^2} - \frac{di_{Sx}}{dt} n_p\omega + \frac{1}{\sigma L_S}\frac{du_{Sy}}{dt} - \Big(R_S/\left(\sigma L_S\right) + \left(\beta M + 1\right)/T_R \Big) \frac{di_{Sy}}{dt}$$
$$- i_{Sy}(\frac{R_S}{T_R}\frac{1}{\sigma L_S}) - i_{Sx} n_p\omega\left(\beta M + 1\right)/T_R + \frac{u_{Sy}}{\sigma L_S T_R}. \tag{13}$$

More compactly, equations (12) and (13) are written in linear regressor form as

$$y(t) = W(t)K \tag{14}$$

with

$$y(t) \triangleq \begin{bmatrix} \frac{d^2 i_{Sx}}{dt^2} - \frac{di_{Sy}}{dt} n_p\omega - \frac{1}{\sigma L_S}\frac{du_{Sx}}{dt} \\ \frac{d^2 i_{Sy}}{dt^2} + \frac{di_{Sx}}{dt} n_p\omega - \frac{1}{\sigma L_S}\frac{du_{Sy}}{dt} \end{bmatrix} \tag{15}$$

and

$$W(t) \triangleq \begin{bmatrix} -\frac{di_{Sx}}{dt}\frac{1}{\sigma L_S} & (\beta M + 1)\left(-\frac{di_{Sx}}{dt} + i_{Sy} n_p\omega\right) + \frac{u_{Sx}}{\sigma L_S} & -\frac{i_{Sx}}{\sigma L_S} \\ -\frac{di_{Sy}}{dt}\frac{1}{\sigma L_S} & (\beta M + 1)\left(-\frac{di_{Sy}}{dt} - i_{Sx} n_p\omega\right) + \frac{u_{Sy}}{\sigma L_S} & -\frac{i_{Sy}}{\sigma L_S} \end{bmatrix} \tag{16}$$

as well as

$$K = \begin{bmatrix} K_1 \\ K_2 \\ K_3 \end{bmatrix} \triangleq \begin{bmatrix} R_S \\ 1/T_R \\ R_S/T_R \end{bmatrix}. \tag{17}$$

This model is over-parameterized in the parameters, that is, they must satisfy the constraint

$$K_3 = K_1 K_2. \tag{18}$$

Replacing K_3 by $K_1 K_2$ in (14) results in a model that is not over-parameterized, but it is no longer linear in the parameters. This issue is considered next.

4 Nonlinear Least-Squares Identification

A discrete-time sampled version of (14) is

$$y(nT) = W(nT)K, \tag{19}$$

where T is the sample period, nT is the time the nth sample is taken, and $K = \begin{bmatrix} K_1 & K_2 & K_3 \end{bmatrix}^T$ is the (over-parameterized) vector of unknown parameters. If the constraint (18) is ignored, then the system is a linear (but over-parameterized) least-squares problem. Theoretically, an exact unique solution for the unknown parameter vector K may be determined after several time instants. However, due to the fact that both $y(nT)$ and $W(nT)$ are measured from signals that are noisy (due to quantization and differentiation), the regressor model (19) is only approximately valid in practice. These sources of error result in an overdetermined system of equations. In order to get around this problem, the solution vector K is specified as that which minimizes a least-squares criterion. Specifically, given $y(nT)$ and $W(nT)$ where $y(nT) = W(nT)K$, one defines

$$E^2(K) = \sum_{n=1}^{N} \left| y(nT) - W(nT)K \right|^2 \tag{20}$$

as the *residual error* associated to a parameter vector K. Then, the least-squares estimate K^* is chosen such that $E^2(K)$ is minimized for $K = K^*$. The function $E^2(K)$ is quadratic and therefore has a unique minimum at the point where $\partial E^2(K)/\partial K = 0$ holds. Solving this expression for K^* yields the least-squares solution to $y(nT) = W(nT)K$ as

$$K^* = \left[\sum_{n=1}^{N} W^T(nT)W(nT) \right]^{-1} \left[\sum_{n=1}^{N} W^T(nT)y(nT) \right]. \tag{21}$$

However, there is no guarantee that the solution of (21) will satisfy the constraint $K_3 = K_1K_2$. Furthermore, the over-parameterized identification model consisting of (17) and (19) results in an ill-conditioned solution for K^*. That is, small changes in the data $W(nT), y(nT)$ can result in large changes in the value computed for K^*. To get around these problems, a *nonlinear* least-squares approach is taken which involves minimizing

$$E^2(K) = \sum_{n=1}^{N} \left| y(nT) - W(nT)K \right|^2 = R_y - 2R_{Wy}^T K + K^T R_W K \tag{22}$$

subject to the constraint $K_3 = K_1K_2$ where

$$R_y \triangleq \sum_{n=1}^{N} y^T(nT)y(nT), R_{Wy} \triangleq \sum_{n=1}^{N} W^T(nT)y(nT)$$
$$R_W \triangleq \sum_{n=1}^{N} W^T(nT)W(nT). \tag{23}$$

On physical grounds, the parameters K_1, K_2 are constrained to the region

$$0 < K_1 < \infty, 0 < K_2 < \infty \tag{24}$$

and the squared error $E^2(K)$ will be minimized in this *open* region. Substituting $K_3 = K_1K_2$ in (22), we obtain a new error function $E_p^2(K_1, K_2)$ as

$$\begin{aligned} E_p^2(K_1, K_2) &\triangleq \sum_{n=1}^{N} \left| y(nT) - W(nT)K \right|^2_{K_3=K_1K_2} \\ &= R_y - 2R_{Wy}^T K \Big|_{K_3=K_1K_2} + \left(K^T R_W K\right)\Big|_{K_3=K_1K_2} . \end{aligned} \tag{25}$$

As the minimum of (25) must occur in the region (24), it follows that the minimum is located at an extremum point. To solve for this minimum thus entails solving simultaneously the two extrema equations

$$p_1(K_1, K_2) \triangleq \frac{\partial E_p^2(K_1, K_2)}{\partial K_1} \tag{26}$$

$$p_2(K_1, K_2) \triangleq \frac{\partial E_p^2(K_1, K_2)}{\partial K_2}, \tag{27}$$

which are *polynomials* in the parameters K_1, K_2. The degrees of the polynomials p_i are given in the table below

	$\deg K_1$	$\deg K_2$
$p_1(K_1, K_2)$	1	2
$p_2(K_1, K_2)$	2	1

These two polynomials are rewritten in the form

$$p_1(K_1, K_2) = a_1(K_2)K_1 + a_0(K_2) \tag{28}$$

$$p_2(K_1, K_2) = b_2(K_2)K_1^2 + b_1(K_2)K_1 + b_0(K_2). \tag{29}$$

A systematic procedure to find all possible solutions to a set of polynomials is provided by elimination theory through the method of resultants [1][2]. However, in this particular example, $p_1(K_1, K_2)$ is of degree 1 in K_1 and can be solved directly. Substituting $K_1 = -a_0(K_2)/a_1(K_2)$ from $p_1(K_1, K_2) = 0$ into $p_2(K_1, K_2) = 0$ and multiplying the result through by $a_1^2(K_2)$, one obtains the (resultant) polynomial

$$r(K_2) = a_0^2(K_2)b_2(K_2) - a_0(K_2)a_1(K_2)b_1(K_2) + a_1^2(K_2)b_0(K_2), \tag{30}$$

where $\deg_{K_2}\{r\} = 5$. The roots of (30) are the only possible candidates for the values of K_2 that satisfy $p_1(K_1, K_2) = p_2(K_1, K_2) = 0$ for some K_1. In the online implementation, the coefficients of the polynomials $a_1(K_2), a_0(K_2)$, $b_2(K_2)$, $b_1(K_2), b_0(K_2)$, whose explicit expressions in terms of the elements of the matrices R_W and R_{Wy} are known a priori vis-a-vis (25), (26), and (27), are computed and stored during data collection. The coefficients of the polynomial $r(K_2)$ are then computed online according to (30). Next, the positive roots K_{2i} of $r(K_2) = 0$ are computed and substituted into $p_1(K_1, K_{2i}) = 0$ which is then solved for its positive roots K_{1j}. By this method of back solving, the finite number of possible candidate solutions (K_{1j}, K_{2i}) are found. The pair that results in the smallest squared error, i.e., the smallest value of $E_p^2(K_1, K_2)$, is chosen.

5 Simulations

The above parameter identification method was studied in simulation using a two-phase equivalent model of an induction machine under closed-loop control. The parameters of the induction machine are (see [13]): $M = 0.0117\,\mathrm{H}$, $L_R = 0.014\,\mathrm{H}$, $L_S = 0.014\,\mathrm{H}$, $R_S = 1.7\,\Omega$, $R_R = 3.9\,\Omega$, $\tau_{L0} = 0.15\,\mathrm{Nm}$, $J = 0.00011$ Kgm^2, and $n_P = 3$. The controller sets the desired rotor speed at $\omega_R = 2\pi \times 75$ rad/s, while the load torque is defined to be $\tau_L \triangleq \tau_{L0} + f\omega$ with $\tau_{L0} = 0.15$ Nm. The data was sampled at $f_S = 4$ kHz which was filtered through a 2nd order low pass Butterworth filter with a cutoff frequency of 70 Hz.

To mimic the ohmic heating of the rotor and stator resistors, in the simulation of the motor model their values were increased by 50% after 3 seconds of operation with the estimator updating the value of T_R every 0.5 seconds. After the update at 3.5 secs the estimator provides the new estimates of R_S and R_R to the controller. Figure 1 below is a plot of $K_2 = 1/T_R$ and its reference versus time showing that after the update the estimator gives the value of K_2 within 2% of the correct value.

To show the importance of having an accurate value of the rotor time constant, the power consumed before and after the rotor time constant update was computed. Figure 2 shows the speed versus time for the simulation. (the transient at $t = 0$ is due to the fact that the flux in the machine is zero so that during the build up of the flux the machine has torque oscillations). Figure 3 below is a plot of the real power $P(t) = u_{Sa}i_{Sa} + u_{Sb}i_{Sb}$ vs time. As the figure shows, the real power jumps up to 66.9 W at 3 sec. After the rotor time constant value is updated to controller at 3.5 seconds, the real power comes down to 63.7 W, which is a 5% decrease. Of course these numbers are small because the simulation was done with a small (a less than kW) machine. In industry where large machines are used, the energy savings would be significant.

As explained above, the rotor time constant $T_R = 1/K_2$ is used to estimate the rotor fluxes which in turn are used to estimate the direct and quadrature currents for use in field oriented control. In field oriented control the motor torque is given by $\tau = \mu\psi_d i_q$ ($\mu = \frac{Mn_p}{JL_R}$) which at constant speed reduces to $\tau = \mu M i_d i_q$. For a given torque, the current magnitude $i_d^2 + i_q^2$ is minimized if $i_d = i_q$ [13]. Thus it is important to estimate the rotor flux angle accurately to have accurate values of the dq currents in order to achieve this minimization.

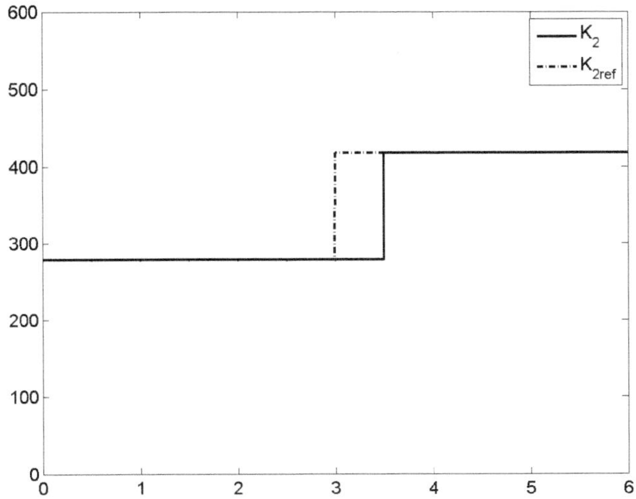

Fig. 1. $K_2 = 1/T_R$ and K_{2ref} vs. time in seconds

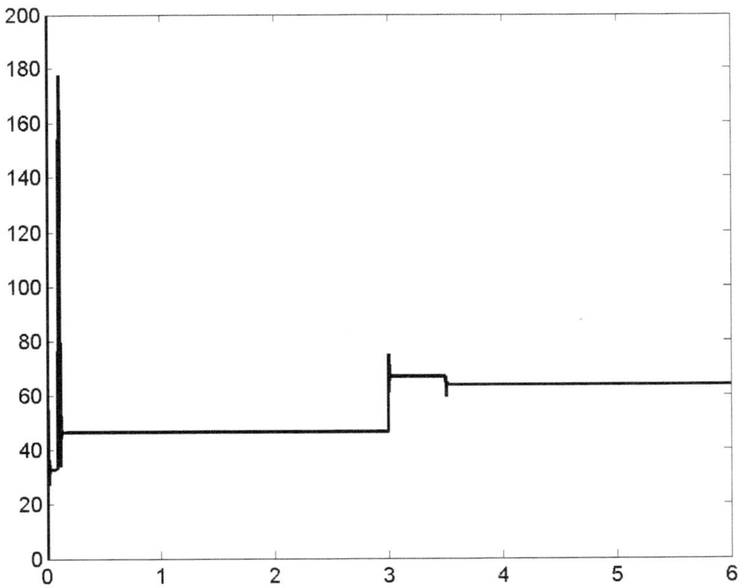

Fig. 2. Speed in radians/sec versus time in seconds

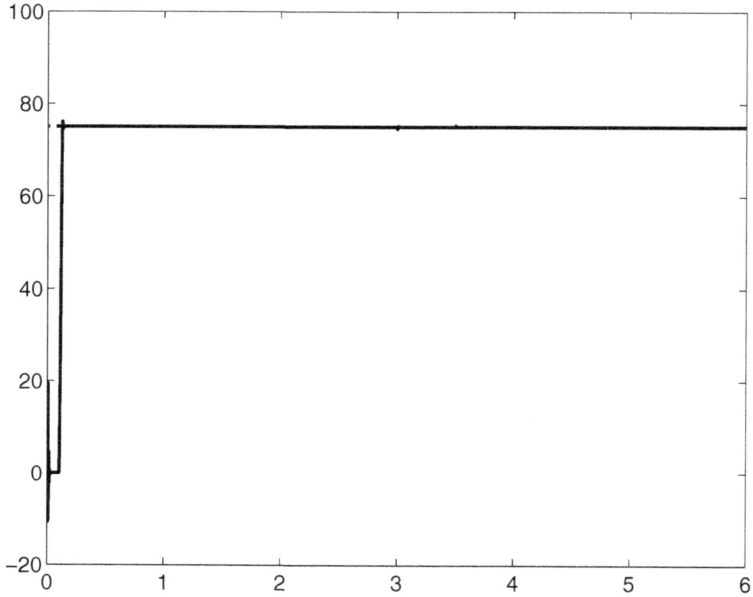

Fig. 3. Real power P in Watts versus time in seconds. (The large transient in the power at the beginning is due to the discontinuity in the acceleration - see the speed trajectory)

6 Conclusions

An approach to solving a nonlinear least-squares parameter identification problem in a *finite* number of steps was presented. This is in contrast to iterative methods which may or may not converge and, even if convergences takes place, it may be to only a local minimum. The method was presented by showing how the rotor time constant of the induction machine can be found online. In this application, the results show that an incorrect value of T_R leads to the controller commanding non-optimum values of the stator currents to the machine which in turn increases the Ohmic losses. That is, a higher power usage is required for the same torque requirement.

References

1. Cox, D., Little, J., O'Shea, D.: Ideals, Varieties, And Algorithms An Introduction to Computational Algebraic Geometry and Commutative Algebra, 2nd edn. Springer, Berlin (1996)
2. von zur Gathen, J., Gerhard, J.: Modern Computer Algebra. Cambridge University Press, Cambridge (1999)

3. Ljung, L.: Challenges of Nonlinear Identification. In: Bode Lecture, IEEE Conference on Decision and Control, Maui HI (2003)
4. Fliess, M., Glad, S.T.: An algebraic approach to linear and nonlinear control. In: Essays on Control: Perspectives in the Theory and Applications, Birkhäuser, pp. 223–267 (1993)
5. Sira-Ramírez, H., Agrawal, S.K.: Differentially Flat Systems. Marcel-Dekker, New York (2004)
6. Diop, S.: Differential-algebraic decision methods and some applications to system theory. Theor. Computer Sci. 98, 137–161 (1992)
7. Ollivier, F.: Le problème de l'identifiabilité structurelle globale: Ètude thèorique, mèthodes effective et bornes de complexè. PhD thesis, École Polytéchnique, Paris, France (1990)
8. Ljung, L., Glad, S.T.: On global identifiability for arbitrary model parameterisations. Automatica 30(2), 265–276 (1994)
9. Margaria, G., Riccomagno, E., Chappell, M.J., Wynn, H.P.: Differential algebra methods for the study of the structural identifiability of biological rational polynomial models (2004) (preprint)
10. Saccomani, M.P.: Some results on parameter identification of nonlinear systems. Cardiovascular Engineering: An Internation Journal 4(1), 95–102 (2004)
11. Diop, S.: Elimination in control theory. MCSS 4, 17–32 (1991)
12. Leonhard, W.: Control of Electrical Drives, 3rd edn. Springer, Berlin (2001)
13. Chiasson, J.: Modeling and High-Performance Control of Electric Machines. John Wiley & Sons, Chichester (2005)
14. Stephan, J., Bodson, M., Chiasson, J.: Real-time estimation of induction motor parameters. IEEE Transactions on Industry Applications 30(3), 746–759 (1994)
15. Stephan, J.: Real-time estimation of the parameters and fluxes of induction motors. Master's thesis, Carnegie Mellon University (1992)
16. Chiasson, J., Bodson, M.: Estimation of the rotor time constant of an induction machine at constant speed. In: Proceedings of the European Control Conference ECC 2007, Kos, Greece, pp. 4673–4678 (July 2007)
17. Oteafy, A., Chiasson, J., Bodson, M.: Online identification of the rotor time constant of an induction machine. In: Proceedings of the American Control Conference, St. Louis MO, pp. 4373–4378 (2009)
18. Wang, K., Chiasson, J., Bodson, M., Tolbert, L.: A nonlinear least-squares approach for estimation of the induction motor parameters. IEEE Transactions on Automatic Control 50(10), 1622–1628 (2005)
19. Wang, K., Chiasson, J., Bodson, M., Tolbert, L.M.: An on-line rotor time constant estimator for the induction machine. In: Proceedings of the IEEE International Electric Machines and Drives Conference, San Antonio TX, pp. 608–614 (May 2005)

Controlling Underactuated Mechanical Systems: A Review and Open Problems

Zhong-Ping Jiang

Abstract. This chapter provides a short review on the popular yet still very important area of controlling underactuated mechanical systems. New solutions to the simultaneous stabilization and tracking problem are proposed for nonholonomic mobile robots using state and output feedback. Some open problems are discussed with a unique objective to solicit fundamentally novel techniques for the further development of modern nonlinear control theory.

1 Introduction

Underactuated mechanical systems refer to those mechanical systems with less number of controls than the degrees of freedom, and arise often from nonholonomic systems with nonintegrable constraints. Examples of underactuated mechanical systems are abundant in our daily life, ranging from spacecraft to ground and marine vehicles such as mobile robots, surface ships and underwater vehicles. Controlling underactuated mechanical systems has been an active research area over the last 25 years. This is because it concerns fundamentally nonlinear control problems which require novel ideas and techniques. One of these challenges in nonholonomic systems is the obstruction to asymptotic stabilization. Indeed, Brockett's necessary condition [4] applied to these inherently nonlinear systems yields a surprising fact that there is no linear, or nonlinear, continuous state-feedback stabilizing control law for this special, but important, class of nonlinear systems. The control systems community has contributed a versatile set of novel ideas and feedback design strategies, including time-varying feedback [41, 40], differential flatness [15, 16, 30], passivity-based control [1, 36, 18], discontinuous or hybrid feedback [2, 3, 39, 5]

Zhong-Ping Jiang
Department of Electrical and Computer Engineering, Polytechnic Institute of New York University, Six Metrotech Center, Brooklyn, NY 11201, U.S.A.
e-mail: zjiang@control.poly.edu

J. Lévine & P. Müllhaupt (Eds.): Adv. in Theory of Control, Signals, LNCIS 407, pp. 77–88.
springerlink.com © Springer-Verlag Berlin Heidelberg 2010

and "time-varying" backstepping [32, 27, 28, 21, 22]. A summary of the earlier research efforts by many researchers can be found in [3, 7, 25] and references therein.

A second feature arising from investigating nonholonomic control problems is that stabilization and tracking are two fundamentally different control problems. Very often, in the traditional literature of control theory, stabilization is regarded as a special case of the tracking problem. Unfortunately, this is not the case for underactuated mechanical systems with nonholonomic constraints. The violation of Brockett's necessary condition for asymptotic stabilization presents a challenge to develop fundamentally new approaches to nonlinear control theory. On the other hand, in the case of trajectory-tracking, there is a local feedback solution if the linearization of the system around the moving trajectory is uniformly controllable. In terms of control terminology, the reference, or to-be-tracked, trajectory must satisfy the persistence excitation (PE) condition to make the tracking control problem tractable. While the stabilization and tracking problems are typically studied as two separate problems, it becomes a natural question to ask: when can we find a (single) control law which can solve both (global) stabilization and (global) tracking simultaneously? We call this problem "*(Global) Simultaneous Stabilization and Tracking*". In this work, we focus on continuous time-varying feedback solutions for underactuated mechanical systems. Due to space limitation, we devote ourselves to the case study of nonholonomic mobile robots.

The purpose of this book chapter is to point out that there is more to be accomplished in this field despite the significant progress made by many researchers. We show by means of the benchmark example of nonholonomic mobile robots that the simultaneous stabilization and tracking problem is largely open for underactuated mechanical systems. Answering this question among other open problems requires the invention of new techniques and methodologies, thereby contributing to the further development of modern nonlinear control theory.

2 Simultaneous Stabilization and Tracking

2.1 Problem Statement

Due to inherent nonlinearity in mechanical systems with nonholonomic constraints, point-stabilization and trajectory-stabilization have often been treated as two separate control problems. One way to unify these two apparently different control problems is to adopt the idea of model reference control. More specifically, consider a nonlinear control system of the form

$$\dot{x} = f(x,u), \quad x \in \Re^n,\ u \in \Re^m \tag{1}$$

and a reference model associated with (1), which describes the desired to-be-tracked trajectory,

$$\dot{x}_d = f(x_d,u_d), \quad x_d \in \Re^n,\ u_d \in \Re^m \tag{2}$$

The central control problem addressed in this work is to identify conditions on the set of admissible reference trajectories x_d and find, if possible, a (single) continuous feedback controller of the form

$$u = u(x, x_d, t) \tag{3}$$

which can stabilize the system (1) to any such x_d, in particular, $x(t) - x_d(t) \to 0$ as $t \to \infty$.

On the one hand, it is worth noting that, of course, this set of so-called feasible trajectories determined by the set $\mathscr{U}$ of admissible reference inputs u_d must be strictly smaller than the set of feasible trajectories determined by *all* piecewise continuous functions. Otherwise, the negative result of [31] implies that, generally speaking, such a controller may not exist. On the other hand, the set $\mathscr{U}$ should be large enough to include a variety of time-varying signals which can generate paths for many practical control applications. Examples of such paths are straight line, circle and a combination of the two.

Clearly, when there exists a set-point and reference input pair $(x_d^\star, u_d^\star)$ such that $f(x_d^\star, u_d^\star) = 0$, the stabilization of set-point $x_d^\star$ is a special case of the simultaneous stabilization and tracking problem by simply setting $u_d \equiv u_d^\star$ and $x_d(0) = x_d^\star$ in (2).

Now, the problem of simultaneous stabilization and tracking is formulated.

Definition 1. For a class of admissible reference trajectories, the problem of simultaneous stabilization and tracking is said to be solvable if one can find a continuous feedback law (3) such that the solutions of the closed-loop system (1)– (3) are bounded over $[0, \infty)$, and in addition, $x(t) - x_d(t)$ goes to zero as time goes to ∞.

Remark 1. When one only has access to part of the state, or a function of the state x, denoted as the output $y = h(x)$, the state-feedback controller (3) cannot be implemented. A solution to circumvent this obstacle is to invoke a dynamic, time-varying, output-feedback control law of the form

$$\dot{\xi} = q(\xi, y, t), \quad u = u(\xi, y, t) \tag{4}$$

In this case, we deal with the problem of simultaneous stabilization and tracking using output-feedback.

Remark 2. A more general, and practically important, control task is to come up with a controller (3), or (4), which enjoys some additional property of disturbance attenuation with respect to modeling errors or external disturbances. The input-to-state stability property [42] is appropriate for quantifying disturbance attenuation.

Remark 3. Although a general answer to the above-mentioned problem is still lacking in the general setting of nonlinear control systems, there are however several continuous feedback solutions to simultaneous stabilization and tracking for nonholonomic mobile robots, surface ships and other underactuated mechanical systems [29, 7, 9, 10, 8, 33]. Some discontinuous feedback solutions may be found in [5] and references therein.

2.2 State-Feedback Control of Nonholonomic Mobile Robots

Consider a two-wheeled nonholonomic mobile robot, as shown in Figure 1, whose dynamics is described by the following ordinary differential equations [9, 10]:

$$\dot{\eta} = J(\eta)\omega, \tag{5}$$

$$M\dot{\omega} + C(\dot{\eta})\omega + D\omega = \tau \tag{6}$$

where $\eta = (x, y, \phi)^T$ denotes the position and the orientation of the robot, $\omega = (\omega_1, \omega_2)^T$ stands for the angular velocities of the rear wheels, $\tau = (\tau_v, \tau_w)^T$ represents the control torques applied to the wheels, and

$$J(\eta) = \frac{r}{2}\begin{bmatrix} \cos\phi & \cos\phi \\ \sin\phi & \sin\phi \\ b^{-1} & -b^{-1} \end{bmatrix}, \; M = \begin{bmatrix} m_{11} & m_{12} \\ m_{12} & m_{11} \end{bmatrix}, \; C(\dot{\eta}) = \begin{bmatrix} 0 & c\dot{\phi} \\ -c\dot{\phi} & 0 \end{bmatrix}, \; D = \begin{bmatrix} d_{11} & 0 \\ 0 & d_{22} \end{bmatrix}$$

with positive constants r, m_{11}, m_{12}, c, d_{11}, and d_{22} related to robot parameters. Notice that the vector field of the robot model is not locally onto, thus Brockett's necessary condition for C^0 time-invariant asymptotic stabilization is violated.

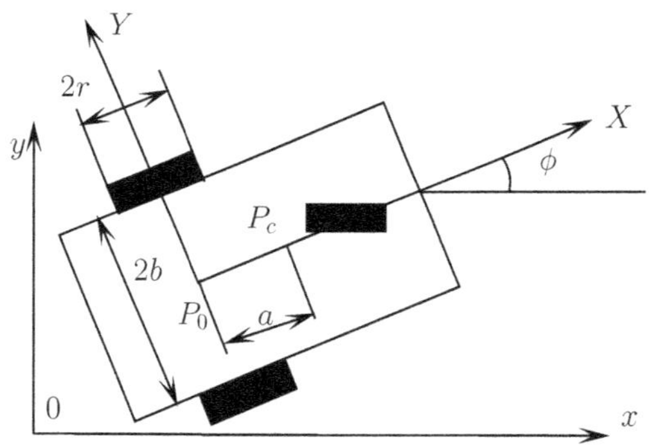

Fig. 1. A Wheeled Mobile Robot

The control objective is to find, if possible, a continuous time-varying feedback law that simultaneously solves stabilization and tracking for a desirable reference trajectory generated by the following virtual robot:

$$\begin{aligned} \dot{x}_d &= v_d \cos\phi_d, \\ \dot{y}_d &= v_d \sin\phi_d, \\ \dot{\phi}_d &= w_d \end{aligned} \tag{7}$$

where $(x_d, y_d, \phi_d)^T$ denotes the position and the orientation of the virtual robot, and v_d, w_d are the linear and angular velocities of the virtual robot, respectively. Assume that v_d and w_d are bounded together with their first and second order derivatives.

Perform the following change of state variables:

$$(v,w)^T = B^{-1}(\omega_1,\omega_2)^T, \quad B = \frac{1}{r}\begin{bmatrix} 1 & b \\ 1 & -b \end{bmatrix}, \tag{8}$$

$$\begin{bmatrix} x_e \\ y_e \\ \phi_e \end{bmatrix} = \begin{bmatrix} \cos\phi & \sin\phi & 0 \\ -\sin\phi & \cos\phi & 0 \\ 0 & 0 & 1 \end{bmatrix} \begin{bmatrix} x - x_d \\ y - y_d \\ \phi - \phi_d \end{bmatrix} \tag{9}$$

As it can be directly checked, the kinematic tracking errors x_e, y_e, ϕ_e satisfy the following differential equations

$$\begin{aligned} \dot{x}_e &= v - v_d \cos\phi_e + y_e w \\ \dot{y}_e &= v_d \sin\phi_e - x_e w \\ \dot{\phi}_e &= w - w_d \, . \end{aligned} \tag{10}$$

As demonstrated in our early work regarding the case-study in backstepping [21], the dynamic ω-subsystem (6) is fully feedback linearizable, or in other words, is feedback equivalent to two integrators. As a consequence, by means of the popular backstepping methodology [26], the desired control laws for the total dynamic model can be derived from the control laws for the kinematic model (10). In the sequel, we will focus our attention on the design of simultaneous stabilizers and trackers for the kinematic model (10).

Toward this end, we will take explicit advantage of the "lower-triangular" structure inside system (10). Specifically, we will use ϕ_e as a virtual control input to stabilize the y_e-subsystem and then apply the backstepping scheme to design a control law for w. A simple adaptation of the design procedure in [21] yields the following virtual control law for ϕ_e, simpler than the ones in [9, 10]:

$$\alpha_{\phi e} = -\arcsin\left(k(t)y_e/\sqrt{1+y_e^2}\right), \quad k(t) = \lambda_1 v_d(t) + \lambda_2 \cos(\lambda_3 t) \tag{11}$$

where λ_i, $1 \le i \le 3$, are positive design parameters so that $\sup_{t\ge 0}|k(t)| < 1$.

Letting $\bar{\phi}_e = \phi_e - \alpha_{\phi e}$, we can rewrite the kinematic tracking error model (10) as

$$\begin{aligned} \dot{x}_e &= v - v_d \cos\phi_e + y_e w \\ \dot{y}_e &= -k(t)v_d y_e/\sqrt{1+y_e^2} - x_e w \\ &\quad + \left[v_d \sin\alpha_{\phi e}(-1+\cos\bar{\phi}_e) + v_d \cos\alpha_{\phi e}\sin\bar{\phi}_e\right] \\ \dot{\bar{\phi}}_e &= w - w_d - \dot{\alpha}_{\phi e} \end{aligned} \tag{12}$$

where

$$\begin{aligned} \dot{\alpha}_{\phi e} &= -\frac{\sqrt{1+y_e^2}}{\sqrt{1+(1-k^2)y_e^2}}\left(\dot{k}(t)\frac{y_e^2}{\sqrt{1+y_e^2}} + k(t)\frac{\dot{y}_e}{(1+y_e^2)^{3/2}}\right) \\ &:= \sigma_1(t,y_e,\phi_e) + \sigma_2(t,y_e)x_e w \end{aligned} \tag{13}$$

Consider the following quadratic function as a Lyapunov function candidate

$$V = \frac{1}{2}x_e^2 + \frac{1}{2}y_e^2 + \frac{1}{2}\bar{\phi}_e^2 \tag{14}$$

Differentiating V along the solutions of (12) leads to

$$\begin{aligned}\dot{V} &= x_e(v - v_d\cos\phi_e - \sigma_2 w) - k(t)v_d\frac{y_e^2}{\sqrt{1+y_e^2}} \\ &+\bar{\phi}_e\left(w - w_d - \sigma_1 + \frac{v_d\sin\alpha_{\phi e}(-1+\cos\bar{\phi}_e) + v_d\cos\alpha_{\phi e}\sin\bar{\phi}_e}{\bar{\phi}_e}\right)\end{aligned} \tag{15}$$

It should be noted that the last term in (15) does not involve any singularity.

This leads us to select the following control laws:

$$v = -c_1x_e + v_d\cos\phi_e + \sigma_2 w, \tag{16}$$

$$w = -c_2\phi_e + w_d + \sigma_1 - \frac{v_d\sin\alpha_{\phi e}(-1+\cos\bar{\phi}_e) + v_d\cos\alpha_{\phi e}\sin\bar{\phi}_e}{\bar{\phi}_e} \tag{17}$$

with $c_1, c_2 > 0$. Thus, (15) becomes

$$\dot{V} \leq -c_1x_e^2 - k(t)v_d\frac{y_e^2}{\sqrt{1+y_e^2}} - c_2\bar{\phi}_e^2\ . \tag{18}$$

By means of the newly developed stability criteria for nonlinear time-varying systems [24], it is not hard to prove that $V(t)$ tends to zero as $t \to \infty$, provided that $v_d(t)$ is PE, that is,

(H1) There exist two positive constants τ_1, τ_2 such that $\int_t^{t+\tau_1} v_d^2(s)ds > \tau_2$ for all $t \geq 0$.

It is worth noting that (H1) is more general than PE conditions of other kind used previously in past literature for studying the tracking problem of underactuated mechanical systems. See, for instance, [7, 9, 10, 21].

When this condition (H1) fails, the following assumption is made:

(H2) The signal v_d is L^1 over $[0, \infty)$.

It is worth noting that (H2) can handle the cases of parking and stabilization.

By application of Lemma 2 in the Appendix and [21, Lemma 2], and following the similar reasoning as in [10], under (H2), we can conclude the convergence to zero of the kinematic tracking errors (x_e, y_e, ϕ_e), or equivalently $(x(t) - x_d(t), y(t) - y_d(t), \phi(t) - \phi_d(t))$.

Summarizing the above, we have

Theorem 1. *Under one of the assumptions (H1) and (H2), the problem of simultaneous stabilization and tracking is solvable for the kinematic model of the nonholonomic mobile robot (5).*

As a direct application of backstepping, following the similar design strategies in our previous work [21, 9, 10], we can generalize Theorem 1 to the dynamic model.

Theorem 2. *Under one of the assumptions (H1) and (H2), the problem of simultaneous stabilization and tracking is solvable for the dynamic model of the nonholonomic mobile robot (5) and (6).*

2.3 Output-Feedback Control of Nonholonomic Mobile Robots

In the previous section, we assume that all state measurements are available for controller design. This may not be the case when the velocity measurements are not perfectly known or when we purposefully want to avoid implementing costly sensors to measure the velocities. In such situations, we often turn to a dynamic, observer-based output feedback controller of the form (4).

In the absence of ω-velocity measurements, the key strategy behind the output-feedback solution to simultaneous stabilization and tracking is to transform the dynamic model (5) and (6) of the robot into a simplified model. The main purpose of this transformation is to turn the *nonlinearities* in the original model into a situation where, in the transformed model, the nonlinearities only depend on the output and the unmeasured states appear linearly. For our control problem, we consider the position and orientation variables η as the (measured) output, and assume the velocity ω as the unmeasured state.

To this end, introduce the following change of coordinates:

$$X = Q(\eta)\omega \tag{19}$$

where, for each fixed η, $Q(\eta)$ is a nonsingular 2×2 matrix.

Direct computation gives:

$$\dot{X} = \left[\dot{Q}(\eta)\omega - Q(\eta)M^{-1}C(\dot{\eta})\omega\right] + Q(\eta)M^{-1}(-D\omega + \tau) \tag{20}$$

Select Q to satisfy the following PDE

$$\dot{Q}(\eta)\omega - Q(\eta)M^{-1}C(\dot{\eta})\omega = 0. \tag{21}$$

One solution to this PDE is [10]:

$$Q(\eta) = \begin{bmatrix} n_{11}\cos(a\Delta\phi) & \Delta\sin(a\Delta\phi) - n_{12}\cos(a\Delta\phi) \\ n_{11}\sin(a\Delta\phi) & -n_{12}\sin(a\Delta\phi) - \Delta\cos(a\Delta\phi) \end{bmatrix}$$

where $n_{11}, n_{12}, a, \Delta$ are appropriate constants (see [10] for the details).

Then, the dynamic model (5) and (6) can be rewritten as

$$\dot{\eta} = J(\eta)Q^{-1}(\eta)X, \tag{22}$$

$$\dot{X} = -Q(\eta)M^{-1}DQ^{-1}(\eta)X + Q(\eta)M^{-1}\tau \tag{23}$$

For this transformed system, we can design a passive, exponential observer of the form

$$\dot{\hat{\eta}} = J(\eta)Q^{-1}(\eta)\hat{X} + K_{01}(\eta - \hat{\eta}), \tag{24}$$
$$\dot{\hat{X}} = -Q(\eta)M^{-1}DQ^{-1}(\eta)\hat{X} + Q(\eta)M^{-1}\tau + K_{02}(\eta - \hat{\eta}) \tag{25}$$

Following [10], by appropriate choice of design functions K_{01}, K_{02}, the observation errors $\tilde{\eta} := \eta - \hat{\eta}$ and $\tilde{X} := X - \hat{X}$ satisfy a globally exponentially stable system.

Accordingly, we can define

$$\hat{\omega} = Q^{-1}(\eta)\hat{X}, \quad \tilde{\omega} = Q^{-1}(\eta)\tilde{X}, \tag{26}$$
$$(\hat{v}, \hat{w})^T = B^{-1}\hat{\omega}^T, \quad (\tilde{v}, \tilde{w})^T = B^{-1}\tilde{\omega}^T . \tag{27}$$

Clearly, these error signals $\tilde{\omega}, \tilde{v}, \tilde{w}$ go to zero at an exponential rate.

Now, the kinematic tracking errors as defined in (9) satisfy, instead of (10),

$$\begin{aligned} \dot{x}_e &= \hat{v} - v_d \cos\phi_e + y_e(\hat{w} + \tilde{w}) + \tilde{v}, \\ \dot{y}_e &= v_d \sin\phi_e - x_e(\hat{w} + \tilde{w}), \\ \dot{\phi}_e &= \hat{w} - w_d + \tilde{w} . \end{aligned} \tag{28}$$

As compared with the state-feedback design procedure in the previous subsection, here we utilize $\hat{v}$ and $\hat{w}$ as the virtual controls for the kinematic and dynamic models. Again using backstepping, by combining the previously introduced design procedure and the one of [10], we can design a dynamic, time-varying output-feedback control law relying upon the measurements of output η that solves simultaneous stabilization and tracking.

Theorem 3. *Under one of the assumptions (H1) and (H2), the problem of simultaneous stabilization and tracking using output feedback is solvable for the nonholonomic mobile robot (5) and (6).*

Proof. similar to the proof of the main result of [10].

3 Open Problems

In spite of significant progress over the past 25 years, the area of controlling underactuated mechanical systems continues to pose challenging problems. Answering these questions certainly contributes to the development of modern nonlinear control theory.

3.1 Adaptive Output Feedback Control

Adaptive output feedback control deals with a situation where the control system in question involves unknown parameters and only the outputs are available for feedback design. To the best of our knowledge, there is no general answer to the question of when an adaptive output feedback controller can be designed for solving simultaneous stabilization and tracking of nonholonomic mobile robots in particular and underactuated mechanical systems in general.

The central difficulty comes from the fact that the design of observers and parameter identifiers and the controller design are intertwined. The presence of nonholonomic constraints prevents from applying earlier adaptive output-feedback schemes [26] directly to this special, but important, class of nonlinear systems. See [9] for a solution to adaptive state-feedback control for the simultaneous stabilization and tracking of nonholonomic mobile robots.

3.2 Control under Saturation Constraints

When the control inputs are subject to saturation constraints, say, $\sup_{t\geq 0}|\tau(t)| \leq \varepsilon$, with $\varepsilon > 0$, the problem of simultaneous stabilization and tracking remains open for nonholonomic mobile robots, let alone more general classes of underactuated mechanical systems. Preliminary results are obtained in [23, 29] for the kinematic model of mobile robots.

3.3 General Models of Underactuated Mechanical Systems

There is very little research accomplished for the simultaneous stabilization and tracking problem for underactuated mechanical systems taking general forms [3, 39], say,

$$\begin{aligned} M_{11}(q)\ddot{q}_1 + M_{12}(q)\ddot{q}_2 + F_1(q,\dot{q}) &= B(q)u, \\ M_{21}(q)\ddot{q}_1 + M_{22}(q)\ddot{q}_2 + F_2(q,\dot{q}) &= 0 \end{aligned} \tag{29}$$

where q is the generalized coordinates and u is the controls. Assuming that $B(q)$ is a full-rank matrix, the second equation of (29) refers to as the nonintegrable constraints on accelerations or the second-order nonholonomic constraints.

New techniques and methodologies are needed to address the simultaneous stabilization and tracking problem for underactuated mechanical systems described by the above general form (29). The challenge will become more daunting when one moves into the exciting area of controlling a group of underactuated mechanical systems, working in a team for complex tasks such as search and rescue, formation control and other military and civilian applications.

Acknowledgements. The author would like to thank Claude Samson, Henk Nijmeijer and Ti-Chun Lee for their insightful remarks on underactuated mechanical systems, and is grateful to his former students K.D. Do and Q. Li for their active participation in this initiative. This work has been supported by NSF grants ECS-0093176, DMS-0504462 and DMS-0906659.

Appendix – Useful Differential Inequalities

The following lemmas on differential inequalities are used, often implicitly, in the controller design for nonholonomic systems; see [19, 21, 8, 9] and [27].

Lemma 1. *Let $V : \Re_+ \to \Re_+$ be a continuously differentiable function satisfying the differential inequality of the form*

$$\dot{V}(t) \leq -a(t)V(t) + (V(t) + \sqrt{V(t)})b(t) \tag{30}$$

where $a(t)$ is PE in the sense of $\liminf_{t\to\infty} \frac{1}{t}\int_{t_0}^{t_0+t} a(\tau)d\tau > 0$ for all $t_0 \geq 0$, and $b(t)$ is an exponentially decaying signal. Then, $V(t)$ is exponentially decaying.

Lemma 2. *Let $V : \Re_+ \to \Re_+$ be continuously differentiable and $W : \Re_+ \to \Re_+$ uniformly continuous satisfying that, for each $t \geq 0$,*

$$\dot{V}(t) \leq -W(t) + p(t) \tag{31}$$

with $p : \Re_+ \to \Re_+$ is L^1. Then, there exists a constant c such that $W(t) \to 0$ and $V(t) \to c$ as $t \to \infty$. Moreover, if $a(t)(\gamma_1(V(t)) - \gamma_2(W(t))) \to 0$ for some continuous and bounded PE-type signal $a(t)$ as in Lemma 1 and two class-K functions γ_1, γ_2, then $V(t) \to 0$ as $t \to \infty$.

References

1. Acosta, J.A., Ortega, R., Astolfi, A., Mahindrakar, A.D.: Interconnection and damping assignment passivity-based control of mechanical systems with underactuation degree one. IEEE Trans. Automat. Contr. 50, 1936–1955 (2005)
2. Astolfi, A.: Discontinuous control of nonholonomic systems. Systems and Control Letters 27, 37–45 (1997)
3. Bloch, A., Baillieul, J., Crouch, P.E., Marsden, J.E.: Nonholonomic Mechanics and Control. Springer, NY (2003)
4. Brockett, R.W.: Asymptotic stability and feedback stabilization. In: Brockett, R.W., Millam, R.S., Sussmann, H.J. (eds.) Differential Geometric Control Theory, pp. 181–191. Birkhäuser, Boston (1983)
5. Buccieri, D., Perritaz, D., Mullhaupt, P., Jiang, Z.P., Bonvin, D.: Velocity-scheduling control for a unicycle mobile robot: theory and experiments. IEEE Trans. Robotics and Automation 25, 451–458 (2009)
6. Coron, J.M.: On the stabilization of some nonlinear control systems: results, tools, and applications. In: Clarke, F.H., Stern, R.J. (eds.) Nonlinear Analysis, Differential Equations and Control, pp. 307–367. Kluwer Academic Publishers, Dordrecht (1999)
7. Dixon, W.E., Dawson, D.M., Zergeroglu, E., Behal, A.: Nonlinear Control of Wheeled Mobile Robots. Springer, New York (2001)
8. Do, K.D., Jiang, Z.P., Pan, J.: Universal controllers for stabilization and tracking of underactuated ships. Systems & Control Letters 47, 299–317 (2002)
9. Do, K.D., Jiang, Z.P., Pan, J.: Simultaneous tracking and stabilization of mobile robots: an adaptive approach. IEEE Trans. Automatic Control 49, 1147–1152 (2004)
10. Do, K.D., Jiang, Z.P., Pan, J.: A global output-feedback controller for simultaneous tracking and stabilization of unicycle-type mobile robots. IEEE Trans. on Robotics and Automation 20, 589–594 (2004)
11. Do, K.D., Pan, J., Jiang, Z.P.: Robust and adaptive path following for underactuated autonomous underwater vehicles. Ocean Engineering 31, 1967–1997 (2004)
12. Do, K.D., Jiang, Z.P., Pan, J.: Robust adaptive path following of underactuated ships. Automatica 40, 929–944 (2004)

13. Egerstedt, M., Hu, X.: A hybrid control approach to action coordination for mobile robots. Automatica 38, 125–130 (2002)
14. Egerstedt, M., Hu, X., Stotsky, A.: Control of mobile platforms using a virtual vehicle approach. IEEE Transaction on Automatic Control 46, 1777–1782 (2001)
15. Fliess, M., Lévine, J., Martin, P., Rouchon, P.: Flatness and defect of nonlinear systems: Introductory theory and examples. Int. J. Control 61, 1327–1361 (1995)
16. Fliess, M., Lévine, J., Martin, P., Rouchon, P.: A Lie-Backlund approach to equivalence and flatness of nonlinear systems. IEEE Trans. Autom. Control 44, 922–937 (1999)
17. Fontaine, I., Lozano, R.: Non-linear Control for Underactuated Mechanical Systems. Springer, London (2001)
18. Fossen, T.I.: Marine Control Systems: Guidance, Navigation and Control of Ships, Rigs and Underwater Vehicles. Marine Cybernetics (2002) ISBN 82-92356-00-2
19. Jiang, Z.P.: Lyapunov design of global state and output feedback trackers for nonholonomic control systems. Int. J. of Control 73, 744–761 (2000)
20. Jiang, Z.P.: Global tracking control of underactuated ships by Lyapunov's direct method. Automatica 38, 301–309 (2002)
21. Jiang, Z.P., Nijmeijer, H.: Tracking control of mobile robots: a case study in backstepping. Automatica 33, 1393–1399 (1997)
22. Jiang, Z.P., Nijmeijer, H.: A recursive technique for tracking control of nonholonomic systems in chained form. IEEE Trans. Automatic Control 44, 265–279 (1999)
23. Jiang, Z.P., Lefeber, E., Nijmeijer, H.: Saturated stabilization and tracking of a nonholonomic mobile robot. Syst. & Control Lett. 42, 327–332 (2001)
24. Jiang, Z.P., Lin, Y., Wang, Y.: Stabilization of nonlinear time-varying systems: A control Lyapunov function approach. J. Syst. Sci. & Complexity 22, 683–696 (2009)
25. Kolmanovsky, I., McClamroch, N.H.: Developments in nonholonomic control problems. IEEE Control Syst. Mag. 15, 20–36 (1995)
26. Krstic, M., Kanellakopoulos, K., Kokotovic, P.V.: Nonlinear and Adaptive Control Design. John Wiley & Sons, New York (1995)
27. Lee, T.C., Jiang, Z.P.: A generalization of Krasovskii-LaSalle theorem for nonlinear time-varying systems: converse results and applications. IEEE Trans. Automatic Control 50, 1147–1163 (2005)
28. Lee, T.C., Jiang, Z.P.: Uniform asymptotic stability of nonlinear switched systems with an application to mobile robots. IEEE Trans. Automatic Control 53, 1235–1252 (2008)
29. Lee, T.C., Song, K.T., Lee, C.H., Teng, C.C.: Tracking control of unicycle-modeled mobile robots using a saturation feedback controller. IEEE Trans. Control Systems Technology 9, 305–318 (2001)
30. Levine, J.: Analysis and Control of Nonlinear Systems: A Flatness-based Approach. Springer, Heidelberg (2009)
31. Lizarraga, D.A.: Obstructions to the existence of universal stabilizers for smooth control systems. Math. Control, Signals, Syst. 16, 255–277 (2003)
32. Malisoff, M., Mazenc, F.: Constructions of Strict Lyapunov Functions. Springer, London (2009)
33. Morin, P., Samson, C.: Control of nonholonomic mobile robots based on the transverse function approach. IEEE Trans. Robotics 25, 1058–1073 (2009)
34. Murray, R.M., Sastry, S.S.: Nonholonomic motion planning: Steering using sinusoids. IEEE Trans. Autom. Control 38, 700–716 (2003)
35. Olfati-Saber, R.: Nonlinear control of underactuated mechanical systems with application to robotics and aerospace vehicles. Ph.D. Thesis, Massachusetts Institute of Technology (2001)

36. Ortega, R., Spong, M., Gomez-Estern, F., Blankenstein, G.: Stabilization of a class of underactuated mechanical systems via interconnection and damping assignment. IEEE Trans. Automat. Contr. 47, 1218–1233 (2002)
37. Qu, Z., Wang, J., Plaisted, C.E., Hull, R.A.: Global-stabilizing near-optimal control design for nonholonomic chained systems. IEEE Trans. Automat. Control 51, 1440–1456 (2006)
38. Refsnes, J.E., Sorensen, A.J., Pettersen, K.Y.: Model-based output feedback control of slender-body underactuated AUVs: Theory and experiments. IEEE Trans. Control Systems Technology 16, 930–946 (2008)
39. Reyhanoglu, M., van der Schaft, A., McClamroch, N.H., Kolmanovsky, I.: Dynamics and control of a class of underactuated mechanical systems. IEEE Trans. Automat. Contr. 44, 1663–1672 (1999)
40. Samson, C.: Velocity and torque feedback control of a nonholonomic cart. In: Canudas de wit, C., et al. (eds.) Advanced Robot Control, pp. 125–151 (1991)
41. Sontag, E.D., Sussmann, H.: Remarks on continuous feedback. In: Proc. IEEE Conf. Decision and Control, Albuquerque, pp. 916–921 (1980)
42. Sontag, E.D.: Input to state stability: Basic concepts and results. In: Nistri, P., Stefani, G. (eds.) Nonlinear and Optimal Control Theory, pp. 163–220. Springer, Berlin (2007)

Time Scaling in Motion Planning and Control of Tree-Like Pendulum Structures

Matthias Krause, Joachim Rudolph, and Frank Woittennek

Abstract. Planar tree-like structures consisting of rigid links with rotational joints are considered. These models can be used to describe the dynamics of planar biped robots, in particular during the single support phase. Another simple structure of this type is the so-called acrobot. In both cases, the base joint is unactuated while motors are available at all other joints. As a result, motion planning and control of such systems remain challenging tasks. It is shown that flatness-based methods can be helpful to their solution if time scaling is taken into account. To this end the known concept of orbital flatness has to be extended. Moreover, controlled time scaling turns out to provide a helpful additional degree of freedom. Motion planning and feedback design are briefly discussed.

1 Introduction

Biped robot locomotion continues to be an interesting field of active research. Starting with fully actuated quasi-static motions, more and more dynamic walking and even running is under examination. A comprehensive view on currently treated topics and results can be found in [10].

Trajectory planning is a major issue in any dynamic locomotion problem, and it is now well known that the flatness concept is particularly helpful in such problems. Anyhow, to the authors' knowledge, the few flatness-based control results in the

Matthias Krause
Lehrstuhl für Verbrennungskraftmaschinen, RWTH Aachen, 52056 Aachen, Germany

Joachim Rudolph
Lehrstuhl für Systemtheorie und Regelungstechnik, Universität des Saarlandes, Campus A5 1, 66123 Saarbrücken, Germany
e-mail: `j.rudolph@lsr.uni-saarland.de`

Frank Woittennek
Institut für Regelungs- und Steuerungstheorie, TU Dresden, 01062 Dresden, Germany
e-mail: `frank.woittennek@tu-dresden.de`

J. Lévine & P. Müllhaupt (Eds.): Adv. in Theory of Control, Signals, LNCIS 407, pp. 89–98.
springerlink.com © Springer-Verlag Berlin Heidelberg 2010

field of biped robots concern only simplified models of specific configurations [7], [8].

Although motions in three dimensions are now a major topic of research, one may still gain insight by studying the planar case. Therefore, here the flatness concept will be exploited to study motion planning and control problems for planar bipeds which have only pointwise contact with the floor. The robots are modeled as pendulum trees, consisting of rigid links with rotational joints equipped with motors. There are no motors at contact points with the ground. The system is, thus, underactuated.

During the single support phase, in addition to a judicious choice of the dependent variables a change of the independent variable, or time scaling, is considered. The use of such time scaling can be considered as a generalization of orbital flatness [2]. Introducing an additional control parameter in the time scaling transformation provides an extra degree of freedom which simplifies motion planning [3, 4]. In particular, it is possible to plan trajectories which avoid impacts when touching the ground. However, due to lack of space this issue cannot be discussed here. Further simplification results from the examination of the acrobot, a basic example of a tree-like structure. Besides motion planning, the consequences of the transformations for the feedback design are also discussed.

The discussion of the double support phase of the biped as well as the ballistic phase are beyond the scope of the present contribution. The same holds for phase transitions. However, similar results are available, which allow for the parametrization and stabilization of complete walking cycles.

2 Mathematical Modeling

The mechanical systems under consideration are planar tree-like structures consisting of $n+1$ rigid bodies, $K_0, \ldots, K_n$, as depicted in Fig. 1(a). The "tree" is a structure of knots and joining edges with unique connections between any two knots. An arbitrary number of bodies can be attached to the body K_0, viewed as the root of the tree. Each body K_k can perform a rotation about a corresponding pivot point P_k. The axes of the joints are collinear and, thus, only planar motion is possible. In Fig. 1(a), one arbitrary body K_k, $k \in \{0, \ldots, n\}$ is shown in detail in order to introduce the corresponding parameters: the length L_k, the mass m_k, and the inertia around the center of mass J_k.

Defining unit vectors $\mathbf{e}^k = \left(\sin\varphi^k, -\cos\varphi^k\right)^T$, $\mathbf{n}^k = \left(\cos\varphi^k, \sin\varphi^k\right)^T$ parallel and orthogonal to the kth edge and using the Einstein summation convention $a_j b^j = \sum_j a_j b^j$, the position of the center of mass $\mathbf{r}^k$ of body K_k can be written as

$$\mathbf{r}^k = \mathbf{r} + l_j^k \mathbf{e}^j, \qquad \dot{\mathbf{r}}^k = \dot{\mathbf{r}} + l_j^k \dot{\varphi}^j \mathbf{n}^j, \qquad j \in \bar{\mathscr{P}}_k. \tag{1}$$

Here, φ^k denotes the absolute orientation angle of K_k w.r.t. the vertical, and $\dot{\varphi}^k = \omega^k$ denotes the corresponding angular velocity. This description can be used

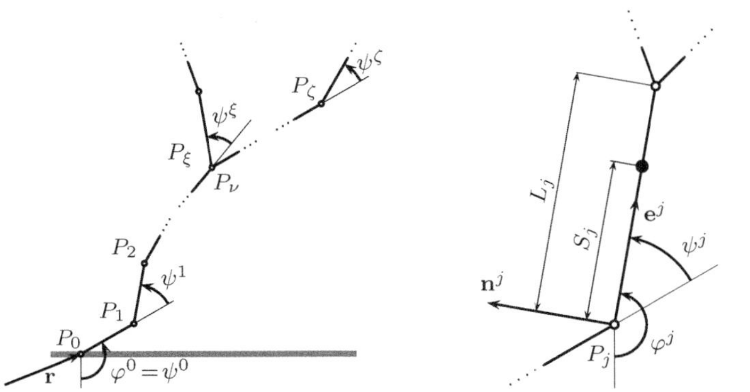

(a) Examplary tree structure. (b) One pendulum in detail.

Fig. 1. Pendulum system considered

if the global root pivot is able to translate. To this end, the vector $\mathbf{r}$ denotes the position of the root pivot in an inertial reference frame. The summation index j runs over the set $\bar{\mathscr{P}}_k = \mathscr{P}_k \cup \{k\}$, where $\mathscr{P}_k$ contains the indices of all bodies lying on the connection between the global root pivot and the link K_k. The center of mass of every body is assumed to lie on the associated edge K_k, at a distance S_k from the pivot P_k. A length parameter l_j^k is defined as $l_j^k = L_j$, for $j \in \mathscr{P}_k$, $l_j^j = S_j$, (i.e., for $k = j$), $l_j^k = 0$, otherwise.

The equations of motion can be derived using d'Alembert's principle in its Lagrangian formulation as

$$
\begin{aligned}
\frac{d}{dt}\left(\frac{\partial T}{\partial \dot{r}_x}\right) - \frac{\partial T}{\partial r_x} + \frac{\partial V}{\partial r_x} &= F_x, \\
\frac{d}{dt}\left(\frac{\partial T}{\partial \dot{r}_y}\right) - \frac{\partial T}{\partial r_y} + \frac{\partial V}{\partial r_y} &= F_y, \\
\frac{d}{dt}\left(\frac{\partial T}{\partial \dot{\psi}^k}\right) - \frac{\partial T}{\partial \psi^k} + \frac{\partial V}{\partial \psi^k} &= \Gamma_k,
\end{aligned}
$$

with $k = 0, \ldots, n$, and with ψ^k denoting the relative angle of K_k w.r.t. its antecessor. Accordingly, $\psi^0 = \varphi^0$ holds because the root pendulum has no antecessor. The generalized force Γ_k is the input torque about the pivot P_k, which is a linear combination of the torques produced by the motors attached at that point. Using the notation (1), the kinetic energy T and the potential energy V can be compactly written as

$$2T = J_i(\dot{\varphi}^i)^2 + m_i\left\langle \dot{\mathbf{r}}^i, \dot{\mathbf{r}}^i \right\rangle = J_i(\dot{\varphi}^i)^2 + m_i\left\langle \dot{\mathbf{r}} + l_j^i \mathbf{n}^j \dot{\varphi}^j, \dot{\mathbf{r}} + l_j^i \mathbf{n}^j \dot{\varphi}^j \right\rangle \quad (2a)$$

$$V = -m_i\left\langle \mathbf{g}, \mathbf{r}^i \right\rangle = -m_i\left\langle \mathbf{g}, \mathbf{r} + l_j^i \mathbf{e}^j \right\rangle, \quad (2b)$$

where $i = 0, \dots, n$ runs over all body indices. Likewise, the index j runs from $j = 0, \dots, n$, although only $j \in \mathscr{P}_i$ is required to get to the center of mass of K_i. This is already taken into account by the definition of the parameter l^i_j which suppresses all bodies K_j with $j \notin \mathscr{P}_i$. The vector $\mathbf{g} = (0, -g)^T$ denotes the gravitational acceleration. Evaluating the equations of motion using (2) yields

$$\frac{d}{dt}\left(m_i\left(\dot{\mathbf{r}} + l^i_j \mathbf{n}^j \dot{\varphi}^j\right)\right) - m\mathbf{g} = \mathbf{F}, \tag{3a}$$

$$J_i \ddot{\varphi}^i + m_i \left\langle \tfrac{d}{dt}\left(\dot{\mathbf{r}} + l^i_j \mathbf{n}^j \dot{\varphi}^j\right), l^i_\mu \mathbf{n}^\mu \right\rangle - m_i \left\langle \mathbf{g}, l^i_\mu \mathbf{n}^\mu \right\rangle = \Gamma_k, \tag{3b}$$

for $i \in \bar{\mathscr{C}}_k$, $\mu \in \{0, \dots, n\} \setminus \mathscr{P}_k$. The summation runs over $\bar{\mathscr{C}}_k = \mathscr{C}_k \cup \{k\}$, with $\mathscr{C}_k$ containing the indices of bodies attached to K_k. Moreover, $m = \sum_{i=0}^n m_i$ denotes the total mass. These equations can be used for arbitrary tree-like pendulum systems. The specific equations depend on the structure considered as described by the sets $\mathscr{P}_k$ and $\mathscr{C}_k$, $k = 0, \dots, n$.

In the biped walking problem, a complete walking cycle may be split into several phases. The most interesting one is likely the single support phase where only one point touches the ground. Assuming that the root pivot has non-sliding contact with the ground it can be modeled as being fixed ($\dot{\mathbf{r}} = \ddot{\mathbf{r}} = \mathbf{0}$) with no actuation torque ($\Gamma_0 = 0$). With these assumptions, (3) yields the following equations for the single support phase:

$$\frac{d}{dt}\left(J_i \dot{\varphi}^i + m_i \left\langle l^i_j \mathbf{n}^j \dot{\varphi}^j, l^i_j \mathbf{n}^j \right\rangle\right) - m_i \left\langle \mathbf{g}, l^i_j \mathbf{n}^j \right\rangle = 0, \tag{4a}$$

$$J_i \ddot{\varphi}^i + m_i \left\langle \tfrac{d}{dt}\left(\dot{\mathbf{r}} + l^i_j \mathbf{n}^j \dot{\varphi}^j\right), l^i_\mu \mathbf{n}^\mu \right\rangle - m_i \left\langle \mathbf{g}, l^i_\mu \mathbf{n}^\mu \right\rangle = \Gamma_k, \tag{4b}$$

for $i \in \bar{\mathscr{C}}_k$, $\mu \in \{0, \dots, n\} \setminus \mathscr{P}_k$, $k = 0, \dots, n$.

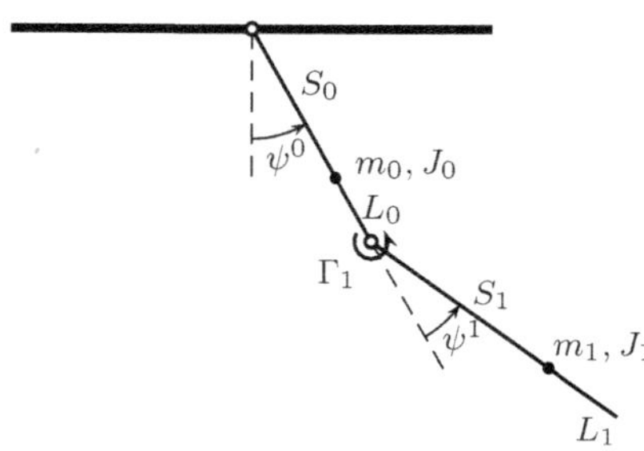

Fig. 2. Sketch of the acrobot

The simplest configuration of the pendulum trees considered here is a two-link planar pendulum with one actuator between the links (see Fig. 2) known as the acrobot [9, 11, 12]. Discussing this simpler system provides valuable additional insight into the time scaling methods proposed.

The equations of motion of this system are well known. In accordance with (4) they can be written as

$$m_{11}(\psi^1)\ddot{\psi}^0 + m_{22}(\psi^1)\ddot{\psi}^1 + c_1(\psi,\dot{\psi}) + g_1(\psi^0,\psi^1) = 0,$$
$$m_{21}(\psi^1)\ddot{\psi}^0 + m_{22}(\psi^1)\ddot{\psi}^1 + c_2(\psi,\dot{\psi}) + g_2(\psi^0,\psi^1) = \Gamma_1,$$

where $\psi = (\psi^0,\psi^1)^T$,

$$m_{11}(\psi^1) = A_1 + A_2 + 2B\cos\psi^1, \quad m_{12}(\psi^1) = A_2 + B\cos\psi^1,$$
$$m_{21}(\psi^1) = A_2 + B\cos\psi^1, \quad m_{22}(\psi^1) = A_2,$$

with $A_1 = J_0 + m_0 S_0^2 + m_1 L_0^2$, $A_2 = J_1 + m_1 S_1^2$, and $B = m_1 L_0 S_1$. Moreover,

$$c_1(\psi,\dot{\psi}) = -(2\dot{\psi}^1\dot{\psi}^0 + (\dot{\psi}^1)^2)B\sin\psi^1, \quad c_2(\psi,\dot{\psi}) = (\dot{\psi}^0)^2 B\sin\psi^1,$$
$$g_1(\psi^0,\psi^1) = (m_0 S_0 + m_1 L_0)g\sin\psi^0 + m_1 S_1\sin(\psi^0+\psi^1)g,$$
$$g_2(\psi^0,\psi^1) = m_1 S_1 g\sin(\psi^0+\psi^1).$$

3 Time Scaling and Flatness Properties

In the sequel, it will be shown that time-scaling transformations can be used to transform the models into flat systems. This is shown in two steps. First, the acrobot structure is considered where the so-called orbital flatness is exhibited. Then, for the general tree-like structure—specifically modeling the biped robot in the single support phase—an extension of the orbital flatness concept is introduced.

3.1 Orbital Flatness of the Acrobot Model

Obviously, m_{ij}, $i,j = 1,2$ and thus the kinetic energy are independent of the absolute angle ψ^0. In [6], this property is called *kinetic symmetry*[1]. The change of coordinates

$$u = \left(m_{22}(\psi^1)\right)^{-1}\left(\Gamma_1 - m_{21}(\psi^1)\ddot{\psi}^0 - c_2(\psi,\dot{\psi}) - g_2(\psi^0,\psi^1)\right),$$

$$z_1 = \psi^0 + \gamma(\psi^1), \quad z_2 = m_{11}(\psi^1)\dot{\psi}^0 + m_{12}(\psi^1)\dot{\psi}^1, \quad \xi_1 = \psi^1, \quad \xi_2 = \dot{\psi}^1, \tag{5}$$

with

$$\gamma(\psi^1) = \int_0^{\psi^1} m_{11}^{-1}(\bar{\psi}^1)m_{12}(\bar{\psi}^1)d\bar{\psi}^1,$$

is introduced to transform the dynamics of the acrobot into a cascade system in strict feedback form

$$\dot{z}_1 = m_{11}^{-1}(\xi_1)z_2, \qquad \dot{z}_2 = -g_1(z_1 - \gamma(\xi_1),\xi_1), \qquad \dot{\xi}_1 = \xi_2, \qquad \dot{\xi}_2 = u. \tag{6}$$

[1] Without including external forces originating from a potential into the Lagrangian, ψ^0 may be considered as a cyclic coordinate, and one may simply speak about a symmetry.

There is only one input, u, and the system can be shown not to be flat.

However, applying the *differential time transformation*

$$\frac{dt}{d\tau} = m_{11}(\psi^1), \tag{7}$$

the system can be parameterized in z_1 and its derivatives. With $\frac{d}{d\tau}z_1 = z_1'$ one obtains

$$z_1' = z_2, \quad z_1'' = -m_{11}(\xi_1)g_1(z_1 - \gamma(\xi_1), \xi_1).$$

Locally, the second equation can be solved for $\xi_1 = \phi(z_1'', z_1)$. Differentiating this expression w.r.t. τ and using (6) in combination with (7) leads to expressions for ξ_2 and, by means of a further differentiation, for u. Thus, z_1 can be interpreted as a flat output of the system with transformed time—i.e., as an orbitally flat output [2].

3.2 Extension to General Tree-Like Structures

Equation (4a) forms a differential relation between the relative angles $\psi^0, \ldots, \psi^n$. Moreover, there are only n input torques. As a consequence, it is not possible to freely assign trajectories to these angles. Thus, one might consider one of the angles as playing a particular role. Clearly, the angle $\psi^0 = \varphi^0$ is a good candidate. For the robot this is the angle between the root and the ground. It is the only angle without an associated torque. The (variable) shape of the structure is then defined by the remaining angles $\psi = (\psi^1, \ldots, \psi^n)$.

The interesting part of the model is the equation (4a)

$$\frac{d}{dt}\left(a^0(\psi)\dot{\varphi}^0 + b_j^0(\psi)\dot{\psi}^j\right) = -\frac{\partial V}{\partial \varphi^0}(\varphi^0, \psi), \quad j \in \{1, \ldots, n\}, \tag{8}$$

not involving any control torque. Applying the *differential time transformation* $\frac{dt}{d\tau} = a^0(\psi)$ and introducing the new variable y yield

$$\frac{d}{d\tau}\left((\varphi^0)' + \frac{b_j^0}{a^0}(\psi^j)'\right) = y'' = -a^0\frac{\partial V}{\partial \varphi^0}, \qquad y = \varphi^0 + \int_{\tau_0}^{\tau} \frac{b_j^0}{a^0}(\psi^j)'\,d\bar{\tau}. \tag{9}$$

This transformation is well-defined, because $a^0(\psi)$ describes an inertia. It is, therefore, always positive and simply denotes a scaling depending on the shape.

Choosing any particular angle $\tilde{\psi}$ from $\{\psi^1, \ldots, \psi^n\}$, one may design a trajectory $\tau \mapsto \tilde{\psi}(\tau)$ and then express the trajectories of the remaining angles as a function of $\tilde{\psi}$. This yields $y = \varphi^0 + \gamma(\tilde{\psi})$ with (j runs over the indices of the angles in $\{\psi^1, \ldots, \psi^n\} \setminus \{\tilde{\psi}\}$)

$$\gamma(\tilde{\psi}) = \int_{\tilde{\psi}_0}^{\tilde{\psi}} \frac{b_j^0}{a^0}\frac{d\psi^j}{d\bar{\tilde{\psi}}}d\bar{\tilde{\psi}}, \qquad y'' = -a^0\frac{\partial V}{\partial \varphi^0} =: f(y - \gamma(\tilde{\psi}), \tilde{\psi}).$$

The latter can be solved for $\tilde{\psi}$. As a result, one may consider $(y, \{\psi^1, \ldots, \psi^n\} \setminus \{\tilde{\psi}\})$ as an *orbitally flat output* of the system.

Since the time transformation involves a functional it slightly generalizes the transformations used to define orbital flatness [2]. The terms *orbitally flat output* and *orbital flatness* might still be used though.

3.3 Controlling the Clock

The parameterization derived so far has one drawback: it does not allow one to assign independent trajectories to all the joint angles and, therefore, motion planning is still difficult. As a result, a generalization may be useful. It is achieved by introducing an extra degree of freedom in the time scaling transformation. To this end, the angles $\psi^1, \ldots, \psi^n$ may also be parameterized by functions $\psi^i = \bar{\psi}^i(s)$, $i = 1, \ldots, n$ of yet another independent variable s. Substituting this expression into (9) leads to

$$y(\tau) = \varphi^0(\tau) + \int_{s(\tau_0)}^{s(\tau)} \frac{\bar{b}_i^0(\sigma)}{\bar{a}^0(\sigma)} \frac{d\bar{\psi}^i}{d\sigma}(\sigma)\, d\sigma =: \varphi^0(\tau) + \bar{\gamma}(s(\tau)),$$

where the bar has been introduced to denote the composite functions. Together with the equation of motion (9) this yields

$$y = \varphi^0 + \bar{\gamma}(s), \qquad y'' = -a^0(\bar{\psi}(s)) \frac{\partial V}{\partial \varphi^0}(\varphi^0, \bar{\psi}(s)) =: \bar{f}(\varphi^0, s). \tag{10}$$

Defining a trajectory $\tau \mapsto y(\tau)$ on a not yet specified interval $[\tau_0, \tau_1]$ allows one to (locally and numerically) solve the second equation in (10),

$$y''(\tau) = \bar{f}\Big(y(\tau) - \bar{\gamma}(s(\tau)), s(\tau)\Big), \tag{11}$$

for the parameter s in order to get a relation $\tau \mapsto s(\tau)$. Furthermore, using (9) and $\psi^i = \bar{\psi}^i(s), i = 1, \ldots, n$, the corresponding trajectories of the angles, $\tau \mapsto \psi^i(\tau)$, $i = 1, \ldots, n$, can also be deduced. Trajectories of derivatives of s w.r.t. τ can be obtained after differentiation of (11).

So far all trajectories have been defined in posture dependent "time" τ, as $\tau \mapsto \psi(\tau)$ with the interval $[\tau_0, \tau_1]$ being chosen arbitrarily. As a consequence, only partial information about the evolution of the motion in physical time t can be deduced directly from the prescribed trajectories. The actual trajectories $t \mapsto \psi(t)$ of the relative pivot angles and the duration of the corresponding motion cycle can be predicted only when the relation

$$t(\tau) = \int_{\tau_0}^{\tau} a^0(\psi(\tilde{\tau}))\, d\tilde{\tau}, \tag{12}$$

following from $dt/d\tau = a^0(\psi)$ has been evaluated.

In transformed time s, the tuple $(y, \psi^1, \ldots, \psi^n)$ forms a set of $n+1$ free parameters. This (vector) fundamental parameter may be interpreted as a generalization of a flat output.

Remark 1. The introduction of an additional free parameter, which in some sense can be viewed as an "extra input", has been used in prior work by other authors [4, 3]—see also [5] for a similar approach.

Of course, for the actual choice of the reference trajectories a number of constraints have to be taken into account. In the case of the biped robot, the free leg cannot pass through the ground, the root leg must keep contact with the ground, torque constraints must be satisfied, singularities must be avoided, and some inverse functions must be determined. Furthermore, one wants to parameterize a "natural looking" motion, and might also wish to ensure that impacts are avoided. Also, a start phase to reach a periodic walking regime from rest as well as a stop phase must be parameterized. Altogether a parameterization of the trajectories of $y, \psi^1, \ldots, \psi^n$ is required with enough freedom, and a detailed numerical study is needed. This work cannot be discussed in detail here but a result is shown below.

3.4 Stabilizing Feedback

A method for the design of feedback laws stabilizing the motion along the reference trajectories designed with the time scaling methods is briefly sketched in the sequel. To this end, the reference (or desired) trajectories planned so far are now denoted as $\varphi_d^0, \psi_d^1, \ldots, \psi_d^n$. Tracking errors of the relative angles are defined as $e^i = \psi^i - \psi_d^i$. They may be stabilized w.r.t. s by defining an error differential equation

$$\frac{d^2 e^i}{ds^2} - (\lambda_1^i + \lambda_2^i)\frac{de^i}{ds} + \lambda_1^i \lambda_2^i e^i = 0. \tag{13}$$

As the definition of the flat output depends on the trajectories, the reference trajectory for y must be corrected by taking the solution of this error system into account:

$$y_d = \varphi_d + \int_0^{s_d} \bar{g}(\sigma) d\sigma, \quad \bar{g}(s) = \frac{\bar{b}_i^0(s)}{\bar{a}^0(s)} \frac{d\psi^i}{ds}(s) = \frac{\bar{b}_i^0(s)}{\bar{a}^0(s)} \Big(\frac{d\psi_d^i}{ds}(s) + \frac{de^i}{ds}(s) \Big). \tag{14}$$

The derivatives of $e_y = y - y_d$ then follow from $s, s', \varphi^0, (\varphi^0)'$ and the reference trajectories as

$$e_y = \varphi - \varphi_d + \int_{s_d}^{s} \bar{g}(\sigma) d\sigma, \tag{15a}$$

$$e_y' = \varphi' - \varphi_d' + \bar{g}(s)s' - \bar{g}(s_d)s_d', \tag{15b}$$

$$e_y'' = \bar{f}(\varphi^0, s) - \bar{f}(\varphi_d^0, s_d), \tag{15c}$$

$$e_y''' = \frac{d}{d\tau} \left(\bar{f}(\varphi^0, s) - \bar{f}(\varphi_d^0, s_d) \right). \tag{15d}$$

A stable error dynamics for e_y (w.r.t. τ) is chosen as

$$e_y'''' + \sum_{i=0}^{3} k_{y,i} \frac{d^i e_y}{d\tau^i} = 0, \tag{16}$$

with an appropriate choice of the coefficients. The control torques are calculated as in the feed-forward control case, except with y'''' replaced by the linear feedback expression following from the error equation. At this point, it is important to notice that the coefficients in the error equations (13) of the relative angles must be chosen depending on the sign of s.

3.5 Simulation Results

A simulation result for the biped structure is presented below. Fig. 3 shows trajectories generated on the basis of the parameterization proposed. The trajectories of the relative joint angles $\psi^i(s), i = 1, \ldots, n$ are shown on the left top subfigure. To its right, the control torques are given (Γ_0, the one at the floor, being zero) and on the upper right, the path of the lower end point of the free leg is drawn. Control constraints from [1] are respected. Fig. 3 also shows a sequence of snaphots taken during this step.

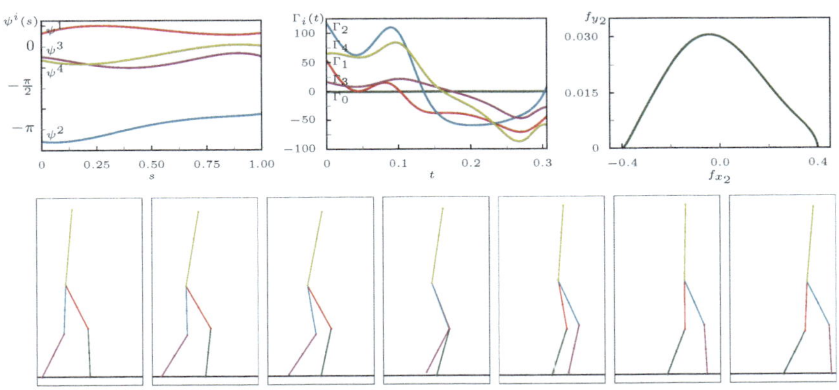

Fig. 3. Simulation results for a step during the single support phase

4 Conclusion

Exploiting the concept of orbital flatness, extending it to transformations involving functionals, and introducing a free time-scale parameter as an “additional input” enables the solution to difficult problems of motion planning and control. This technique has been shown for general planar tree-like pendulum structures. The method can be used in particular to plan walking motions of robots with models that fall into this class of systems. Specifically, the single support phase of biped walking robots is considered here. Additionally, the phases with two point floor contact and the ballistic phase (with no ground contact) can be treated on a flatness basis. In the first case, this treatment is rather standard since the system is overactuated. In the ballistic phase, the methods introduced here can be applied on the reduced system obtained after elimination of the invariants of motion.

References

1. Chevallereau, C., Abba, G., Aoustin, Y., Plestan, F., Westervelt, E.R., Canudas de Wit, C., Grizzle, J.W.: Rabbit: A testbed for advanced control theory. IEEE Contr. Syst. Mag. 23, 57–78 (2003)
2. Fliess, M., Lévine, J., Martin, P., Rouchon, P.: A Lie-Bäcklund approach to equivalence and flatness of nonlinear systems. IEEE Trans. Autom. Contr. 44, 922–937 (1999)
3. Kiss, B., Szádesczky-Kardoss, E.: Tracking control of the orbitally flat kinematic car with a new time-scaling input. In: Proc. 46th IEEE Conf. on Decision and Control, New Orleans, LA (2007)
4. Lévine, J.: On the synchronization of a pair of independent windshield wipers. IEEE Trans. Contr. Syst. Technol. 5, 787–795 (2004)
5. Aguiar, A.P., Hespanha, J.P., Kokotović, P.V.: Path-Following for Non-Minimum Phase Systems Removes Performance Limitations. IEEE Trans. Autom. Contr. 50, 234–239 (2005)
6. Olfati-Saber, R.: Nonlinear control of underactuated mechanical systems with application to robotics and aerospace vehicles. Ph.D. diss. Massachusetts Inst. of Techn., Boston (2001)
7. Rouchon, P., Sira-Ramirez, H.: Control of the walking toy: A flatness approach. In: Proc. American Contr. Conf. (2003)
8. Sangwan, V., Agrawal, S.K.: Differentially flat design of bipeds ensuring limit-cycles. In: Proc. IEEE Int. Conf. on Robotics and Automation (2007)
9. Spong, M.: The swing up control problem for the acrobot. IEEE Control Systems Magazine 15, 49–55 (1995)
10. Westervelt, E.R., Grizzle, J.W., Chevallereau, C., Choi, J.H., Morris, B.: Feedback Control of Dynamic Bipedal Robot Locomotion. Taylor & Francis, Abington (2007)
11. Willson, S.S., Mullhaupt, P., Bonvin, D.: Quotient method for controlling the acrobot. In: Proc. 48th IEEE Conf. on Decision and Control, Shanghai, China (2009)
12. Xin, X., Kaneda, M.: New analytical results of the energy based swinging up control of the acrobot. In: Proc. 43rd IEEE Conf. on Decision and Control (2004)

Mechanical Version of the CRONE Suspension

Alain Oustaloup and Xavier Moreau

Abstract. This paper deals with an application of the non integer differentiation in vehicle suspension area: the CRONE suspension, French acronym of *suspension á Comportement Robuste d'Ordre Non Entier*. This suspension results from a traditional suspension system whose order 1 dashpot is replaced by a non-integer order dashpot. The different steps, from the concept to its practical realisation, are presented. A quarter-car model is used to illustrate the performances. The frequency and time responses, for various values of the vehicle load, reveal a great stability robustness: the resonance in the frequency domain and the damping ratio in the time domain remain almost constant whatsoever the load variations are.

1 Introduction

The history of the non integer (so-called fractional) differentiation can be dated back to 1695, when L'Hospital and Leibniz were communicating whether it made sense to define an operator d^n/dt^n for $n = 0.5$ [1]. In the $18th$ century there were only few contributions to this topic and it was Euler who again raised the question of a derivative of order n for n being a fraction. Later, Liouville attempted to give a logical definition of fractional derivatives. One can state that the whole theory of fractional derivatives and integrals was established in the $2nd$ half of the $19th$ century [2], [3], [4]. During the $20th$ century, fractional-order systems, or systems described using fractional derivatives and integrals, have been studied by many in the engineering area [5], [6], [7], [8]. It should be noted that there is a growing number of physical systems whose behaviour can be compactly described using fractional-order system theory [9], [10], [11], [12]. Of specific interest to engineers are electrochemical processes, long lines, dielectric polarization, colour noise, chaos and viscoelastic materials. After this brief introduction, part 2 presents some basic definitions to

Alain Oustaloup · Xavier Moreau
Bordeaux University – IMS, 351, cours de la Libération, 33405 Talence Cedex, France
e-mail: {firstname.lastname}@ims-bordeaux.fr

J. Lévine & P. Müllhaupt (Eds.): Adv. in Theory of Control, Signals, LNCIS 407, pp. 99–112.
springerlink.com © Springer-Verlag Berlin Heidelberg 2010

introduce non integer differentiation from integer differentiation. Then, an application of non integer differentiation in vehicle suspension area is presented: the CRONE suspension [13]. The CRONE approach leads to a robustness of stability degree versus load variation. Finally, the last part summarises all the main points developed in this paper.

2 From Integer Differentiation to Non Integer Differentiation

2.1 *Towards Non Integer Differentiation*

A natural approach of the non integer derivative of a causal or not causal function $f(t)$, using a generalization of the well known real integer order derivative definition, consists in:

- expressing the order 1 derivative under an adequate form,
- also expressing the order 2 derivative under an analogous form,
- then extending to the non integer case the generic form resulting,
- to obtain thus a generalization to the integer and non integer case.

2.2 *Generic Form of the Order 1 and 2 Derivatives*

The order 1 left derivative is defined by:

$$D^1 f(t) = \lim_{h\to 0} \frac{f(t) - f(t-h)}{h}. \tag{1}$$

A discretization of t to the sampling interval h, namely $t = Kh$, is translated by:

$$D^1 f(t) = \lim_{h\to 0} \frac{f(Kh) - f((K-1)h)}{h}. \tag{2}$$

The introduction of the delay operator q^{-1} applicable to a concrete function and defined by

$$q^{-1} f(Kh) = f((K-1)h), \tag{3}$$

allows to write:

$$D^1 f(t) = \lim_{h\to 0} \frac{1 - q^{-1}}{h} f(Kh). \tag{4}$$

A similar calculation carried out for an order 2 derivative, leads to:

$$D^2 f(t) = \lim_{h\to 0} \frac{\left(1 - q^{-1}\right)^2}{h^2} f(Kh). \tag{5}$$

2.3 *Generalization to the Integer and Non Integer Case*

The generalization to any (integer or non integer, real or complex) order is immediate and leads to the definition proposed by Grünwald in 1867, that is to say:

$$D^n f(t) = \lim_{h \to 0} \frac{\left(1 - q^{-1}\right)^n}{h^n} f(Kh), \tag{6}$$

namely, developping $(1 - q^{-1})^n$ by the Newton binomial formula:

$$D^n f(t) = \lim_{h \to 0} \frac{1}{h^n} \left[\sum_{k=0}^{\infty} (-1)^k \frac{n(n-1)(n-2)\ldots(n-k+1)}{k!} q^{-k} \right] f(Kh), \tag{7}$$

or

$$D^n f(t) = \lim_{h \to 0} \frac{1}{h^n} \sum_{k=0}^{\infty} (-1)^k \frac{n(n-1)(n-2)\ldots(n-k+1)}{k!} q^{-k} f(Kh), \tag{8}$$

or even, given that

$$q^{-k} f(Kh) = f((K-k)h) = f(t-kh) : \tag{9}$$

$$D^n f(t) = \lim_{h \to 0} \frac{1}{h^n} \sum_{k=0}^{\infty} (-1)^k \frac{n(n-1)(n-2)\ldots(n-k+1)}{k!} f(t-kh), \tag{10}$$

or even, under a more condensed writing:

$$D^n f(t) = \lim_{h \to 0} \frac{1}{h^n} \sum_{k=0}^{\infty} a_k f(t-kh), \tag{11}$$

by putting:

$$a^k = (-1)^k \frac{n(n-1)(n-2)\ldots(n-k+1)}{k!}, \tag{12}$$

or, using the usual symbolism: $a_k = (-1)^k \binom{n}{k}$ with $a_k = 1$ for $k = 0$.

2.4 *Local and Global Characterization*

Through the function $f(t-kh)$ which introduces terms in $f(t), f(t-h)$, $f(t-2h), \ldots$, thus past samples, the formula

$$D^n f(t) = \lim_{h \to 0} \frac{1}{h^n} \sum_{k=0}^{\infty} a_k f(t-kh), \tag{13}$$

shows that, contrary to integer derivative, the non integer derivative of a function at a given instant t takes into account the values of this function at all the past instants.

So, if integer derivative gives a local characterization of the function (slope of the tangent to the curve at the instant t for the order 1 derivative), non integer derivative turns out to give a global characterization.

2.5 Memory Notion

Taking into account all the past of the function through a weighted sum of the samples which expresses (through the weighting coefficients) a different weight according to the sample, non integer differentiation introduces a memory notion through a lessening or a stressing of the past in conformity with the weighting coefficients which are such as:

- for $n = -1$ (which corresponds to an integration), the past is not weighted (figure 1);
- for $n > -1$ (which corresponds to less than an integration and to a differentiation), the past is lessened (with weighting coefficients which keep the same sign for $-1 < n < 1$ and which change sign for $n > 1$ (figure 2));
- for $n < -1$ ((figure 3) which corresponds to more than an integration), the past is stressed (with weighting coefficients which keep the same sign).

Such as characterized, this memory notion evokes a memory subtle form through which the recollection of an event depends on the nature of this event:

- for usual events, the recollection of more ancient events is less important than the recollection of more recent events;
- on the other hand, for unusual events, the recollection of more ancient events can be more important than the recollection of more recent events.

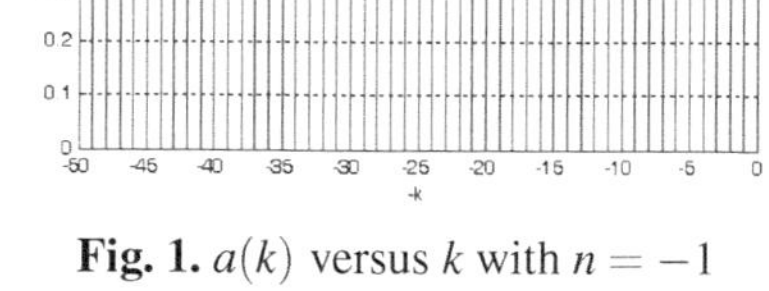

Fig. 1. $a(k)$ versus k with $n = -1$

3 The CRONE Approach in Vehicle Suspension Area: The CRONE Suspension

In dynamics, for any vehicle, the suspension system must perform two main functions [13]. Firstly, it provides a high degree of isolation for the vehicle body from the loads applied between the wheels and roads to ensure passenger comfort; and, secondly, it keeps the wheels in close contact with the road surface to ensure an adequate roadholding when accelerating, braking or cornering. These functions must then be optimised within several constraints: first, the minimum value of the relative body/wheel workspace usage; second, control of vehicle attitude in manoeuvring;

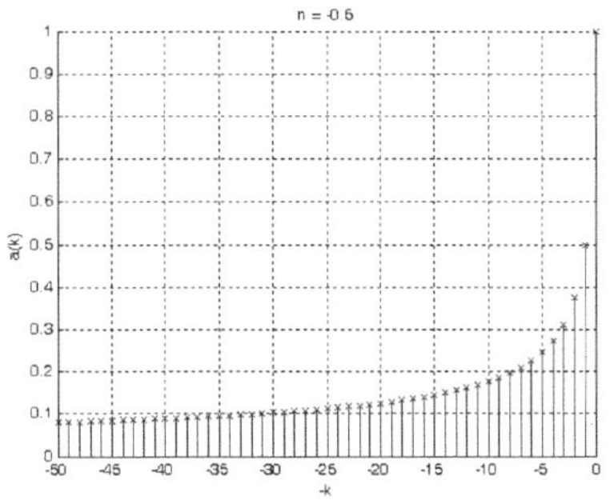

(a) $n = -0.5$

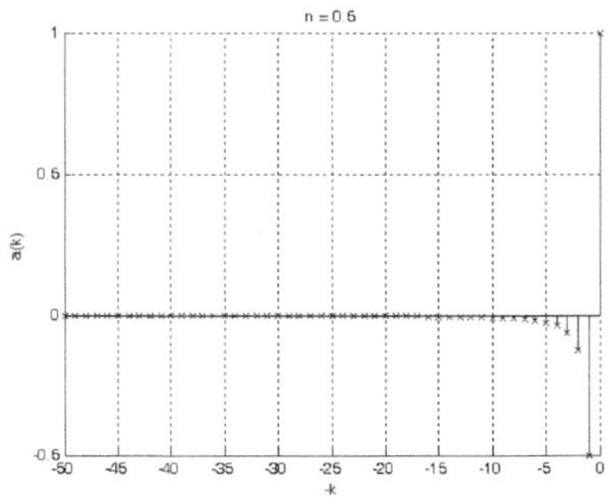

(b) $n = 0.5$

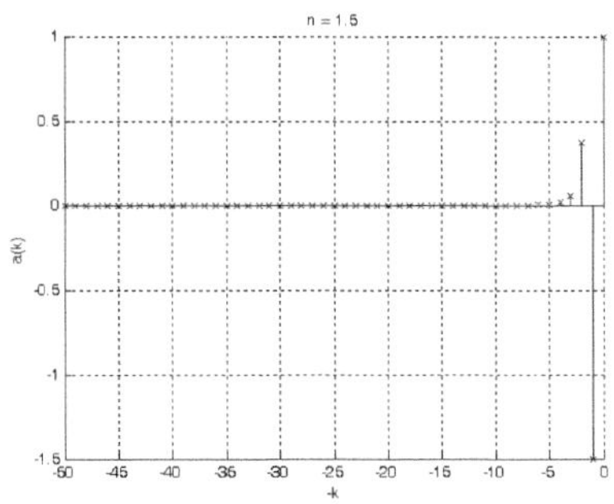

(c) $n = 1.5$

Fig. 2. $a(k)$ versus k for different values of n

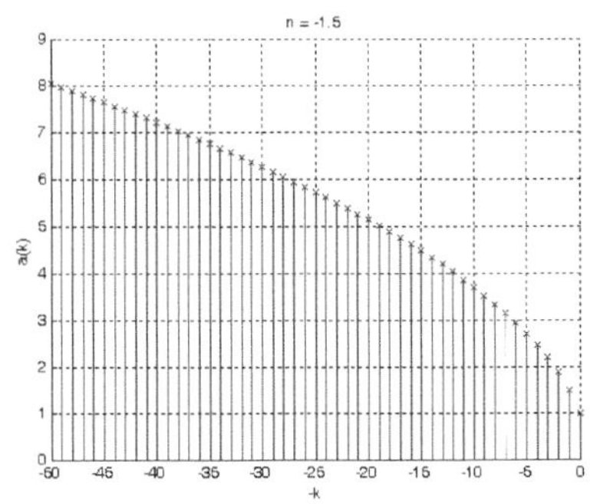

Fig. 3. $a(k)$ versus k with $n < -1$

third, minimum value of the power consumption. In fact, the conflict between these various aspects of vehicle behaviour is the main problem in suspension system design.

In statics, the vehicle bodywork weight is compensated:

- by the self leveller device, in addition to the suspension, in a hydropneumatic technology;

- by the spring of the suspension itself in a mechanical technology, which implies that the CRONE suspension mechanical version is, like the usual suspension
  - of type mass-spring-dashpot
  - apart that the aforementioned dashpot is actually a damping device of another kind.

Just like the hydropneumatic version of the CRONE suspension, the aim is to ensure simultaneously to the vehicle bodywork

- a better vibration isolation by a reduction of the vertical accelerations (in performance terms);
- and a better robustness of stability degree in relation to the carried load (in robustness terms).

Contrary to the effective strategies that consist in reparametrizing an usual suspension through a modification (linked or not)

- of the spring *stiffness*
- and of the dashpot *viscous damping* (or *viscous friction coefficient*),

our strategy consists in restructuring an usual suspension through a modification of the *order* of the dashpot (defined by a force-displacement transfer), which comes to replace

- the usual dashpot (of order 1)
- by a dashpot of non integer order.

4 Principle and Modeling of the CRONE Suspension

Let a suspension be of mass-spring-dashpot type (figure 4) , in which:

- M represents the bodywork mass supported by each wheel,
- k denotes the spring stiffness,
- C denotes the dashpot viscous damping,
- $x(t)$ and $y(t)$ denote the vertical displacements of the wheel and of the bodywork respectively.

The usual suspension uses a dashpot which develops a force proportional to its relative (or differential) speed, that is to say to the first derivative of its relative (or differential) displacement.

The principle of the CRONE suspension consists in replacing the order 1 dashpot so defined by a non integer order dashpot. This dashpot develops a force proportional to the non integer derivative of its relative displacement, namely:

$$F(t) = C\left(\frac{d}{dt}\right)^m [x(t) - y(t)], \quad \text{with} \quad 0 < m < 1. \tag{14}$$

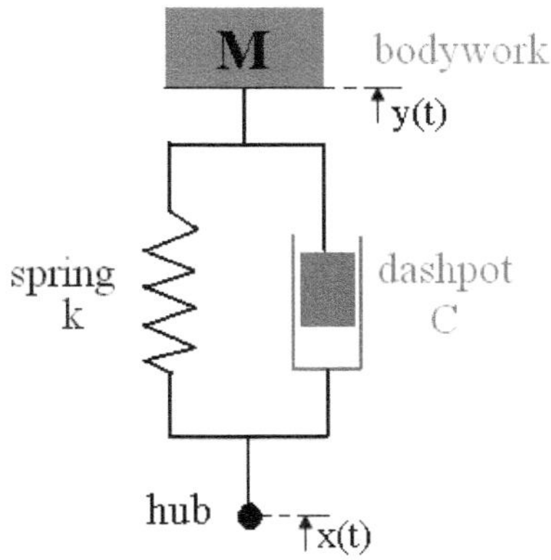

Fig. 4. Suspension scheme

Applying the *Newton's second principle* to the usual and CRONE suspensions, makes it possible to establish the differential equations:

$$M\frac{d^2y(t)}{dt^2} = -k[y(t)-x(t)] - C\left(\frac{d}{dt}\right)[y(t)-x(t)] \tag{15}$$

and

$$M\frac{d^2y(t)}{dt^2} = -k[y(t)-x(t)] - C\left(\frac{d}{dt}\right)^m[y(t)-x(t)], \tag{16}$$

namely:

$$M\frac{d^2y(t)}{dt^2} + C\frac{dy(t)}{dt} + ky(t) = C\frac{dx(t)}{dt} + kx(t) \tag{17}$$

and

$$M\frac{d^2y(t)}{dt^2} + C\left(\frac{d}{dt}\right)^m y(t) + ky(t) = C\left(\frac{d}{dt}\right)^m x(t) + kx(t), \tag{18}$$

from which one draws the transmittances:

$$H(s) = \frac{Y(s)}{X(s)} = \frac{k+Cs}{k+Cs+Ms^2} \tag{19}$$

for the *usual suspension (m=1)*, and

$$H(s) = \frac{Y(s)}{X(s)} = \frac{k+Cs^m}{k+Cs^m+Ms^2} \tag{20}$$

for the *CRONE suspension ($0 < m < 1$).*

4.1 *Initial Behaviour : No Initial Acceleration for the CRONE Suspension*

The deformation (by crushing) of a tyre while climbing a pavement which constitutes an extreme test case in vibratory isolation by impact absorption, makes that the wheel hub describes a profile that looks more like a ramp than a step.

So, the ramp function turns out to be a model of elementary deterministic solicitation sufficiently representative of reality, so enabling to express the vertical displacement of the hub under the form:

$$x(t) = tu(t), \tag{21}$$

where $u(t)$ denotes the unit step function. Concerning the response $y(t)$ to this ramp try, it is possible to write, from the initial value theorem

$$\lim_{t \to 0} \ddot{y}(t) = \lim_{s \to +\infty} s[s^2 H(s) X(s)], \tag{22}$$

particularized by

$$\lim_{t \to 0} \ddot{y}(t) = \lim_{s \to +\infty} sH(s) \quad \text{for} \quad X(s) = \frac{1}{s^2} \ : \tag{23}$$

$$\ddot{y}(0^+) = \lim_{s \to +\infty} \frac{ks + Cs^2}{k + Cs + Ms^2} = \frac{C}{M} \tag{24}$$

for the *usual suspension (m=1)*, and

$$\ddot{y}(0^+) = \lim_{s \to +\infty} \frac{k + Cs^{1+m}}{k + Cs^m + Ms^2} = 0 \tag{25}$$

for the *CRONE suspension (0 < m < 1)*.

The comparison of these relations shows that the initial acceleration of the bodywork is finite for the usual suspension and nil for the CRONE suspension.

Independently of the robustness of the CRONE suspension that we are going to show, this property is already remarkable. It indeed ensures a better comfort for the passengers.

4.2 *Stability Degree Robustness*

4.2.1 Robustness Tests

For different parametric states of the usual and CRONE suspensions, namely

$$k = 17055\, N/m \quad \text{and} \quad C = 5500\, Ns/m \quad \text{for} \quad m = 1$$

and

$$k = 4264\, N/m \quad \text{and} \quad C = 9000\, Ns^{0.8}/m \quad \text{for} \quad m = 0.8,$$

these states ensuring approximately the same dynamics to the two suspensions for a mass of 300 kg, the tests on the degree stability robustness in relation to the carried load are completed for different values of the mass M, namely 150 kg, 300 kg, 600 kg and 900 kg.

4.2.2 Stability Degree Measure

Stability degree is quantified (or measured):

- either by the resonance ratio, defined from the frequency response (figure 5);
- either by the first overshoot, defined from the step response (figure 6);
- or by the damping ratio, defined from the roots of the characteristic equation (figure 7).

That amounts to saying that the robustness tests as defined rely successively on

- the frequency and step responses of the two suspensions
- and also the roots of their characteristic equation.

4.2.3 Frequency and Setp Responses: Robustness of the Resonance Ratio and of the First Overshoot for the CRONE Suspension

Figures 5 and 6 present the frequency and step responses for both usual and CRONE suspensions.

These responses given for the different values of the vehicle load, reveal the robustness of stability degree for the CRONE suspension through the constancy – of the resonance ratio in frequency domain – and of the first overshoot in time domain.

4.2.4 Characteristic Equation Roots: Robustness of the Damping Ratio for the CRONE Suspension

The algorithm, that the "***Generalized characteristic equation***" module of the CRONE software uses, is applied to find the roots of the characteristic equation

- of the usual suspension on one hand, namely ($m = 1$)

$$Ms^2 + Cs + k = 0 \tag{26}$$

- of the CRONE suspension on the other hand, namely ($m = 0.8$)

$$Ms^2 + Cs^{0.8} + k = 0. \tag{27}$$

For the parametric states of the usual and CRONE suspensions as previously defined, figure 7 illustrates the image of the roots (or the poles) so obtained for the different mass values defined by the test conditions. Knowing that the damping ratio is given by the cosine of the half centre angle that the conjugate complex pole pair forms, the angular quasi-constancy of the CRONE suspension poles is representative of its robustness.

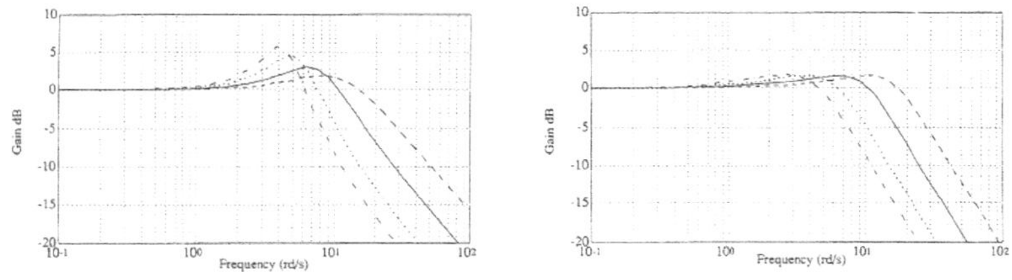

(a) Gain diagrams of the usual suspension (b) Gain diagrams of the CRONE suspension

Fig. 5. Frequency response: – $M = 300\ Kg$; … $M = 600\ Kg$

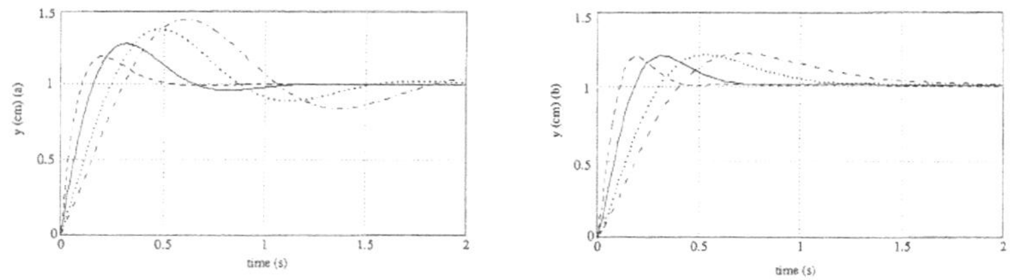

(a) Step responses of the usual suspension (b) Step responses of the CRONE suspension

Fig. 6. Step response: – $M = 300\ Kg$; … $M = 600\ Kg$

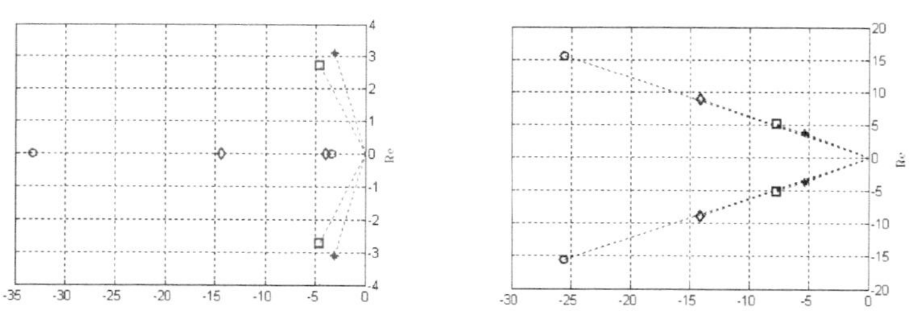

(a) Image of the poles of the usual suspension (b) Image of the poles of the CRONE suspension

Fig. 7. Image of the poles of the usual and CRONE suspensions for different loads: $\lozenge\ M = 300\ Kg$; $\square\ M = 600\ Kg$

5 Idea of the Synthesis of a Non Integer Order Dashpot

The synthesis of a non integer order dashpot is based on the non-stationarity of an usual dashpot through the time variation of the viscous damping. Indeed, the force developed by a non integer order dashpot, namely

$$F(t) = C\left(\frac{d}{dt}\right)^m [x(t) - y(t)], \tag{28}$$

can be interpreted as resulting from an usual dashpot with variable viscous damping, namely

$$F(t) = C(t)\left(\frac{d}{dt}\right)[x(t) - y(t)], \tag{29}$$

from where one draws, by identifying these two equations:

$$C(t) = C\,\frac{\left(\frac{d}{dt}\right)^m [x(t) - y(t)]}{\left(\frac{d}{dt}\right)[x(t) - y(t)]}. \tag{30}$$

So, the solution consists:

- firstly, in computing (in conformity with this relation) the instantaneous viscous damping $C(t)$ from the relative displacement recorded by a position sensor;
- secondly, in computing the instantaneous section $s(t)$ of the dashpot hole from the value of $C(t)$ so obtained.

6 Active Character of the CRONE Suspension

The time variation of the viscous damping $C(t)$ for a mass of $300kg$ and for step and harmonic responses as shown in figure 8, presents positive and negative values that translate the existence of dissipative and active phases, so expressing that the damping device which is at stake is in fact

- a continuously controlled dashpot for the dissipative phases
- relayed by a hydraulic actuator controlled in force for the active phases (which makes active the CRONE suspension).

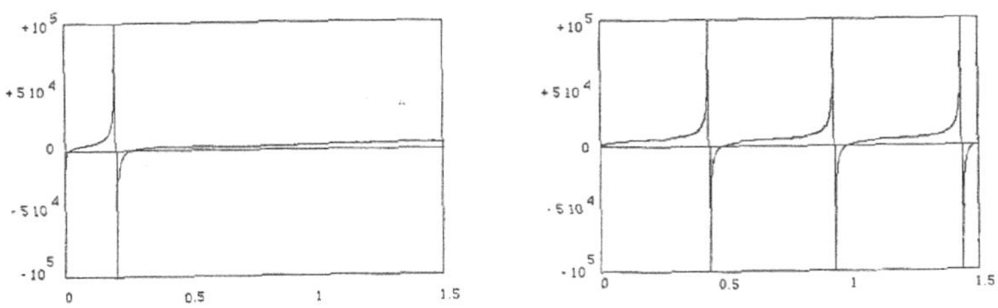

(a) Variation of $C(t)$ for a step response (b) Variation of $C(t)$ for a harmonic response

Fig. 8. Variation of $C(t)$ for step and harmonic responses

7 Piloted Passive CRONE Suspension

Given that the energies at stake during the active phases are negligible in front of those corresponding to the dissipative phases (quantified study in sinusoidal state), which confers a quasi dissipative character to the dashpot of non integer order (so justifying its dashpot denomination), it is convenient to adopt the following process:

- a priori, cancel $C(t)$ when it becomes negative;
- a posteriori, verify that the cancellation of $C(t)$ so defined does not practically affect the performances (by means of an eventual reparametrization of the suspension).

The modification of $C(t)$ in conformity with a new viscous damping, denoted by $C^*(t)$, such as

$$C^*(t) \begin{cases} 0 \quad \text{when} \quad C(t) \leq 0, \\ C\,\dfrac{\left(\frac{d}{dt}\right)^m [x(t)-y(t)]}{\left(\frac{d}{dt}\right)[x(t)-y(t)]} \quad \text{when} \quad C(t) > 0, \end{cases} \tag{31}$$

defines the CRONE suspension said modified or, in other terms, the piloted passive CRONE suspension which constitutes the most economical solution as it needs no exterior energy supply and which approximatively keeps the same performances (figure 9) by means of the following reparametrization:

$$k = 10000\,N/m, \quad C = 9000\,Ns^{0.7}/m \quad \text{and} \quad m = 0.7. \tag{32}$$

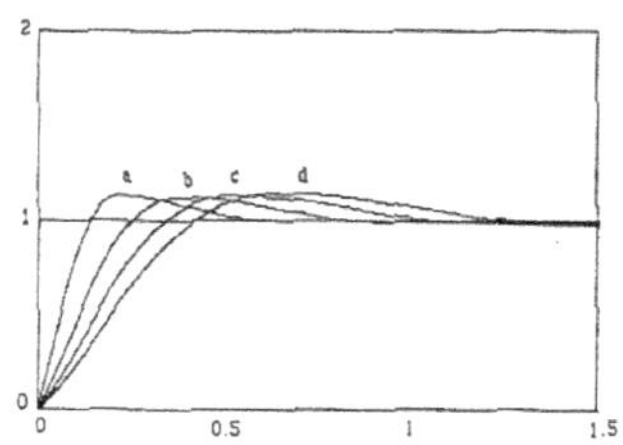

Fig. 9. Step responses of the modified CRONE suspension: (a) 300 kg; (b) 600 kg; (c) 900 kg; (d) 1200 kg

8 Contract Collaboration

The industrial partner is the Peugeot-Citroën (PSA) company.

The contribution of the collaboration is:

- the conception and the achievement of the non-integer order dashpot (figure 10) (according to the patents 90 046 13 and 95 050 84 registered respectively in 1990 and 1995)

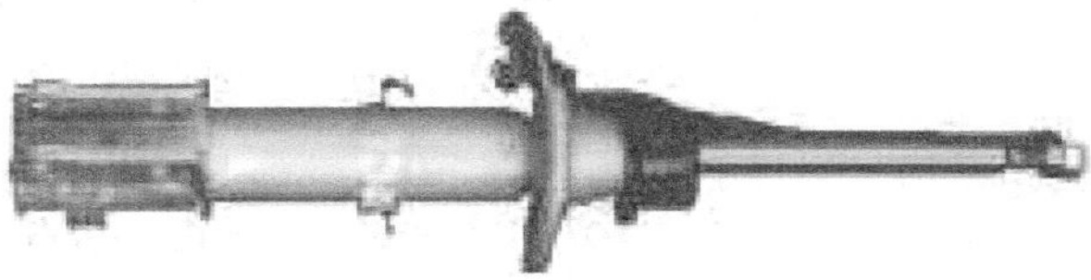

Fig. 10. Non-integer order dashpot

(a) In the front

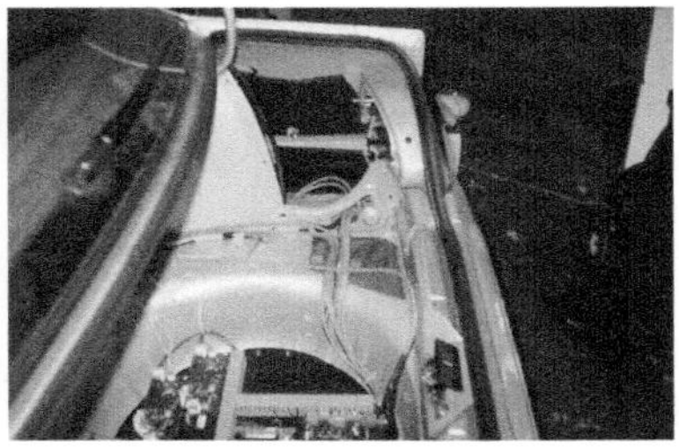

(b) At the rear

Fig. 11. Non-integer order Dashpot implementation

- and also its implementation in a 406 Peugeot (figure 11) in 1998 with the help of the Coverplant company.

9 Conclusion

This paper has shown that the CRONE suspension provides remarkable performance: better robustness of stability degree versus load variations of the vehicle. Robustness is illustrated by the frequency and time responses obtained for different values of the load. From the concept of the CRONE suspension, a technological solution has been developed with PSA. This solution, called the passive CRONE suspension, uses a continuously controlled damper. Its design permits it to be manufactured at the same cost as a traditional automobile damper. Bench test on a prototype have validated the theoretical expectations.

References

1. Dugowson, S.: Les différentielles métaphysiques: histoire et philosophie de la généralisation de l'ordre de dérivation PHD thesis, Université Paris Nord (1994)
2. Oldham, K.B., Spanier, J.: The fractional calculus. Academic Press, New-York (1974)
3. Miller, K.S., Ross, B.: An introduction to the fractional calculus and fractional differential equations. A Wiley-Interscience Publication, Hoboken (1993)
4. Samko, S.G., Kilbas, A.A., Marichev, O.I.: Fractional integrals and derivatives: theory and applications. Gordon and Breach Science Publishers (1993)

5. Oustaloup, A.: La dérivation non entière: théorie, synthèse et applications. Edition Hermès, Paris (1995)
6. Lin, J.: Modélisation et identification de systèmes d'ordre non entier PHD thesis, Université de Poitiers (2001)
7. Cois, O.: Systèmes linéaires non entiers et identification par modèle non entier: application en thermique PHD thesis, Université Bordeaux 1 (2002)
8. Moreau, X., Ramus-Serment, C., Oustaloup, A.: Fractional Differentiation in Passive Vibration Control. Journal of Nonlinear Dynamics 29, 343–362 (2002)
9. Trigeassou, J.C., Poinot, T., Lin, J., Oustaloup, A., Levron, F.: Modeling and identification of a non integer order system. In: Proceedings of ECC 1999, European Control Conference, Karlsruhe, Germany (1999)
10. Ramus-Serment, C., Moreau, X., Nouillant, M., Oustaloup, A., Levron, F.: Generalised approach on fractional response of fractal networks. Journal of Chaos, Solitons and Fractals 14, 479–488 (2002)
11. Petras, I., Vinagre, B.M., O'Leary, P., Dorcak, L.: Analogue Realizations of Fractional-Order Controllers. Journal of Nonlinear Dynamics 29, 281–296 (2002)
12. Aoun, M., Malti, R., Levron, F., Oustaloup, A.: Numerical Simulations of Fractional Systems: An Overview of Existing Methods and Improvements. Journal of Nonlinear Dynamics 38, 117–131 (2004)
13. Moreau, X., Altet, O., Oustaloup, A.: The CRONE Suspension: Management of Comfort-Road Holding Dilemma. Journal of Nonlinear Dynamics 38, 461–484 (2004)
14. Dauphin-Tanguy G.: Les bond-graphs. Edition Hermès, Paris (2000)
15. Poinot, T., Trigeassou, J.C.: Identification of Fractional Systems Using an Output-Error Technique. Journal of Nonlinear Dynamics 38, 133–154 (2004)
16. Serrier, P., Moreau, X., Oustaloup, A.: Advances in Fractional Calculus. In: Limited-Bandwidth Fractional Differentiator: Synthesis and Application in Vibration Isolation, pp. 287–302. Springer, Heidelberg (2007)

Electrostatic MEMS: Modelling, Control, and Applications

Guchuan Zhu

Abstract. This paper addresses issues related to the modelling and the control of electrostatic microelectromechanical systems (MEMS) in applications requiring high accuracy positioning, wide operation range, and high control bandwidth. A particular emphasis is put on the choice of control system architecture and its influence on potential performance in different practical operation conditions.

1 Introduction

In the last years, there has been a surge of interest in MEMS fabrication and applications. Currently available MEMS fabrication techniques enable the construction of devices with high-precision displacement and high-quality optical surface. However, most of the known applications take advantage of only the native properties of MEMS in building miniaturized systems for sensing or actuation operating in an open-loop manner. This would not allow fully exploiting the promises the MEMS technology can offer. MEMS requiring high accuracy positioning, wide operation range, and high control bandwidth include, just to name a few, all-optical switches, optical scanning mirrors, recoverable photopic crystal devices, tunable optical filters, and deformable micro-mirrors, which are key components in such systems as optical communications, astronomical telescopes, medical and biological instruments, and many other scientific and engineering applications. Constructing highly-integrated microsystems for high-functionality and high-performance applications represents one of the future trends in MEMS technology.

Among diverse actuation mechanisms, electrostatic actuation is the most popular one because of its simple structural geometry, flexible operation, and easy fabrication from standard and well-understood materials [1]. However, this actuation

Guchuan Zhu
Department of Electrical Engineering, École Polytechnique de Montréal, C.P. 6079, Succursale centre-ville, Montreal, QC, Canada H3C 3A7
e-mail: guchuan.zhu@polymtl.ca

J. Lévine & P. Müllhaupt (Eds.): Adv. in Theory of Control, Signals, LNCIS 407, pp. 113–123.
springerlink.com ©Springer-Verlag Berlin Heidelberg 2010

scheme results in highly nonlinear dynamics, giving rise to a saddle-node bifurcation, called "pull-in," which limits the stable open-loop operation to a small portion of the whole physically available range [2]. Extending the stable operation range and further enhancing the performance of electrostatic MEMS constitute one of the central topics in the field of MEMS which has motivated the work on the application of a variety of nonlinear control techniques to this problem (see, e.g., [3, 4, 5, 6, 7, 8]). Robust and adaptive control of MEMS in the presence of parasitics and parametric uncertainties has also been addressed (see, e.g., [9, 10, 11]).

This paper aims at getting an overview of diverse MEMS control strategies and an insight into the potential performance one can expect regarding the architecture of control system, in particular the choice of control variables. For simplicity, only one degree-of-freedom (1DOF) electrostatic parallel-plate actuators will be considered and the model of such a system is presented in Section 2. The stabilizability of typical control strategies is presented in Section 3-5, while the issue related to modelling and control of MEMS devices in the presence of modelling errors due to parametric uncertainties and parasitics is discussed in Section 6. Finally, some concluding remarks are given in Section 7.

2 Dynamic Model of Electrostatic Parallel-Plate MEMS

The schematic representation of 1DOF electrostatic parallel-plate actuators is shown in Fig. 1 where m is the mass of the moveable plate, k is the stiffness coefficient, b is the damping coefficient, G is the air gap, G_0 is the zero-voltage gap, p is the normalized displacement, Δ is the insulating layer thickness, R is the loop resistance, V_a is the voltage across the actuator, and V_s and I_s are the source voltage and the source current, respectively. When the moveable plate is supposed to be a rigid body without deformation and only the main electric field within the gap is considered, the actuator capacitance can be computed by $C_a = \varepsilon A/G$ [2] where A is the effective area of electrodes and ε is the permittivity in the gap. The electrostatic force on the moving plate corresponding to bias voltage V_a is then given by [2]

$$F_e = \frac{V_a^2}{2}\frac{\partial C_a}{\partial G} = -\frac{2\varepsilon A V_a^2}{G^2} = -\frac{Q_a^2}{2\varepsilon A} \tag{1}$$

where $Q_a = V_a/C_a$ is the charge on the actuator. The equation of motion of the actuator is then given by (see, e.g., [2]):

$$m\ddot{G} + b\dot{G} + k(G - G_0) = F_e. \tag{2}$$

By scaling system variables with respect to a critical bias voltage V_{pi} (the so-called pull-in voltage) [12, 5]:

$$p = 1 - \frac{G}{G_0},\ q = \frac{3Q_a}{2C_0 V_{pi}},\ u = \frac{V_s}{V_{pi}},\ i = \frac{I_s}{V_{pi}\omega_0 C_0},\ r = \omega_0 C_0 R, \tag{3}$$

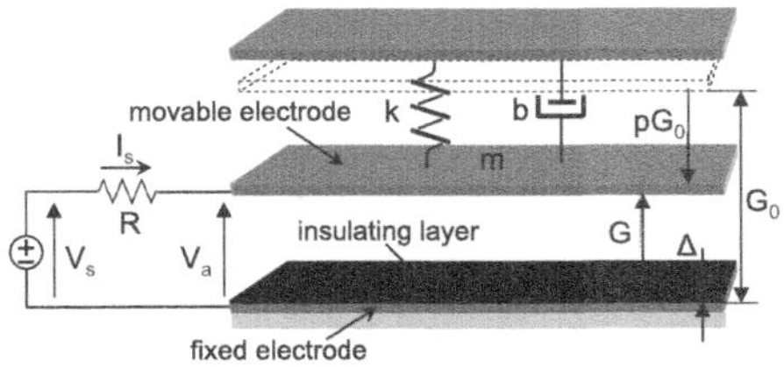

Fig. 1. Schematic representation of 1DOF parallel-plate electrostatic actuator. The top structure is fixed for sustaining the moveable plate.

where $C_0 := \varepsilon A/G_0$ is the capacitance at zero-voltage position, $\omega_0 := \sqrt{k/m}$ is the undamped natural frequency, and $\zeta := b/2m\omega_0$ is the damping ratio, the normalized bias voltage can be expressed by $u_a = 3q(1-p)/2$ and the dynamics of normalized charge become $\dot{q} = 2i/3$. Denoting $x = (x_1, x_2, x_3)^T = (p, \dot{p}, q)^T$, the state-space model of the system in a normalized time scale with respect to ω_0 is given by [5]:

$$\dot{x}_1 = x_2, \tag{4a}$$

$$\dot{x}_2 = -2\zeta x_2 - x_1 + \frac{1}{3}x_3^2, \tag{4b}$$

$$\dot{x}_3 = \frac{1}{r}x_3(x_1 - 1) + \frac{2}{3r}u, \tag{4c}$$

which is defined on a restricted state space $\mathscr{X} = \{x \in \mathbb{R}^3 \mid x_1 < 1 - \delta\}$ with $\delta = \Delta/G_0$ being the normalized thickness of the insulating layer. Note that for simplicity, contact dynamics on the boundary of $\mathscr{X}$ are not considered in this paper. Consequently, the closed-loop stability obtained by the presented control laws holds, in essence, locally.

3 Voltage Control and Capacitive Feedback

An intuitive choice of system output is the voltage across the device

$$y = u_a = \frac{3}{2}x_3(1 - x_1) = h_1(x) \tag{5}$$

which is the most commonly used control variable in open-loop control schemes. We then have[1]

$$\dot{y} = \underbrace{-\frac{3}{2r}x_3\left((1-x_2)^2 + rx_2\right)}_{L_f h_1} + \underbrace{\frac{1-x_1}{r}}_{L_g h_1}u. \tag{6}$$

As $L_g h_1 \neq 0$ for all $x_1 < 1$, System (4) with u_a as output has uniform relative degree 1 [13]. The input-output linearization can be obtained by the following control

[1] $L_f h(x)$ is the Lie derivative of $h(x)$ along the vector field f defined as $L_f h(x) = \frac{\partial h(x)}{\partial x} f(x)$.

$$u = (L_g h_1)^{-1} \left(\tilde{u} - L_f h_1\right), \tag{7}$$

and is of the form

$$\dot{y} = \tilde{u}, \tag{8a}$$
$$\dot{z} = \eta(z, y) \tag{8b}$$

where $\tilde{u}$ is a new control signal, $z = (x_1, x_2)^T$, and

$$\eta(z, y) = \begin{pmatrix} x_2 \\ -2\zeta x_2 - x_1 + \dfrac{4}{27}\dfrac{y^2}{(1-x_1)^2} \end{pmatrix}. \tag{9}$$

Clearly, the zero-dynamics (8b) coincide with that of the mechanical subsystem.

An obvious control is

$$\tilde{u} = -k(y - \bar{y}), \quad k > 0, \tag{10}$$

where $\bar{y}$ is the reference signal corresponding to a set-point. Hence, any equilibrium of the zero-dynamics must satisfy $x_2 = 0$ and $\bar{y}^2 = 27\bar{x}_1(1-\bar{x}_1)^2/4$. Denoting $\Delta x_1 = x_1 - \bar{x}_1$ and $\Delta x_2 = x_2 - \bar{x}_2$, the Jacobian linearization of the zero-dynamics around the equilibrium point is given by

$$\frac{d}{dt}\begin{pmatrix} \Delta x_1 \\ \Delta x_2 \end{pmatrix} = \begin{pmatrix} 0 & 1 \\ -1 + \dfrac{2\bar{x}_1}{1-\bar{x}_1} & -2\zeta \end{pmatrix} \begin{pmatrix} \Delta x_1 \\ \Delta x_2 \end{pmatrix}$$

which is stable if and only if $\bar{x}_1 \in [0, 1/3)$. It is straightforward to show that $\bar{x}_1 = 1/3$ is a saddle-node bifurcation point at which $\bar{y} = \bar{u}_a = 1$, corresponding to V_{pi}.

A control scheme, called capacitive feedback, has been proposed by [14] that can extend the stable operational range beyond the pull-in position by inserting a serial capacitor (see Fig 2(a)). It can be seen that

$$V_a + V_{C_f} = V_a\left(1 + C_a/C_f\right) = V_a\left(1 + \rho_f C_a/C_0\right)$$

where $\rho_f := C_0/C_f$, representing the scale of C_f with respect to C_0. Therefore in equilibrium, the relationship between actuation and bias voltages is given in normalized coordinates by $\overline{u}_a = \overline{u}_s(1-\overline{x}_1)/(1-\overline{x}_1+\rho_f)$. Noting that in any equilibrium $\overline{x}_3^2 = 3\overline{x}_1$, we obtain

$$\overline{u}_s^2 = \frac{27}{4}\overline{x}_1\left(1 - \overline{x}_1 + \rho_f\right)^2. \tag{11}$$

It can be seen from actuation curves in equilibrium points shown in Fig 2(b) that the insertion of a serial capacitor has the effect of pushing the pull-in position away and that for $\rho_f > 2$ or equivalently $C_f < C_0/2$, the saddle-node bifurcation will be removed [14]. However, the insertion of serial capacitor will increase the footprint.

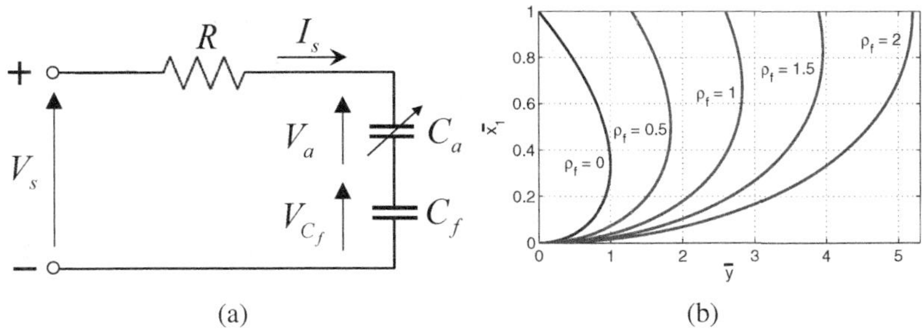

Fig. 2. Capacitive feedback: (a) scheme; (b) effect of the size of C_f to actuation curve in equilibrium

Capacitive feedback can also be implemented by closed-loop control without using physical serial capacitor. In fact, if the system output is chosen as

$$y = \frac{3}{2} x_3 \left(1 - x_1 + \rho_f\right) = h_2(x), \tag{12}$$

System (4) has still uniform relative degree 1 and its input-output linearization is of the form given in (8) with

$$\eta(z, y) = \begin{pmatrix} x_2 \\ -2\zeta x_2 - x_1 + \dfrac{4}{27} \dfrac{y^2}{(1 - x_1 + \rho_f)^2} \end{pmatrix} \tag{13}$$

which is obtained by the control

$$u = \left(L_g h_2\right)^{-1} \left(\tilde{u} - L_f h_2\right), \tag{14}$$

where

$$L_f h_2 = -\frac{3}{2r} x_3 \left((1 - x_1 + \rho_f)(1 - x_1) + r x_2\right), \; L_g h_2 = \frac{1 - x_1 + \rho_f}{r}. \tag{15}$$

Therefore, (11) holds in any equilibrium point and the corresponding zero-dynamics are stable at any position within the air gap if $\rho_f > 2$. Moreover, by choosing the linear control as the one given in (10), it is possible to make the electrical subsystem arbitrarily fast. The performance of the system will then be dominated by that of the zero-dynamics, that is the mechanical subsystem.

4 Charge Feedback Control

Another way for displacement control is to drive the device by prescribed charge. With charge as control variable, we have from (4) that at equilibrium $\bar{x}_3^2 = 3\bar{x}_1$. This implies that every equilibrium be unique in the whole gap range and be stable, hence

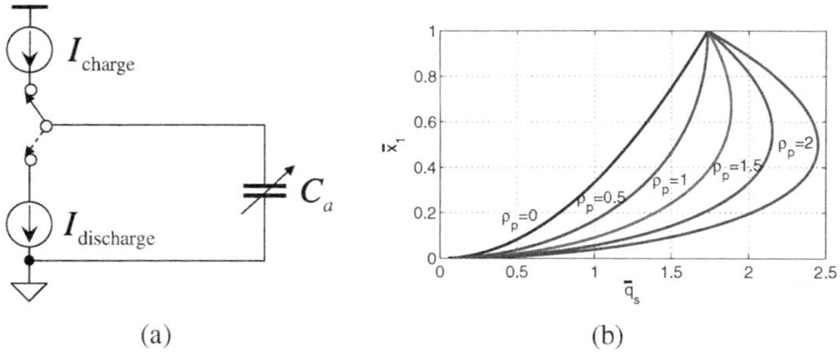

Fig. 3. Charge control: (a) control scheme implemented by ideal current sources; (b) effect of parallel capacitance C_p to actuation curve in equilibrium

the voltage pull-in will not occur [15]. Figure 3(a) illustrates a basic charge control scheme which is one of the diverse implementations proposed in the literature [12]. Compared to voltage control, implementation of current source requires more complicated circuits and bigger footprint.

Charge feedback can be implemented by closed-loop nonlinear control using voltage source [3, 4]. In fact, if the charge is chosen as system output

$$y = q = x_3 = h_3(x), \tag{16}$$

the input-output linearization of the systems is of the form of (8) with

$$\eta(z,y) = \begin{pmatrix} x_2 \\ -2\zeta x_2 - x_1 + \frac{1}{3}y^2 \end{pmatrix}. \tag{17}$$

The linearizing control can be derived directly from (4) which is given by

$$u = \frac{3r}{2}\left(\tilde{u} - \frac{1}{r}x_3(x_1 - 1)\right). \tag{18}$$

Clearly the zero-dynamics verify $\bar{x}_3^2 = 3\bar{x}_1$ in any equilibrium and are stable in the whole operational range. Furthermore, if the electrical-subsystem is rended arbitrarily fast by an appropriate control described in (10), then the performance of the system will be dominated by mechanical-subsystem. Thus, the performance may be poor if the natural damping of the devices is too low or too high. A solution for improving system performance is proposed in [3] by using passivity-based dynamic feedback.

Charge control may also suffer from instability, called charge pull-in, when there are capacitors parallel to the device, represented by C_p [15, 16]. This is due to the fact that the charge for driving the device including parallel capacitance is given by

$$Q_s = Q_a + Q_p = Q_a(1 + C_p/C_a) = Q_a(1 + \rho_p C_0/C_a)$$

where $\rho_p := C_p/C_0$, representing the size of C_p with respect to C_0. Therefore, the relationship between the charge on the device, q_s, and that for driving the actuator, q, is given in normalized coordinates by $q_s = q(1+\rho_p(1-x_1))$ and the equilibrium satisfies

$$\bar{q}_s^2 = 3\bar{x}_1 \left(1+\rho_p\left(1-\bar{x}_1\right)\right)^2. \tag{19}$$

Actuation curves in equilibrium points corresponding to different size of C_p, represented by ρ_p, are shown in Fig 3(b). It can be shown that equilibrium points in the whole physically feasible gap are all stable if and only if $\rho_p < 1/2$ or equivalently $C_p < C_0/2$ [15, 16].

5 Position Feedback Control

Position feedback is the most popular scheme employed in electrostatic MEMS control due to the fact that it allows removing the phenomenon of pull-in. When the position is chosen as system output, $y = x_1$, we obtain by a direct computation:

$$\begin{aligned}
\dot{y} &= x_2,\\
\ddot{y} &= -2\zeta\dot{y} - y + \frac{1}{3}x_3^2\\
y^{(3)} &= -2\zeta\ddot{y} - \dot{y} + \frac{2}{3}x_3\dot{x}_3
\end{aligned}$$

which leads to

$$x_3 = \pm\sqrt{3\left(\ddot{y}+2\zeta\dot{y}+y\right)}, \tag{21a}$$

$$u = \frac{9r}{4x_3}\left(y^{(3)}+2\zeta\ddot{y}+\dot{y}\right)+\frac{3}{2}x_3(1-y) \tag{21b}$$

Since the states as well as the input can be expressed by output and its derivatives, we can conclude that System (4) is differentially flat, except for $x_3 = 0$, with $y = x_1$ as flat output [17, 5]. The system is therefore exactly linearizable by state feedback and a diffeomorphism [17]. This explains why position feedback can completely remove pull-in. Furthermore, as the system is equivalent to $y^{(3)} = \tilde{u}$, where $\tilde{u}$ is a new input, the tracking control can be carried out easily in the framework of flat systems. Note that being treated as tracking problem, the desired behavior can be specified through reference trajectories. Therefore, one can expect a high performance.

Nevertheless, position feedback control suffers from a singularity related to the controllability. It is straightforward to show that due to the quadratic term q^2 (or x_3^2 in (4)) appearing in the mechanical subsystem, the Jacobian linearization of such a system is not controllable at the points where $q = 0$. Consequently, controls derived from feedback linearization with position as output will usually explode as the trajectory of the system is approaching these points. To avoid the singularity due to uncontrollable linearization, direct nonlinear designs should be considered.

Examples of such designs include passivity-based control [4], Lyapunov approach [8], and backstepping on nonlinear term q^2 [18].

6 Dealing with Modelling Errors

To obtain a higher performance, one may need to use more accurate models. However, due to particular physical property of MEMS, accurate modelling may result in very complex mathematical model. For example, the dynamics of micro-actuators are significantly affected by the pressure due to the surrounding air which cannot escape immediately as the moveable plate moves against the fixed one, creating the so-called squeezed film damping force. To reduce this effect, the moveable plate is typically patterned with circular holes allowing the flow of air. In this case, the squeezed film damping coefficient in (2) is given by [19]

$$b(x) = \frac{12\mu A^2}{n\pi G_0^3 x^3}\left(\frac{\lambda}{2} - \frac{\lambda^2}{8} - \frac{\ln\lambda}{4} - \frac{3}{8}\right) \tag{22}$$

where μ is the gas viscosity, n is the total number of holes in the moveable plate, and λ is the fraction occupied by the holes. Obviously, squeezed film damping is a state dependent quantity.

Modelling phenomena such as fringing fields and deformations leads also to complex dynamics described by partial differential equations (see, e.g., [20]). Other sources of uncertainty, such as parasitics, complicate the modelling issue even further. In addition, most of the existing formulations do not precisely represent the behavior of real devices due to dielectrics, imperfect conductors, rough sidewalls, rounded corners, etc.

If the objective is one of modelling the plant for the purpose of controlling it, then the model sought only needs to capture the essential dynamical behavior of the system and yet be simple enough to make the control design less complicated to implement. One of such a model is shown in Fig. 4. This model is based on the simplified one given in Section 2 while presenting diverse unmodeled phenomena by serial and parallel capacitors, denoted by C_s and C_p respectively, which can be expressed in terms of *capacitive influence coefficients*, defined as:

$$\rho_s := C_0/C_s, \quad \rho_p := C_p/C_0. \tag{23}$$

Note that this model captures a quite wide range of physical phenomena. For example, the serial capacitor can characterize the effect of fringing fields and deformations, and the parallel capacitor can represent the parasitics due to thin layer-out of micro structures [21, 20]. Following the same line of reasoning, we can model devices with more complex configuration. The electrical dynamics of the system shown in Fig. 4 are given by [22]

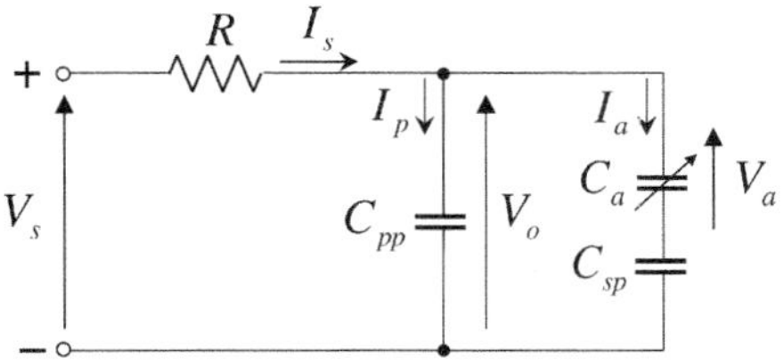

Fig. 4. Equivalent circuit of a generic model. I_a: current through the actuator; I_p: current through C_p; V_p: voltage across C_p.

$$\dot{Q}_a = \frac{1}{R\left(1+\rho_p\rho_s+\rho_p\dfrac{G}{G_0}\right)}\left(V_s - \left(\frac{G}{\varepsilon A}+\rho_s\frac{G_0}{\varepsilon A}+R\rho_p\frac{\dot{G}}{G_0}\right)Q_a\right). \tag{24}$$

The electrical-subsystem in (4c) then becomes:

$$\dot{x}_3 = \frac{1}{r(1+\rho_p(1-x_1)+\rho_p\rho_s)}\left(\frac{2}{3}u-(1-x_1)x_3-\rho_s x_3+r\rho_p x_2 x_3\right). \tag{25}$$

Since the nominal plant is an ideal rigid body, the mechanical subsystem still follows the same dynamics described in (4a-4b). Therefore only the dynamics of electrical subsystem are affected.

Note that ρ_p and ρ_s are both bounded and that under realistic operation conditions the term $(1+\rho_p(1-x_1)+\rho_p\rho_s)$ is positive and bounded. It is also reasonable to suppose that uncertainty of system parameters m, b, k, and r is bounded. Hence, the control of micro-actuator in the presence of modelling errors and parametric uncertainty can be handled in the framework of input-to-state stability (ISS) [23]. With ISS approach, robust control system design amounts to finding a control for which the closed-loop system is stable with respect to disturbances, considered now as inputs to the system. Though ISS-based robust control usually only guarantees boundedness of the closed-loop signals, it can handle arbitrary time-varying uncertainties. This feature is particularly suitable for dealing with state-dependent modelling errors, as those considered in the above discussion. Note that the ISS framework can also deal with other types of uncertainties, such as sensor noise and disturbance rejection, allowing one to address more generic and practical control problems in microsystems.

7 Concluding Remarks

The capability of achieving full gap stable operation regarding different MEMS control strategies is closely related to the architecture of control system, or more specifically the choice of system output. Native voltage control suffers from instability due to pull-in, whereas more judicious choice of system output leads to capacitive feedback, charge feedback, or position feedback which allow a full-gap stabilization. It is also seen that several intuitive control schemes can also be implemented by nonlinear control algorithms, making it possible to enhance the performance of existing systems. However, care must be taken in order to get rid of singularities and performance issues related to diverse practical operation conditions. Finally, the techniques presented in this paper can eventually be extended to devices with more complex structure, such as torsional micro-mirrors and multi-DOF MEMS.

Acknowledgements. The author gratefully acknowledges Professors Jean Lévine, Laurent Praly, Lahcen Saydy, Yves-Alain Peter, and Muthukumaran Packirisamy for their invaluable collaboration in carrying out related research work presented in this paper.

References

1. Kovacs, G.T.A.: Micromachined Transducers Sourcebook. McGraw-Hill, New York (1998)
2. Senturia, S.D.: Microsystem Design. Kluwer Academic Publishers, Norwell (2002)
3. Maithripala, D.H.S., Berg, J.M., Dayawansa, W.P.: ASME Journal of Dynamic Systems, Measurement and Control 127, 443 (2005)
4. Maithripala, D.H.S., Kawade, B.D., Berg, J.M., Dayawansa, W.P.: Int. J. Robust Nonlinear Control 15, 839 (2005)
5. Zhu, G., Lévine, J., Praly, L., Peter, Y.A.: J. Microelectromech. Syst. 15(5), 1165 (2006)
6. Zhu, G., Lévine, J., Praly, L.: Proc. of the 44th IEEE CDC and ECC 2005, Seville, Spain, pp. 7534–7539 (2005)
7. Malisoff, M., Mazenc, F., de Queiroz, M.: Int. J. Robust Nonlinear Control 18(18), 1637 (2008)
8. Zhu, G., Lévine, J., Praly, L.: Proc. of the 446th IEEE Conference on Decision and Control, New Orleans, LA, USA, pp. 2433–2438 (2007)
9. Zhu, G., Penet, J., Saydy, L.: ASME Journal of Dynamic Systems, Measurement and Control 129(6), 786 (2007)
10. Tee, K.P., Ge, S.S., Tay, E.H.: IEEE Trans. Contr. Syst. Technol. 17(2), 304 (2009)
11. Bastani, Y., de Queiroz, M.S.: ASME Journal of Dynamic Systems, Measurement and Control 131(1), 014503.1 (2009)
12. Pont-Nin, J., Rodríguez, A., Castañer, L.: J. Microelectromech. Syst. 11(3), 196 (2002)
13. Isidori, A.: Nonlinear Control Systems, 3rd edn. Springer, London (1995)
14. Seeger, J.I., Crary, S.B.: Tech. Dig. 9th Int. Conf. Solid-State Sensors and Actuators (Transducers 1997), pp. 1133–1136 (1997)
15. Seeger, J., Boser, B.: J. Microelectromech. Syst. 12(5), 656 (2003)
16. Wickramasinghe, I.P.M., Maithripala, D.H.S., Kawade, B.D., Berg, J.M., Dayawansa, W.P.: IEEE Trans. Contr. Syst. Technol. 17(1), 249 (2009)
17. Lévine, J.: Analysis and Control of Nonlinear Systems: A Flatness-Based Approach. Springer, Berlin (2009)
18. Younis, M., Gao, F., de Queiroz, M.S.: Proc. of the 2007 American Control Conference, New York, NY, pp. 3180–3185 (2007)
19. Bergqvist, J., Rudolf, F., Maisano, J., Parodi, F., Rossi, M.: Int. Conf. Solid-State Sensors Actuators Digest, New York, NY, pp. 266–269 (1991)

20. Zhu, G., Saydy, L., Hosseini, M., Chianetta, J.F., Peter, Y.A.: A robustness approach for handling modeling errors in parallel-plate electrostatic MEMS control. J. Microelectromech. Syst. 17(6), 1302 (2008)
21. Chan, E.K., Dutton, R.W.: J. Microelectromech. Syst. 9(3), 321 (2000)
22. Zhu, G., Penet, J., Saydy, L.: ASME Journal of Dynamic Systems, Measurement, and Control 129(6), 786 (2007)
23. Sontag, E.D.: Nonlinear Control in the Year 2000, vol. 2, pp. 443–468. Springer, Berlin (2000)

Part II
Mathematical Tools

Flatness Characterization: Two Approaches

Felix Antritter and Jean Lévine

Abstract. We survey two approaches to flatness necessary and sufficient conditions and compare them on examples.

1 Introduction

In this survey we consider underdetermined implicit systems of the form

$$F(x, \dot{x}) = 0 \tag{1}$$

with $x \in X$, X being an inifnitely differentiable manifold of dimension n, whose tangent bundle is denoted by $\mathrm{T}X$, and $F : \mathrm{T}X \to \mathbb{R}^{n-m}$ regular in the sense that $\mathrm{rk}\,\frac{\partial F}{\partial \dot{x}} = n - m$ in a suitable open dense subset of $\mathrm{T}X$. Differential flatness, or more shortly, flatness was introduced in 1992 [20, 11]. In the setting of implicit control systems it may be roughly described as follows: there exists a smooth mapping $x = \varphi(y, \dot{y}, \ldots, y^{(r)})$ with $y = (y_1, \ldots, y_m)^T$ of dimension m, $r = (r_1, \ldots, r_m)^T \in \mathbb{N}^m$, such that

$$F(\varphi(y, \dot{y}, \ldots, y^{(r)}), \dot{\varphi}(y, \dot{y}, \ldots, y^{(r+1)})) \equiv 0 \tag{2}$$

with φ invertible in the sense that there exists a locally defined smooth mapping ψ and a multi-index s such that $y = \psi(x, \dot{x}, \ldots, x^{(s)})$.

The vector y is called a *flat output*.

Felix Antritter
Automatisierungs- und Regelungstechnik, Universität der Bundeswehr München, Werner-Heisenberg-Weg 37, DE-85579 Neubiberg, Germany
e-mail: `felix.antritter@unibw.de`

Jean Lévine
Mines ParisTech, CAS- Centre Automatique et Systèmes, Mathématiques et Systèmes, 35, rue Saint-Honoré, 77305, Fontainebleau, France
e-mail: `jean.levine@mines-paristech.fr`

J. Lévine & P. Müllhaupt (Eds.): Adv. in Theory of Control, Signals, LNCIS 407, pp. 127–139.
springerlink.com © Springer-Verlag Berlin Heidelberg 2010

This concept has inspired an important literature. See [10, 21, 19, 26, 27, 31] for surveys on flatness and its applications. Various formalisms have been introduced: finite dimensional differential geometric approaches [4, 14, 30], [32, 28], differential algebra and related approaches [12, 3, 15], infinite dimensional differential geometry of jets and prolongations [13, 33, 19, 6, 7, 23], [22, 24], which is adopted here. The interested reader may refer to [1, 13, 16], [19, 23, 34] for more details.

The first part of the paper recalls the mathematical setting. In Section 3 the approch introduced in [19, 2] for the characterization of differentially flat systems is recalled. Then, in Section 4, we introduce a novel characterization using the so-called Generalized Euler-Lagrange Operator. We conclude the paper with examples.

2 Implicit Control Systems on Manifolds of Jets of Infinite Order

Given an infinitely differentiable manifold X of dimension n, we denote its tangent space at $x \in X$ by $\mathrm{T}_x X$, and its tangent bundle by $\mathrm{T}X$. Let F be a meromorphic function from $\mathrm{T}X$ to $\mathbb{R}^{n-m}$. We consider an underdetermined implicit system of the form (1) regular in the sense that $\mathrm{rk}\, \frac{\partial F}{\partial \dot{x}} = n - m$ in a suitable open dense subset of $\mathrm{T}X$.

Following [17, 18], we consider the infinite dimensional manifold $\mathfrak{X}$ defined by $\mathfrak{X} \stackrel{\text{def}}{=} X \times \mathbb{R}^n_\infty \stackrel{\text{def}}{=} X \times \mathbb{R}^n \times \mathbb{R}^n \times \ldots$, made of an infinite (but countable) number of copies of $\mathbb{R}^n$, with the global infinite set of coordinates[1] $\overline{x} = \left(x, \dot{x}, \ldots, x^{(k)}, \ldots,\right)$, endowed with the product topology. Recall that, in this topology, a function φ from $\mathfrak{X}$ to $\mathbb{R}$ is *continuous* (resp. *differentiable*) if φ depends only on a finite (but otherwise arbitrary) number of variables and is continuous (resp. differentiable) with respect to these variables. C^∞ or analytic or meromorphic functions from $\mathfrak{X}$ to $\mathbb{R}$ are then defined as in the usual finite dimensional case since they only depend on a finite number of variables. We endow $\mathfrak{X}$ with the so-called trivial Cartan field ([16, 34]) $\tau_{\mathfrak{X}} = \sum_{i=1}^n \sum_{j \geq 0} x_i^{(j+1)} \frac{\partial}{\partial x_i^{(j)}}$. We also denote by $L_{\tau_{\mathfrak{X}}} \gamma = \sum_{i=1}^n \sum_{j \geq 0} x_i^{(j+1)} \frac{\partial \gamma}{\partial x_i^{(j)}} = \frac{d\gamma}{dt}$ the Lie derivative of a differentiable function γ along $\tau_{\mathfrak{X}}$ and $L^k_{\tau_{\mathfrak{X}}} \gamma$ its kth iterate. Since $\frac{d}{dt} x_i^{(j)} \stackrel{\text{def}}{=} \dot{x}_i^{(j)} = x_i^{(j+1)}$, the Cartan field acts on coordinates as a shift to the right. $\mathfrak{X}$ is thus called *manifold of jets of infinite order.*

A regular implicit control system is defined as a triple $(\mathfrak{X}, \tau_{\mathfrak{X}}, F)$ *with* $\mathfrak{X} = X \times \mathbb{R}^n_\infty$, $\tau_{\mathfrak{X}}$ *its associated trivial Cartan field, and* F *meromorphic from* $\mathrm{T}X$ *to* $\mathbb{R}^{n-m}$) *satisfying* $\mathrm{rk}\, \frac{\partial F}{\partial \dot{x}} = n - m$ *in a suitable open subset of* $\mathrm{T}X$.

We next consider the cotangent space $T^*_{\overline{x}}\mathfrak{X}$ with $dx_i^{(j)}$, $i = 1, \ldots, n$, $j \geq 0$ as basis, dual to the $\frac{\partial}{\partial x_i^{(j)}}$'s. 1-forms on $\mathfrak{X}$ are then defined in the usual way.

[1] From now on, $\overline{x}$ $\overline{y}$, ...stand for the sequences of jets of infinite order of x, y,...

The set of 1-forms is noted $\Lambda^1(\mathfrak{X})$. We also denote by $\Lambda^p(\mathfrak{X})$ the module of all the p-forms on $\mathfrak{X}$.

2.1 Flatness

We recall the following definitions and result [17, 18, 19]:

Given two regular implicit control systems $(\mathfrak{X}, \tau_{\mathfrak{X}}, F)$, with $\mathfrak{X} = X \times \mathbb{R}^n_\infty$, $\dim X = n$ and $\mathrm{rk}\, \frac{\partial F}{\partial \dot{x}} = n - m$, and $(\mathfrak{Y}, \tau_{\mathfrak{Y}}, G)$, with $\mathfrak{Y} = Y \times \mathbb{R}^p_\infty$, $\dim Y = p$, $\tau_{\mathfrak{Y}}$ its trivial Cartan field, and $\mathrm{rk}\, \frac{\partial G}{\partial \dot{y}} = p - q$, we set $\mathfrak{X}_0 = \{\overline{x} \in \mathfrak{X} | L^k_{\tau_{\mathfrak{X}}} F(\overline{x}) = 0,\ \forall k \geq 0\}$ and $\mathfrak{Y}_0 = \{\overline{y} \in \mathfrak{Y} | L^k_{\tau_{\mathfrak{Y}}} G(\overline{y}) = 0,\ \forall k \geq 0\}$. They are endowed with the topologies and differentiable structures induced by $\mathfrak{X}$ and $\mathfrak{Y}$ respectively.

Definition 1. *The control systems* $(\mathfrak{X}, \tau_{\mathfrak{X}}, F)$ *and* $(\mathfrak{Y}, \tau_{\mathfrak{Y}}, G)$ *are said* locally Lie-Bäcklund equivalent *(or shortly L-B equivalent) in a neighbourhood* $\mathcal{X}_0 \times \mathcal{Y}_0$ *of the pair* $(\overline{x}_0, \overline{y}_0) \in \mathfrak{X}_0 \times \mathfrak{Y}_0$ *if and only if*

(i) there exists a one-to-one meromorphic mapping $\Phi = (\varphi, \dot{\varphi}, \ldots)$ *from* $\mathcal{Y}_0$ *to* $\mathcal{X}_0$ *satisfying* $\Phi(\overline{y}_0) = \overline{x}_0$ *and such that* $\Phi_* \tau_{\mathfrak{Y}} = \tau_{\mathfrak{X}}$*;*
(ii) there exists Ψ *one-to-one and meromorphic from* $\mathcal{X}_0$ *to* $\mathcal{Y}_0$*, with* $\Psi = (\psi, \dot{\psi}, \ldots)$*, such that* $\Psi(\overline{x}_0) = \overline{y}_0$ *and* $\Psi_* \tau_{\mathfrak{X}} = \tau_{\mathfrak{Y}}$*.*

The mappings Φ *and* Ψ *are called* mutually inverse Lie-Bäcklund isomorphisms *at* $(\overline{x}_0, \overline{y}_0)$*.*

Definition 2. *The implicit system* $(\mathfrak{X}, \tau_{\mathfrak{X}}, F)$ *is* locally flat *in a neighborhood of* $(\overline{x}_0, \overline{y}_0) \in \mathfrak{X}_0 \times \mathbb{R}^m_\infty$ *if and only if it is locally L-B equivalent around* $(\overline{x}_0, \overline{y}_0)$ *to the trivial implicit system* $(\mathbb{R}^m_\infty, \tau_{\mathbb{R}^m_\infty}, 0)$*. In this case, the mutually inverse L-B isomorphisms* Φ *and* Ψ *are called* inverse trivializations.

Theorem 1. *The system* $(\mathfrak{X}, \tau_{\mathfrak{X}}, F)$ *is locally flat at* $(\overline{x}_0, \overline{y}_0) \in \mathfrak{X}_0 \times \mathbb{R}^m_\infty$ *if and only if there exists a local meromorphic invertible mapping* Φ *from* $\mathbb{R}^m_\infty$ *to* $\mathfrak{X}_0$*, with meromorphic inverse, satisfying* $\Phi(\overline{y}_0) = \overline{x}_0$*, and such that*[2]

$$\Phi^* dF = 0. \tag{3}$$

3 Necessary and Sufficient Conditions: Generalized Moving Frame Structure Equations

3.1 Algebraic Characterization of the Differential of a Trivialization

Consider the following matrix, polynomial with respect to the differential operator $\frac{d}{dt}$ (we use indifferently $\frac{d}{dt}$ for $L_{\tau_{\mathfrak{X}}}$ or $L_{\tau_{\mathbb{R}^m_\infty}}$, the context being unambiguous):

[2] Note that if Φ is a meromorphic mapping from $\mathfrak{Y}$ to $\mathfrak{X}$, the (backward) image by Φ of a 1-form is defined in the same way as in the finite dimensional context.

$$P(F) = \frac{\partial F}{\partial x} + \frac{\partial F}{\partial \dot{x}} \frac{d}{dt}, \quad P(\varphi) = \sum_{j \geq 0} \frac{\partial \varphi}{\partial y^{(j)}} \frac{d^j}{dt^j} \tag{4}$$

with $P(F)$ (resp. $P(\varphi)$) of size $(n-m) \times n$ (resp. $n \times m$).

Equation (3)) reads:

$$\Phi^* dF = P(F)P(\varphi)dy = 0. \tag{5}$$

Clearly, the entries of the matrices in (4) are polynomials in the differential operator $\frac{d}{dt}$ with meromorphic coefficients from $\mathfrak{X}$ to $\mathbb{R}$. We denote by $\mathfrak{K}$ the field of meromorphic functions from $\mathfrak{X}$ to $\mathbb{R}$ and by $\mathfrak{K}[\frac{d}{dt}]$ the (non-commutative) principal ideal ring of polynomials in $\frac{d}{dt}$ with coefficients in $\mathfrak{K}$. For $r, s \in \mathbb{N}$, let us denote by $\mathcal{M}_{r,s}[\frac{d}{dt}]$ the module of $r \times s$ matrices over $\mathfrak{K}[\frac{d}{dt}]$ (see e.g. [8]). Matrices whose inverse belong to $\mathcal{M}_{r,r}[\frac{d}{dt}]$ are called *unimodular matrices* . They form a multiplicative group denoted by $\mathcal{U}_r[\frac{d}{dt}]$.

Every matrix $M \in \mathcal{M}_{r,s}[\frac{d}{dt}]$ admits a *Smith decomposition* (or diagonal reduction)

$$VMU = (\Delta, 0_{r,s-r}) \text{ if } r \leq s, \text{ and} \begin{pmatrix} \Delta \\ 0_{r-s,s} \end{pmatrix} \text{ if } s \leq r \tag{6}$$

with $V \in \mathcal{U}_r[\frac{d}{dt}]$ and $U \in \mathcal{U}_s[\frac{d}{dt}]$ and Δ diagonal (see e.g. [8]). U and V are indeed non unique. We say that $U \in \mathsf{R}-\mathsf{Smith}\,(M)$ and $V \in \mathsf{L}-\mathsf{Smith}\,(M)$.

A matrix $M \in \mathcal{M}_{r,s}[\frac{d}{dt}]$ is said *hyper-regular* if and only if its Smith decomposition leads to $\Delta = I$. An interpretation of this property in terms of controllability in the sense of [9], may be found in [18].

From now on, we assume that $P(F)$ is hyper-regular in a neighborhood of $\overline{x}_0$. In place of (5), we first solve the matrix equation:

$$P(F)\Theta = 0 \tag{7}$$

where $\Theta \in \mathcal{M}_{n,m}[\frac{d}{dt}]$ is not supposed to be of the form $P(\varphi)$. It may be verified that matrices $\Theta \in \mathcal{M}_{n,m}[\frac{d}{dt}]$ satisfying (7) have the structure

$$\Theta = U \begin{pmatrix} 0_{n-m,m} \\ I_m \end{pmatrix} W \tag{8}$$

with $U \in \mathsf{R}-\mathsf{Smith}\,(P(F))$ and $W \in \mathcal{U}_m[\frac{d}{dt}]$ arbitrary. Clearly Θ is itself hyper-regular and admits the Smith decomposition

$$Q\Theta Z = QU \begin{pmatrix} 0_{n-m,m} \\ I_m \end{pmatrix} WZ = Q\hat{U}R = \begin{pmatrix} I_m \\ 0_{n-m,m} \end{pmatrix} \tag{9}$$

with $Q \in \mathcal{U}_n[\frac{d}{dt}]$, $Z \in \mathcal{U}_m[\frac{d}{dt}]$, $R = WZ$ and $\hat{U} = U \begin{pmatrix} 0_{n-m,m} \\ I_m \end{pmatrix}$.

3.2 Integrability

We denote by ω the m-dimensional vector 1-form defined by

$$\omega(\overline{x}) = \begin{pmatrix} \omega_1(\overline{x}) \\ \vdots \\ \omega_m(\overline{x}) \end{pmatrix} = (I_m, 0_{m,n-m})\, Q(\overline{x}) dx_{|\mathfrak{X}_0} \tag{10}$$

with Q given by (9), the restriction to $\mathfrak{X}_0$ meaning that $\overline{x} \in \mathfrak{X}_0$ satisfies $L^k_{\tau_\mathfrak{X}} F = 0$ for all k and that the $dx_j^{(k)}$ are such that $dL^k_{\tau_\mathfrak{X}} F = 0$ in $\mathfrak{X}_0$ for all k. Since Q is hyper-regular, the forms $\omega_1, \ldots, \omega_m$ are independent by construction.

Theorem 2. *A necessary and sufficient condition for system (1) to be locally flat around* $(\overline{x}_0, \overline{y}_0)$ *is that there exist* $U \in \mathsf{R}-\mathsf{Smith}\,(P(F))$, $Q \in \mathsf{L}-\mathsf{Smith}\left(\hat{U}\right)$, *with* $\hat{U}$ *given by (9) and a matrix* $M \in \mathfrak{U}_m[\frac{d}{dt}]$ *such that* $d(M\tau) = 0$.

We denote by $(\Lambda^p(\mathfrak{X}))^m$ the space of m-dimensional vector p-forms on $\mathfrak{X}$, by $(\Lambda(\mathfrak{X}))^m$ the space of m-dimensional vector forms of arbitrary degree on $\mathfrak{X}$, and by $\mathcal{L}_q\left((\Lambda(\mathfrak{X}))^m\right) = \bigcup_{p\geq 1} \mathcal{L}\left((\Lambda^p(\mathfrak{X}))^m, (\Lambda^{p+q}(\mathfrak{X}))^m\right)$ the space of linear operators from $(\Lambda^p(\mathfrak{X}))^m$ to $(\Lambda^{p+q}(\mathfrak{X}))^m$ for all $p \geq 1$, where $\mathcal{L}(\mathcal{P}, \mathcal{Q})$ denotes the set of linear mappings from a given space $\mathcal{P}$ to a given space $\mathcal{Q}$.

In order to develop the expression $d(\mu\kappa)$ for $\mu \in \mathcal{L}_q\left((\Lambda(\mathfrak{X}))^m\right)$ and for all $\kappa \in (\Lambda^p(\mathfrak{X}))^m$ and all $p \geq 1$, we define the operator $\mathfrak{d}$ by:

$$\mathfrak{d}(\mu)\,\kappa = d(\mu\,\kappa) - (-1)^q \mu\, d\kappa. \tag{11}$$

Note that (11) uniquely defines $\mathfrak{d}(\mu)$ as an element of $\mathcal{L}_{q+1}\left((\Lambda(\mathfrak{X}))^m\right)$.

Theorem 3. *The system* $(\mathfrak{X}, \tau_\mathfrak{X}, F)$ *is locally flat iff there locally exists* $\mu \in \mathcal{L}_1\left((\Lambda(\mathfrak{X}))^m\right)$, *and a matrix* $M \in \mathfrak{U}_m[\frac{d}{dt}]$ *such that*

$$d\omega = \mu\,\omega, \qquad \mathfrak{d}(\mu) = \mu^2, \qquad \mathfrak{d}(M) = -M\mu. \tag{12}$$

with the notation $\mu^2 = \mu\mu$ *and where* ω *is defined by (10). In addition, if (12) holds true, a flat output* y *is obtained by integration of* $dy = M\omega$.

Remark 1. *Note that the two first conditions of (12) are comparable to conditions (A) and (B) of [6, 7]. However, the last condition of (12) is different from condition (C) of [6, 7] and is easier to check.*

Note also that conditions (12) may be seen as a generalization in the framework of manifolds of jets of infinite order of Cartan's well-known moving frame structure equations (see e.g. [5]).

3.3 A Sequential Procedure

We start with $P(F)$ hyper-regular and compute the vector 1-form ω defined by (10).

1. We identify the operator μ such that $d\omega = \mu\omega$ componentwise. It is proven in [19] that such μ always exists.
2. Among the possible μ's, only those satisfying $\mathfrak{d}(\mu) = \mu^2$ are kept. It is shown in [19] that such μ always exists.
3. We then identify M such that $\mathfrak{d}(M) = -M\mu$ componentwise.
4. If, among such M's, there is a unimodular one, the system is flat and a flat output is obtained by integration of $dy = M\omega$. Otherwise the system is not flat.

More details and examples may be found in [18, 19].

4 Necessary and Sufficient Conditions Using the Generalized Euler-Lagrange Operator

Another way of analysing (3) consists in characterizing the change of coordinates corresponding to the mapping Φ in (3). More precisely (3) reads

$$\sum_{j=1}^{m}\sum_{k=0}^{r_j}\left(\frac{\partial F}{\partial x}\frac{\partial \varphi}{\partial y_j^{(k)}}dy_j^{(k)} + \frac{\partial F}{\partial \dot{x}}\frac{d}{dt}\left(\frac{\partial \varphi}{\partial y_j^{(k)}}\right)dy_j^{(k)} + \frac{\partial F}{\partial \dot{x}}\frac{\partial \varphi}{\partial y_j^{(k)}}dy_j^{(k+1)}\right) = 0 \tag{13}$$

Since the one forms $dy_1, \ldots, dy_1^{(r_1)}, \ldots, dy_m, \ldots, dy_m^{(r_m)}$ are independent by assumption, (13) yields, for every $j = 1, \ldots, m$,

$$\begin{cases} \dfrac{\partial F}{\partial \dot{x}}\dfrac{\partial \varphi}{\partial y_j^{(r_j)}} = 0 \\ \dfrac{\partial F}{\partial x}\dfrac{\partial \varphi}{\partial y_j^{(k)}} + \dfrac{\partial F}{\partial \dot{x}}\dfrac{d}{dt}\left(\dfrac{\partial \varphi}{\partial y_j^{(k)}}\right) + \dfrac{\partial F}{\partial \dot{x}}\dfrac{\partial \varphi}{\partial y_j^{(k-1)}} = 0, \quad \forall k = 1, \ldots, r_j \\ \dfrac{\partial F}{\partial x}\dfrac{\partial \varphi}{\partial y_j} + \dfrac{\partial F}{\partial \dot{x}}\dfrac{d}{dt}\left(\dfrac{\partial \varphi}{\partial y_j}\right) = 0 \end{cases} \tag{14}$$

The Generalized Euler-Lagrange operator $\mathcal{E}_F$ associated to F is defined by

$$\mathcal{E}_F = \frac{\partial F}{\partial x} - \frac{d}{dt}\left(\frac{\partial F}{\partial \dot{x}}\right) \tag{15}$$

In the case $n - m = 1$, it is well-known that the curves that extremize the cost function $J = \int_0^T F(x, \dot{x})dt$ are those satisfying the Euler-Lagrange equation $\mathcal{E}_F = 0$, which justifies our terminology.

Using (15) and elementary calculus, (14) yields:

Theorem 4. *A necessary and sufficient condition for (1) to be diffferentially flat is that there exist* $(r_1, \ldots, r_m)$ *with* $\sum_{i=1}^m r_i + m \geq n$ *and a solution* φ *of the following triangular system of PDEs in an open dense subset of* $\mathfrak{X}$

$$\begin{cases} \frac{\partial F}{\partial \dot{x}} \frac{\partial \varphi}{\partial y_j^{(r_j)}} = 0 \\ \frac{\partial F}{\partial \dot{x}} \frac{\partial \varphi}{\partial y_j^{(l)}} = \sum_{k=0}^{r_j - l - 1} (-1)^{k+1} \frac{d^k}{dt^k} \left(\mathcal{E}_F \frac{\partial \varphi}{\partial y_j^{(l+k+1)}} \right), \quad \forall l = 0, \ldots, r_j - 1 \ , \\ 0 = \sum_{k=0}^{r_j} (-1)^k \frac{d^k}{dt^k} \left(\mathcal{E}_F \frac{\partial \varphi}{\partial y_j^{(k)}} \right) \end{cases} \tag{16}$$

satisfying $d\varphi_1 \wedge \ldots \wedge d\varphi_n \neq 0$.

Remark 2. *If there exists a coordinate transformation* φ *that satisfies the conditions of Theorem 4 with given* $r_1, \ldots, r_m$*, meaning that the system is flat, then* $g_j = \sum_{i=1}^n \frac{\partial \varphi_i}{\partial y_j^{(r_j)}} \frac{\partial}{\partial \dot{x}_i}$ *, if non zero, defines a ruled direction [32, 25, 19].*

5 Examples

5.1 An Academic Example: Generalized Moving Frame Approach

We consider the 3-dimensional system with 2 inputs:

$$\dot{x}_1 = u_1, \quad \dot{x}_2 = u_2, \quad \dot{x}_3 = \sin\left(\frac{u_1}{u_2}\right) \tag{17}$$

or, in implicit form:

$$F(x_1, x_2, x_3, \dot{x}_1, \dot{x}_2, \dot{x}_3) \triangleq \dot{x}_3 - \sin\left(\frac{\dot{x}_1}{\dot{x}_2}\right) = 0. \tag{18}$$

It is readily seen that $P(F) = \left[-\cos(\frac{\dot{x}_1}{\dot{x}_2})\dot{x}_2^{-1}\frac{d}{dt} \middle| \dot{x}_1 \cos(\frac{\dot{x}_1}{\dot{x}_2})\dot{x}_2^{-2}\frac{d}{dt} \middle| \frac{d}{dt} \right]$ and that $VP(F)U = (1\ 0\ 0)$ with

$$V = 1, \quad U = \left(\begin{array}{c|c|c} \frac{\dot{x}_1}{a\dot{x}_2} & 1 + \frac{\dot{x}_1}{a(\dot{x}_2)^2} \cos\left(\frac{\dot{x}_1}{\dot{x}_2}\right) \frac{d}{dt} & \frac{\dot{x}_1}{a\dot{x}_2} \frac{d}{dt} \\ \frac{1}{a} & \frac{1}{a\dot{x}_2} \cos\left(\frac{\dot{x}_1}{\dot{x}_2}\right) \frac{d}{dt} & -\frac{1}{a} \frac{d}{dt} \\ 0 & 0 & 1 \end{array} \right) \tag{19}$$

where $a = -\frac{1}{\dot{x}_2}\cos\left(\frac{\dot{x}_1}{\dot{x}_2}\right)\left(\frac{\ddot{x}_1\dot{x}_2-\dot{x}_1\ddot{x}_2}{(\dot{x}_2)^2}\right)$. Then, $Q\hat{U}R = \begin{pmatrix}1 & 0\\ 0 & 1\\ 0 & 0\end{pmatrix}$ is computed with

$$Q = \left(\begin{array}{c|c|c} 1 & -\frac{\dot{x}_1}{\dot{x}_2} & 0 \\ 0 & 0 & 1 \\ -\frac{1}{a\dot{x}_2}\cos\left(\frac{\dot{x}_1}{\dot{x}_2}\right)\frac{d}{dt} & \frac{\dot{x}_1}{a(\dot{x}_2)^2}\cos\left(\frac{\dot{x}_1}{\dot{x}_2}\right)\frac{d}{dt} & \frac{1}{a}\frac{d}{dt}\end{array}\right), \quad R = \begin{pmatrix}1 & 0\\ 0 & 1\end{pmatrix} \tag{20}$$

So, $(\omega_1\ \omega_2)^T = \begin{pmatrix}1 & 0 & 0\\ 0 & 0 & 1\end{pmatrix} Qdx = \left(dx_1 - \frac{\dot{x}_1}{\dot{x}_2}dx_2\ \ dx_3\right)^T$ and $d\omega = \left(\frac{1}{\sqrt{1-(\dot{x}_3)^2}}dx_2 \wedge dx_3\ \ 0\right)^T$. According to section 3.3, step 1,

$$\mu = \begin{pmatrix} 0 & \left(-\frac{\dot{x}_3}{(1-(\dot{x}_3)^2)^{\frac{3}{2}}}dx_2 \wedge d\dot{x}_3 + \eta d\dot{x}_3\right)\wedge \frac{d}{dt} \\ 0 & 0 \end{pmatrix}. \tag{21}$$

Step 2 yields $\eta = \frac{x_2\dot{x}_3}{(1-\dot{x}_3)^{\frac{3}{2}}} + \sigma(\dot{x}_3)$. For step 3 we set $M = \begin{pmatrix}1 & m_{12}\frac{d}{dt}\\ 0 & 1\end{pmatrix}$ which yields $m_{12} = -\left(\frac{x_2}{\sqrt{1-(\dot{x}_3)^2}} + \sigma_1(\dot{x}_3)\right)$ with σ_1 a primitive of σ. Thus, $d(M\omega) = 0$ and setting $(dy_1\ dy_2)^T = M\omega$, one obtains

$$y_1 = x_1 - \frac{\dot{x}_1}{\dot{x}_2}x_2 + \sigma_2(\dot{x}_3), \quad y_2 = x_3 \tag{22}$$

where $\sigma_2(\dot{x}_3)$ is an arbitrary meromorphic function (a primitive of σ_1). By inversion of (22) we get

$$\begin{aligned} x_1 &= y_1 - \arcsin(\dot{y}_2)\frac{\sqrt{1-(\dot{y}_2)^2}}{\ddot{y}_2}(\dot{y}_1 - \sigma_1(\dot{y}_2)\ddot{y}_2) - \sigma_2(\dot{y}_2) \\ x_2 &= -\frac{\sqrt{1-(\dot{y}_2)^2}}{\ddot{y}_2}(\dot{y}_1 - \sigma_1(\dot{y}_2)\ddot{y}_2) \\ x_3 &= y_2 \end{aligned} \tag{23}$$

5.2 Academic Example: Euler-Lagrange Operator

We consider once more the example (18). We have

$$\frac{\partial F}{\partial \dot{x}} = \left(-\dot{x}_2^{-1}\cos\left(\frac{\dot{x}_1}{\dot{x}_2}\right), \dot{x}_1\dot{x}_2^{-2}\cos\left(\frac{\dot{x}_1}{\dot{x}_2}\right), 1\right), \ \mathcal{E}_F = (\eta_1, \eta_2, 0) \tag{24}$$

with $\eta_1 = -\frac{\ddot{x}_2}{\dot{x}_2^2}\cos\left(\frac{\dot{x}_1}{\dot{x}_2}\right) - \frac{\ddot{x}_1\dot{x}_2 - \dot{x}_1\ddot{x}_2}{\dot{x}_2^3}\sin\left(\frac{\dot{x}_1}{\dot{x}_2}\right)$ and
$\eta_2 = -\frac{\ddot{x}_1\dot{x}_2 - 2\dot{x}_1\ddot{x}_2}{\dot{x}_2^3}\cos\left(\frac{\dot{x}_1}{\dot{x}_2}\right) + \frac{\dot{x}_1(\ddot{x}_1\dot{x}_2 - \dot{x}_1\ddot{x}_2)}{\dot{x}_2^4}\sin\left(\frac{\dot{x}_1}{\dot{x}_2}\right)$.

The first two equations of (16), with $r_1 = r_2 = 2$, read

$$-\frac{1}{\dot{x}_2}\cos\left(\frac{\dot{x}_1}{\dot{x}_2}\right)\left(\frac{\partial\varphi_1}{\partial\dot{y}_j} - \frac{\dot{x}_1}{\dot{x}_2}\frac{\partial\varphi_2}{\partial\dot{y}_j}\right) + \frac{\partial\varphi_3}{\partial\dot{y}_j} = 0, \quad j = 1,2 \tag{25}$$

If we assume that $\frac{\partial\varphi_3}{\partial\dot{y}_j} = \frac{\partial\varphi_3}{\partial\ddot{y}_j} = 0, \quad j = 1,2$ and introduce the variable

$$\psi = \frac{\dot{x}_1}{\dot{x}_2} \tag{26}$$

with $\frac{\partial}{\partial\ddot{y}}\psi = 0$ we obtain from (25)

$$\frac{\partial\varphi_1}{\partial\ddot{y}_j} - \psi\frac{\partial\varphi_2}{\partial\ddot{y}_j} = \frac{\partial}{\partial\ddot{y}_j}(\varphi_1 - \psi\varphi_2) = 0, \quad j = 1,2$$

Setting $\kappa(y,\dot{y}) = \varphi_1 - \psi\varphi_2$, we get

$$\dot{\kappa} = \dot{\varphi}_1 - \psi\dot{\varphi}_2 - \dot{\psi}\varphi_2 = -\dot{\psi}\varphi_2 \tag{27}$$

Using the definition of κ and (27) we obtain:

$$\varphi_1 = \kappa - \frac{\dot{\kappa}\sqrt{1-\dot{\varphi}_3}}{\ddot{\varphi}_3}\arcsin(\dot{\varphi}_3), \quad \varphi_2 = -\frac{\dot{\kappa}}{\ddot{\varphi}_3}\sqrt{1-\dot{\varphi}_3}, \quad \varphi_3 = \varphi_3(y) \tag{28}$$

Choosing $\varphi_3 = y_2, \quad \kappa = y_1$, we arrive at the invertible transformation

$$x_1 = \varphi_1 = y_1 - \frac{\dot{y}_1}{\ddot{y}_2}\sqrt{1-\dot{y}_2^2}\arcsin(\dot{y}_2),\ x_2 = \varphi_2 = -\frac{\dot{y}_1}{\ddot{y}_2}\sqrt{1-\dot{y}_2^2},$$

with $x_3 = \varphi_3 = y_2$, which gives the same formula as (23) with $\sigma_1 = \sigma_2 = 0$. Hence (y_1, y_2) is indeed a flat output, which implies that the remaining equations of (16) are satisfied.

5.3 An Example Proposed by P. Rouchon

Consider the implicit control system

$$F(x,\dot{x}) = \dot{x}_1\dot{x}_3 - (\dot{x}_2)^2 = 0. \tag{29}$$

We thus have $\frac{\partial F}{\partial x} = (0 \quad 0 \quad 0)$, $\frac{\partial F}{\partial \dot{x}} = (\dot{x}_3 \quad -2\dot{x}_2 \quad \dot{x}_1)$ and

$$\mathcal{E}_F = \frac{\partial F}{\partial x} - \frac{d}{dt}\left(\frac{\partial F}{\partial \dot{x}}\right) = -\frac{d}{dt}\left(\frac{\partial F}{\partial \dot{x}}\right) = (-\ddot{x}_3 \quad 2\ddot{x}_2 \quad -\ddot{x}_1).$$

The lowest possible choice of (r_1, r_2) in Theorem 4 is $r_1 = r_2 = 1$. However, there is no solution of (16) for these values, and we choose $r_1 = r_2 = 2$. The two first equations of (16) read

$$\dot{\varphi}_3 \frac{\partial \varphi_1}{\partial \ddot{y}_j} - 2\dot{\varphi}_2 \frac{\partial \varphi_2}{\partial \ddot{y}_j} + \dot{\varphi}_1 \frac{\partial \varphi_3}{\partial \ddot{y}_j} = 0, \quad j = 1, 2 \tag{30}$$

We divide (30) by $\dot{\varphi}_3$ to obtain

$$\frac{\partial \varphi_1}{\partial \ddot{y}_j} - 2\psi \frac{\partial \varphi_2}{\partial \ddot{y}_j} + \psi^2 \frac{\partial \varphi_3}{\partial \ddot{y}_j} = 0, \quad j = 1, 2 \tag{31}$$

where, taking account of the system equation (29),

$$\psi = \frac{\dot{\varphi}_2}{\dot{\varphi}_3} = \sqrt{\frac{\dot{\varphi}_1}{\dot{\varphi}_3}}. \tag{32}$$

If we assume that ψ doesn't depend on $\ddot{y}_1$ and $\ddot{y}_2$, equation (31) reads $\frac{\partial}{\partial \ddot{y}_j}\left(\varphi_1 - 2\psi\varphi_2 + \psi^2\varphi_3\right) = 0$, for $j = 1, 2$. In other words, there exists a function κ satisfying $\frac{\partial \kappa}{\partial \ddot{y}_j} = 0$ for $j = 1, 2$, such that

$$\varphi_1 - 2\psi\varphi_2 + \psi^2\varphi_3 = \kappa \tag{33}$$

Differentiating the latter relation with respect to t, and taking into account the relation $\dot{\varphi}_1 - 2\psi\dot{\varphi}_2 + \psi^2\dot{\varphi}_3 = 0$ obtained from (29) and (32), we get

$$\varphi_2 - \psi\varphi_3 = -\frac{\dot{\kappa}}{2\dot{\psi}}. \tag{34}$$

We again differentiate the latter relation with respect to t to obtain

$$\varphi_3 = \frac{\ddot{\kappa}\dot{\psi} - \dot{\kappa}\ddot{\psi}}{2\dot{\psi}^3} \tag{35}$$

thanks to $\dot{\varphi}_2 - \psi\dot{\varphi}_3 = 0$ from (32). Thus, solving the system (33)–(35), we immediately obtain

$$\begin{aligned} \varphi_1 &= \kappa - \psi\frac{\dot{\kappa}}{\dot{\psi}} + \psi^2\left(\frac{\ddot{\kappa}\dot{\psi} - \dot{\kappa}\ddot{\psi}}{2\dot{\psi}^3}\right) \\ \varphi_2 &= -\frac{\dot{\kappa}}{2\dot{\psi}} + \psi\left(\frac{\ddot{\kappa}\dot{\psi} - \dot{\kappa}\ddot{\psi}}{2\dot{\psi}^3}\right) \\ \varphi_3 &= \frac{\ddot{\kappa}\dot{\psi} - \dot{\kappa}\ddot{\psi}}{2\dot{\psi}^3} \end{aligned} \tag{36}$$

where κ and ψ are arbitrary functions of $y_1, y_2, \dot{y}_1, \dot{y}_2$.

Note that choosing $\kappa = y_1$ and $\psi = y_2$ yields, after inversion of (36) with (32):

$$y_1 = x_1 - 2x_2\frac{\dot{x}_2}{\dot{x}_3} + x_3\frac{\dot{x}_1}{\dot{x}_3}, \quad y_2 = \frac{\dot{x}_2}{\dot{x}_3},$$

which is similar to the solution obtained by F. Ollivier[3].

Similarly, the solution of K. Schlacher and M. Schöberl [29] may be recovered by posing $\kappa = y_1 - y_2\frac{\dot{y}_1}{\dot{y}_2}$ and $\psi = \frac{\dot{y}_1}{2\dot{y}_2}$ which, again after inversion of (36) with (32), yields:

$$y_1 = x_1 - x_3\frac{\dot{x}_1}{\dot{x}_3}, \quad y_2 = x_2 - x_3\frac{\dot{x}_2}{\dot{x}_3}.$$

6 Conclusion

In this survey we presented two dual approaches to flatness necessary and sufficient conditions, one based on the integration of 1-forms and the second based on the integration of a set of PDEs involving a generalized Euler-Lagrange operator. Their complexity is compared on examples.

Acknowledgements. This work has been partially supported by a PROCOPE program of EGIDE "Algorithmique en Calcul Formel pour les Systèmes Différentiellement Plats", N. 20146UH and DAAD N. 50018800 "Implementierung notwendiger und hinreichender Kriterien für differentielle Flacheit mittels Computer Algebra".

References

1. Anderson, R.L., Ibragimov, N.H.: Lie-Bäcklund Transformations in Applications. SIAM, Philadelphia (1979)
2. Antritter, F., Lévine, J.: Towards a computer algebraic algorithm for flat output determination. In: Proc. ISSAC 2008 (2008)
3. Aranda-Bricaire, E., Moog, C., Pomet, J.-B.: A linear algebraic framework for dynamic feedback linearization. IEEE Trans. Automat. Control 40(1), 127–132 (1995)
4. Charlet, B., Lévine, J., Marino, R.: Sufficient conditions for dynamic state feedback linearization. SIAM J. Control Optimiz. 29(1), 38–57 (1991)
5. Chern, S., Chen, W., Lam, K.: Lectures on Differential Geometry. Series on University Mathematics, vol. 1. World Scientific, Singapore (2000)
6. Chetverikov, V.: New flatness conditions for control systems. In: Proceedings of NOLCOS 2001, St. Petersburg, pp. 168–173 (2001)
7. Chetverikov, V.: Flatness conditions for control systems. Preprint DIPS (2002), `http://www.diffiety.ac.ru`
8. Cohn, P.: Free Rings and Their Relations. Academic Press, London (1985)
9. Fliess, M.: A remark on Willems' trajectory characterization of linear controllability. Systems & Control Letters 19, 43–45 (1992)

[3] Personal communication.

10. Fliess, M., Lévine, J., Martin, P., Ollivier, F., Rouchon, P.: Controlling nonlinear systems by flatness. In: Byrnes, C., Datta, B., Gilliam, D., Martin, C. (eds.) Systems and Control in the Twenty-First Century, pp. 137–154. Birkhäuser, Boston (1997)
11. Fliess, M., Lévine, J., Martin, P., Rouchon, P.: Sur les systèmes non linéaires différentiellement plats. C.R. Acad. Sci. Paris I(315), 619–624 (1992)
12. Fliess, M., Lévine, J., Martin, P., Rouchon, P.: Flatness and defect of nonlinear systems: introductory theory and examples. Int. J. Control 61(6), 1327–1361 (1995)
13. Fliess, M., Lévine, J., Martin, P., Rouchon, P.: A Lie-Bäcklund approach to equivalence and flatness of nonlinear systems. IEEE Trans. Automat. Control 44(5), 922–937 (1999)
14. Franch, J.: Flatness, Tangent Systems and Flat Outputs. PhD thesis, Universitat Politècnica de Catalunya Jordi Girona (1999)
15. Jakubczyk, B.: Invariants of dynamic feedback and free systems. In: Proc. ECC 1993, Groningen, pp. 1510–1513 (1993)
16. Krasil'shchik, I.S., Lychagin, V.V., Vinogradov, A.M.: Geometry of Jet Spaces and Nonlinear Partial Differential Equations. In: Gordon and Breach, New York (1986)
17. Lévine, J.: On necessary and sufficient conditions for differential flatness. In: Proc. of IFAC NOLCOS 2004 Conference, Stuttgart (2004)
18. Lévine, J.: On necessary and sufficient conditions for differential flatness. arXiv:math.OC/0605405 (2006), `http://www.arxiv.org`
19. Lévine, J.: Analysis and Control of Nonlinear Systems: A Flatness-based Approach. Mathematical Engineering Series. Springer, Heidelberg (2009)
20. Martin, P.: Contribution à l'Étude des Systèmes Différentiellement Plats. PhD thesis, École des Mines de Paris (1992)
21. Martin, P., Murray, R., Rouchon, P.: Flat systems. In: Bastin, G., Gevers, M. (eds.) Plenary Lectures and Minicourses, Proc. ECC 1997, Brussels, pp. 211–264 (1997)
22. Pereira da Silva, P., Filho, C.C.: Relative flatness and flatness of implicit systems. SIAM J. Control and Optimization 39(6), 1929–1951 (2001)
23. Pomet, J.-B.: A differential geometric setting for dynamic equivalence and dynamic linearization. In: Jakubczyk, B., Respondek, W., Rzeżuchowski, T. (eds.) Geometry in Nonlinear Control and Differential Inclusions, pp. 319–339. Banach Center Publications, Warsaw (1993)
24. Rathinam, M., Murray, R.: Configuration flatness of Lagrangian systems underactuated by one control. SIAM J. Control Optimiz. 36(1), 164–179 (1998)
25. Rouchon, P.: Necessary condition and genericity of dynamic feedback linearization. J. Math. Systems Estim. & Control 4(2), 257–260 (1994)
26. Rudolph, J.: Flatness Based Control of Distributed Parameter Systems. Shaker Verlag, Aachen (2003)
27. Rudolph, J., Winkler, J., Woittenek, F.: Flatness Based Control of Distributed Parameter Systems: Examples and Computer Exercises from Various Technological Domains. Shaker Verlag, Aachen (2003)
28. Schlacher, K., Schöberl, M.: Construction of flat outputs by reduction and elimination. In: Proc. 7th IFAC Symposium on Nonlinear Control Systems, Pretoria, South Africa, pp. 666–671 (August 2007)

29. Schlacher, K., Schöberl, M.: Construction of flat outputs by reduction and elimination. In: Lévine, J., Müllhaupt, P. (eds.) Advances in the Theory of Control, Signals and Systems, with Physical Modeling, Springer, Heidelberg (2010)
30. Shadwick, W.: Absolute equivalence and dynamic feedback linearization. Systems & Control Letters 15, 35–39 (1990)
31. Sira-Ramirez, H., Agrawal, S.: Differentially Flat Systems. Marcel Dekker, New York (2004)
32. Sluis, W.: A necessary condition for dynamic feedback linearization. Systems & Control Letters 21, 277–283 (1993)
33. van Nieuwstadt, M., Rathinam, M., Murray, R.: Differential flatness and absolute equivalence of nonlinear control systems. SIAM J. Control Optim. 36(4), 1225–1239 (1998)
34. Zharinov, V.: Geometrical Aspect of Partial Differential Equations. World Scientific, Singapore (1992)

Nonholonomic Mechanics, Dissipation and Quantization*

Anthony M. Bloch

Abstract. In this review paper we consider some of the basics of nonholonomic systems, considering in particular how it is possible do derive nonholonomic equations of motion as a limit of a Lagrangian system subject to dissipation. This in then extended to show how dissipation may be induced from a Hamiltonian field with a view to quantization of the system.

Keywords: Nonholonomic Systems, Dissipation, Quantization.

1 Introduction

In this (mainly) review paper we consider some of the basics of nonholonomic systems, considering in particular how it is possible do derive nonholonomic equations of motion as a limit of Lagrangian system subject to dissipation. This is then extended to show how dissipation may be induced from a Hamiltonian field, thus keeping the full system Hamiltonian, with a view to quantization of the system. Some of the basic ideas in nonholonomic systems theory may be found in [Bloch, Krishnaprasad, Marsden, and Murray(1996)] and [Bloch(2003)] (see also e.g. [Bullo and Lewis(2005)]) which thus give further background on the ideas described here. Below we give various references which link systems with nonholonomic constraints to the limit of infinite

A. Bloch
Department of Mathematics
University of Michigan
Ann Arbor, MI 48109
Tel.: +734-647-4980
Fax: +734-763-0937
e-mail: abloch@umich.edu

* Support from NSF grants DMS-0604307 and DMS-0907949 is gratefully acknowledged.

J. Lévine & P. Müllhaupt (Eds.): Adv. in Theory of Control, Signals, LNCIS 407, pp. 141–152.
springerlink.com © Springer-Verlag Berlin Heidelberg 2010

friction. The key idea goes back to Caratheodory and there have been various interesting contributions since then including those by Fufaev and Kozlov among others.

2 Vertical Disk

We being by discussing a key example which is useful for illustrating many of the key idea in nonholonomic mechanics and control, the vertical disk (see [Bloch(2003)]). In this example the configuration space: $Q = \mathbb{R}^2 \times S^1 \times S^1$, parameterized by coordinates $q = (x, y, \theta, \varphi)$.

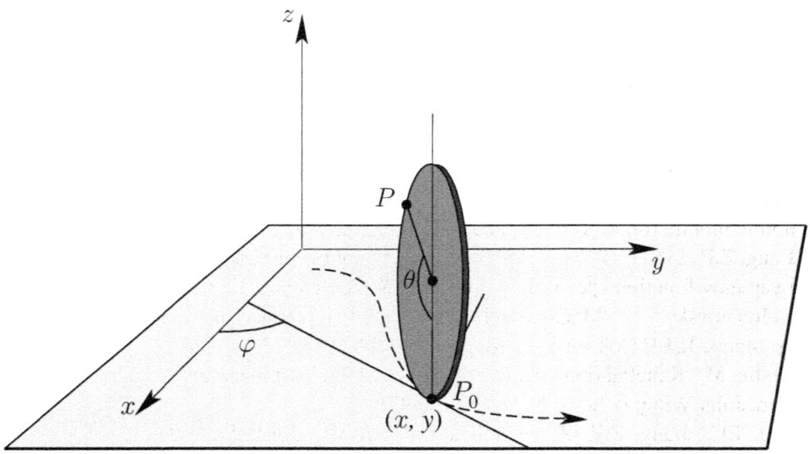

Fig. 1. The geometry for the rolling disk

The Lagrangian for the system is simply the kinetic energy

$$L(x, y, \theta, \phi, \dot{x}, \dot{y}, \dot{\theta}, \dot{\phi}) = \frac{1}{2}m(\dot{x}^2 + \dot{y}^2) + \frac{1}{2}I\dot{\theta}^2 + \frac{1}{2}J\dot{\varphi}^2.$$

If R is the radius of the disk, the nonholonomic constraints of rolling without slipping are

$$\begin{aligned} \dot{x} &= R(\cos\varphi)\dot{\theta} \\ \dot{y} &= R(\sin\varphi)\dot{\theta}, . \end{aligned}$$

Dynamics of the Controlled Disk. We consider the case where we have two controls, one that can steer the disk and another that determines the roll torque. We obtain the Lagrange d'Alembert equations

$$\frac{d}{dt}\left(\frac{\partial L}{\partial \dot{q}}\right) = u_1 X_1 + u_2 X_2 + \lambda_1 W_1 + \lambda_2 W_2,$$

where

$$\frac{\partial L}{\partial \dot{q}} = (m\dot{x}, m\dot{y}, I\dot{\theta}, J\dot{\varphi})^T,$$

$$X_1 = (0,0,1,0)^T, X_2 = (0,0,0,1)^T,$$

and

$$W_1^T = (1, 0, -R\cos\varphi, 0), \quad W_2^T = (0, 1, -R\sin\varphi, 0)^T,$$

together with the constraint equations.

Here u_1, u_2 are natural controls. We call the variables θ and ϕ "base" or "controlled" variables and the variables x and y "fiber" variables. While θ and φ are controlled directly, the variables x and y are controlled indirectly via the constraints. This a special case of a general construction on bundles (see [Bloch(2003)]).

It is clear here that the base variables are controllable in any sense we can imagine. Moreover the full system is controllable also by virtue of the nonholonomic (nonintegrable) nature of the constraints.

Also of interest is the so called ***Kinematic Controlled Disk***. In this case we imagine we have direct control over velocities rather than forces and, accordingly, we consider the most general first order system satisfying the constraints or lying in the "constraint distribution".

In this case the system is

$$\dot{q} = u_1 \overline{X}_1 + u_2 \overline{X}_2$$

where $\overline{X}_1 = (\cos\varphi, \sin\varphi, 1, 0)^T$ and $\overline{X}_2 = (0,0,0,1)^T$.

Interesting problems related to this system including motion planning and stabilization. Aspects of this are discussed in [Bloch(2003)] and references therein.

Nonholonomic Equations of Motion. We now discuss the nonholonomic equations of motion in general: see e.g [Bloch(2003)].

The Lagrange-d'Alembert Principle. Consider a system with a configuration space Q, local coordinates q^i and m nonintegrable constraints

$$\dot{s}^a + A^a_\alpha(r,s)\dot{r}^\alpha = 0$$

where $q = (r,s) \in \mathbb{R}^{n-p} \times \mathbb{R}^p$, which we write as $q^i = (r^\alpha, s^a)$, where $1 \leq \alpha \leq n-p$ and $1 \leq a \leq p$.

We also assume we have a Lagrangian $L(q^i, \dot{q}^i)$. The equations of motion given by Lagrange-d'Alembert principle.

Definition 1. The ***Lagrange-d'Alembert equations of motion*** for the system are those determined by

$$\delta \int_a^b L(q^i, \dot{q}^i)\, dt = 0,$$

where we choose variations $\delta q(t)$ of the curve $q(t)$ that satisfy $\delta q(a) = \delta q(b) = 0$ and $\delta q(t)$ satisfies the constraints for each t where $a \leq t \leq b$.

This principle is supplemented by the condition that the curve itself satisfies the constraints. Note that we take the variation *before* imposing the constraints; that is, we do not impose the constraints on the family of curves defining the variation.

This leads to the equations of motion

$$\left(\frac{d}{dt} \frac{\partial L}{\partial \dot{r}^\alpha} - \frac{\partial L}{\partial r^\alpha} \right) = A^a_\alpha \left(\frac{d}{dt} \frac{\partial L}{\partial \dot{s}^a} - \frac{\partial L}{\partial s^a} \right), \qquad \alpha = 1, \ldots, n-m. \tag{1}$$

The equations (1) combined with the constraint equations

$$\dot{s}^a = -A^a_\alpha \dot{r}^\alpha, \qquad a = 1, \ldots, m, \tag{2}$$

give a complete description of the *equations of motion* of the system. Notice that they consist of $n-m$ second-order equations and m first-order equations.

We remark that this is in contrast to so called variational nonholonomic systems (sometimes called vakonomic systems) where we constrain the class of curves over which we take variations. Such constrained variational problems may be solved by appending the constraints to the Lagrangian via Lagrange multipliers (for details and background see [Bloch(2003)].

3 Chaplygin Sleigh

One of the striking feature of nonholonomic systems is that while they conserve energy they need not conserve volume in the phase space (or momentum, even in the presence of symmetries). For more on this see [Bloch(2003)], [Zenkov and Bloch(2003)] and [Bloch, Marsden, and Zenkov(2009)].

Here we describe the Chaplygin sleigh, perhaps the simplest mechanical system which illustrates the possible dissipative nature of energy preserving nonholonomic systems.

If v denotes the velocity of the system along the direction of the blade and ω its angular velocity one can show that the equations of motion reduce to

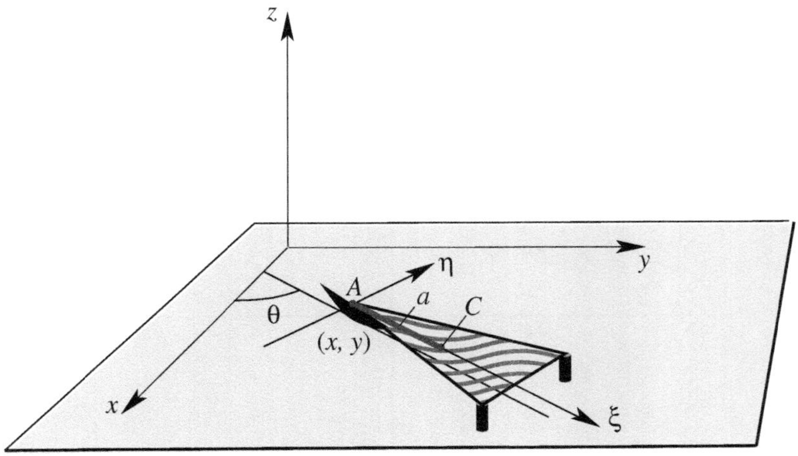

Fig. 2. The Chaplygin sleigh is a rigid body moving on two sliding posts and one knife edge

$$\dot{v} = a\omega^2$$
$$\dot{\omega} = -\frac{ma}{I + ma^2} v\omega$$

The equations have a family of relative equilibria given by $(v, \omega)|v =$ const, $\omega = 0$.

Linearizing about any of these equilibria one finds one zero eigenvalue and one negative eigenvalue. In fact the solution curves are ellipses in $v - \omega$ plane with the positive v-axis attracting all solutions.

This is a special case of the so-called *Euler-Poincaré-Suslov* equations, an important special case of the reduced nonholonomic equations.

Another example is the *Euler-Poincaré-Suslov Problem* on $SO(3)$. In this case the problem can be formulated as the standard Euler equations

$$I\dot{\omega} = I\omega \times \omega$$

where $\omega = (\omega_1, \omega_2, \omega_3)$ are the system angular velocities in a frame where the inertia matrix is of the form $I = \text{diag}(I_1, I_2, I_3)$ and the system is subject to the constraint

$$a \cdot \omega = 0$$

where $a = (a_1, a_2, a_3)$. The nonholonomic equations of motion are then given by

$$I\dot{\omega} = I\omega \times \omega + \lambda a$$

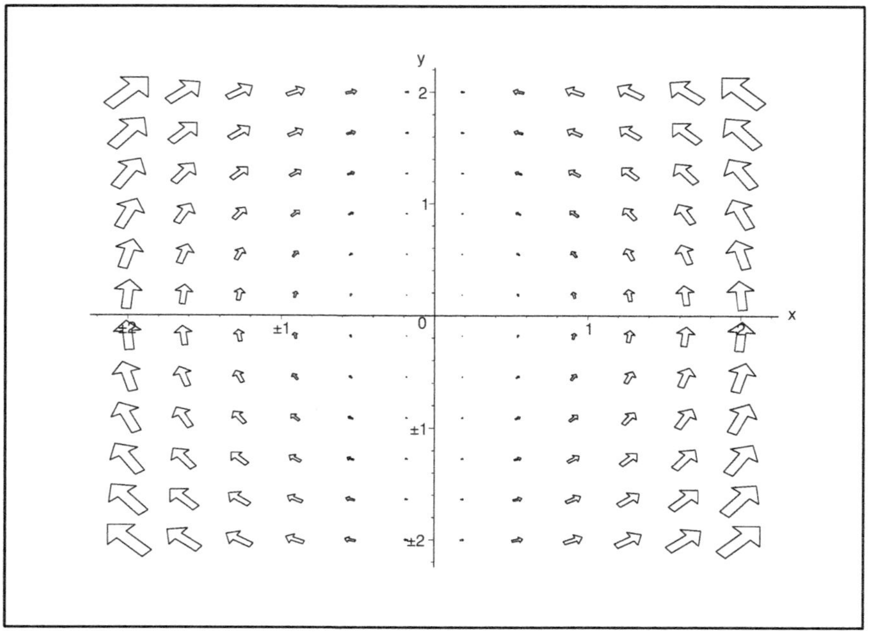

Fig. 3. Chaplygin Sleigh phase portrait

subject to the constraint. Solving for λ we get

$$\lambda = -\frac{I^{-1}a \cdot (I\omega \times \omega)}{I^{-1}a \cdot a} .$$

If a is an eigenvector of the moment of inertia tensor the flow is measure preserving.

We can extend to the general Euler-Poincaré-Suslov equations on a Lie algebra $\mathfrak{g}$ where the system is characterized by the Lagrangian $L = \frac{1}{2}\mathbb{I}_{AB}\Omega^A\Omega^B$ and the left-invariant constraint

$$\langle a, \Omega \rangle = a_A \Omega^A = 0. \tag{3}$$

Here $a = a_A e^A \in \mathfrak{g}^*$ and $\Omega = \Omega^A e_A$, where e_A, $A = 1, \ldots, k$, is a basis for $\mathfrak{g}$ and e^A is its dual basis. Multiple constraints may be imposed as well. The classical examples of such systems are the systems just discussed: the *Chaplygin Sleigh* and the *Suslov problem* introduced by Chaplygin in 1895 and Suslov in 1902, respectively.

4 Lamb Model of Damping

Our goal here is to implement the constraints in the sleigh model by an external field which in turn imposes dissipative motion on the sleigh. The model of dissipation that we use goes back to Lamb in 1900 (see [Lamb(1900)] and was discussed in detail in [Bloch, Hagerty and Weinstein (2004)] The original Lamb model is an oscillator physically coupled to a string. The vibrations of the oscillator transmit waves into the string and are carried off to infinity. Hence the oscillator loses energy and is effectively damped by the string.

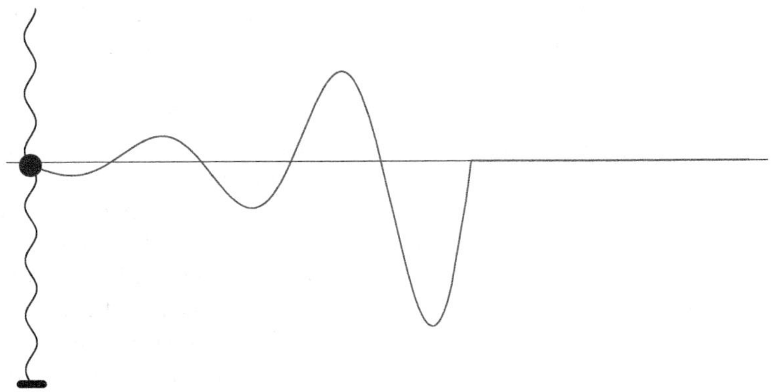

Fig. 4. Lamb model of an oscillator coupled to a string

Let $w(x,t)$ denote displacement of a string. with mass density ρ, tension T. Assuming a singular mass density at $x = 0$, we couple to this an oscillator with position q and mass M (see figure 4) yielding the dynamics:

$$\begin{aligned} \frac{\partial^2 w}{\partial t^2} &= c^2 \frac{\partial^2 w}{\partial x^2} \\ M\ddot{q} + Vq &= T[w_x]_{x=0} \\ q(t) &= w(0,t). \end{aligned}$$

$[w_x]_{x=0} = w_x(0+,t) - w_x(0-,t)$ is the jump discontinuity of the slope of the string. Note that this is a Hamiltonian system.

We can now solve for w and reduce (via elementary Fourier analysis) to obtain a reduced form of the dynamics describing the explicit motion of the oscillator subsystem,

$$M\ddot{q} + \frac{2T}{c}\dot{q} + Vq = 0.$$

The coupling term arises explicitly as a Rayleigh dissipation term $\frac{2T}{c}\dot{q}$ in the dynamics of the oscillator.

5 Nonholonomic Systems as Limit

There is an interesting history behind the question of whether the Lagrange–d'Alembert equations can be obtained by starting with an unconstrained system subject to appropriately chosen dissipative forces, and then letting these forces go to infinity in an appropriate manner.

Nonholonomic constraints can be regarded in some sense as due to "infinite" friction. Several authors have asked if this can be quantified. Interestingly this goes back at least to the work of Caratheodory who asked if the limiting case of such friction could explain the motion of Chaplygin's sleigh. Caratheodory claimed this could not be done, but Fufaev in [Fufaev(1964)] showed that this was indeed possible. The general case was considered by Kozlov, [Kozlov(1983)] and Karapetyan [Karapetyan(1983)].

Kozlov ([Kozlov(1992)] showed also that variational nonholonomic equations (i.e. solutions of a constrained variational problem such as an optimal control problem, see [Bloch(2003)]) can be obtained as the result of another limiting process: He added a parameter-dependent "inertial term" to the Lagrangian of the constrained system, and then showed that the unconstrained equations approach the variational equations as the parameter approaches infinity.

The key idea in the nonholonomic setting is to take a nonlinear Rayleigh dissipation function of the form

$$F = -\frac{1}{2}k\sum_{j=1}^{m}\left(\sum_{i=1}^{n} a_i^{(j)}(\mathbf{q})\dot{q}_i\right)^2 \tag{4}$$

where $\sum_{i=1}^{n} a_i^{(j)}(\mathbf{q})\dot{q}_i = 0$, $i = 1 \ldots m$ are the constraints and $k >$ is a positive constant. Taking the limit as k goes to zero and using Tikhonov's theorem yields the nonholonomic dynamics.

However, the system in this setting is still not Hamiltonian. The goal here is to keep the system in the class of Hamiltonian systems by emulating the dissipation by coupling to an external field. We shall consider this issue in the next section.

Now consider again the Chaplygin sleigh which illustrates in very nice fashion the approach to limiting friction.

This mechanical system has three coordinates, two for the center of mass (x_C, y_C) and one "internal" angular variable θ for the rotation with respect to the knife edge located at $(x, y) = (x_C + a\cos\theta, y_C + a\sin\theta)$. The system can rotate freely around (x, y) but is only allowed to translate in the direction $(\cos\theta, \sin\theta)$: if we choose our coordinates as $\mathbf{q} = (x, y, \theta)$ there is a single constraint given by

$$\dot{x}\sin\theta - \dot{y}\cos\theta = 0, \tag{5}$$

or, $\mathbf{a}^{(1)} = (\sin\theta, -\cos\theta, 0)$.

The equations of motion can be also obtained using the virtual force method starting with the unconstrained Lagrangian

$$L_0 = \frac{m}{2}\left[\left(\dot{x} - a\dot{\theta}\sin\theta\right)^2 + \left(\dot{y} + a\dot{\theta}\cos\theta\right)^2\right] + \frac{I}{2}\dot{\theta}^2, \tag{6}$$

and using a Lagrange multiplier in the equations of motion:

$$\begin{aligned} m\frac{d}{dt}\left(\dot{x} - a\dot{\theta}\sin\theta\right) &= -\lambda\sin\theta, \\ m\frac{d}{dt}\left(\dot{y} + a\dot{\theta}\cos\theta\right) &= \lambda\cos\theta, \\ (I + ma^2)\ddot{\theta} + ma\dot{\theta}(\dot{x}\cos\theta + \dot{y}\sin\theta) &= 0. \end{aligned} \tag{7}$$

Carathedory and Fufaev added a viscous friction force of the from

$$R = -Nu \tag{8}$$

to the sleigh equations.where u is the velocity in the direction perpendicular to the blade. (Note that we interchange u and v compared to the original paper of Fufaev.)

Setting

$$k^2 = \frac{m}{I + ma^2}, \quad \epsilon = \frac{I}{Na^2} \tag{9}$$

the equations with dissipation become

$$u = \epsilon a\dot{\omega} \tag{10}$$

$$\dot{v} = a\omega^2 + \epsilon a\omega\dot{\omega} \tag{11}$$

$$ak^2\dot{\omega} + v\omega = -\epsilon a\ddot{\omega} \tag{12}$$

It is clear that as ϵ goes to zero one recovers the original equations. Cartheodory incorrectly argued however that since no matter how small ϵ is these equations yield trajectories which differ from that of the original system, dissipation cannot yield the nonholonomic constraints.

Fufaev realized this is not correct since the system degenerates from a system of three to two equations and thus there is a singularity. Setting $\mu = \epsilon a$ and $\sigma = \dot{\omega}$ we then get

$$\dot{\omega} = \sigma \tag{13}$$

$$\dot{v} = a\omega^2 + \mu\omega\sigma \tag{14}$$

$$\mu\dot{\sigma} = -ak^2\sigma - v\omega\,. \tag{15}$$

Then as $\mu \to 0$ we get rapid motion except for the surface

$$ak^2\sigma + \mu\omega = 0. \tag{16}$$

The slow motion of this surface onto the v-ω plane then gives the correct equations of motion.

6 Dissipation and Quantization

One can show ([Bloch and Rojo (2008)]) that the sleigh equations can be obtained from a variational principle as reduced equations of motion after the system is coupled to an environment described by an $U(1)$ infinite field of the form $\mathbf{a}(\mathbf{z},t) \equiv [\cos\alpha(\mathbf{z},t), \sin\alpha(\mathbf{z},t)]$. For the Lagrangian of the free field we choose

$$L_{\mathrm{F}} = \frac{K}{2}\int d^2\mathbf{z}\,\dot{\mathbf{a}}^2, \tag{17}$$

and we couple the sleigh and the field with a term of the form

$$L_1 = \int d^2\mathbf{z}\,\delta(\mathbf{z}-\mathbf{x})\,\left[\gamma\dot{\mathbf{x}}\cdot\mathbf{a} + \mu\cos\left(\alpha(\mathbf{z},t)-\theta\right)\right]. \tag{18}$$

The first term in square brackets corresponds to a minimal coupling that favors $\dot{\mathbf{x}}$ in the direction of $\mathbf{a}$; the second has the form of a potential coupling that favors an alignment of the internal variable θ with the local direction of $\mathbf{a}$.

The total action is $S = \int dt(L_0 + L_F + L_1)$ where L_0 is the Lagrangian of the free sleigh

$$L_0 = \frac{m}{2}\left[\left(\dot{x} - a\dot{\theta}\sin\theta\right)^2 + \left(\dot{y} + a\dot{\theta}\cos\theta\right)^2\right] + \frac{I}{2}\dot{\theta}^2, \tag{19}$$

and can be regarded as a full "microscopic" theory of the sleigh coupled to an environment.

The equations of motion of the combined system are now obtained from a variational principle, $\delta S = 0$.

Now take the limit $\mu \to \infty$ and use singular perturbation theory. For very large μ we can show that we have a very slow dynamics on the right hand side of the equations of motion., which amounts to saying that in the $\mu \to \infty$ limit the variables $\alpha(\mathbf{x},t)$ and θ are pinned to the same value. We also obtain

$$\dot{x}\sin\alpha(\mathbf{x},t) - \dot{y}\cos\alpha(\mathbf{x},t) = \dot{x}\sin\theta - \dot{y}\cos\theta = 0, \tag{20}$$

which means that the constraint is satisfied and one can show the full equations are given also.

One can now consider quantization of the system. in the case $a = 0$.

The Hamiltonian in this limit has the form

$$H = \frac{1}{2m}\left[p_x - \lambda\cos\alpha(\mathbf{x})\right]^2 + \frac{1}{2m}\left[p_y - \lambda\sin\alpha(\mathbf{x})\right]^2 + \frac{1}{2I}p_\theta^2 \tag{21}$$

$$+ \frac{1}{2K}\int d\mathbf{z}\,\Pi^2\left(\alpha(\mathbf{z})\right) + \mu\cos\left[\theta - \alpha(\mathbf{x})\right]. \tag{22}$$

For the quantization of H we proceed with the usual replacements

$$\mathbf{p} = -i\hbar(\partial_x, \partial_y), \quad p_\theta = -i\hbar\partial_\theta, \quad \Pi\left(\alpha(\mathbf{z})\right) = -i\hbar\partial_{\alpha(\mathbf{z})}. \tag{23}$$

We can then analyze the corresponding Schroedinger equation. In the quasiclassical limit the fluctuations of the angle are small and centered around given eigenstates $\theta = \theta_{\mathbf{k}}$. This means that, up to small quantum fluctuations, the knife edge is pointing in the direction defined by the classical constraint. Details may be found in [Bloch and Rojo (2008)].

We note also that an alternate approach to quantization can be obtained using the inverse problem of the calculus of variations (see [Bloch, Fernandez and Mestdag (2009)]). In this setting one obtains an associated system which give the nonholonomic equations on invariant manifolds. This system can shown to be variational using the inverse problem and can then be quantized.

We note finally that control of such nonholonomic systems with internal dissipation is of interest and the dissipation leads to interesting controlled dynamics. We are currently pursuing work in this area with Luis Narnanjo and Dmitry Zenkov. See also [Osborne and Zenkov(2005)].

References

[Arnold, Kozlov, and Neishtadt(1988)] Arnold, V.I., Kozlov, V.V., Neishtadt, A.I.: Dynamical Systems III. Encyclopedia of Math., vol. 3. Springer, Heidelberg (1988)

[Bloch(2003)] Bloch, A.M.: Nonholonomic Mechanics and Control. Interdisciplinary Applied Mathematics. Springer, Heidelberg (2003)

[Bloch, Fernandez and Mestdag (2009)] Bloch, A.M., Fernandez, O., Mestdag, T.: Hamiltonization of Nonholonomic Systems and the Inverse Problem of the Calculus of Variations. Reports on Mathematical Physics 63, 225–249 (2009)

[Bloch and Rojo (2008)] Bloch, A.M., Rojo, A.: Quantization of a nonholonomic system. Phys. Rev. Letters 101, 030404 (2008)

[Bloch, Krishnaprasad, Marsden, and Murray(1996)] Bloch, A.M., Krishnaprasad, P.S., Marsden, J.E., Murray, R.: Nonholonomic mechanical systems with symmetry. Arch. Rat. Mech. An. 136, 21–99 (1996)

[Bloch, Krishnaprasad, Marsden, and Ratiu(1996)] Bloch, A.M., Krishnaprasad, P.S., Marsden, J.E., Ratiu, T.S.: The Euler–Poincaré equations and double bracket dissipation. Comm. Math. Phys. 175, 1–42 (1996)

[Bloch, Hagerty and Weinstein (2004)] Bloch, A.M., Hagerty, P., Weinstein, M.: Radiation induced instability. SIAM J. Applied Math. 64, 484–524 (2004)

[Bloch, Marsden, and Zenkov(2009)] Bloch, A.M., Marsden, J.E., Zenkov, D.E.: Quasivelocities and symmetries in nonholonomic systems. Dynamical Systems 24, 187–222 (2009)

[Bullo and Lewis(2005)] Bullo, F., Lewis, A.D.: Geometric control of mechanical systems. Texts in Applied Mathematics, vol. 49. Springer, New York (2005) (Modeling, analysis, and design for simple mechanical control systems)

[Carathéodory(1933)] Carathéodory, C.: Der Schlitten. Z. Angew. Math. und Mech. 13, 71–76 (1933)

[Cendra, Marsden, and Ratiu(2001b)] Cendra, H., Marsden, J.E., Ratiu, T.S.: Geometric mechanics, Lagrangian reduction and nonholonomic systems. In: Enquist, B., Schmid, W. (eds.) Mathematics Unlimited-2001 and Beyond, pp. 221–273. Springer, Heidelberg (2001)

[Fufaev(1964)] Fufaev, N.A.: On the possibility of realizing a nonholonomic constraint by means of viscous friction forces. Prikl. Math. Mech 28, 513–515 (1964)

[Karapetyan(1983)] Karapetyan, A.V.: On realizing nonholonomic constraints by viscous friction forces and Celtic stones stability. Prikl. Matem. Mekhan. USSR 45, 30–36 (1983)

[Kozlov(1983)] Kozlov, V.V.: Realization of nonintegrable constraints in classical mechanics. Sov. Phys. Dokl. 28, 735–737 (1983)

[Kozlov(1992)] Kozlov, V.V.: The problem of realizing constraints in dynamics. J. Appl. Math. Mech. 56(4), 594–600 (1992)

[Lamb(1900)] Lamb, H.: On the peculiarity of wave-system due to the free vibrations of a nucleus in an extended medium. Proc. London Math, Society 32, 208–211 (1900)

[Marsden and Ratiu(1999)] Marsden, J.E., Ratiu, T.S.: Introduction to Mechanics and Symmetry. Texts in Applied Mathematics, vol. 17. Springer, Heidelberg (1999), First Edition (1994), Second Edition (1999)

[Neimark and Fufaev(1972)] Neimark, J.I., Fufaev, N.A.: Dynamics of Nonholonomic Systems. Translations of Mathematical Monographs, vol. 33. AMS, Providence (1972)

[Osborne and Zenkov(2005)] Osborne, J.M., Zenkov, D.V.: Steering the Chaplygin sleigh by a moving mass. In: Proc. IEEE CDC, vol. 44, pp. 1114–1118 (2005)

[Zenkov and Bloch(2003)] Zenkov, D.V., Bloch, A.M.: Invariant measures of nonholonomic flows with internal degrees of freedom. Nonlinearity 16, 1793–1807 (2003)

Controlled Lagrangians

Dong Eui Chang

Abstract. We report our recent progress on the method of controlled Lagrangians. We present the following: a set of new matching conditions for controlled Lagrangian systems with external forces including velocity-independent forces; a criterion for energy shaping and exponential stabilizability by dissipation for all linear controlled Lagrangian systems; and a criterion for energy shaping and exponential stabilizability by dissipation for all controlled Lagrangian systems with one degree of underactuation. We illustrate the criteria with examples.

1 Introduction

The main idea of the method of controlled Lagrangians is as follows. Given an unstable mechanical system such as an inverted pendulum, we transform it via feedback to a stable mechanical system such as a hanging pendulum and then achieve asymptotic stability by dissipative feedback forcing. One usually requires that the energy function of the transformed system should have a non-degenerate minimum at an equilibrium point of interest, due to which the method of controlled Lagrangians is sometimes called the energy shaping method. One can also transform an external force via feedback to a dissipative force, and this is called force shaping.

Although the requirement of non-degeneracy of a minimum of the "shaped" energy function has been well accepted by the community, it has been overlooked that this is equivalent to considering the linearization of a system. Hence, the study of stabilization for linear mechanical systems is important not only by itself but also for understanding stabilization of nonlinear mechanical systems. In this paper we address only non-degenerate energy shaping and call it simply energy shaping for convenience. We note that degenerate energy shaping is much more difficult.

Dong Eui Chang
Applied Mathematics, University of Waterloo, 200 University Ave. W., Waterloo, Ontario, N2L 3G1, Canada
e-mail: dechang@math.uwaterloo.ca

J. Lévine & P. Müllhaupt (Eds.): Adv. in Theory of Control, Signals, LNCIS 407, pp. 153–159.
springerlink.com © Springer-Verlag Berlin Heidelberg 2010

The following gives a brief history of the energy shaping method. Potential shaping was initiated in [1] to stabilize a fully-actuated mechanical system. A kinetic energy shaping technique was introduced in [2] to stabilize a rotational motion of an *underactuated* rigid body. A total energy shaping method for underactuated mechanical systems was first presented in [3, 4]. For a complete bibliography on the history, we refer the readers to [5, 6, 7, 8] and references therein.

In the present paper we report some of the recent results obtained by the author. First, we give new matching conditions for controlled Lagrangian systems with external forces including velocity-independent forces. Second, we provide conditions for matching and force shaping. Third, we completely solve the problem of energy shaping for linear mechanical control systems. Last, we give a necessary and sufficient condition for energy shaping and exponential stabilizability by dissipation for the class of all controlled Lagrangian systems with one degree of underactuation. There are two examples to illustrate the main results. The present paper improves [5] because we take into account both velocity-independent and velocity-dependent external forces.

2 The Method of Controlled Lagrangians

Basic Notions on the Theory of Controlled Lagrangians

Let Q be a configuration manifold of dimension n. One may assume that Q is an open subset of $\mathbb{R}^n$ because we are interested mainly in local stability. A controlled Lagrangian (CL) system on TQ is a triple (L,F,W): L is a Lagrangian function of the form

$$L(q,\dot{q}) = \frac{1}{2}m(\dot{q},\dot{q}) - V(q)$$

where $m \in \Gamma(S^2(T^*Q))$ is non-degenerate and V is a function on Q; $F : TQ \to T^*Q$ is a fiber-preserving map called external force; and W is a subbundle of T^*Q called control bundle. Every control force is W-valued. When a map u is W-valued, we simply denote it by $u \in W$. Finally, let $W^0 = \{v \in TQ \mid \langle \alpha, v\rangle = 0\, \forall \alpha \in W\}$ be the annihilator of W.

The equations of motion of a controlled Lagrangian system (L,F,W) with a control $u \in W$ are given by

$$m^\flat(\nabla_{\dot{q}}\dot{q}) + \mathbf{d}V = F + u \tag{1}$$

where ∇ is the symmetric affine connection induced from the (pseudo-Riemannian) metric m. In coordinates,

$$m_{ij}\ddot{q}^j + [jk,i]\dot{q}^j\dot{q}^k + \frac{\partial V}{\partial q^i} = F_i + u_i, \quad i = 1,\dots,n$$

where $[jk,i] = \frac{1}{2}\left(\frac{\partial m_{ij}}{\partial q^k} + \frac{\partial m_{ki}}{\partial q^j} - \frac{\partial m_{jk}}{\partial q^i}\right)$.

Define the energy function $E : TQ \to \mathbb{R}$ of (L, F, W) by

$$E(q, \dot{q}) = \frac{1}{2} m(\dot{q}, \dot{q}) + V(q).$$

Thus, along the trajectory of (1),

$$\frac{dE}{dt} = \langle F, \dot{q} \rangle + \langle u, \dot{q} \rangle$$

where the quantity on the right-hand side is called the power generated by F and u.

A force F is called dissipative if $\langle F(q,\dot{q}), \dot{q} \rangle \leq 0$ for all $(q, \dot{q}) \in TQ$. In particular, F is called gyroscopic if $\langle F(q,\dot{q}), \dot{q} \rangle = 0$ for all $(q, \dot{q}) \in TQ$. When one is interested in local stability it suffices to consider locally dissipative forces or locally gyroscopic forces, i.e., one requires $\langle F(q,\dot{q}), \dot{q} \rangle \leq 0$ or $\langle F(q,\dot{q}), \dot{q} \rangle = 0$ only in a neighborhood of an equilibrium of interest. However, in this paper we do not deal with locally dissipative forces or locally gyroscopic forces because we will not be addressing force shaping in detail. We instead refer the readers to [5] for supplementary work on (local) force shaping.

The linearization of a controlled Lagrangian system (L, F, W) at an equilibrium point $(q, \dot{q}) = (q_e, 0)$ is a linear controlled Lagrangian system (L^ℓ, F^ℓ, W^ℓ) given by $L^\ell = \frac{1}{2} m_{ij}(q_e) \dot{q}^i \dot{q}^j - \frac{1}{2} \frac{\partial^2 V}{\partial q^i \partial q^j}(q_e)(q^i - q_e^i)(q^j - q_e^j)$; $F^\ell = \frac{\partial F}{\partial q^i}(q_e, 0)(q^i - q_e^i) + \frac{\partial F}{\partial \dot{q}^i}(q_e, 0)\dot{q}^i$; and $W^\ell = W(q_e)$.

Matching and Force Shaping

Consider two controlled Lagrangian systems (L, F, W) and $(\widehat{L}, \widehat{F}, \widehat{W})$ defined by

$$\begin{aligned} L &= \frac{1}{2} m(\dot{q}, \dot{q}) - V(q), \quad \widehat{L} = \frac{1}{2} \widehat{m}(\dot{q}, \dot{q}) - \widehat{V}(q), \\ F &= F_0 + F_1 + F_2, \quad \widehat{F} = \widehat{F}_0 + \widehat{F}_1 + \widehat{F}_2 \end{aligned}$$

where $F_0, \widehat{F}_0 : Q \to T^*Q$ are velocity-independent forces defined by $F_0 = \alpha$ and $\widehat{F}_0 = \widehat{\alpha}$ for some one-forms α and $\widehat{\alpha}$ on Q; $F_1, \widehat{F}_1 : TQ \to T^*Q$ are forces linear in velocity defined by

$$\langle F_1(v), u \rangle = A(v, u), \quad \langle \widehat{F}_1(v), u \rangle = \widehat{A}(v, u)$$

for some $A, \widehat{A} \in \Gamma(T^*Q \otimes T^*Q)$; and $F_2, \widehat{F}_2 : TQ \to T^*Q$ are forces quadratic in velocity defined by

$$\langle F_2(v), u \rangle = B(v, v, u), \quad \langle \widehat{F}_2(v), u \rangle = \widehat{B}(v, v, u)$$

for some $B, \widehat{B} \in \Gamma(S^2(T^*Q) \otimes T^*Q)$. Here, we consider velocity-dependent forces of degree 2, in order to match the terms on the left-hand side of (1) that are quadratic in velocity. Let ∇ and $\widehat{\nabla}$ denote the metric connections of m and $\widehat{m}$, respectively.

Theorem 1 (Matching). *In order that the two CL systems $(L, F = F_0 + F_1 + F_2, W)$ and $(\widehat{L}, \widehat{F} = \widehat{F}_0 + \widehat{F}_1 + \widehat{F}_2, \widehat{W})$ are feedback equivalent, it is necessary and sufficient that*

$$
\begin{aligned}
&\langle \mathbf{d}V - \alpha - m^\flat \widehat{m}^\sharp (\mathbf{d}\widehat{V} - \widehat{\alpha})), Z\rangle = 0,\\
&\widehat{A}(X, \widehat{m}^\sharp m^\flat Z) = A(X,Z),\\
&\widehat{B}(X,Y,\widehat{m}^\sharp m^\flat Z) = \widehat{K}(X,Y,\widehat{m}^\sharp m^\flat Z) + B(X,Y,Z),\\
&\widehat{W} = \widehat{m}^\flat m^\sharp W
\end{aligned}
$$

for every $X, Y \in TQ$ and $Z \in W^0$, where $\widehat{K} \in \Gamma(S^2(T^*Q) \otimes T^*Q)$ is defined by

$$
\widehat{K}(X,Y,Z) = \widehat{m}(\widehat{\nabla}_X Y - \nabla_X Y, Z)
$$

for all $X, Y, Z \in TQ$.

Theorem 2 (Matching and Force Shaping). *Consider a CL system $(L, F = F_0 + F_1 + F_2, W)$. In order to find a CL system $(\widehat{L}, \widehat{F} = \widehat{F}_1 + \widehat{F}_2, \widehat{W})$ with dissipative $\widehat{F}$ that is feedback equivalent to (L, F, W), it is necessary and sufficient to find a non-degenerate $\widehat{m} \in \Gamma(S^2(T^*Q))$ and a function $\widehat{V} : Q \to \mathbb{R}$ that satisfy*

$$
\begin{aligned}
&(\mathbf{d}V - \alpha - m^\flat \widehat{m}^\sharp \mathbf{d}\widehat{V})|_{W^0} = 0,\\
&\mathrm{Sym}(A^\flat \widehat{m}^\sharp m^\flat)|_{(W^0)^{\otimes 2}} \preceq 0,\\
&\mathrm{Sym}(\widehat{R})|_{(W^0)^{\otimes 3}} = 0
\end{aligned}
$$

where $\widehat{R} \in \Gamma(S^2(T^*Q) \otimes T^*Q)$ is defined by

$$
\widehat{R}(X,Y,Z) = \frac{1}{2}(\nabla_{\widehat{m}^\sharp m^\flat Z}\widehat{m})(\widehat{m}^\sharp m^\flat X, \widehat{m}^\sharp m^\flat Y) + B(\widehat{m}^\sharp m^\flat X, \widehat{m}^\sharp m^\flat Y, Z)
$$

for all $X, Y, Z \in TQ$.

Sometimes it is computationally easier to find a $\widehat{T} = m\widehat{m}^{-1}m \in \Gamma(S^2(T^*Q))$ instead of finding an $\widehat{m}$ directly. The following corollary considers this case.

Corollary 1. *Let $\widehat{T} = m\widehat{m}^{-1}m \in \Gamma(S^2(T^*Q))$. Then, the three matching conditions in Theorem 2 are equivalent to the following:*

$$
\begin{aligned}
&(\mathbf{d}V - \alpha - \widehat{T}^\flat m^\sharp \mathbf{d}\widehat{V})|_{W^0} = 0,\\
&\mathrm{Sym}(A^\flat m^\sharp \widehat{T}^\flat)|_{(W^0)^{\otimes 2}} \preceq 0,\\
&\mathrm{Sym}(\widehat{J})|_{(W^0)^{\otimes 3}} = 0
\end{aligned}
\tag{2}
$$

where $\widehat{J} \in \Gamma(S^2(T^*Q) \otimes T^*Q)$ *is defined by*

$$\widehat{J}(X,Y,Z) = \frac{1}{2}(\nabla_{m^\sharp \widehat{T}^\flat Z}\widehat{T})(X,Y) - B(m^\sharp \widehat{T}^\flat X, m^\sharp \widehat{T}^\flat Y, Z)$$

for all $X,Y,Z \in TQ$.

Energy Shaping for Linear Controlled Lagrangian Systems

For linear controlled Lagrangian systems without any external forces we have a complete understanding of energy shaping and stabilization by dissipation after shaping energy.

Theorem 3 (Energy Shaping). *A linear CL system* $(L,0,W)$ *is feedback equivalent to a linear CL system* $(\widehat{L},0,\widehat{W})$ *with positive definite energy, if and only if the uncontrollable dynamics of* $(L,0,W)$*, if it exists, is oscillatory.*[1] *Moreover, if* $(L,0,W)$ *is controllable, then* $(\widehat{L},0,\widehat{W})$ *can be exponentially stabilized by any* $\widehat{W}$*-valued linear symmetric dissipative feedback* $\widehat{u}$ *with* $\mathrm{rank}\,\widehat{u} = \dim \widehat{W}$. *If* $(L,0,W)$ *is not controllable, then it cannot be exponentially stabilized by any controller.*

Let us apply Theorem 3 to the system $(L,0,W)$ with $L = \frac{1}{2}(\dot{x}^2 + \dot{y}^2 + \dot{z}^2) - \frac{1}{2}(x^2 + 2xy + 3y^2 + 2\varepsilon yz)$ and $W = \mathrm{span}\{(0,0,1)\}$ with a parameter $\varepsilon \in \mathbb{R}$. If $\varepsilon \neq 0$, then the system is controllable, so it is possible to shape energy and then exponentially stabilize the energy-shaped system with any linear symmetric dissipative feedback having rank 1. If $\varepsilon = 0$, then the system is not controllable, but its uncontrollable dynamics is oscillatory. Hence, it is still possible to shape the energy function although it is impossible to exponentially stabilize the system.

Energy Shaping for Controlled Lagrangian Systems with One Degree of Underactuation

When the degree of underactuation is one, i.e., # (configuration variables) − # (controls) = 1, we have a necessary and sufficient condition for energy shaping, and a necessary and sufficient condition for exponential stabilization by dissipation after energy shaping.

Theorem 4 (Energy Shaping and Exponential Stabilization by Dissipative Forcing). *Let* $(L,0,W)$ *be a CL system with one degree of under-actuation. Let* $(L^\ell,0,W^\ell)$ *denote its linearization at an equilibrium point of interest. Then, there exists a CL system* $(\widehat{L},\widehat{F},\widehat{W})$ *feedback equivalent to* $(L,0,W)$*, with a gyroscopic force* $\widehat{F}$ *quadratic in velocity, and its energy* $\widehat{E}$ *having a non-degenerate minimum at the equilibrium, if and only if the uncontrollable dynamics of* $(L^\ell,0,W^\ell)$*, if it exists, is oscillatory. In particular, if* $(L^\ell,0,W^\ell)$ *is controllable, then* $(\widehat{L},\widehat{F},\widehat{W})$ *can be exponentially stabilized by any* $\widehat{W}$*-valued linear symmetric dissipative feedback* $\widehat{u} \in \Gamma(S^2(T^*Q))$ *with* $\mathrm{rank}\,\widehat{u} = \dim \widehat{W}$. *If* $(L^\ell,0,W^\ell)$ *is not controllable, then* $(\widehat{L},\widehat{F},\widehat{W})$ *cannot be exponentially stabilized by any dissipative feedback.*

---

[1] A linear dynamics $\dot{x} = Ax$ on $\mathbb{R}^n$ is called oscillatory if A is diagonalizable over $\mathbb{C}$ and every eigenvalue of A is a non-zero purely imaginary number.

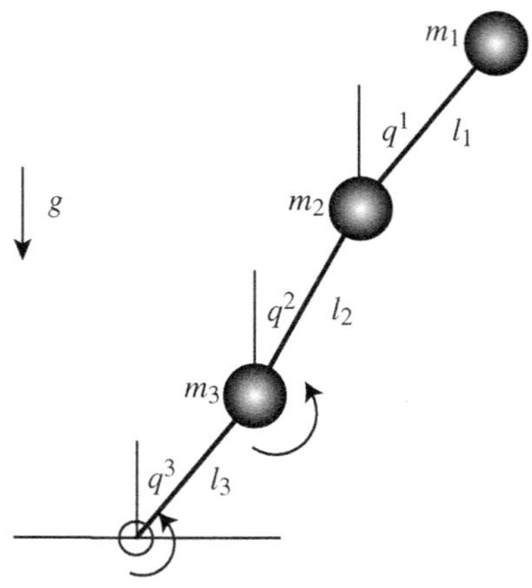

Fig. 1. A 3-link robot arm with two actuators

Let us apply Theorem 4 to stabilize the 3-link robot arm with two actuators in Fig. 1. The coordinates $q=(q^1,q^2,q^3)$ denote the angles between the arms and the vertical line. In these coordinates, the mass matrix is given by

$$m=\begin{bmatrix} a_{11} & a_{12}\cos(q^1-q^2) & a_{13}\cos(q^1-q^3)\\ a_{21}\cos(q^1-q^2) & a_{22} & a_{23}\cos(q^2-q^3)\\ a_{31}\cos(q^1-q^3) & a_{32}\cos(q^2-q^3) & a_{33}\end{bmatrix}$$

where

$$\begin{aligned} a_{kk} &= \sum_{i=1}^{k} m_i(l_i)^2, \quad k=1,2,3;\\ a_{1k} &= a_{k1} = m_1 l_1 l_k, \quad k=2,3;\\ a_{23} &= a_{32} = (m_1+m_2)l_2 l_3. \end{aligned}$$

The potential energy V is given by

$$V=b_1\cos q^1+b_2\cos q^2+b_3\cos q^3$$

where

$$b_k = g l_k \sum_{i=1}^{k} m_i, \quad i=1,2,3.$$

Assume that there is no friction at any joint. This system is a CL system $(L,0,W)$ where $L=\frac{1}{2}m(\dot q,\dot q)-V(q)$ and $W=\mathrm{span}\{\mathbf{d}q^2,\mathbf{d}q^3\}$. The control objective is to stabilize the equilibrium $(q,\dot q)=(0,0)$. One can easily verify that the linearization of this system at $(0,0)$ is controllable, so by Theorem 4 it is possible to shape the energy function of the system and then exponentially stabilize the equilibrium by any linear symmetric dissipative feedback force with rank 2.

3 Conclusion

We have presented criteria for "energy-shapability" and exponential stabilizability by dissipation after energy shaping for the class of all linear controlled Lagrangian systems without external forces and for the class of all one-degree-of-underactuation controlled Lagrangian systems without external forces. These criteria are easy to verify in advance before any concrete controller design.

We note that when there is an external force, the matching conditions in Theorem 2 or Corollary 1 can be further refined by locally shaping the force and energy simultaneously. We refer the readers to [5] for more on this topic.

References

1. Arimoto, S., Miyazaki, F.: Stability and robustness of PID feedback control for robot manip- ulators of sensory capability. In: Brady, M., Paul, R.P. (eds.) Robotics Research, First International Symposium, pp. 783–799. MIT Press, Cambridge (1983)
2. Bloch, A.M., Krishnaprasad, P.S., Marsden, J.E., Sanchez de Alvarez, G.: Stabilization of rigid body dynamics by internal and external torques. Automatica 28, 745–756 (1992)
3. Bloch, A.M., Leonard, N.E., Marsden, J.E.: Stabilization of mechanical systems using controlled Lagrangians. In: Proc. 36th IEEE Conference on Decision and Control, San Diego, CA, USA, pp. 2356–2361 (1997)
4. Bloch, A.M., Leonard, N.E., Marsden, J.E.: Controlled Lagrangians and the stabilization of mechanical systems I: The first matching theorem. IEEE Trans. Automat. Contr. 45, 2253–2270 (2000)
5. Chang, D.E.: The method of controlled Lagrangians: Energy plus force shaping. In: Proc. 48th IEEE Conference on Decision and Control, Shanghai, China, pp. 3329–3334 (2009)
6. Chang, D.E.: Generalization of the IDA-PBC method for stabilization of mechanical systems. In: Proc. 18th Mediterranean Conference on Control and Automation, Marrakech, Morocco (2010)
7. Chang, D.E.: Stabilizability of controlled Lagrangian systems of two degrees of freedom and one degree of under-actuation by the energy shaping method. IEEE Trans. Automat. Contr. (in press)
8. Ortega, R., Garcia Canseco, E.: Interconnection and damping assignment passivity-based control: A survey. European J. of Control 10, 432–450 (2004)

Compensation of Input Delay for Linear, Nonlinear, Adaptive, and PDE Systems

Miroslav Krstic

Abstract. We present a tutorial introduction to methods for stabilization of systems with long input delays. The methods are based on techniques originally developed for boundary control of partial differential equations. We start with a consideration of linear systems, first with a known delay and then subject to a small uncertainty in the delay. Then we study linear systems with constant delays that are completely unknown, which requires an adaptive control approach. For linear systems, we also present a method for compensating arbitrarily large but known time-varying delays. Finally, we consider nonlinear control problems in the presence of arbitrarily long input delays.

An enormous wealth of knowledge and research results exists for control of systems with state delays and input delays. Problems with long input delays, for unstable plants, represent a particular challenge. In fact, they were the first challenge to be tackled, in Otto J. M. Smith's article [1], where the compensator known as the Smith predictor was introduced five decades ago. The Smith predictor's value is in its ability to compensate for a long input or output delay in set point regulation or constant disturbance rejection problems. However, its major limitation is that, when the plant is unstable, it fails to recover the stabilizing property of a nominal controller when delay is introduced.

A substantial modification to the Smith predictor, which removes its limitation to stable plants was developed three decades ago in the form of finite spectrum assignment (FSA) controllers [2, 3, 4]. More recent treatment of this subject can also be found in the books [5, 6]. In the FSA approach, the system

$$\dot{X}(t) = AX(t) + BU(t-D), \tag{1}$$

where X is the state vector, U is the control input (scalar in our consideration here), D is an arbitrarily long delay, and (A,B) is a controllable pair, is stabilized with the

Miroslav Krstic
Department of Mechanical and Aerospace Engineering, University of California, San Diego, La Jolla, CA 92093-0411, USA
e-mail: krstic@ucsd.edu

J. Lévine & P. Müllhaupt (Eds.): Adv. in Theory of Control, Signals, LNCIS 407, pp. 161–178.
springerlink.com © Springer-Verlag Berlin Heidelberg 2010

infinite-dimensional predictor feedback

$$U(t) = K\left[e^{AD}X(t) + \int_{t-D}^{t} e^{A(t-\theta)}BU(\theta)d\theta\right], \tag{2}$$

where the gain K is chosen so that the matrix $A+BK$ is Hurwitz. The word 'predictor' comes from the fact that the bracketed quantity is actually the future state $X(t+D)$, expressed using the current state $X(t)$ as the initial condition and using the controls $U(\theta)$ from the past time window $[t-D,t]$. Concerns are raised in [7] regarding the robustness of the feedback law (2) to digital implementation of the distributed delay (integral) term but are resolved with appropriate discretization schemes [8, 9].

One can view the feedback law (2) as being given implicitly, since U appears both on the left and on the right, however, one should observe that the input memory $U(\theta), \theta \in [t-D,t]$ is actually a part of the state of the overall infinite-dimensional system, so the control law is actually given by an explicit full-state feedback formula. The predictor feedback (2) actually represents a particular form of boundary control, commonly encountered in the context of control of partial differential equations.

Motivated by our recent efforts in solving boundary control problems for various classes of partial differential equations (PDEs) using the continuum version of the backstepping method [10, 11], we review in this article various extensions to the predictor feedback design that we have recently developed, particularly for nonlinear and PDE systems. These extensions are the subject of our new book [12]. They include the extension of predictor feedback to nonlinear systems and PDEs with input delays, various robustness and inverse optimality results, a delay-adaptive design, an extension to time-varying delays, and observer design in the presence of sensor delays and PDE dynamics. This article is a tutorial introduction to these design tools and concludes with a brief review of some open problems and research opportunities.

1 Lyapunov Functional and Its Immediate Benefits

The key to various extensions to the predictor feedback that we present here is the observation that the invertible backstepping transformation

$$w(x,t) = u(x,t) - \int_0^x Ke^{A(x-y)}Bu(y,t)dy - Ke^{Ax}X(t), \tag{3}$$

$$u(x,t) = w(x,t) + \int_0^x Ke^{(A+BK)(x-y)}Bw(y,t)\,dy + Ke^{(A+BK)x}X(t), \tag{4}$$

where

$$u(x,t) = U(t+x-D), \tag{5}$$

can transform the system (1), (2) into the *target system*

$$\dot{X}(t) = (A+BK)X(t)+Bw(0,t)\,, \tag{6}$$
$$w_t(x,t) = w_x(x,t)\,, \tag{7}$$
$$w(D,t) = 0\,, \tag{8}$$

which is a cascade of an undriven transport PDE w-subsystem and the exponentially stable X-system.

We shoe the equivalence of the closed-loop system (1), (2) and the target system (6), (7), (8) as follows. With (3), (5), $w(x,t)$ can be alternatively written as

$$w(x,t) = U(t+x-D) - \int_0^x Ke^{A(x-y)}BU(t+y-D)dy - Ke^{Ax}X(t), \tag{9}$$

By noting that

$$w(0,t) = U(t-D) - KX(t), \tag{10}$$

we obtain

$$\dot{X}(t) = AX(t)+BU(t-D) \tag{11}$$
$$= (A+BK)X(t)+B\Big(U(t-D) - KX(t)\Big) \tag{12}$$
$$= (A+BK)X(t)+Bw(0,t)\,. \tag{13}$$

We further calculate

$$\begin{aligned}
\frac{\partial}{\partial t}w(x,t) &\triangleq U'(t+x-D) - \int_0^x Ke^{A(x-y)}BU'(t+y-D)dy \\
&\quad -Ke^{Ax}\Big(AX(t)+BU(t-D)\Big) \\
&= U'(t+x-D) - \int_0^x \frac{\partial}{\partial y}\Big(Ke^{A(x-y)}BU(t+y-D)\Big)dy \\
&\quad + \int_0^x \frac{\partial(Ke^{A(x-y)})}{\partial y}BU(t+y-D)dy - Ke^{Ax}\Big(AX(t)+BU(t-D)\Big) \\
&= U'(t+x-D) - \Big(KBU(t+x-D) - Ke^{-Ax}BU(t-D)\Big) \\
&\quad - \int_0^x KAe^{A(x-y)}BU(t+y-D)dy - Ke^{Ax}\Big(AX(t)+BU(t-D)\Big) \\
&= U'(t+x-D) - KBU(t+x-D) \\
&\quad - \int_0^x KAe^{A(x-y)}BU(t+y-D)dy - Ke^{Ax}AX(t)
\end{aligned} \tag{14}$$

and

$$\frac{\partial}{\partial x}w(x,t) \triangleq U'(t+x-D) - KBU(t+x-D)$$

$$-\int_0^x KAe^{A(x-y)}BU(t+y-D)dy - KAe^{Ax}X(t). \tag{15}$$

Thus, we obtain

$$\frac{\partial}{\partial t}w(x,t) = \frac{\partial}{\partial x}w(x,t), \tag{16}$$

which establishes (7). Finally, for $x = D$ it follows that

$$\begin{aligned} w(D,t) &= U(t) - \int_0^D Ke^{A(D-y)}BU(t+y-D)dy - Ke^{AD}X(t) \\ &= U(t) - K\left(\int_{t-D}^t e^{A(t-\theta)}BU(\theta)dy + e^{AD}X(t)\right). \end{aligned} \tag{17}$$

With (2), we obtain (8).

Since the undriven transport PDE (7), (8) is exponentially stable, the overall cascade is exponentially stable. This fact is established with a Lyapunov functional

$$V(t) = X(t)^T PX(t) + 2\frac{|PB|^2}{\lambda_{\min}(Q)}\int_0^D (1+x)w(x,t)^2 dx, \tag{18}$$

where P is the solution of the Lyapunov equation

$$P(A+BK) + (A+BK)^{\mathrm{T}}P = -Q. \tag{19}$$

Taking the derivative of the Lyapunov function, we calculate

$$\begin{aligned} \frac{dV}{dt} &= \Big((A+BK)X(t) + Bw(0,t)\Big)^T PX(t) \\ &\quad + X^T(t)P\Big((A+BK)X(t) + Bw(0,t)\Big) \\ &\quad + \frac{2|PB|^2}{\lambda_{\min}(Q)}\int_0^D (1+x)2w\frac{\partial w}{\partial t}dx \\ &= X^T(t)(A+BK)^T PX(t) + (Bw(0,t))^T PX(t) + X^T(t)P(A+BK)X(t) \\ &\quad + X^T(t)PBw(0,t) + \frac{2|PB|^2}{\lambda_{\min}(Q)}\int_0^D (1+x)2w\frac{\partial w}{\partial x}dx \\ &= -X^T(t)QX(t) + w^T(0,t)B^T PX(t) + X^T(t)PBw(0,t) \\ &\quad + \frac{2|PB|^2}{\lambda_{\min}(Q)}\int_0^D \left(\frac{\partial}{\partial x}\Big((1+x)w^2\Big) - \frac{\partial(1+x)}{\partial x}w^2\right)dx \\ &= -X^T(t)QX(t) + w^T(0,t)B^T PX(t) + X^T(t)PBw(0,t) \\ &\quad + \frac{2|PB|^2}{\lambda_{\min}(Q)}(1+D)\underbrace{w^2(D,t)}_{0} - \frac{2|PB|^2}{\lambda_{\min}(Q)}w^2(0,t) \\ &\quad - \frac{2|PB|^2}{\lambda_{\min}(Q)}\int_0^D w^2(x,t)dx \end{aligned}$$

$$
\begin{aligned}
&= -X^T(t)QX(t) + w^T(0,t)B^T PX(t) + X^T(t)PBw(0,t) \\
&\quad -\frac{2|PB|^2}{\lambda_{\min}(Q)} w^2(0,t) - \frac{2|PB|^2}{\lambda_{\min}(Q)} \int_0^D w^2(x,t)dx.
\end{aligned} \tag{20}
$$

Thus

$$
\begin{aligned}
\frac{dV}{dt} &= -\begin{bmatrix} X(t) & w(0,t) \end{bmatrix} \begin{bmatrix} Q & -PB \\ -B^T P^T & \dfrac{2|PB|^2}{\lambda_{\min}(Q)} \end{bmatrix} \begin{bmatrix} X(t) \\ w(0,t) \end{bmatrix} \\
&\quad -\frac{2|PB|^2}{\lambda_{\min}(Q)} \int_0^D w^2(x,t)dx.
\end{aligned} \tag{21}
$$

Since the matrix

$$
\begin{bmatrix} Q & -PB \\ -B^T P^T & \dfrac{2|PB|^2}{\lambda_{\min}(Q)} \end{bmatrix} \tag{22}
$$

is positive definite, it follows that the equilibrium $X = 0, w(x) \equiv 0$ is exponentially stable in the sense of the euclidean norm on X and the L_2 norm of w. With further analysis, exponential stability of the equilibrium $X = 0, u(x) \equiv 0$ is also established, obtaining the following theorem.

Theorem 1. *There exist positive constants G and g such that the solutions of the closed-loop system (1), (2) satisfy* $\Gamma(t) \leq G e^{-gt}\Gamma(0)$ *for all* $t \geq 0$, *where*

$$
\Gamma(t) = |X(t)|^2 + \int_0^D u(x,t)^2 dx. \tag{23}
$$

In the literature on delay systems the representation through the transport PDE state (5) is somewhat non-standard. The constructions provided in the transport PDE notation can also be expressed in the delay notation, such that the Lyapunov functional (18) is written as

$$
V(t) = X(t)^T PX(t) + 2\frac{|PB|^2}{\lambda_{\min}(Q)} \int_{t-D}^t (1 + \theta + D - t)W(\theta)^2 d\theta, \tag{24}
$$

and the backstepping transformation (3) is

$$
W(\theta) = U(\theta) - K\left[\int_{t-D}^{\theta} e^{A(\theta-\sigma)} BU(\sigma)d\sigma + e^{A(\theta+D-t)}X(t)\right], \tag{25}
$$

with $-D \leq t - D \leq \theta \leq t$. We pursue the PDE notation for delay systems so we can seamlessly transition to PDE problems in the latter sections of the article.

From this point on, the ability to construct a Lyapunov functional can be exploited in various ways, including deriving disturbance attenuation estimates when the system (1) is subject to an additive disturbance, proving robustness to a small actuator lag, and conducting an inverse optimal redesign of the predictor feedback. We consider these three problems in the current section. In subsequent sections, we present further, more substantial, benefits of constructing a Lyapunov functional and a backstepping transformation. These benefits are the establishment of robustness to a small error in D, where the error is allowed to be either positive or negative, the design of adaptive controllers in the presence of a completely unknown and arbitrarily long D, the design of stabilizing predictor feedback for time varying delays, and the design of predictor feedback for some classes of nonlinear and PDE systems.

We now consider the system

$$\dot{X}(t) = AX(t) + BU(t-D) + B_1 d(t), \tag{26}$$

where $d(t)$ is an unmeasurable disturbance which is bounded but its bound is unknown, and the controller

$$U(t) = \frac{c}{s+c}\left\{K\left[\mathrm{e}^{AD}X(t) + \int_{t-D}^{t} \mathrm{e}^{A(t-\theta)}BU(\theta)d\theta\right]\right\}, \tag{27}$$

where $c > 0$, and where we use the transfer function representation for compactness of notation.

We introduce the Lyapunov functional

$$V(t) = X(t)^T PX(t) + 2\frac{|PB|^2}{\lambda_{\min}(Q)}\int_0^D (1+x)w(x,t)^2 dx + \frac{1}{2}w(D,t)^2. \tag{28}$$

Note that, due to the change in the control law from (2) to (27), the quantity $w(D,t)$, which is zero in (8), is not zero in (28). The following result is established with (28).

Theorem 2. *There exists a positive constant c^* such that for all $c > c^*$, the feedback system (26), (27) is L_∞-stable, that is, there exist positive constants $\beta_1, \beta_2, \gamma_1$ such that*

$$N(t) \leq \beta_1 \mathrm{e}^{-\beta_2 t} N(0) + \gamma_1 \sup_{\tau\in[0,t]} |d(\tau)|, \tag{29}$$

where

$$N(t) = \left(|X(t)|^2 + \int_{t-D}^{t} U(\theta)^2 d\theta + U(t)^2\right)^{1/2}. \tag{30}$$

Furthermore, there exists a constant $c^{**} > c^*$ such that for all $c \geq c^{**}$ the feedback (27) minimizes the cost functional

$$J = \sup_{d\in\mathscr{D}} \lim_{t\to\infty} \left[2cV(t) + \int_0^t \left(Q(\tau) + \dot{U}(t)^2 - c\gamma_2 d(\tau)^2\right) d\tau\right] \tag{31}$$

for each

$$\gamma_2 \geq \gamma_2^{**} = 8\frac{|PB|^2}{\lambda_{\min}(Q)}, \tag{32}$$

where $Q(t) \geq \mu N(t)^2$ for some $\mu(c,\gamma_2) > 0$, which is such that $\mu(c,\gamma_2) \to \infty$ as $c \to \infty$, and $\mathscr{D}$ is the set of linear scalar-valued functions of X.

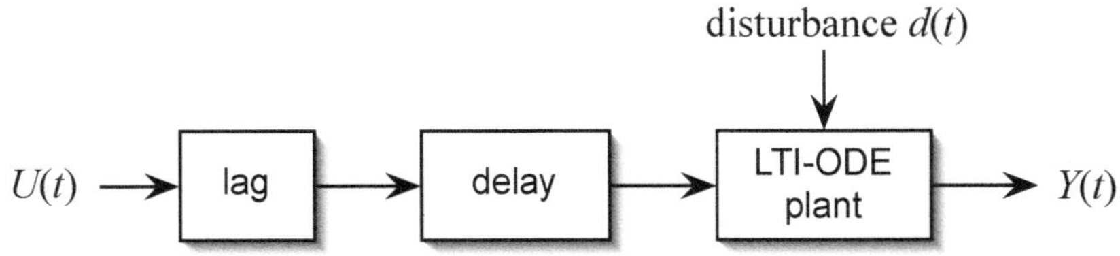

Fig. 1. An ODE with input delay and with an unmodeled input lag and additive disturbance. A suitable form of robustness holds with respect to both perturbations under predictor feedback (2), as stated in Theorem 7.

The following four special cases can be inferred from Theorem 2. First, the predictor feedback (2) is robust to the introduction of a lag $\frac{c}{s+c}$ for sufficiently high c. The lag can be either a part of the control law, as in (27), or an unmodeled part of the system dynamics, as shown in Figure 1. This robustness to input lag may not be surprising, but it is not obvious, in the light of various negative results on robustness of hyperbolic PDEs to infinitesimal perturbations. Second, the system under predictor feedback (2), as well as under feedback (27) with sufficiently high c, has a finite L_∞ gain relative to an additive disturbance. Third, the feedback (27) is an inverse optimal stabilizer for sufficiently high but finite c, in the absence of the disturbance d. This result is not so easy to see intuitively. It is obtained by writing the feedback law in terms of $\dot{U}(t)$ as the control input, in which case the feedback law is of the (Lie derivative) form '$-L_gV$' [13]. Fourth, in the presence of the disturbance, the feedback (27) with sufficiently high c is an inverse optimal solution to a differential game problem [14] with a positive definite penalty on the state and control, and a negative-definite penalty on the disturbance.

2 Delay-Robustness, Delay-Adaptivity, and Time-Varying Delays

In control systems with input delay, the length of the delay is the most significant possible uncertainty, both in the sense of robustness to a small mismatch in the delay D when designing constant predictor feedback and in the sense of designing delay-adaptive predictor feedback for a large uncertainty in the delay D.

2.1 Robustness to Delay Mismatch

We first discuss the problem of robustness to delay mismatch, as depicted in Figure 2, and consider the feedback system

$$\dot{X}(t) = AX(t) + BU(t - D_0 - \Delta D), \tag{33}$$

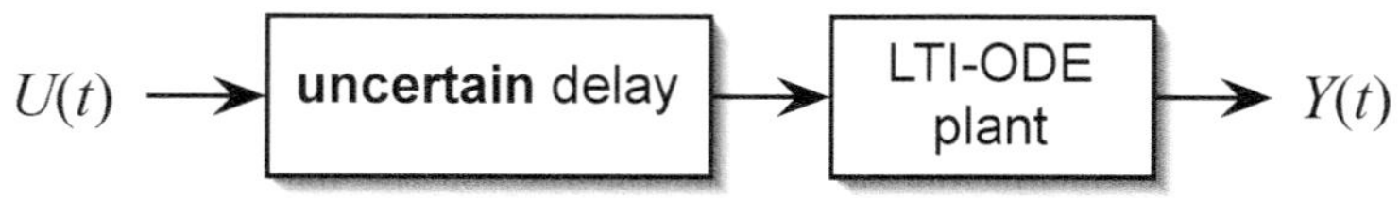

Fig. 2. An ODE with input delay which is known up to a small mismatch error ΔD, which can be either positive or negative. Stability is preserved under predictor feedback (34) for sufficiently small $|\Delta D|$ but arbitrarily large D, as stated in Theorem 3.

$$U(t) = K\left[e^{AD_0}X(t) + \int_{t-D_0}^{t} e^{A(t-\theta)}BU(\theta)d\theta\right]. \tag{34}$$

The actuator delay mismatch ΔD can be either positive or negative relative to the assumed actuator delay $D_0 > 0$. However, the actual delay must be nonnegative, $D_0 + \Delta D \geq 0$. For the study of robustness to a small ΔD, we use two different Lyapunov functionals, one for $\Delta D > 0$, which is the easier of the two cases, and another for $\Delta D < 0$, in which case we employ

$$V(t) = X(t)^T PX(t) + \frac{a}{2}\int_0^{D_0+\Delta D}(1+x)w(x,t)^2dx + \frac{1}{2}\int_{\Delta D}^{0}(D_0+x)w(x,t)^2dx \tag{35}$$

with a sufficiently large a.

Theorem 3. *There exists a positive constant δ such that for all $\Delta D \in (-\delta,\delta)$ there exist positive constants G and g such that the solutions of the closed-loop system (33), (34) satisfy $\Gamma(t) \leq Ge^{-gt}\Gamma(0)$ for all $t \geq 0$, where*

$$\Gamma(t) = |X(t)|^2 + \int_{t-\bar{D}}^{t} U(\theta)^2 d\theta \tag{36}$$

and where

$$\bar{D} = D_0 + \max\{0, \Delta D\}. \tag{37}$$

The significance of this robustness result can be assessed based on the intuition drawn from existing results. For example, finite-dimensional feedback laws for finite-dimensional plants are robust to small delays [15], however, this result does not apply to our infinite-dimensional problem. The delay perturbation to predictor feedback incorporates the possibility of two different classes of perturbations, depending on whether ΔD is positive or negative, so off-the-shelf results cannot be used.

The result of Theorem 3 may be surprising in light of Datko's negative result on delay-robustness for certain examples of hyperbolic PDEs with boundary control [16]. Even though the input-delay problem also involves a hyperbolic PDE, such a negative result does not hold for predictor feedback because of a significant difference between first-order and second-order hyperbolic PDEs. The second-order hyperbolic PDEs in Datko's work have infinitely many eigenvalues on the imaginary axis, whereas is not the case with an ODE with input delay, even when the ODE is unstable, only a finite number of open-loop eigenvalues may be in the closed right-half plane.

2.2 Delay-Adaptive Control

Now we turn our attention from robustness to small delay mismatch to adaptivity for large delay uncertainty. Several results exist on adaptive control of systems with input delays, including [17, 18]. However, existing results deal with parametric uncertainties in the ODE plant, whereas the key challenge is uncertainty in the delay.

Let us consider the plant (1) but with a transport PDE representation of the input delay given as

$$\dot{X}(t) = AX(t) + Bu(0,t), \tag{38}$$

$$Du_t(x,t) = u_x(x,t), \tag{39}$$

$$u(1,t) = U(t). \tag{40}$$

Here, the actuator state is defined as

$$u(x,t) = U(t + D(x-1)), \qquad x \in [0,1] \tag{41}$$

instead of the definition (5) with $x \in [0,D]$. We take the predictor feedback in the certainty equivalence form

$$U(t) = K\left[e^{A\hat{D}(t)}X(t) + \hat{D}(t)\int_0^1 e^{A\hat{D}(t)(1-y)}Bu(y,t)dy\right], \tag{42}$$

where the update law for the estimate $\hat{D}(t)$ is designed as

$$\dot{\hat{D}}(t) = \gamma \mathrm{Proj}_{[0,\bar{D}]}\{\tau(t)\}, \tag{43}$$

$$\tau(t) = -\frac{\int_0^1 (1+x)w(x,t)Ke^{A\hat{D}(t)x}dx\,(AX(t)+Bu(0,t))}{1+X(t)^T PX(t) + b\int_0^1 (1+x)w(x,t)^2 dx}, \tag{44}$$

$$w(x,t) = u(x,t) - \hat{D}(t)\int_0^x Ke^{A\hat{D}(t)(x-y)}Bu(y,t)dy - Ke^{A\hat{D}(t)x}X(t), \tag{45}$$

where $b \geq \frac{4\bar{D}|PB|^2}{\lambda_{\min}(Q)}$ and where $\bar{D}$ is an a priori known upper bound on D. The standard projection operator projects $\hat{D}(t)$ into the interval $[0,\bar{D}]$. The structure of the adaptive control system is shown in Figure 3. The choice of the update law (43)–(45) is motivated by a rather subtle Lyapunov analysis, resulting in a normalization of the update law, without the use of any filters or overparametrization.

Theorem 4. *Consider the closed-loop adaptive system (38)–(45). There exists* $\gamma^* >$ 0 *such that for all* $\gamma \in (0,\gamma^*)$ *there exist positive constants* R *and* ρ *(independent of the initial conditions) such that for all initial conditions satisfying* $(X_0, u_0, \hat{D}_0) \in$

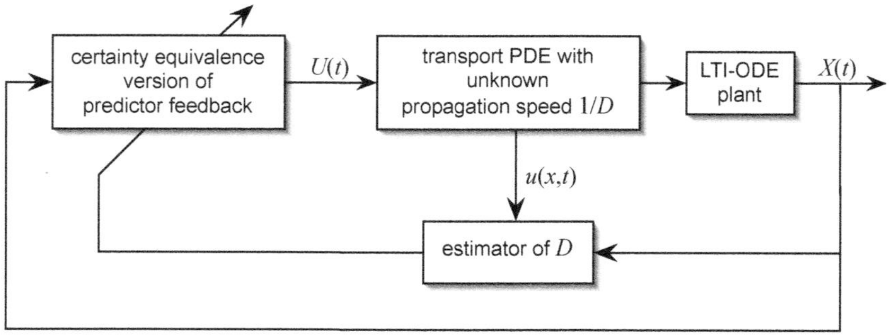

Fig. 3. Delay-adaptive predictor feedback for a true delay D varying in a broad range from 0 to a possibly large value $\bar{D}$. The certainty-equivalence controller (42) is combined with the update law (43)–(45). Global stability and regulation of the state and control are achieved, as specified in Theorem 4.

$\mathbb{R}^n \times L_2[0,1] \times [0,\bar{D}]$, *the norm of the solutions obeys an exponential bound relative to the norm of initial conditions, namely*

$$\Upsilon(t) \le R\left(e^{\rho\Upsilon(0)} - 1\right), \quad \text{for all } t \ge 0, \tag{46}$$

where

$$\Upsilon(t) = |X(t)|^2 + \int_0^1 u(x,t)^2 dx + \left(D - \hat{D}(t)\right)^2. \tag{47}$$

Furthermore

$$\lim_{t\to\infty} X(t) = 0, \quad \lim_{t\to\infty} U(t) = 0. \tag{48}$$

Example 1. We illustrate the delay-adaptive design for the example plant

$$X(s) = \frac{e^{-s}}{(s-0.75)} U(s) \tag{49}$$

with the simulation results given in Figure 4. The period up to 1 sec is the dead time, the parameter estimation is active until about 3 sec, the control evolution is exponential (corresponding to a predominantly LTI system) after 3 sec, and the state evolves exponentially after 4 sec. The adaptive controller is successful both with $\hat{D}(0) = 0$ and with $\hat{D}(0) = 2D$ (100% parameter error in both cases). ■

The controller (42)–(45) uses full state measurement of the transport PDE state. In the absence of such measurement, a slightly different design guarantees local stability, which is the strongest result achievable in that case due to a nonlinear parametrization of the operator e^{-Ds}.

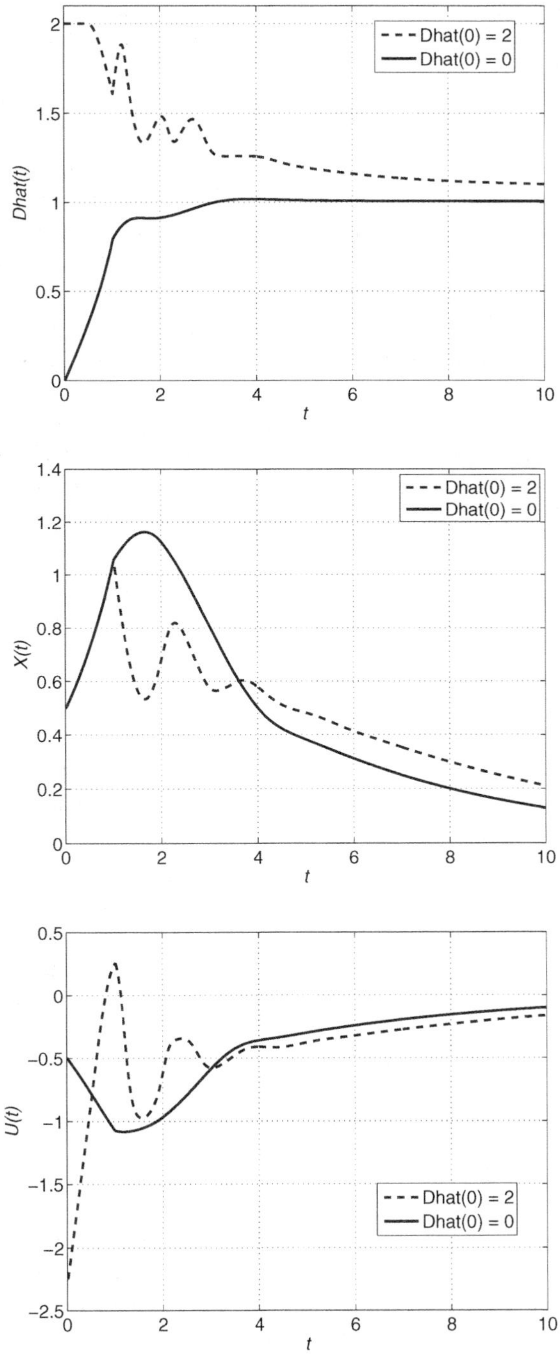

Fig. 4. Time responses of $\hat{D}(t)$, $X(t)$, and $U(t)$ under delay-adaptive predictor feedback for an unstable first-order plant. Stabilization is achieved both with $\hat{D}(0) = 0$ and with a $\hat{D}(0)$ that heavily overestimates the true D.

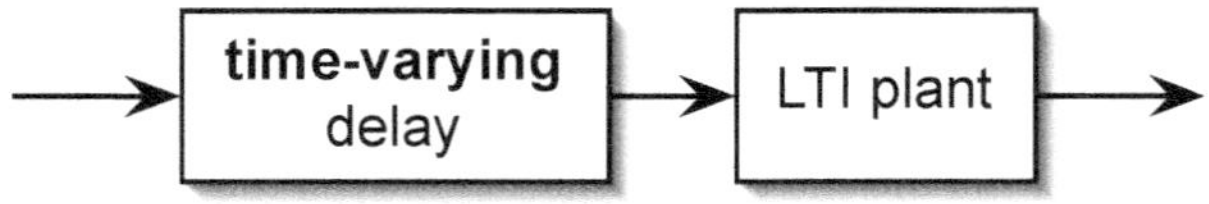

Fig. 5. Linear system $\dot{X}(t) = AX(t) + BU(\phi(t))$ with time-varying actuator delay $\delta(t) = t - \phi(t)$. The predictor feedback (51) with compensation of the time-varying delay achieves exponential stabilization in the sense of Theorem 5.

2.3 *Time-Varying Input Delay*

Before we close this section on uncertain delays, let us briefly turn our attention to the problem of *time-varying* known input delays, which is depicted in Figure 5. We consider the system

$$\dot{X}(t) = AX(t) + BU(\phi(t)). \tag{50}$$

A predictor feedback for this system is

$$U(t)=K\left[e^{A(\phi^{-1}(t)-t)}X(t) + \int_{\phi(t)}^{t} e^{A(\phi^{-1}(t)-\phi^{-1}(\theta))}B\frac{U(\theta)}{\phi'(\phi^{-1}(\theta))}d\theta\right], \text{ for all } t \geq 0. \tag{51}$$

With rather extensive effort, going through a transport PDE representation with $u(x,t) = U\left(\phi\left(t + x\left(\phi^{-1}(t) - t\right)\right)\right)$ and the time-varying backstepping transformation

$$w(x,t) = u(x,t) - Ke^{Ax(\phi^{-1}(t)-t)}X(t) - K\int_0^x e^{A(x-y)(\phi^{-1}(t)-t)}Bu(y,t)\left(\phi^{-1}(t) - t\right)dy \tag{52}$$

into the target system

$$\dot{X}(t) = (A+BK)X(t) + Bw(0,t), \tag{53}$$

$$w_t(x,t) = \pi(x,t)w_x(x,t), \tag{54}$$

$$w(1,t) = 0, \tag{55}$$

where the variable speed of propagation of the transport equation w is given by

$$\pi(x,t) = \frac{1 + x\left(\frac{d\left(\phi^{-1}(t)\right)}{dt} - 1\right)}{\phi^{-1}(t) - t}, \tag{56}$$

we obtain the following stabilization result.

Theorem 5. *Consider the closed-loop system (50), (51). Let the delay function $\delta(t) = t - \phi(t)$ be strictly positive and uniformly bounded from above. Let the delay rate function $\delta'(t)$ be strictly smaller than 1 and uniformly bounded from below.*

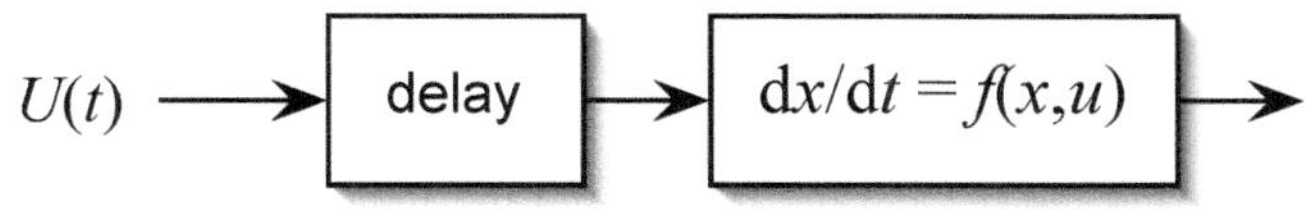

Fig. 6. Nonlinear control in the presence of arbitrarily long input delay. Global stabilization is achieved with the predictor feedback (59)–(61) if the plant is forward complete and globally asymptotically stabilizable in the absence of delay, as stated in Theorem 6.

There exist positive constants G and g (the latter one being independent of ϕ) such that

$$|X(t)|^2+\int_{\phi(t)}^{t}U^2(\theta)d\theta\le Ge^{-gt}\left(|X_0|^2+\int_{\phi(0)}^{0}U^2(\theta)d\theta\right),\quad \textit{for all } t\ge 0. \tag{57}$$

3 Predictor Feedback for Nonlinear Systems

In the area of robust nonlinear control various types of uncertainties are considered—unmeasurable disturbances, static nonlinear functional perturbations, dynamic perturbations on the state, and dynamic perturbations on the input. The unmodeled *input* dynamics are represent the greatest challenge in robust nonlinear control. It is for this reason no surprise that long delays at the input of nonlinear systems, as depicted in Figure 6, have remained an unsolved challenge in nonlinear control. Considerable success has been achieved in recent years with control of nonlinear systems with state delay [19, 20, 21, 22], however, only one result exists where input delay of arbitrary length is being addressed [23]. Systematic *compensation* of input delays of arbitrary length is non-existent.

A conceptually easy and natural way to compensate input delays in nonlinear control is through an extension of predictor feedback to nonlinear systems, which we present next. Consider the general class of nonlinear systems

$$\dot{X}(t)=f(X(t),U(t-D)),\quad f(0,0)=0, \tag{58}$$

and assume that a feedback law $U=\kappa(X)$ with $\kappa(0)=0$ is known which globally asymptotically stabilizes the system at the origin when $D=0$. Denote the initial conditions as $Z_0=Z(0)$ and $U_0(\theta)=U(\theta),\theta\in[-D,0]$. A predictor feedback is given by

$$U(t)=\kappa(P(t)), \tag{59}$$

where the predictor is defined as

$$P(t)=\int_{t-D}^{t}f(P(\theta),U(\theta))d\theta+Z(t),\quad t\ge 0, \tag{60}$$

$$P(\theta) = \int_{-D}^{\theta} f(P(\sigma), U_0(\sigma))d\sigma + Z_0, \quad \theta \in [-D, 0]. \tag{61}$$

A key feature to note about the predictor $P(t)$ is that it is defined implicitly, through a nonlinear integral equation, rather than explicitly, through matrix exponentials and the variation of constants formula, as is the case when the plant is linear. The lack of an explicit formula for $P(t)$ is not necessarily an obstacle numerically, since $P(t)$ is defined in terms of its past values.

The more serious question is conceptual, does a solution for $P(t)$ always exist? Fortunately, the answer to this question is rather simple. Since control has no effect for D seconds after it has been applied, the system can indeed exhibit finite escape over that period, resulting in a finite escape for $P(t)$ since the predictor is governed by the same model as the plant. Hence, a natural way to ensure global existence of the predictor state is to assume that the plant is *forward complete*.

A system is said to be forward complete if, for all initial conditions and all locally bounded input signals, its solutions exist for all time. This definition does not require the solutions to be *uniformly* bounded. They can be growing to infinity as time goes to infinity. For example, all LTI systems, stable or unstable, driven by inputs of exponential growth, are forward complete. The same is true of nonlinear systems with globally Lipschitz right-hand sides, but also of many systems that are neither globally Lipschitz nor stable but contain super-linear nonlinearities that induce limit cycles, rather than finite escape.

The nonlinear predictor design is developed for two classes of systems. For the broad class of *forward complete* systems, that is, systems that do not exhibit a finite escape time for any initial condition and any input signals that remain finite over finite time intervals, which includes many mechanical and other systems, predictor feedback is developed which achieves global asymptotic stability, as long as the system without delay is globally asymptotically stabilizable. However, the predictor requires the solution of a nonlinear integral equation, or a nonlinear DDE, in real time.

Theorem 6. *Let $\dot{X} = f(X,U)$ be forward complete and $\dot{X} = f(X,\kappa(X))$ be globally asymptotically stable at $X = 0$. Consider the closed-loop system (58)–(61). There exists a function $\hat{\beta} \in \mathscr{K}\mathscr{L}$ such that*

$$|Z(t)| + \|U\|_{L_\infty[t-D,t]} \leq \hat{\beta}\left(|Z(0)| + \|U_0\|_{L_\infty[-D,0]}, t\right) \tag{62}$$

for all $(Z_0, U_0) \in \mathbb{R}^n \times L_\infty[-D,0]$ and for all $t \geq 0$.

As we have mentioned above, the only weakness of predictor feedback laws is that $P(t)$ may not be explicitly computable. Fortunately, a significant class of nonlinear system exists which are not only forward complete and globally stabilizable, but where $P(t)$ is also explicitly computable. This is the class of *strict-feedforward* systems [13].

Example 2. We illustrate the explicit computability of the predictor, and thus of the feedback law, for an example of a strict-feedforward system. Consider the third-order system.

$$\dot{X}_1(t) = X_2(t) + X_3^2(t), \tag{63}$$
$$\dot{X}_2(t) = X_3(t) + X_3(t)U(t-D), \tag{64}$$
$$\dot{X}_3(t) = U(t-D), \tag{65}$$

which is not feedback linearizable and is in the strict-feedforward class. The globally asymptotically stabilizing predictor feedback for this system is given by

$$\begin{aligned} U(t) &= -P_1(t) - 3P_2(t) - 3P_3(t) - \frac{3}{8}P_2^2(t) \\ &\quad + \frac{3}{4}P_3(t)\left(-P_1(t) - 2P_2(t) + \frac{1}{2}P_3(t) + \frac{P_2(t)P_3(t)}{2}\right. \\ &\quad \left. + \frac{5}{8}P_3^2(t) - \frac{1}{4}P_3^3(t) - \frac{3}{8}\left(P_2(t) - \frac{P_3^2(t)}{2}\right)^2\right), \end{aligned} \tag{66}$$

where the D-second-ahead predictor of $(X_1(t), X_2(t), X_3(t))$ is given explicitly by

$$\begin{aligned} P_1(t) &= X_1(t) + DX_2(t) + \frac{1}{2}D^2X_3(t) + DX_3^2(t) + 3X_3(t)\int_{t-D}^{t}(t-\theta)U(\theta)d\theta \\ &\quad + \frac{1}{2}\int_{t-D}^{t}(t-\theta)^2U(\theta)d\theta + \frac{3}{2}\int_{t-D}^{t}\left(\int_{t-D}^{\theta}U(\sigma)d\sigma\right)^2 d\theta, \end{aligned} \tag{67}$$

$$\begin{aligned} P_2(t) &= X_2(t) + DX_3(t) + X_3(t)\int_{t-D}^{t}U(\theta)d\theta + \int_{t-D}^{t}(t-\theta)U(\theta)d\theta \\ &\quad + \frac{1}{2}\left(\int_{t-D}^{t}U(\theta)d\theta\right)^2, \end{aligned} \tag{68}$$

$$P_3(t) = X_3(t) + \int_{t-D}^{t}U(\theta)d\theta. \tag{69}$$

Note that the nonlinear infinite-dimensional feedback operator employs a finite Volterra series in $U(\theta)$. ■

4 Delay-PDE Cascades

When a plant with an input delay is a PDE, such as, for example, in Figure 7, special challenges arise in the design of predictor feedback, particularly if the PDE is actuated through boundary control, which makes the input operator (commonly denoted as B) unbounded. In [12] we consider two benchmark delay-PDE cascades, one where the plant is a parabolic PDE and the other where the plant is a second-order hyperbolic PDE. We review here the parabolic case, where the plant is an unstable reaction-diffusion equation with an arbitrarily large number of unstable eigenvalues in open loop.

Consider the PDE system

$$u_t(x,t) = u_{xx}(x,t) + \lambda u(x,t), \tag{70}$$

Fig. 7. Control of an unstable parabolic PDE with input delay, that is, of a boundary controlled cascade of a transport PDE and a reaction-diffusion PDE. Explicit gains are derived for the predictor feedback (73). As stated in Theorem 7, stability is achieved in a somewhat non-standard Sobolev norm, rather than in the basic L_2 norm of the state of the PDE cascade.

$$u(0,t) = 0, \tag{71}$$
$$u(1,t) = U(t-D), \tag{72}$$

where λ is an arbitrary constant. We derive a stabilizing feedback law in the explicit form

$$\begin{aligned} U(t) = {} & 2\sum_{n=1}^{\infty}\int_0^1 \sin(\pi n\xi)\lambda\xi \frac{I_1\left(\sqrt{\lambda(1-\xi^2)}\right)}{\sqrt{\lambda(1-\xi^2)}}d\xi \\ & \times\left(-e^{(\lambda-\pi^2n^2)D}\int_0^1 \sin(\pi n y)u(y,t)dy\right. \\ & \left. +\pi n(-1)^n\int_{t-D}^t e^{(\lambda-\pi^2n^2)(t-\theta)}U(\theta)d\theta\right), \end{aligned} \tag{73}$$

where $I_1(\cdot)$ is a Bessel function.

Theorem 7. *Consider the closed-loop system (70)–(73). There exists a positive continuous function $\rho : \mathbb{R}^2 \to \mathbb{R}_+$ such that, for all initial conditions $(u_0, U_0) \in L_2[0,1] \times H_1[0,D]$, and for all $c > 0$, the solutions are bounded in the following sense*

$$\Upsilon(t) \le \rho(D,\lambda)e^{cD}\Upsilon(0)e^{-\min\{2,c\}t}, \qquad \textit{for all } t \ge 0, \tag{74}$$

where

$$\Upsilon(t) = \int_0^1 u^2(x,t)dx + \int_{t-D}^t \left(U^2(\theta) + \dot{U}^2(\theta)\right)d\theta. \tag{75}$$

Two elements of this result are of significance and they arise in any application of predictor feedback to PDEs with boundary control. First, the feedback law (73) is derived explicitly. The explicit determination of the control gains is made possible by first deriving the control gain for $D = 0$ explicitly, which was achieved in [24], and then by solving the undriven version of the PDE system (70)–(72) with an initial condition given by the control gain for $D = 0$. In more specific terms, we solve the PDE systems

$$k_{xx}(x,y) = k_{yy}(x,y) + \lambda k(x,y), \qquad 0 \le y \le x \le 1, \tag{76}$$
$$k(x,0) = 0, \tag{77}$$

$$k(x,x) = -\frac{\lambda}{2}x, \tag{78}$$

and

$$\gamma_x(x,y) = \gamma_{yy}(x,y) + \lambda\gamma(x,y), \qquad (x,y) \in [1, 1+D] \times (0,1), \tag{79}$$
$$\gamma(x,0) = 0, \tag{80}$$
$$\gamma(x,1) = 0, \tag{81}$$
$$\gamma(1,y) = k(1,y). \tag{82}$$

Note that the k-system is hyperbolic and defined on a triangular domain, whereas the γ-system is parabolic and defined on a rectangular semi-infinite domain, as well as that the solution to the k-system acts as an initial condition to the γ-system, as given by (82). The process of explicitly solving for $\gamma(x,y)$ is the PDE equivalent of analytically finding the vector $K\mathrm{e}^{AD}$ in (2).

Second, when dealing with boundary control of a PDE with input delay, we are facing the problem of control of two PDEs from different classes, such as a parabolic PDE and a first-order hyperbolic PDE in the case covered here, where the PDEs are interconnected through a boundary. While for each one of the two PDEs individually a natural system norm may be the standard L_2 norm, for the interconnected system this may not be the case and a higher order norm may have to be used for one of the subsystems, as is the case in (75).

5 Conclusions

The PDE backstepping approach is a potentially powerful tool in advancing the design techniques for systems with input and output delays. Two ideas presented in this article may be of interest to researchers in delay systems. The first idea is the construction of backstepping transformations that allow one to deal with delays and PDE dynamics at the input, as well as in the main line of applying control action, such as in the chain of integrators for systems in triangular forms. The second idea is the construction of Lyapunov functionals and explicit stability estimates, with the help of direct and inverse backstepping transformations.

References

1. Smith, O.J.M.: A controller to overcome dead time. ISA 6, 28–33 (1959)
2. Manitius, A.Z., Olbrot, A.W.: Finite spectrum assignment for systems with delays. IEEE Trans. on Automatic Control 24, 541–553 (1979)
3. Kwon, W.H., Pearson, A.E.: Feedback stabilization of linear systems with delayed control. IEEE Trans. on Automatic Control 25, 266–269 (1980)
4. Artstein, Z.: Linear systems with delayed controls: a reduction. IEEE Trans. on Automatic Control 27, 869–879 (1982)
5. Zhong, Q.-C.: Robust Control of Time-delay Systems. Springer, Heidelberg (2006)

6. Michiels, W., Niculescu, S.-I.: Stability and Stabilization of Time-Delay Systems: An Eigenvalue-Based Approach. SIAM, Philadelphia (2007)
7. Mondie, S., Michiels, W.: Finite spectrum assignment of unstable time-delay systems with a safe implementation. IEEE Trans. on Automatic Control 48, 2207–2212 (2003)
8. Zhong, Q.-C.: On distributed delay in linear control laws—Part I: Discrete-delay implementation. IEEE Transactions on Automatic Control 49, 2074–2080 (2006)
9. Zhong, Q.-C., Mirkin, L.: Control of integral processes with dead time—Part 2: Quantitative analysis. IEE Proc. Control Theory & Appl. 149, 291–296 (2002)
10. Krstic, M., Smyshlyaev, A.: Boundary Control of PDEs: A Course on Backstepping Designs. SIAM, Philadelphia (2008)
11. Vazquez, R., Krstic, M.: Control of Turbulent and Magnetohydrodynamic Channel Flows. Birkhauser, Basel (2007)
12. Krstic, M.: Delay Compensation for Nonlinear, Adaptive, and PDE Systems. Birkhauser, Basel (2009)
13. Sepulchre, R., Jankovic, M., Kokotovic, P.: Constructive Nonlinear Control. Springer, Heidelberg (1997)
14. Krstic, M., Deng, H.: Stabilization of Nonlinear Uncertain Systems. Springer, Heidelberg (1998)
15. Teel, A.R.: Connections between Razumikhin-type theorems and the ISS nonlinear small gain theorem. IEEE Transactions on Automatic Control 43, 960–964 (1998)
16. Datko, R.: Not all feedback stabilized hyperbolic systems are robust with respect to small time delays in their feedbacks. SIAM Journal on Control and Optimization 26, 697–713 (1988)
17. Ortega, R., Lozano, R.: Globally stable adaptive controller for systems with delay. Internat. J. Control 47, 17–23 (1988)
18. Niculescu, S.-I., Annaswamy, A.M.: An Adaptive Smith-Controller for Time-delay Systems with Relative Degree $n* \geq 2$. Systems and Control Letters 49, 347–358 (2003)
19. Jankovic, M.: Control Lyapunov-Razumikhin functions and robust stabilization of time delay systems. IEEE Trans. on Automatic Control 46, 1048–1060 (2001)
20. Germani, A., Manes, C., Pepe, P.: Input-output linearization with delay cancellation for nonlinear delay systems: the problem of the internal stability. International Journal of Robust and Nonlinear Control 13, 909–937 (2003)
21. Karafyllis, I.: Finite-time global stabilization by means of time-varying distributed delay feedback. SIAM Journal of Control and Optimization 45, 320–342 (2006)
22. Mazenc, F., Bliman, P.-A.: Backstepping design for time-delay nonlinear systems. IEEE Transactions on Automatic Control 51, 149–154 (2006)
23. Mazenc, F., Mondie, S., Francisco, R.: Global asymptotic stabilization of feedforward systems with delay at the input. IEEE Trans. Automatic Control 49, 844–850 (2004)
24. Smyshlyaev, A., Krstic, M.: Closed form boundary state feedbacks for a class of 1D partial integro-differential equations. IEEE Trans. on Automatic Control 49(12), 2185–2202 (2004)

Boundary Value Problems and Convolutional Systems over Rings of Ultradistributions

Hugues Mounier, Joachim Rudolph, and Frank Woittennek

Abstract. One dimensional boundary value problems with lumped controls are considered. Such systems can be modeled as modules over a ring of Beurling ultradistributions with compact support. This ring appears naturally from a corresponding Cauchy problem. The heat equation with different boundary conditions serves for illustration.

1 Introduction

The design of feedforward and feedback control for finite dimensional systems and delay systems is largely simplified by flatness based control, respectively freeness. This has been shown in numerous academic case studies and industrial applications. A central part in the control design (the importance of which has often been under-estimated) is trajectory planning.

It is particularly useful for distributed parameter systems with lumped control inputs, a class of systems the models of which include partial differential equations. In the linear case, as for delay systems, a module-theoretic framework has been established, and the trajectory planning is based on the use of a module basis, which plays a role similar to the one of a flat output in finite-dimensional flat systems.

Examples of distributed parameter systems that have been studied are heat conductors, elastic piezo-beams and plates, elastic robot arms, ropes, electric lines,

Hugues Mounier
Laboratoire des Signaux et Systèmes, Supélec, 3, rue Joliot Curie, 91192 Gif-sur-Yvette, France

Joachim Rudolph
Lehrstuhl für Systemtheorie und Regelungstechnik, Universität des Saarlandes, Campus A5 1, 66123 Saarbrücken, Germany
e-mail: j.rudolph@lsr.uni-saarland.de

Frank Woittennek
Institut für Regelungs- und Steuerungstheorie, TU Dresden, 01062 Dresden, Germany
e-mail: frank.woittennek@tu-dresden.de

J. Lévine & P. Müllhaupt (Eds.): Adv. in Theory of Control, Signals, LNCIS 407, pp. 179–188.
springerlink.com © Springer-Verlag Berlin Heidelberg 2010

tubular chemical reactors, and heat exchangers (see, e.g., [16, 17]). Although many of the problems considered are linear with fixed boundary, some nonlinear and free boundary value problems have been solved, too.

Here, based on the example of the linear heat equation the choice of the ring used to represent the system as a module is further discussed. It is shown that a suitable ring is $\mathscr{R} = \mathbb{C}(\partial_t)[\mathfrak{S}] \cap \mathscr{E}'^*$, where ∂_t stands for time derivation, $\mathfrak{S}$ is a collection of spatially dependant hyperbolic functions, and $\mathscr{E}'^*$ is a ring of Beurling ultradistributions.

2 Motivating Example: The Heat Equation

The one dimensional heat equation might be viewed as one of the simplest problems of the class considered in the sequel. It will, therefore, be used for motivation. Moreover, this discussion is based on elementary calculations, which allow one to capture the idea of the approach without entering into deeper mathematical considerations.

Consider the system

$$\partial_x^2 w(x,t) = \partial_t w(x,t), \qquad x \in [0,1],\ t \in \mathbb{R} \tag{1a}$$

$$\partial_x w(0,t) = 0, \qquad w(1,t) = u(t) \tag{1b}$$

with homogeneous initial conditions. These equations model the heat conduction in a rod of unit length, where $w(x,t)$ denotes the temperature at the point x at time t. The first boundary condition means that there is no heat flux at $x = 0$, the second one means that the temperature at $x = 1$ is considered as a control input $u(t)$.

2.1 *Symbolic Viewpoint*

Use the Laplace transform w.r.t. t to obtain

$$s\widehat{w}(x,s) = \partial_x^2 \widehat{w}(x,s) \tag{2}$$

from (1a). (Mikusiński's oprational calculus would lead to similar formulae.) The characteristic equation associated with (2) reads $\zeta^2 - s = 0$, i.e. $\zeta = \pm\sqrt{s}$, and the general solution of (1a) can, thus, be written as $\widehat{w}(x,s) = e^{x\sqrt{s}}\gamma_1(s) + e^{-x\sqrt{s}}\gamma_2(s)$ or

$$\widehat{w}(x,s) = \cosh(x\sqrt{s})\lambda_1(s) + \frac{\sinh(x\sqrt{s})}{\sqrt{s}}\lambda_2(s). \tag{3}$$

The second formulation is easier to handle, because with

$$\widehat{C}_0(x) = \cosh(x\sqrt{s}), \quad \widehat{C}_1(x) = \frac{\sinh(x\sqrt{s})}{\sqrt{s}} \tag{4}$$

one has the relations $\partial_x\widehat{C}_0(x) = s\widehat{C}_1(x), \partial_x\widehat{C}_1(x) = \widehat{C}_0(x)$. Furthermore, as $\widehat{C}_0(0) = 1$ and $\widehat{C}_1(0) = 0$, the parameters λ_1 and λ_2 admit a direct interpretation through

$\lambda_1(s) = \hat{w}(0,s)$ and $\lambda_2(s) = \partial_x \widehat{w}(0,s)$. The general form of the solution and its first derivative can thus be written

$$\widehat{w}(x,s) = \widehat{C}_0(x)\lambda_1(s) + \widehat{C}_1(x)\lambda_2(s)$$
$$\partial_x \widehat{w}(x,s) = s\widehat{C}_1(x)\lambda_1(s) + \widehat{C}_0(x)\lambda_2(s).$$

The boundary conditions (1b) yield

$$\lambda_2(s) = 0, \qquad \widehat{C}_0(1)\lambda_1(s) = \widehat{u}(s),$$

and the equation $\cosh(\sqrt{s})\,\widehat{w}(x,s) = \cosh(x\sqrt{s})\,\widehat{u}(s)$, or

$$\widehat{C}_0(1)\widehat{w}(x,s) = \widehat{C}_0(x)\,\widehat{u}(s).$$

As a result one has a parametrization in $\lambda_1(s)$:

$$\widehat{u}(s) = \widehat{C}_0(1)\lambda_1(s) \tag{5a}$$
$$\widehat{w}(x,s) = \widehat{C}_0(x)\lambda_1(s). \tag{5b}$$

The free parameter λ_1 may, therefore, be considered as a *flat or basic output*. In a module theoretic framework on an appropriate ring (to be defined) it would form a basis of a corresponding free module.

Formally, write $\cosh(\sqrt{s}) = \sum_{i \geqslant 0} s^i/((2i)!)$, and introduce $\omega(t) = w(0,t)$ to denote the function corresponding to λ_1 in the time domain. Then, in the time domain

$$w(x,t) = \sum_{i \geqslant 0} \frac{x^{2i}}{(2i)!}\,\omega^{(i)}(t), \quad u(t) = \sum_{i \geqslant 0} \frac{1}{(2i)!}\,\omega^{(i)}(t). \tag{6}$$

Convergence of the above series can be shown (see, e.g., [6, 11, 12]) provided $t \mapsto \omega(t)$ is a Beurling ultradifferentiable function of Gevrey order 2 (cf. the app.).

2.2 *Temporal Viewpoint*

A different look on the problem is based on a Cauchy-Kowaleski form of the system:

$$\partial_x^2 w(x,t) = \partial_t w(x,t), \qquad x \in [0,1], t \in [0,\infty[\tag{7a}$$
$$\partial_x w(0,t) = 0, \quad w(0,t) = \omega(t), \tag{7b}$$

which allows one to search for a formal solution

$$w(x,t) = \sum_{i \geqslant 0} a_i(t) \frac{x^i}{i!}$$

where the functions a_i are infinitely differentiable. A formal check based upon (7) gives $a_{i+2}(t) = \dot{a}_i(t), i \geqslant 0, a_1(t) = 0, a_0(t) = \omega(t)$. Thus, for $i \geqslant 0$, one has $a_{2i}(t) = \omega^{(i)}(t), a_{2i+1}(t) = 0$, which implies (6).

3 Module Theoretic Formulation over Appropriate Rings

Generalizing the ideas of the introductory example, this section describes how boundary value problems can be reformulated as linear systems of equations over rings of ultradistributions. These equations serve as the defining relations for the module representing the system under consideration. The question of the appropriate choice of the coefficient rings of this module is brought up because its particular choice may play an important rhole in whether the system module is free. The latter property essentially simplifies trajectory planning and control design.

3.1 Class of Models Considered

In order to keep the exposition simple, in the sequel the following particular class of systems, with distributed variables $w_1, \ldots, w_l$ and lumped variables $u = (u_1, \ldots, u_m)$ is considered:

$$\begin{aligned} \partial_x w_i = A_i w_i + B_i u, \quad & w_i : \Omega_i \to \mathscr{F}^p, \quad u \in \mathscr{F}^m \\ & A_i \in (\mathbb{R}[\partial_t])^{p_i \times p_i}, \quad B_i \in (\mathbb{R}[\partial_t])^{p_i \times m}, \quad i \in \{1, \ldots, l\} \end{aligned} \tag{8a}$$

where $\mathscr{F}$ represents an appropriate space $\mathscr{E}^*(\mathbb{R})$ of smooth functions or (ultra-) distributions $\mathscr{D}'^*(\mathbb{R})$ to be specified in Sect. 3.2 below. The intervals $\Omega_1, \ldots, \Omega_l$ are open neighborhoods of $\tilde{\Omega}_i = [x_{i,0}, x_{i,1}]$. Without loss of generality, assume $x_{i,0} = 0$.

A key hypothesis will be the following: The characteristic polynomials of the matrices $A_1, \ldots, A_l$ can be written

$$P_i(\lambda) := \det(\lambda I - A_i) = \sum_{\nu=0}^{p_i} a_{i,\nu} \lambda^\nu, \quad a_{i,\nu} = \sum_{\mu \le p_i - \nu} a_{i,\nu,\mu} \partial_t^\mu \tag{8b}$$

with $a_{i,\nu,\mu} \in \mathbb{R}$, $a_{i,p_i,0} = 1$. Moreover, their principal parts $\sum_{\mu+\nu=p_i} a_{i,\nu,\mu} \partial_t^\mu \lambda^\nu$ are hyperbolic w.r.t. the time t, i.e., the roots of $\sum_{\mu+\nu=p_i} a_{i,\nu,\mu} \lambda^\nu$ are real.

The models are completed by boundary conditions

$$\sum_{i=1}^{l} L_i w_i(0) + R_i w_i(\ell_i) + Du = 0 \tag{8c}$$

with $D \in (\mathbb{R}[\partial_t])^{q \times m}$ and $L_i, R_i \in (\mathbb{R}[\partial_t])^{q \times p_i}$.

Remark 1. Note that the above assumptions apply to a large class of spatially one-dimensional boundary controlled evolution equations, including Euler-Bernoulli or Timoshenko beam equations, more general parabolic diffusion-reaction-convection equations, damped and undamped wave-equations etc. An exception are the models of internally damped mechanical systems.

Example 1. Consider an example similar to (1). The model is given by

$$\partial_x^2 w(x,t) = \partial_t w(x,t), \qquad x \in [0,\ell],\ t \in [0,+\infty[\tag{9a}$$

$$\partial_x w(0,t) = 0, \quad \partial_x w(\ell,t) = u(t), \tag{9b}$$

which may be rewritten in the form (8a), (8c) as

$$\partial_x \begin{pmatrix} w(x,t) \\ \partial_x w(x,t) \end{pmatrix} = \begin{pmatrix} 0 & 1 \\ \partial_t & 0 \end{pmatrix} \begin{pmatrix} w(x,t) \\ \partial_x w(x,t) \end{pmatrix} \tag{10a}$$

$$\begin{pmatrix} 0 & 1 \\ 0 & 0 \end{pmatrix} \begin{pmatrix} w(0,t) \\ \partial_x w(0,t) \end{pmatrix} + \begin{pmatrix} 0 & 0 \\ 0 & 1 \end{pmatrix} \begin{pmatrix} w(\ell,t) \\ \partial_x w(\ell,t) \end{pmatrix} = \begin{pmatrix} 0 \\ 1 \end{pmatrix} u(t). \tag{10b}$$

The characteristic polynomial $P(\lambda) = \lambda^2 - \partial_t$ of the coefficient matrix in (10a) has the principal part λ^2 which is clearly hyperbolic w.r.t. the time axis.

3.2 *Solution of the Cauchy Problem*

Some properties of the solution of the Cauchy problem (8a) with initial conditions given at $x = \xi$, i.e.

$$\partial_x w = Aw + Bu, \quad w(\xi) = w_\xi \tag{11}$$

with $A \in (\mathbb{R}[\partial_t])^{p\times p}$, $B \in (\mathbb{R}[\partial_t])^{p\times q}$ as assumed in the previous section for A_i, B_i, will be used. The notation of the previous section is used in what follows, dropping the index $i \in \{1,\dots,l\}$.

Choose[1] $\mathscr{E}^*(\mathbb{R}) = \mathscr{E}^{(p/(p-1))}(\mathbb{R})$ (resp. $\mathscr{D}'^*(\mathbb{R}) = \mathscr{D}'^{(p/(p-1))}(\mathbb{R})$) which corresponds to Beurling ultradifferentiable functions (resp. ultradistributions) of Gevrey order $p/(p-1)$ introduced in the appendix.

Consider the initial value problem

$$P(\partial_x)v(x) = 0, \quad (\partial_x^j v)(0) = v_j \in \mathscr{F}, \quad j = 0,\dots,p-1 \tag{12}$$

associated with the characteristic polynomial

$$P(\lambda) := \det(\lambda I - A) = \sum_{j=0}^{p} a_j \lambda^j, \quad a_j = \sum_{\mu \le p-j} a_{j,\mu} \partial_t^\mu .$$

Conformal with [8, Thrm. 12.5.6] or [15, Thrm 2.5.2, Prop. 2.5.6] the initial value problem (12) has a unique solution. This solution may be written as

$$v(x) = \sum_{j=0}^{p-1} C_j(x) v_j,$$

[1] Depending on the particular p.d.e. under consideration, choosing larger spaces $\mathscr{E}^*$ of smooth functions and smaller spaces $\mathscr{D}'^*$ of ultradistributions and even distributions may be possible.

where juxtaposition of symbols means convolution and $C_0,\dots,C_{p-1}$ are smooth functions[2] mapping Ω to the space of compactly supported Beurling ultradistributions $\mathscr{E}'^*(\mathbb{R}) := \mathscr{E}'^{(p/(p-1))}(\mathbb{R})$ of Gevrey order $p/(p-1)$. The functions $C_0,\dots,C_{p-1}$ satisfy $(k,j \in \{0,\dots,p-1\})$

$$\partial_x^k C_j(0) = \begin{cases} 1, & k=j \\ 0, & k \neq j \end{cases} \tag{13}$$

and

$$\partial_x C_j = C_{j-1} - a_j C_{p-1}, \quad j=1,\dots,p-1, \quad \partial_x C_0 = -a_0 C_{p-1}. \tag{14}$$

With these preparatory steps, the unique solution $x \mapsto \Phi(x,\xi)$ of the initial value problem (11) can be expressed as

$$w(x) = \Phi(x,\xi) w_\xi + \Psi(x,\xi) u. \tag{15}$$

Therein, $\Phi(x,\xi) \in \mathscr{E}'^*(\mathbb{R})^{p\times p}$ and $\Psi(x,\xi) \in \mathscr{E}'^*(\mathbb{R})^{p\times m}$ are given by

$$\Phi(x,\xi) = \sum_{j=0}^{p-1} A^j C_j(x-\xi), \quad \Psi(x,\xi) = \int_\xi^x \Phi(x,\zeta) d\zeta B. \tag{16}$$

That (15) with the matrices given in (16) is indeed a solution of (11) can be checked by plugging it into the p.d.e. in (11) and then employing (14) in combination with the Cayley-Hamilton theorem. Moreover, observe that $\Psi(\xi,\xi) = 0$ while $\Phi(\xi,\xi)$ is the identity. As a consequence, the restriction of $x \mapsto w(x)$ to $x = \xi$ indeed equals w_ξ.

Uniqueness of the solution (15) can be led back to the uniqueness of the scalar problem (12). To this end assume the existence of two different solutions of (11) which, by linearity, implies the existence of a non-zero solution of the homogeneous p.d.e. $\partial_x \tilde{w}(x) = A\tilde{w}(x)$ with data $\tilde{w}(\xi) = 0$. Differentiating this latter differential equation $p-1$ times w.r.t. x and using the Cayley-Hamilton theorem, one observes that all components of $\tilde{w}$ satisfy (12) with zero data $\tilde{w}(\xi) = \cdots = \partial_x^{p-1}\tilde{w}(\xi) = 0$.

Remark 2. As in the example introduced in sec. 2 the solution of the Cauchy problem (11) can be achieved either by direct computations in the time domain (cf. sec. 2.2) or, alternatively, by means of the Laplace transform (cf. sec. 2.1). According to the classical theory of ordinary differential equations, the solution of the Cauchy problem (11) in the Laplace domain always exists even if the characteristic polynomial of A does not satisfy the conditions formulated in section 3.1. However, these conditions are necessary in order to ensure the existence of time-domain interpretations of such solutions as compactly-supported ultradistributions. More specifically, they ensure particular growth bounds (w.r.t. the complex Laplace variable s) of the partial Laplace transforms $\widehat{C}_0(x),\dots,\widehat{C}_{p-1}(x)$ w.r.t. time of $C_0(x),\dots,C_{p-1}(x)$.

[2] A function $C : \Omega \to \mathscr{E}'^*$ is called of class C^∞ if it defines a map $\mathscr{D}^* \to C^\infty(\Omega)$, i.e., for any test function $\varphi \in \mathscr{D}^*$ the function $\Omega \ni x \mapsto C(x)[\varphi]$ belongs to $C^\infty(\Omega)$. It can be shown that this mapping is continuous.

These bounds are specified in the appropriate Paley-Wiener theorems for ultradistributions (see, e.g., [9, 10, 15]) and distributions (see, e.g., [7]).

Example 2 (Ex. 1 continued). As $p = 2$, for every fixed $x \in \Omega$, $C_0(x), C_1(x)$ are ultradistributions of Gevrey order 2 (elements of $\mathscr{E}'^{(2)}$). Clearly, for this simple example $C_0(x)$, $C_1(x)$ can be given explicitly: While their Laplace transforms simply correspond to (4), in the time domain one gets for all $v_0, v_1 \in \mathscr{E}^{(2)}(\mathbb{R})$ (cf. (6))

$$C_0(x)v_0 = \sum_{k=0}^{\infty} \frac{x^{2k}}{(2k)!}\partial_t^k v_0, \quad C_1(x)v_1 = \sum_{k=0}^{\infty} \frac{x^{2k+1}}{(2k+1)!}\partial_t^k v_1.$$

According to (15) and (16) the solutions of the (spatial) Cauchy problem with data $w(\xi) = c = (c_1, c_2)^T$ is given by

$$w(x) = \Phi(x,\xi)c, \quad \Phi(x,\xi) = \begin{pmatrix} C_0(x-\xi) & C_1(x-\xi) \\ \partial_t C_1(x-\xi) & C_0(x-\xi) \end{pmatrix}. \tag{17}$$

In particular, one has $w(x) = C_0(x-\xi)c_1 + C_1(x-\xi)c_2$.

3.3 System Module

Using the solutions of the initial value problem in the boundary conditions (8c), one obtains

$$w_i(x) = \Phi_i(x,\xi_i)w_i(\xi_i) + \Psi_i(x,\xi_i)u, \quad i = 1,\dots,l, \quad P_\xi c_\xi = 0 \tag{18}$$

Here $\xi = (\xi_1,\dots,\xi_l)$ is *arbitrary but fixed*, $c_\xi^T = (w_1^T(\xi_1),\dots,w_l^T(\xi_l),u^T)$, $P_\xi = (P_{\xi,1},\dots,P_{\xi,l+1})$ with

$$P_{\xi,i} = L_i\Phi_i(0,\xi_i) + R_i\Phi_i(\ell_i,\xi_i), \quad i = 1,\dots,l$$

$$P_{\xi,l+1} = D + \sum_{i=1}^{l} L_i\Psi_i(0,\xi_i) + R_i\Psi_i(\ell_i,\xi_i).$$

The system will be represented by a module generated by c_ξ, u with the presentation given in (18) — cf. [4, 3, 2, 13]. The ring of coefficients must contain at least the entries of $\Phi_i(x,\xi_i)$, $\Psi_i(x,\xi_i)$, $i = 1,\dots,l$, and the entries of P_ξ, which consist of values of functions $C_{i,j}$, $j = 1,\dots,p_i$, $i = 1,\dots,l$ from $\mathbb{R}$ in $\mathscr{E}'^{\star}$. Moreover, the matrices may also contain values of spatial integrals of $C_{i,j}$. A possible choice for the ring of coefficients is, thus, $\mathscr{R}^I = \mathbb{C}[\partial_t, \mathfrak{S}, \mathfrak{S}^I] \subset \mathscr{E}'^{*}$ with

$$\mathfrak{S} = \{C_{i,j}(x) | x \in \mathbb{R}; i = 1,\dots,l; j = 0,\dots,p_i - 1\},$$
$$\mathfrak{S}^I = \{C_{i,j}^I(x) | x \in \mathbb{R}; i = 1,\dots,l; j = 0,\dots,p_i - 1\}$$

and

$$C^I_{i,j}(x) = \int_0^x C_{i,j}(\zeta)d\zeta, \quad i = 1,\ldots,l, \quad j = 0,\ldots,p_i - 1.$$

This ring is isomorphic to a subring of $\mathscr{E}'^*$.

Following [14, 1, 5], in order to simplify the analysis of the module properties instead of $\mathscr{R}^I$, the larger ring $\mathscr{R} = \mathbb{C}(\partial_t)[\mathfrak{S}] \cap \mathscr{E}'^*$ may be considered.

Definition 1. The *convolutional system* Σ associated with the boundary value problem (8) is the module generated by the components of c_ξ and u over $\mathscr{R}$, with the presentation matrix P_ξ.

One may check that Σ is independent of the choice of ξ (cf. [19, Sect. 3.3] and [18, Remark 4]).

Example 3 (Ex. 2 continued). Substituting (17) into the boundary conditions (10b) one obtains $L\Phi(0,\xi)c + R\Phi(\ell,\xi)c - Du = 0$ or, even more explicitly,

$$\begin{pmatrix} 0 & 1 \\ 0 & 0 \end{pmatrix} \begin{pmatrix} C_0(-\xi) & C_1(-\xi) \\ \partial_t C_1(-\xi) & C_0(-\xi) \end{pmatrix} c + \begin{pmatrix} 0 & 0 \\ 0 & 1 \end{pmatrix} \begin{pmatrix} C_0(\ell-\xi) & C_1(\ell-\xi) \\ \partial_t C_1(\ell-\xi) & C_0(\ell-\xi) \end{pmatrix} c - \begin{pmatrix} 0 \\ 1 \end{pmatrix} u = 0.$$

As a result, one has

$$\begin{pmatrix} -\partial_t C_1(\xi) & C_0(\xi) & 0 \\ \partial_t C_1(\ell-\xi) & C_0(\ell-\xi) & -1 \end{pmatrix} \begin{pmatrix} c_1 \\ c_2 \\ u \end{pmatrix} = 0, \quad w(x) = \Phi(x,\xi) \begin{pmatrix} c_1 \\ c_2 \end{pmatrix},$$

the first equation of which may be written

$$P_\xi \begin{pmatrix} c \\ u \end{pmatrix} = 0 \qquad \text{with} \qquad P_\xi = \begin{pmatrix} -\partial_t C_1(\xi) & C_0(\xi) & 0 \\ \partial_t C_1(\ell-\xi) & C_0(\ell-\xi) & -1 \end{pmatrix}.$$

Thus, the *convolutional system* Σ associated with the boundary value problem (10) is the module generated by c_1, c_2, and u over $\mathscr{R} = \mathbb{C}(\partial_t)[\{C_0(x), C_1(x) | x \in \mathbb{R}\}] \cap \mathscr{E}'^*$, with the above defined presentation matrix P_ξ. Alternatively, instead of starting with a module over $\mathscr{R}$ one may directly pass to $\mathscr{E}'^*$.

4 Conclusion

A ring has been exhibited over which systems of one dimensional boundary controlled distributed parameter systems may be viewed as convolutional systems. It appears that this ring is well suited for controllability studies, especially when one is interested in the relations between algebraic and trajectory related controllabillity properties. For a particular subclass of the class of models considered here, it is established in [20], through Bézout ring properties, that torsion freeness and freeness are equivalent over such types of rings for systems in which the p.d.e.'s are of second order only. However, known results for the rings of entire functions of Paley-Wiener type (which are isomorphic to $\mathscr{E}'^*$ via the Laplace transform) suggest that in some situations it may be advantageous to consider systems over even larger subrings of $\mathscr{E}'^*$ to obtain similar results.

Appendix: Ultradistributions and Ultradifferentiable Functions

Some basic definitions about Gevrey functions and the corresponding classes of ultradistributions are recalled here.

Definition 2 (see, e.g. [9],[8, Def. 12.7.3, p. 137]). An infinitely differentiable function $f : \Omega \to \mathbb{C}$ (with $\Omega \subset \mathbb{R}^n$ open) belongs to the small Gevrey class $\mathscr{E}^{(\alpha)}(\Omega)$ (or the space of Beurling ultradifferentiable functions of Gevrey class α) if for all $M \in \mathbb{R}^+$ and all compact sets $K \subset \Omega$ there exists $C_{K,M}$ such that

$$\sup_{t \in \Omega, k \geq 0} |\partial_t^{(k)} f(t)| \leq C_{K,M} M^k (k!)^\alpha.$$

A sequence (f_n), $n \in \mathbb{N}$, $f_n \in \mathscr{E}^{(\alpha)}(\Omega)$ converges to $f \in \mathscr{E}^{(\alpha)}(\Omega)$, if for all compact $K \subset \Omega$ and all $M \in \mathbb{R}^+$

$$\lim_{n \to \infty} \sup_{t \in \Omega, k \geq 0} \frac{|\partial_t^{(k)} (f_n(t) - f(t))|}{M^k (k!)^\alpha} = 0.$$

The space of compactly supported functions in $\mathscr{E}^{(\alpha)}$ is denoted by $\mathscr{D}^\alpha(\Omega)$. A sequence (f_n), $f_n \in \mathscr{D}^{(\alpha)}(\Omega)$, $n \in \mathbb{N}$ converges in $\mathscr{D}^{(\alpha)}(\Omega)$ if it converges in $\mathscr{E}^{(\alpha)}(\Omega)$ and, moreover, $\cup_{n \in \mathbb{N}} \mathrm{supp} f_n$ is compact. The space $\mathscr{D}'^{(\alpha)}(\mathbb{R})$ (resp. $\mathscr{E}'^{(\alpha)}(\mathbb{R})$) of Beurling ultradistributions (resp. Beurling ultradistributions with compact support) of Gevrey order α is the space of linear continuous functionals on $\mathscr{D}^{(\alpha)}(\mathbb{R})$ (resp. $\mathscr{E}^{(\alpha)}(\mathbb{R})$).

The Laplace transform of an ultradistribution $f \in \mathscr{E}'^*$ is given by $\widehat{f}(s) = f(g_\xi)$ with $g_s(t) = \mathrm{e}^{-st}$. The isomorphism between the two convolution rings of ultradistributions with compact support and their Laplace transforms is given by a Paley-Wiener type theorem which can be found in [10].

References

1. Brethé, D., Loiseau, J.J.: A result that could bear fruit for the control of delay-differential systems. In: Proc. 4th IEEE Mediterranean Symp. Control Automation, Chania, Greece, pp. 168–172 (1996)
2. Fliess, M., Mounier, H.: Controllability and observability of linear delay systems: an algebraic approach. ESAIM: COCV (Control, Optimisation and Calculus of Variations) 3, 301–314 (1998)
3. Fliess, M., Mounier, H.: Tracking control and π-freeness of infinite dimensional linear systems. In: Picci, G., Gilliam, D.S. (eds.) Dynamical Systems, Control, Coding, Computer Vision, pp. 45–68. Birkhäuser, Basel (1999)

4. Fliess, M., Mounier, H.: An algebraic framework for infinite dimensional linear systems. e-STA (Sciences et Technologies de l'Automatique) 1(1) (2004)
5. Glüsing-Lüerßen, H.: A behavioral approach to delay-differential systems. SIAM J. Control Optim. 35(2), 480–499 (1997)
6. Hill, C.D.: Parabolic equations in one space variable and the non-characteristic Cauchy problem. Communications on Pure and Applied Mathematics XX, 619–633 (1967)
7. Hörmander, L.: The Analysis of Linear Partial Differential Operators I: Distribution Theory and Fourier Analysis. In: Grundlehren der mathematischen Wissenschaften, vol. 256, Springer, Heidelberg (1983)
8. Hörmander, L.: The Analysis of Linear Partial Differential Operators II: Differential Operators with Constant Coefficients. In: Grundlehren der mathematischen Wissenschaften, 2nd edn., vol. 257, Springer, Heidelberg (1990)
9. Komatsu, H.: Ultradistributions. I. Structure theorems and a characterization. J. Fac. Sci. Univ. Tokyo Sect. IA Math. 20, 25–105 (1973)
10. Komatsu, H.: Ultradistributions II. The kernel theorem and ultradistributions with support in a manifold. J. Fac. Sci. Univ. Tokyo Sect. IA Math. 24, 607–628 (1973)
11. Laroche, B., Martin, P., Rouchon, P.: Motion planning for the heat equation. Int. J. Robust and Nonlinear Control 10, 629–643 (2000)
12. Lynch, A.F., Rudolph, J.: Flatness-based boundary control of a class of quasilinear parabolic distributed parameter systems. Internat. J. Control 75(15), 1219–1230 (2002)
13. Mounier, H.: Propriétés structurelles des systèmes linéaires à retards : aspects théoriques et pratiques. Thèse de Doctorat, Université Paris-Sud, Orsay (1995)
14. Mounier, H.: Algebraic interpretations of the spectral controllability of a linear delay system. Forum Math. 10, 39–58 (1998)
15. Rodino, L.: Linear Partial Differential Operators in Gevrey Spaces. World Scientific, Singapore (1993)
16. Rudolph, J.: Flatness based control of distributed parameter systems. In: Berichte aus der Steuerungs- und Regelungstechnik, Shaker Verlag, Aachen (2003)
17. Rudolph, J., Winkler, J., Woittennek, F.: Flatness based control of distributed parameter systems: Examples and computer exercises from various technological domains. In: Berichte aus der Steuerungs- und Regelungstechnik, Shaker Verlag, Aachen (2003)
18. Rudolph, J., Woittennek, F.: Motion planning and open loop control design for linear distributed parameter systems with lumped controls. Internat. J. Control 81(3), 457–474 (2008)
19. Woittennek, F.: Beiträge zum Steuerungsentwurf für lineare, örtlich verteilte Systeme mit konzentrierten Stelleingriffen. In: Berichte aus der Steuerungs- und Regelungstechnik, Shaker Verlag, Aachen (2007)
20. Woittennek, F., Mounier, H.: Controllability of networks of spatially one-dimensional second order p.d.e. – an algebraic approach. SIAM J. Control Optim. 48(6), 3882–3902 (2010)

Wei-Norman Technique for Control Design of Bilinear ODE Systems with Application to Quantum Control

Markku Nihtilä

Abstract. A two-level quantum system model describing population transfer driven by a laser field is studied. A four-dimensional real-variable differential equation model is first obtained from the complex-valued two-level model describing the wave function of the system. Due to bilinearity in the control and the states Lie-algebraic techniques can be applied for constructing the state transition matrix of the system. The Wei-Norman technique is used in the construction. The exponential representation of the transition matrix includes three base functions, two of which serves as the parameter functions, which can be chosen freely. This corresponds to considering the overall control system as an underdetermined differential system. In this framework the initial and final states can be defined corresponding to the two levels of the original system model. Then flatness-based design is applied for explicitly calculating the parameter functions, which in turn give the desired input–output pairs. This input then drives the state of the system from the given initial state to the given final state in a finite time.

Keywords: Quantum control, Control design, Open-loop control, Wei-Norman technique, Two-level systems, Ordinary differential equation models.

1 Introduction

In quantum computation a two-level system, so-called qubit, forms a basic element for building up multi-qubit computing elements of future quantum computers, see [15], [22] & [10]. Then a key problem is to drive the qubit from one stable level to another, much like the classical commutation of an ordinary bit from 0 to 1.

Markku Nihtilä
University of Eastern Finland,
Department of Physics and Mathematics,
Kuopio Campus, POB 1627,
FI-70211 Kuopio, Finland
e-mail: markku.nihtila@uef.fi
http://www.uef.fi

J. Lévine & P. Müllhaupt (Eds.): Adv. in Theory of Control, Signals, LNCIS 407, pp. 189–199.
springerlink.com © Springer-Verlag Berlin Heidelberg 2010

Molecular excitation, i.e. driving of an ensemble of molecules from one locally stable steady state to another is one alternative for a qubit structure. This type of systems are controlled by using coherent light. Based on laser technology shorter and shorter coherent pulses can be generated for controlling molecular excitation, see [15]–[20]. The goal is to direct molecular reactions towards unprobable but desirable direction [2]-[5]. Then nonlinear and more and more sophisticated control methods are needed for properly designing durations and forms of the control pulses. In classical N-level problems the system to be controlled can be modelled by using ordinary $2N$-dimensional differential equation systems. Due to femto- and picosecond scale pulses, feedback is not in general applicable in the control design for these systems. Flatness-based control [6]–[9] is then an ideal methodology for open-loop design needed in quantum control problems. Then, for obtaining smaller maximum values for the controls, due to e.g. technical limitations, the transition time has to be increased. This offers a way to obtain a feasible compromise between the strenght and lenght of the control signals applied.

This two-level quantum control problem and some related studies have been carried out by several authors earlier, too, see [1],[11], [12], & [21]. Especially, in [21] a very similar approach as ours is used. A main difference there is that the authors use quaternion-type framework in the state-variable representation instead of a standard $\mathbb{R}^n$-representation.

However, we start from the basic definition of differential flatness. The system

$$\frac{dx}{dt} = f(x,u); \quad x(t) \in \mathbb{R}^n, \quad u(t) \in \mathbb{R}^m \tag{1}$$

is called differentially flat if there exists algebraic functions $\mathcal{A}$, $\mathcal{B}$, $\mathcal{C}$, and finite integers α, β, and γ such that for any pair (x,u) of inputs and controls, satisfying the dynamics (1), there exists a function z, called a flat (or linearizing) output, such that the following equations are satisfied

$$x(t) = \mathcal{A}(z, \dot{z}, \ldots, z^{(\alpha)}) \tag{2}$$
$$u(t) = \mathcal{B}(z, \dot{z}, \ldots, z^{(\beta)}) \tag{3}$$
$$z(t) = \mathcal{C}(x, u, \dot{u}, \ldots, u^{(\gamma)})\,. \tag{4}$$

Here we study the problem of a two-level quantum system from the flatness viewpoint. We start with the standard finite-state Schrödinger equation of two energy levels and transform it into a control theoretic form. A four-dimensional real-variable differential equation model is then obtained from the complex-valued two-level model. Due to bilinearity in the control and the states Lie-algebraic techniques can be applied for constructing the state transition matrix of the system. The Wei-Norman technique is used in the construction according to [23]. The exponential representation of the transition matrix includes three base functions, two of which serve as the parameter functions, which can be chosen freely. In this framework the initial and final states can be defined corresponding to the two levels of the original system model. Then flatness-based design is applied for explicitly calculating the parameter functions, which in turn give the desired input–output pairs. These inputs then drive the state of the system from the given initial state to the given final state in a finite time and produce the corresponding outputs via parametrization.

2 Model Conversion

At first, we repeat here shortly as a background the derivation of our dynamics. This was given e.g. in [3]. Population transfer in a two-level quantum system can be described by the time-dependent Schrödinger equation, i.e. by the dynamics

$$\mathrm{i}\frac{d\tilde{\psi}}{dt} = \tilde{H}(t)\,\tilde{\psi}, \quad \tilde{H}(t) = \begin{bmatrix} E_1 & \Omega(t) \\ \Omega^*(t) & E_2 \end{bmatrix}, \tag{5}$$

where the modified Planck's constant $\hbar = \frac{h}{2\pi}$ has been scaled to $\hbar = 1$, and $\mathrm{i} = \sqrt{-1}$. The wavefunction $\tilde{\psi} : \mathbb{R} \to \mathbb{C}^2$ has the probabilistic interpretation, in the sense that

$$\|\tilde{\psi}(t)\|^2 = |\tilde{\psi}_1(t)|^2 + |\tilde{\psi}_2(t)|^2 = 1, \forall\, t \in \mathbb{R}, \tag{6}$$

where $\tilde{\psi} = (\tilde{\psi}_1, \tilde{\psi}_2)$. The control is given by $\Omega : \mathbb{R} \to \mathbb{C}$, and Ω^* is the complex conjugate of Ω. E_1 and E_2 are the energy levels. The unitary transformation $\tilde{\psi} \mapsto \psi$ and $\Omega \mapsto u$ by

$$\tilde{\psi}(t) = U(t)\,\psi(t), \quad U(t) = \begin{bmatrix} e^{-\mathrm{i}E_1 t} & 0 \\ 0 & e^{-\mathrm{i}E_2 t} \end{bmatrix}, \tag{7}$$

$$u(t) = e^{-\mathrm{i}(E_2 - E_1)t}\,\Omega(t) \tag{8}$$

transforms (5) to

$$\mathrm{i}\frac{d\psi}{dt} = H(t)\,\psi, \quad H(t) = \begin{bmatrix} 0 & u(t) \\ u^*(t) & 0 \end{bmatrix}. \tag{9}$$

The componentwise representation

$$\psi(t) = \psi_1(t)\begin{bmatrix} 1 \\ 0 \end{bmatrix} + \psi_2(t)\begin{bmatrix} 0 \\ 1 \end{bmatrix} \tag{10}$$

converts (9) to the dynamics

$$\begin{aligned}\dot{\psi}_1 &= -\mathrm{i}\, u\, \psi_2, \\ \dot{\psi}_2 &= -\mathrm{i}\, u^*\, \psi_1.\end{aligned} \tag{11}$$

By using the real-valued decompositions

$$\begin{cases}\psi_1 = x_1 + \mathrm{i}\, x_2 \\ \psi_2 = x_3 + \mathrm{i}\, x_4 \\ \;u \;= u_1 + \mathrm{i}\, u_2\end{cases} \tag{12}$$

one obtains a state-variable representation in standard control theoretic form

$$\begin{bmatrix}\dot{x}_1\\ \dot{x}_2\\ \dot{x}_3\\ \dot{x}_4\end{bmatrix} = \begin{bmatrix} x_4 & x_3\\ -x_3 & x_4\\ x_2 & -x_1\\ -x_1 & -x_2\end{bmatrix}\begin{bmatrix}u_1\\ u_2\end{bmatrix}, \quad \text{or} \quad \frac{\mathrm{d}x}{\mathrm{d}t} = \big(u_1F_1 + u_2F_2\big)x, \tag{13}$$

$$F_1 = \begin{bmatrix} 0 & 0 & 0 & 1\\ 0 & 0 & -1 & 0\\ 0 & 1 & 0 & 0\\ -1 & 0 & 0 & 0\end{bmatrix}, \quad F_2 = \begin{bmatrix} 0 & 0 & 1 & 0\\ 0 & 0 & 0 & 1\\ -1 & 0 & 0 & 0\\ 0 & -1 & 0 & 0\end{bmatrix}. \tag{14}$$

The constraint equation (6) is converted into the constraint

$$\sum_{k=1}^{4} x_k^2 = 1.$$

Remark 1. It can be shown that F_1 and F_2 together with their Lie product $2F_3 = [F_1, F_2] = F_1F_2 - F_2F_1$ form a Lie algebra with some isomorfic "brothers". This can be used as a basis for differential geometric considerations of the control system (13), see [3]. However, the elementary approach applied in this paper is sufficient for our parametrization purposes.

3 Wei-Norman Formulation of State Transition Matrix

The Lie algebra of the matrices F_1, F_2, and F_3 is three-dimensional with the relations

$$[F_1, F_2] = 2F_3, \quad [F_2, F_3] = 2F_1, \quad [F_3, F_1] = 2F_2, \tag{15}$$

$$F_3 = \begin{bmatrix} 0 & -1 & 0 & 0\\ 1 & 0 & 0 & 0\\ 0 & 0 & 0 & 1\\ 0 & 0 & -1 & 0\end{bmatrix}. \tag{16}$$

Due to the linear structure of the system model (13) with respect to the state x, the state transition matrix of the system, denoted by Φ, and which relates the values of the state according to

$$x(t) = \Phi(t,0)\, x(0) \tag{17}$$

can be written as a product of exponentials

$$\Phi(t,0) = \mathrm{e}^{g_1 F_1}\, \mathrm{e}^{g_2 F_2}\, \mathrm{e}^{g_3 F_3} \tag{18}$$

where the exponentials are defined by

$$\mathrm{e}^{g_i F_i} = \sum_{k=0}^{\infty} \frac{1}{k!}\, g_i^k F_i^k\,, \quad i = 1, 2, 3. \tag{19}$$

The state transition matrix satisafies the following initial-value problem (IVP1)

$$\frac{\partial}{\partial t}\Phi(t,0) = F(t)\,\Phi(t,0); \quad \Phi(0,0) = I, \tag{20}$$

$$F(t) = u_1(t)F_1 + u_2(t)F_2 + 0 \cdot F_3. \tag{21}$$

The technique we are using is nowadays called Wei-Norman technique according to the paper of Wei and Norman [23]. Substitution of (18) to the IVP1 gives

$$\begin{aligned}\frac{\partial}{\partial t}\Phi &= \dot{g}_1 F_1\, \Phi + \dot{g}_2\, \mathrm{e}^{g_1 F_1} F_2\, \mathrm{e}^{-g_1 F_1}\, \Phi \\ &\quad + \dot{g}_3\, \mathrm{e}^{g_1 F_1} \mathrm{e}^{g_2 F_2} F_3\, \mathrm{e}^{-g_2 F_2} \mathrm{e}^{-g_1 F_1}\, \Phi\end{aligned} \tag{22}$$

By using (several times) the Campbell-Baker-Hausdorff formula for square matrices A and B of the same dimension

$$\begin{aligned}\mathrm{e}^{A}\, B\, \mathrm{e}^{-A} &= B + [A,B] + [A,[A,B]]/2! \\ &\quad + [A,[A,[A,B]]]/3! + \cdots\end{aligned} \tag{23}$$

in the equation (22) it can be represented (after some tedious calculations) in the form

$$\frac{\partial}{\partial t}\Phi = [f_1(t)F_1 + f_2(t)F_2 + f_3(t)F_3]\; \Phi \tag{24}$$

$$f_1(t) = \dot{g}_1 + \dot{g}_3 \sin(2g_2) \tag{25}$$

$$f_2(t) = \dot{g}_2 \cos(2g_1) - \dot{g}_3 \cos(2g_2)\, \sin(2g_1) \tag{26}$$

$$f_3(t) = \dot{g}_2 \sin(2g_1) + \dot{g}_3 \cos(2g_2)\, \cos(2g_1) \tag{27}$$

The various series representations of the matrix exponentials can be represented in closed form due to specific forms of the matrices F_1, F_2, and F_3 as it is described in Appendix. By comparing the coefficients of the F_i's in (24) and (20)–(21) one finally obtains a differential relation between the g_i's and the controls u_1 and u_2 in the form of a matrix equation

$$\begin{bmatrix} 1 & 0 & \sin(2g_2) \\ 0 & \cos(2g_1) & -\cos(2g_2)\sin(2g_1) \\ 0 & \sin(2g_1) & \cos(2g_2)\cos(2g_1) \end{bmatrix} \begin{bmatrix} \dot{g}_1 \\ \dot{g}_2 \\ \dot{g}_3 \end{bmatrix} = \begin{bmatrix} u_1 \\ u_2 \\ 0 \end{bmatrix} \tag{28}$$

4 Parametrization of the System Model

Because the system has two (scalar) controls we can choose two of the three base functions g_i freely corresponding to free selection of the two controls. The third base function has to be determined from the last equation of (28). Parametrization actually means that the input–output pairs can be determined from the parameter functions without explicitly solving of the system equations.

Due to the flatness-based design idea, computation of the third base function as well as of the controls must not include integrations as given by the equations (2). Only differentiations are allowed. Consequently, based on the third equation in (28), the base functions g_2 and g_3 are chosen as parameter functions. Then these are also so-called flat outputs, see [9], denoted by $z = (z_1, z_2) = (g_2, g_3)$. The parametrization obtained in this way for g_1 and the controls are given by

$$g_1 = \frac{1}{2}\arctan\left[-\cos(2g_2)\,\frac{\dot{g}_3}{\dot{g}_2} \right] \tag{29}$$

$$u_1 = \dot{g}_1 + \dot{g}_3 \sin(2g_2) \tag{30}$$

$$u_2 = \sqrt{\dot{g}_2^2 + \dot{g}_3^2 \cos^2(2g_2)}\,. \tag{31}$$

The state variables are calculated by using the state transition matrix equations (17) or (18)

$$x(t) = \Phi(t,0)x(0) = \mathrm{e}^{g_1 F_1}\,\mathrm{e}^{g_2 F_2}\,\mathrm{e}^{g_3 F_3}\,x(0).$$

5 State Transfer

In population transfer problems from the level 1 corresponding to the situation

$$|\psi_1(0)|^2 = x_1(0)^2 + x_2(0)^2 = 0 \tag{32}$$

to the level 2, where

$$|\psi_2(T)|^2 = x_3(T)^2 + x_4(T)^2 = 0\,, \tag{33}$$

where T is the transfer time, we can parametrize the partial trajectory by using a sufficiently smooth, but otherwise arbitrarily chosen, parametrization $x_1, x_2 : [0, T] \to \mathbb{R}$ with the boundary conditions

$$x_1(0)^2 + x_2(0)^2 = 0, \tag{34}$$

$$x_1(T)^2 + x_2(T)^2 = 1. \tag{35}$$

By partitioning the state vector for two two-dimensional parts as

$$x(t) = (\,w(t), v(t)\,)\,, \quad w(t) = (\,x_1(t), x_2(t)\,)\,, \quad v(t) = (\,x_3(t), x_4(t)\,) \tag{36}$$

we can represent the task of driving the state from the initial one to the final one in a finite time T as follows

$$x(0) = \begin{bmatrix} 0 \\ v_0 \end{bmatrix} = \begin{bmatrix} 0 \\ 0 \\ x_{30} \\ x_{40} \end{bmatrix} \to x(T) = \begin{bmatrix} w_T \\ 0 \end{bmatrix} = \begin{bmatrix} x_{1T} \\ x_{2T} \\ 0 \\ 0 \end{bmatrix} \tag{37}$$

$$\begin{bmatrix} 0 \\ 0 \\ \sin\alpha \\ \cos\alpha \end{bmatrix} \to \begin{bmatrix} \cos\beta \\ \sin\beta \\ 0 \\ 0 \end{bmatrix} \tag{38}$$

We have chosen a specific parametrization for the initial and final values of the state, because the sum of the squares of the nonzero state components must be equal to 1 at the both ends of the planned trajectory.

The state transition equation $x(T) = \Phi(T, 0)x(0)$ can now be written in the form

$$\begin{bmatrix} w_T \\ 0 \end{bmatrix} = \begin{bmatrix} A & B \\ C & D \end{bmatrix} \begin{bmatrix} 0 \\ v_0 \end{bmatrix}, \tag{39}$$

where A, B, C, and D are 2×2-blocks of the 4×4-dimensional state transition matrix $\Phi(T, 0)$. Consequently,

$$w_T = Bv_0\,, \quad \text{and} \quad 0 = Dv_0\,. \tag{40}$$

In Appendix one alternative for choosing the feasible final values for g_2 and g_3 is derived. The final value of g_1 depends on the derivatives of g_2 and g_3. This means that we have to adjust these derivatives via the equation (29) to agree with the requirement $g_1(T) = \pi/2$.

6 Conclusion

This paper is among the first ones in the series where our results on parametrization of finite-state quantum control problems are represented. Earlier we have been concentrating on studying parametrization of systems described by partial differential control systems, see [13] & [14]. Flatness-based ideas, originally developed by Michel Fliess and his co-workers, see the seminal paper [9], have been developed for open-loop control design. These, however, form a means for control design of a large class of nonlinear differential systems. In quantum control problems, where laser pulses are used for the control, the dynamics is so fast that, at least at the present level of the speed of possible computations, feedback control seems to be impossible to implement.

Here we studied a two-level population transfer problem, which is described by a 4-dimensional ordinary differential system, bilinear in the two scalar controls. Without more advanced differential geometric considerations, which might be helpful in understanding quantum phenomena in general, we use the formulation found generally in the literature, to obtain our basic driftless system model of the form $\dot{x} = g(x)u$, where g is linear in the state x.

Acknowledgment. This work was supported in part by the European Commission, in Marie Curie programme's Transfer of Knowledge project *Parametrization in the Control of Dynamic Systems* (PARAMCOSYS, MTKD-CT-2004-509223), which is greatly acknowledged.

Appendix: Computation of the State Transition Matrix

The state transition matrix

$$\Phi(t, 0) = \mathrm{e}^{g_1 F_1}\, \mathrm{e}^{g_2 F_2}\, \mathrm{e}^{g_3 F_3} \tag{41}$$

where the exponentials are defined by absolutely convergent infinite series

$$\mathrm{e}^{g_i F_i} = \sum_{k=0}^{\infty} \frac{1}{k!} g_i^k F_i^k \,, i = 1, 2, 3 \tag{42}$$

we obtain for the exponent functions the representations in closed form

$$\mathrm{e}^{g_i F_i} = \cos g_i\, I + \sin g_i\, F_i \tag{43}$$

due to the fact that $F_i^2 = -I$, $i = 1, 2, 3$, where I is 4×4 identity matrix. Then the product of the three exponent functions is of the form

$$\Phi = (c_1 I + s_1 F_1)(c_2 I + s_2 F_2)(c_3 I + s_3 F_3) \tag{44}$$

$$c_i = \cos g_i, \quad s_i = \sin g_i, \quad i = 1, 2, 3. \tag{45}$$

Now the D-part and B-part of the transfer matrix Φ are given by

$$D = \begin{bmatrix} d_1 & d_2 \\ d_3 & d_4 \end{bmatrix} = \begin{bmatrix} c_1c_2c_3 - s_1s_2s_3 & c_1c_2s_3 + s_1s_2c_3 \\ -c_1c_2s_3 - s_1s_2c_3 & c_1c_2c_3 - s_1s_2s_3 \end{bmatrix} \tag{46}$$

$$= c_1c_2 \begin{bmatrix} c_3 & s_3 \\ -s_3 & c_3 \end{bmatrix} - s_1s_2 \begin{bmatrix} s_3 & -c_3 \\ c_3 & s_3 \end{bmatrix}, \tag{47}$$

$$B = \begin{bmatrix} b_1 & b_2 \\ b_3 & b_4 \end{bmatrix} = \begin{bmatrix} c_1s_2c_3 - s_1c_2s_3 & c_1s_2s_3 + s_1c_2c_3 \\ -c_1s_2s_3 - s_1c_2c_3 & c_1s_2c_3 - s_1c_2s_3 \end{bmatrix} \tag{48}$$

$$= c_1s_2 \begin{bmatrix} c_3 & s_3 \\ -s_3 & c_3 \end{bmatrix} - s_1c_2 \begin{bmatrix} s_3 & -c_3 \\ c_3 & s_3 \end{bmatrix}. \tag{49}$$

We must have $D = 0$ due to the requirement $Dv_0 = 0$ for any $v_0 = (\, x_{30}, x_{40}\,)$ satisfying the requirement $x_{30}^2 + x_{40}^2 = 1$. Then we have two alternatives in (47):

$$\begin{cases} a) & c_1 = s_2 = 0 \\ b) & s_1 = c_2 = 0 \end{cases} \Rightarrow \quad D = 0 \quad \therefore \quad Dv_0 = 0. \tag{50}$$

These conditions give two possibilities

$$a) \begin{cases} \cos g_1(T) = 0\,, & g_1(T) = \dfrac{\pi}{2} \\ \sin g_2(T) = 0\,, & g_2(T) = 0, \end{cases} ; \quad b) \begin{cases} \sin g_1(T) = 0\,, & g_1(T) = 0 \\ \cos g_2(T) = 0\,, & g_2(T) = \dfrac{\pi}{2}. \end{cases} \tag{51}$$

In the case of the first alternative *a)* we have

$$B = -s_1c_2 \begin{bmatrix} s_3 & -c_3 \\ c_3 & s_3 \end{bmatrix} ; \quad w_T = Bv_0 = \begin{bmatrix} \sin g_3 & -\cos g_3 \\ \cos g_3 & \sin g_3 \end{bmatrix} \begin{bmatrix} \sin\alpha \\ \cos\alpha \end{bmatrix} \tag{52}$$

$$= \begin{bmatrix} \sin g_3 \sin\alpha - \cos g_3 \cos\alpha \\ \cos g_3 \sin\alpha + \sin g_3 \cos\alpha \end{bmatrix} = \begin{bmatrix} \cos(g_3 + \alpha) \\ \sin(g_3 + \alpha) \end{bmatrix} = \begin{bmatrix} \cos\beta \\ \sin\beta \end{bmatrix} \tag{53}$$

$$g_3(T) = \beta - \alpha. \tag{54}$$

In the same way the alternative *b)* can be solved. Due to trigonometric functions in the equations there are also other possibilities for the final values of g_2 and g_3 deviating by the multiples of π or 2π. These details are not considered here.

References

1. D'Alessandro, D., Dahleh, M.: Optimal control of two-level quantum systems. IEEE Trans. Autom. Contr. 46, 866–876 (2001)
2. Bandrauk, A.D., Delfour, M.C., LeBris, C. (eds.): Quantum Control: Mathematical and Numerical Challenges. CRM Proc. & Lecture Notes, vol. 33. American Mathematical Society, Rhode Island (2002)
3. Boscain, H., Charlot, G., Gauthier, J.-P., Guérin, S., Jauslin, H.-.R.: Optimal control in laser-induced population transfer for two- and three-level quantum systems. J. Math. Phys. 43, 2107–2132 (2002)
4. Brown, E., Rabitz, H.: Some mathematical and algorithmic challenges in the control of quantum dynamics phenomena. J. Math. Chem. 31, 17–63 (2002)
5. Le Bris, C.: Control theory applied to quantum chemistry: Some tracks. In: ESAIM: Proceedings of Contrôle des Systèmes Gouvernés par des Équations aux Dérivées Partielles, vol. 8, pp. 77–94 (2000)
6. Lévine, J.: On necessary and sufficient conditions for differential flatness (December 2006), http://arxiv.org/abs/math.OC/0605405
7. Fliess, M.: Generalized controller canonical forms for linear and nonlinear dynamics. IEEE Trans. Autom. Contr. 35, 994–1001 (1990)
8. Fliess, M., Lévine, J., Martin, P., Rouchon, P.: On differentially flat nonlinear systems. In: Fliess, M. (ed.) Proc. IFAC Symp. Nonlinear Control Systems Design, Bordeaux, France, pp. 408–412 (1992)
9. Fliess, M., Lévine, J., Martin, P., Rouchon, P.: Flatness and defect of non-linear systems: Introductory theory and applications. Int. J. Contr. 61, 1327–1361 (1995)
10. Fuchs, G.D., Dobrovitski, V.V., Toyli, D.M., Heremans, F.J., Awschalom, D.D.: Gigahertz Dynamics of a Strongly Driven Single Quantum Spin. Science 326, 1520–1522 (2009)
11. Graichen, K., Zeitz, M.: Feedforward control design for finite-time transition problems of nonlinear systems with input and output constraints. IEEE Trans. Autom. Contr. 53, 1273–1278 (2008)
12. Grivopoulos, S., Bamieh, B.: Optimal population transfers in a quantum system for large transfer time. IEEE Trans. Autom. Contr. 53, 980–992 (2008)
13. Nihtilä, M., Tervo, J., Kokkonen, P.: Control of Burgers' system via parametrization. In: Allgöwer, F. (ed.) Preprints of 6th IFAC Symp. Nonlinear Control Systems, NOLCOS 2004, Stuttgart, Germany, vol. I, pp. 423–428 (2004)
14. Nihtilä, M., Tervo, J., Kokkonen, P.: Pseudo-differential operators in parametrization of boundary-value control systems. In: CD-ROM Proceedings of the 43rd IEEE Conf. Decision and Control, CDC 2004 (IEEE Catalog number 04CH37601C, ISBN 0-7803-8683-3), Paradise Islands, The Bahamas, pp. 1958–1963 (2004)
15. Press, D., Ladd, T.D., Zhang, B., Yamamoto, Y.: Complete quantum control of a single quantum dot spin using ultrafast optical pulses. Nature 456, 218–221 (2008)
16. Pople, J.A.: Nobel lecture: Quantum chemical models. Rev. Mod. Phys. 71, 1267–1274 (1999)
17. Rabitz, H., de Vivie-Riedle, R., Motzkus, M., Kompa, K.: Whither the future of controlling quantum phenomena? Science 288, 824–828 (2000)

18. Shapiro, M., Brumer, P.: Principles of Quantum Control of Molecular Processes. John Wiley & Sons Inc., Hoboken (2003)
19. Wang, X., Schirmer, S.: Analysis of Lyapunov control for Hamiltonian quantum systems. In: Sixth EUROMECH (European Mechanics Society) Nonlinear Dynamics Conference, ENOC 2008, St. Petersburg, Russia (2008)
20. Schirmer, S.: Implementation of quantum gates via optimal control. J. Mod. Optics (2009), DOI: 10.1080/09500340802344933
21. Pereira da Silva, P.S., Rouchon, P.: Flatness-based control of a single qubit gate. IEEE Trans. Autom. Contr. 53, 775–779 (2008)
22. Vedral, V., Plenio, M.B.: Basics of quantum computation. Progress in Quantum Electronics 20, 1–39 (1998)
23. Wei, J., Norman, E.: On global representation of the solutions of linear differential equations as product of exponentials. In: Proc. AMS, vol. 15, pp. 327–334 (1964)

Interval Methods for Verification and Implementation of Robust Controllers

Andreas Rauh and Harald Aschemann

Abstract. In recent years, powerful interval arithmetic tools have been developed for the computation of guaranteed enclosures of the sets of all reachable states of dynamical systems. In such simulations, uncertainties in initial conditions and parameters are taken into account by worst-case bounds. The resulting enclosures are verified in the sense that all reachable states are guaranteed to be included. This is achieved by taking into account both the influence of the above-mentioned uncertainties as well as numerical inaccuracies arising from computer implementations using finite-precision floating-point arithmetic. In this contribution, a computational framework for both offline and online applications of interval tools in control design is presented. Verified computational procedures and their applications to the solution of initial value problems for both ordinary differential equations and differential algebraic equations are summarized. These algorithms are employed for verified feedforward control design as well as state and disturbance estimation for a distributed heating system.

1 Introduction

The basis of the procedures presented in this contribution are interval techniques which have been developed to quantify rounding errors in finite-precision floating-point arithmetic as well as to determine the influence of uncertainties in mathematical system models [12, 6]. For technical applications, these models are either given by sets of algebraic equations, difference equations, ordinary differential equations (ODEs), or differential-algebraic equations (DAEs).

Andreas Rauh
Chair of Mechatronics, University of Rostock
e-mail: `Andreas.Rauh@uni-rostock.de`

Harald Aschemann
Chair of Mechatronics, University of Rostock
e-mail: `Harald.Aschemann@uni-rostock.de`

J. Lévine & P. Müllhaupt (Eds.): Adv. in Theory of Control, Signals, LNCIS 407, pp. 201–211.
springerlink.com © Springer-Verlag Berlin Heidelberg 2010

Software libraries which implement basic interval arithmetic functionalities such as evaluation of arithmetic operations and functions (e.g. trigonometric and other transcendental functions) are, for instance, implemented in the C++ toolbox PROFIL/BIAS [7]. In addition, most verified computational algorithms make use of partial derivatives of the first and higher orders as well as coefficients of Taylor series. Such derivatives are obtained efficiently with the help of algorithmic differentiation. The C++ library that is used for this purpose is FADBAD++ [2].

On the basis of these software libraries, tools for verified integration of initial value problems (IVPs) for sets of ODEs have been developed. Examples for tools operating on the principle of interval arithmetic are VNODE, VNODE-LP [13], and VALENCIA-IVP [1]. In addition, program packages such as VSPODE [10] and COSY VI [3] make use of Taylor model arithmetic in order to reduce the influence of overestimation which might lead to extremely conservative enclosures of the exact solution sets if naive implementations of interval algorithms are applied.

On the one hand, these software packages are the basis for approaches aiming at offline verification, design, stability analysis and optimization of robust open-loop and closed-loop control strategies (cf. [16, 17, 15, 18]). On the other hand, they are also applicable under certain prerequisites to the online computation of feedforward control strategies as well as for the computation of state and disturbance estimates.

In offline applications, interval tools are used to quantify the effects of uncertainties which result from, for example, manufacturing tolerances or measurement errors occurring unavoidably in any technical application. In the offline design and proof of feasibility, verified enclosures of *all possibly admissible* solutions of control synthesis are determined after computation of verified enclosures of *all reachable* states. In this case, computing time is of minor importance. However, in online applications, we have to fulfill real-time requirements by computing only one guaranteed admissible solution taking into account the influence of all possible uncertainties. This solution must not violate any constraints on state and control variables.

Verified simulation algorithms for sets of ODEs and DAEs are summarized in the Sections 2 and 3, respectively. In Section 4, DAE-based solution procedures for feedforward control as well as state and disturbance estimation are presented for finite-dimensional system models. These strategies are applied in real-time to a finite volume representation of a distributed heating system in Section 5. Conclusions and an outlook on future work are given in Section 6.

2 Verified Simulation of ODEs in VALENCIA-IVP

In this section, we consider the verified solution of IVPs to the set of ODEs

$$\dot{x}(t) = f(x(t), t) \ , \ \ x \in \mathbb{R}^{n_x} \tag{1}$$

with the uncertain initial conditions $x(0) \in [x(0)] := [\underline{x}(0) \ ; \ \overline{x}(0)]$, $\underline{x}_i(0) \leq \overline{x}_i(0)$ for all $i = 1, \ldots, n_x$ with the help of the verified solver VALENCIA-IVP.

In the basic version of VALENCIA-IVP, time-varying state enclosures

$$[x_{encl}(t)] := x_{app}(t) + [R(t)] \tag{2}$$

are computed iteratively which consist of a non-verified approximate solution $x_{app}(t)$ with guaranteed error bounds $[R(t)]$. For the sake of simplicity, we specify the iteration formulas for the ODE (1) in the time interval $0 \leq t \leq T$. In this case, an interval containing the derivatives $[\dot{R}(t)]$ of the desired error bounds $[R(t)]$ can be computed by

$$\begin{aligned}\left[\dot{R}^{(\kappa+1)}(t)\right] &= -\dot{x}_{app}(t) + f\left(\left[x_{encl}^{(\kappa)}(t)\right], t\right) \\ &= -\dot{x}_{app}(t) + f\left(x_{app}(t) + \left[R^{(\kappa)}(t)\right], t\right) =: r\left(\left[R^{(\kappa)}(t)\right], t\right)\end{aligned} \tag{3}$$

if

$$\left[\dot{R}^{(\kappa+1)}(t)\right] \subseteq \left[\dot{R}^{(\kappa)}(t)\right] \tag{4}$$

holds with

$$\left[R^{(\kappa+1)}(t)\right] \subseteq \left[R^{(\kappa+1)}(0)\right] + t \cdot r\left(\left[R^{(\kappa)}([0\,;\,t])\right], [0\,;\,t]\right) \tag{5}$$

and $t = T$ as well as $[x(0)] \subseteq x_{app}(0) + \left[R^{(\kappa+1)}(t)\right]$.

In addition, exponential state enclosures can be determined to prevent the growth of interval diameters especially in simulations of asymptotically stable systems. The basic idea is to use the ansatz

$$[x_{encl}(t)] := \exp([\Lambda] \cdot t) \cdot [x_{encl}(0)] \tag{6}$$

with

$$[\Lambda] := \operatorname{diag}\{[\lambda_i]\} \tag{7}$$

for the state enclosures, where the coefficients $[\lambda_i]$ are computed iteratively by

$$\left[\lambda_i^{(\kappa+1)}\right] := \frac{f_i\left(\exp\left(\left[\Lambda^{(\kappa)}\right] \cdot [0\,;\,T]\right) \cdot [x_{encl}(0)], [0\,;\,T]\right)}{\exp\left(\left[\lambda_i^{(\kappa)}\right] \cdot [0\,;\,T]\right) \cdot [x_{encl,i}(0)]} \tag{8}$$

for all $i = 1, \ldots, n_x$ in the case of convergence, that means, for $\left[\lambda_i^{(\kappa+1)}\right] \subseteq \left[\lambda_i^{(\kappa)}\right]$.

The iteration formula (8) is only admissible if the value zero does not belong to the set of all reachable states in the time interval $[0\,;\,T]$. This property is checked by the guaranteed enclosures obtained in the formulas (3)–(5) before the exponential state enclosures are evaluated. A detailed derivation of the iteration formulas of VALENCIA-IVP can, for example, be found in [1, 15]. To further tighten the computed state enclosures, consistency tests are available which exclude physically unreasonable domains resulting from overestimation by constraint propagation based on conservation properties derived from suitable balance equations such as energy balances for mechanical systems [5]. In control systems, the above-mentioned algorithms can be used to prove whether the enclosures of all reachable states remain within given bounds for known control strategies. In the case of feedback control, it is moreover possible to show whether the resulting control $u(x(t))$ (given by analytic expressions) matches the corresponding physical input constraints.

3 Verified Solution of IVPs for DAEs in VALENCIA-IVP

In this section, we consider semi-explicit DAEs

$$\dot{x}(t) = f(x(t), y(t), t) \tag{9}$$

$$0 = g(x(t), y(t), t) \tag{10}$$

with $f : D \mapsto \mathbb{R}^{n_x}$, $g : D \mapsto \mathbb{R}^{n_y}, D \subset \mathbb{R}^{n_x} \times \mathbb{R}^{n_y} \times \mathbb{R}^1$, and the consistent initial conditions $x(0)$ and $y(0)$. As for the ODEs in Section 2, these DAEs may further depend on uncertain parameters p. To simplify the notation, the dependency on p is not explicitly denoted. However, all presented results are also applicable to $p_i \in \left[\underline{p}_i \; ; \; \overline{p}_i\right]$ with $\underline{p}_i < \overline{p}_i$, $i = 1, \ldots, n_p$. The basis for the following applications is the computation of guaranteed enclosures for both consistent initial conditions and solutions to IVPs for DAEs. The enclosures for the differential and algebraic variables $x_i(t)$ and $y_j(t)$, respectively, are defined by

$$[x_i(t)] := x_{app,i}(t_k) + (t - t_k) \cdot \dot{x}_{app,i}(t_k) + [R_{x,i}(t_k)] + (t - t_k) \cdot [\dot{R}_{x,i}(t)] \quad \text{and} \tag{11}$$

$$[y_j(t)] := y_{app,j}(t_k) + (t - t_k) \cdot \dot{y}_{app,j}(t_k) + [R_{y,j}(t)] \tag{12}$$

with $i = 1, \ldots, n_x$, $j = 1, \ldots, n_y$, and $t \in [t_k \; ; \; t_{k+1}]$, $t_0 \leq t \leq t_f$.

In (11) and (12), t_k and t_{k+1} are two subsequent points of time between which guaranteed state enclosures are determined. For $t = t_0$, the conditions

$$[x(t_0)] = x_{app}(t_0) + [R_x(t_0)] \qquad \text{and} \qquad [y(t_0)] = y_{app}(t_0) + [R_y(t_0)] \tag{13}$$

have to be fulfilled with non-verified approximate solutions $x_{app}(t)$ and $y_{app}(t)$. They are computed, for example, by the non-verified DAE solver DAETS [14].

The following three-stage algorithm allows us to determine guaranteed state enclosures for a system of DAEs using the Krawczyk iteration [9] which solves nonlinear algebraic equations in a verified way.

Step 1. Compute hidden constraints that have to be fulfilled for the verified enclosures of the initial conditions $x(0)$ and $y(0)$ as well as for the time responses $x(t)$ and $y(t)$ by considering algebraic equations $g_i(x)$ which do not depend explicitly on y. Differentiation with respect to time leads to

$$\frac{d^j g_i(x)}{dt^j} = \left(\frac{\partial L_f^{j-1} g_i(x)}{\partial x}\right)^T \cdot f(x, y) = L_f^j g_i(x) = 0 \tag{14}$$

with $L_f^0 g_i(x) = g_i(x)$. The Lie derivatives $L_f^j g_i(x)$ are computed by algorithmic differentiation using FADBAD++ [2] up to the smallest order $j > 0$ for which $L_f^j g_i(x)$ depends on at least one component of y.

Step 2. Compute initial conditions for the equations (9) and (10) such that the constraints (10) and (14) are fulfilled using the Krawczyk iteration.

Step 3. Substitute the state enclosures (11) and (12) for the vectors $x(t)$ and $y(t)$ in (9) and (10) and solve the resulting equations for $[\dot{R}_x(t)]$ and $[R_y(t)]$ with the help of the Krawczyk iteration. Consider the hidden constraints (14) to restrict the set of feasible solutions.

4 DAEs for Verified Feedforward Control and State Estimation

Besides simulation of systems with known control inputs, VALENCIA-IVP can be employed for *trajectory planning* and computation of *feedforward control* strategies for ODE and DAE systems. In the case of trajectory planning, reference signals $w(t)$ of open-loop controllers (S is open in Fig. 1) or closed-loop controllers (S is closed in Fig. 1) are calculated in such a way that the output $y(t)$ follows a desired time response $y_d(t)$ within given tolerances. For closed-loop control, the structure and parameters of $u(\hat{x}, w)$ are assumed to be determined beforehand using classical techniques for control synthesis. If state estimation techniques are employed in Fig. 1 to reconstruct non-measured components of x_s, p, and q, the estimate $\hat{x}$ is fed back as a substitute for the unknown quantities in the closed-loop control $u(\hat{x}, w)$.

To determine feedforward control strategies (and reference signals, resp.), we compute the inputs $u(t)$ (and $w(t)$, resp.) as components of the vector $y(t)$ of algebraic state variables in the DAEs (9), (10) after describing the desired system outputs by the algebraic equations

$$0 = h(x_s(t), u(t), q(t), t) - (y_d(t) + y_{tol}(t)) \tag{15}$$

with worst-case interval bounds $[y_{tol}(t)]$ for the tolerances $y_{tol}(t)$ between the actual and desired outputs $y(t)$ and $y_d(t)$. Note that all parameter vectors p and q may contain interval uncertainties. The resulting DAE system is solved by VALENCIA-IVP for the control sequence $u(t)$ and the consistent states $x(t)$.

Compared to design approaches based on symbolic formula manipulation which can be applied to feedforward control of nonlinear exactly input-to-state linearizable sets of ODEs (as a special case of differentially flat systems) [4, 11], numerical, interval-based approaches are more flexible. First, uncertainties and robustness requirements can be taken into account directly in the constraints (15). In addition, the verified approach can also handle differentially non-flat systems if stability of the internal dynamics can be guaranteed by techniques published, for example, in [18]. For most of these systems, the output $y(t)$ does not coincide exactly with $y_d(t)$. However, verified techniques still allow us to compute control sequences (if they exist) for which the tolerances $[y_{tol}(t)] \neq [0\,;\,0]$ in (15) are not violated.

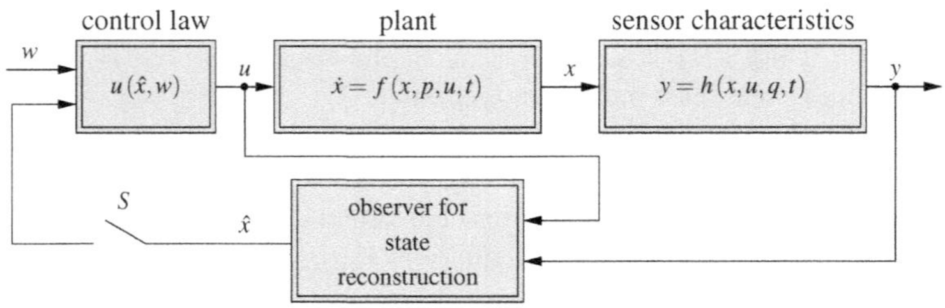

Fig. 1. Observer-based closed-loop control of nonlinear dynamical systems

Since most control structures rely on information on *estimates* for non-measured states, parameters, and disturbances, an extension of the DAE approach is considered. In classical interval observers, a two-stage approach is used for reconstruction of the non-measured quantities in a filter step by solving the measurement equations for the same number of variables as linearly independent measurements (cf. [8]). In a second stage, this information is predicted over time with the techniques from Sections 2 and 3 up to the point at which the next measured data are available.

In contrast, the DAE-based solution procedure employs a one-stage approach. To estimate non-measured quantities, the output equation $y_m(t) = h(x(t))$ is included as a further time-dependent algebraic constraint with interval uncertainties of the measured variables and their derivatives. Here, the Lie derivatives of $y_m(t) = h(x(t))$ coincide directly with the hidden constraints (14) which are evaluated in each time interval in which VALENCIA-IVP is used to integrate the dynamical system model by solving the corresponding IVP to the set of DAEs.

5 Control of a Distributed Heating System

To visualize the practical applicability of verified DAE solvers for feedforward control as well as state and disturbance estimation, we consider the distributed heating system in Fig. 2. The controlled variable of this system is the temperature at a given position of the rod. Control and disturbance inputs are provided by four Peltier elements and cooling units. The temperature $\vartheta(z,t)$ of the rod depends both on the spatial variable z and on the time t. The temperature distribution is given by the parabolic partial differential equation

$$\frac{\partial\vartheta(z,t)}{\partial t} - \frac{\lambda}{\rho c_p}\frac{\partial^2\vartheta(z,t)}{\partial z^2} + \frac{\alpha}{h\rho c_p}\vartheta(z,t) = \frac{\alpha}{h\rho c_p}\vartheta_U \tag{16}$$

which is discretized in its spatial coordinate using a finite volume discretization to obtain a model for offline simulation as well as online state and disturbance estimation. Balancing of heat exchange between four volume elements leads to the ODEs

$$\begin{bmatrix} \dot{x}_1(t) \\ \dot{x}_2(t) \\ \dot{x}_3(t) \\ \dot{x}_4(t) \end{bmatrix} = \begin{bmatrix} a_{11} & a_{12} & 0 & 0 \\ a_{12} & a_{22} & a_{12} & 0 \\ 0 & a_{12} & a_{22} & a_{12} \\ 0 & 0 & a_{12} & a_{11} \end{bmatrix} \cdot \begin{bmatrix} x_1(t) \\ x_2(t) \\ x_3(t) \\ x_4(t) \end{bmatrix} + \frac{1}{m_s c_p} \begin{bmatrix} 1 \\ 0 \\ 0 \\ 0 \end{bmatrix} u(t) + \frac{\alpha A}{m_s c_p} \begin{bmatrix} e_1(t) \\ e_2(t) \\ e_3(t) \\ e_4(t) \end{bmatrix} \tag{17}$$

for the temperatures $x_i(t)$ in the segments $i = 1, \ldots, n = 4$ with the coefficients

$$a_{11} = -\frac{\alpha A l_s + \lambda_s bh}{l_s m_s c_p}, \quad a_{12} = \frac{\lambda_s bh}{l_s m_s c_p}, \quad \text{and} \quad a_{22} = -\frac{\alpha A l_s + 2\lambda_s bh}{l_s m_s c_p}. \tag{18}$$

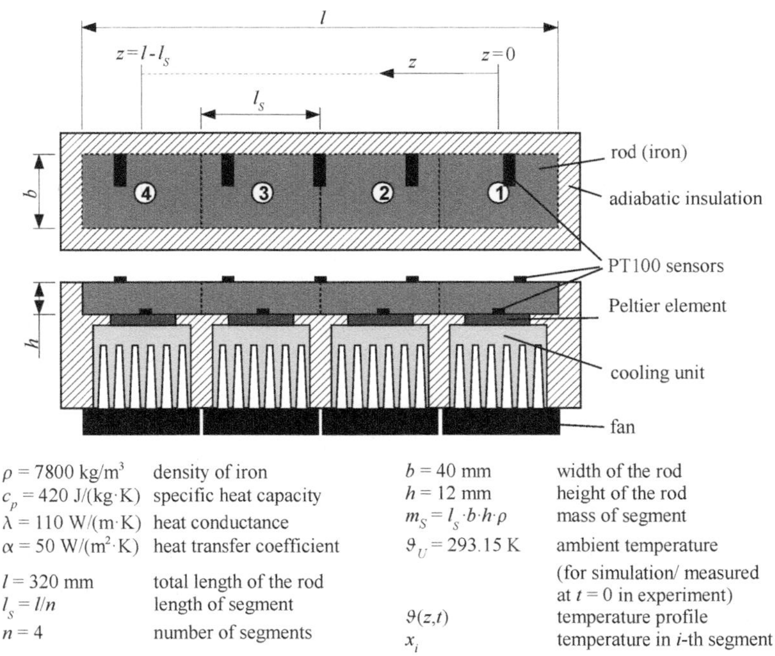

Fig. 2. Experimental setup of a distributed heating system

In (17), the input signal $u(t)$ corresponds to the heat flow into the first segment of the rod. The goal of feedforward control (determined numerically by a DAE solver) is the computation of an input $u(t) = u_1(t)$ in such a way that the output temperature in an arbitrary segment follows the specification

$$y_d(t) = \vartheta_0 + \frac{(\vartheta_f - \vartheta_0)}{2}\left(1 + \tanh\left(k\left(t - \frac{3600\,\mathrm{s}}{2}\right)\right)\right) \tag{19}$$

with $\vartheta_0 = \vartheta_U(0)$, $\vartheta_f = \vartheta_0 + 10\,\mathrm{K}$, and $k = 0.0015$ exactly. The prediction time horizon for the DAE solver is $t_{k+1} - t_k = 1\,\mathrm{s}$. To select a specific value from the control intervals, the definition $u_1(t) = 0.5 \cdot (\underline{u}_1(t) + \overline{u}_1(t))$ with $t \in [t_k; t_{k+1})$ is used.

The additive terms $e_i(t)$, $i = 1, \ldots, n = 4$ summarize errors resulting from the discretization of the PDE and unmodeled disturbances which are estimated by a Luenberger observer and the novel DAE-based approach, see the experimental results in Fig. 3. The interval observer detects the point of time from which on the Luenberger observer yields consistent estimates. Both estimators make use of the measured temperatures $y_1 = x_1$ and $y_2 = x_4$. If model errors are neglected, all e_i are equal to the ambient temperature $\vartheta_U(0)$. For the implementation of the disturbance observer, the ODEs (17) are extended by $\dot{e} = 0$ with $e = e_1 = \ldots = e_4$. To quantify the influence of measurement errors, the uncertainties $x_i \in y_j + [-1\,;\,1]\,\mathrm{K}$, $\dot{x}_i \in [-0.5\,;\,1.5]\,\dot{y}_j$, $i \in \{1,4\}$, $j \in \{1,2\}$ are considered by the DAE solver. To compensate model errors and disturbances, output feedback $u_2(t)$ is introduced in addition to the feedforward control $u_1(t)$ by a PI controller compensating the largest time constant of the plant (17). Therefore, the total control input is given by $u(t) = u_1(t) + u_2(t)$.

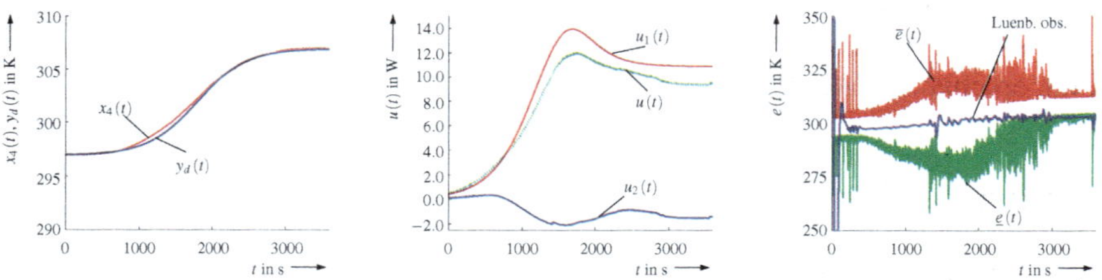

(a) Output temperature $x_4(t)$. **(b)** Control variable $u(t)$. **(c)** Disturbance estimate $e(t)$.

Fig. 3. Experimental results for closed-loop control of the heating system

For specification of the *flat* output $g(x,t) = x_4(t) - y_d(t) = 0$, the *structural analysis* performed in VALENCIA-IVP provides the following result:

The Lie derivative $L_f^4 g$ corresponds to the smallest order of the derivative of the output equation $g(x,t)$ which is influenced directly by the control input u. Since the number of unknowns and the number of hidden constraints is identical in this case, the equations $L_f^1 g = 0, \ldots, L_f^4 g = 0$ can be solved directly by interval Newton techniques for the consistent states x_1, x_2, and x_3, as well as the control input u. The value of x_4 is known a-priori from $g = L_f^0 g = 0$ for each point of time t.

For specification of a *non-flat* output, for example $g(x,t) = x_3(t) - y_d(t) = 0$, the order δ is now lower than the number of unknown variables, that is, the relative degree δ of the system is smaller than the dimension of the state vector. Thus, the equations $L_f^1 g = 0, \ldots, L_f^\delta g = 0$ cannot be solved directly. Information on the initial conditions of the system has to be included in the following two-stage procedure. In the first stage, a set of ODEs or DAEs is identified automatically which includes the system's output and can be solved as an IVP by specification of corresponding initial conditions. If these equations result in a set of DAEs, the initial conditions are computed consistently using the output equation $g = L_f^0 g = 0$ and, if necessary, the lower-order constraints $L_f^1 g = 0, \ldots, L_f^\tau g = 0$, $\tau < \delta$. In the second stage, the solution to this IVP is substituted for the corresponding state variables in $L_f^{\tau+1} g = 0, \ldots, L_f^\delta g = 0$. These purely algebraic equations are now solved for the remaining states and the control input $u(t)$ using interval Newton techniques.

For specification of x_3 as the output, it is at least necessary to know the initial temperature $x_4(0)$, see the following result of the structural analysis with $\tau = 0$:

	x_1	x_2	x_3	x_4	t	u
$\dot{x}_1$	•	•				•
$\dot{x}_2$	•	•	◇			
$\dot{x}_3$		•	◇	○		
$\dot{x}_4$			◇	○		
$g(x,t)$			◇		◇	

	x_1	x_2	x_3	x_4	t	u
$L_f^0 g$			◇		◇	
$L_f^1 g$		•	◇	○	◇	
$L_f^2 g$	•	•	◇	○	◇	
$L_f^3 g$	•	•	◇	○	◇	•

Legend:
◇ a-priori known
○ determined via IVP solver (ODE/ DAE) (stage 1)
• determined via algebraic constraints of DAE (stage 2)

6 Conclusions and Outlook on Future Research

In this paper, interval-based approaches for the verification and implementation of robust control strategies have been presented and applied to a finite volume representation of a distributed heating system. For this system, the online computation of feedforward control using VALENCIA-IVP is extended by classical output feedback for compensation of model and parameter uncertainties and neglected disturbances. Furthermore, a verified estimation procedure for internal system states and disturbances has been described which is implemented by a one-stage approach instead of the classical two-stage procedure used by other interval observers. This observer can be applied to verify the admissibility and reliability of classical non-verified observers such as Luenberger-type observers by comparison of their estimates with the verified error bounds obtained in the interval approach.

In future work, further relations between reachability of states and the controllability of uncertain dynamical systems on the one hand and the solvability of DAEs describing feedforward control problems on the other hand will be investigated. Moreover, generalizations of the routine implemented in VALENCIA-IVP for the detection of hidden algebraic constraints will be studied to extend the presented automated feedforward control to multiple-input multiple-output systems for which desired output trajectories are prescribed for non-flat outputs and for which ambiguities in the solution might exist. Finally, combinations with verified tools for stability analysis based on interval evaluation of Lyapunov functions will be discussed to prove stability of non-observable or non-controllable internal dynamics.

References

1. Auer, E., Rauh, A., Hofer, E.P., Luther, W.: Validated Modeling of Mechanical Systems with SMARTMOBILE: Improvement of Performance by VALENCIA-IVP. In: Hertling, P., Hoffmann, C.M., Luther, W., Revol, N. (eds.) Real Number Algorithms. LNCS, vol. 5045, pp. 1–27. Springer, Heidelberg (2008)
2. Bendsten, C., Stauning, O.: FADBAD++, Version 2.1 (2007), `http://www.fadbad.com`
3. Berz, M., Makino, K.: COSY INFINITY Version 8.1. User's Guide and Reference Manual. Tech. Rep. MSU HEP 20704, Michigan State University (2002)
4. Fliess, M., Lévine, J., Martin, P., Rouchon, P.: Flatness and Defect of Nonlinear Systems: Introductory Theory and Examples. International Journal of Control 61, 1327–1361 (1995)
5. Freihold, M., Hofer, E.P.: Derivation of Physically Motivated Constraints For Efficient Interval Simulations Applied to the Analysis of Uncertain Dynamical Systems. Special Issue of the International Journal of Applied Mathematics and Computer Science AMCS, Verified Methods: Applications in Medicine and Engineering 19(3), 485–499 (2009)
6. Jaulin, L., Kieffer, M., Didrit, O., Walter, É.: Applied Interval Analysis. Springer, London (2001)
7. Keil, C.: PROFIL/BIAS, Version 2.0.8 (2008), `http://www.ti3.tu-harburg.de/keil/profil/`
8. Kletting, M., Rauh, A., Aschemann, H., Hofer, E.P.: Interval Observer Design Based on Taylor Models for Nonlinear Uncertain Continuous-Time Systems. In: CD-Proc. of the 12th GAMM-IMACS International Symposium on Scientific Computing, Computer Arithmetic, and Validated Numerics SCAN 2006. IEEE Computer Society, Duisburg (2007)
9. Krawczyk, R.: Newton-Algorithmen zur Bestimmung von Nullstellen mit Fehlerschranken. Computing 4, 189–201 (1969) (In German)
10. Lin, Y., Stadtherr, M.A.: Validated solution of initial value problems for ODEs with interval parameters. In: NSF Workshop Proceeding on Reliable Engineering Computing, Savannah GA (2006)
11. Marquez, H.J.: Nonlinear Control Systems. John Wiley & Sons, Inc., New Jersey (2003)
12. Moore, R.E.: Interval Arithmetic. Prentice-Hall, Englewood Cliffs (1966)
13. Nedialkov, N.S.: Interval Tools for ODEs and DAEs. In: CD-Proc. of the 12th GAMM-IMACS International Symposium on Scientific Computing, Computer Arithmetic, and Validated Numerics SCAN 2006. IEEE Computer Society, Duisburg (2007)
14. Nedialkov, N.S., Pryce, J.D.: DAETS — Differential-Algebraic Equations by Taylor Series (2008), `http://www.cas.mcmaster.ca/~nedialk/daets/`
15. Rauh, A., Brill, M., Günther, C.: A Novel Interval Arithmetic Approach for Solving Differential-Algebraic Equations with VALENCIA-IVP. Special Issue of the International Journal of Applied Mathematics and Computer Science AMCS, Verified Methods: Applications in Medicine and Engineering 19(3), 381–397 (2009)
16. Rauh, A., Hofer, E.P.: Interval Methods for Optimal Control. In: Frediani, A., Buttazzo, G. (eds.) Proc. of the 47th Workshop on Variational Analysis and Aerospace Engineering 2007, Erice, Italy. School of Mathematics, pp. 397–418. Springer, Heidelberg (2009)

17. Rauh, A., Minisini, J., Hofer, E.P.: Towards the Development of an Interval Arithmetic Environment for Validated Computer-Aided Design and Verification of Systems in Control Engineering. In: Cuyt, A., Krämer, W., Luther, W., Markstein, P. (eds.) Numerical Validation in Current Hardware Architectures. LNCS, vol. 5492, pp. 175–188. Springer, Heidelberg (2008)
18. Rauh, A., Minisini, J., Hofer, E.P.: Verification Techniques for Sensitivity Analysis and Design of Controllers for Nonlinear Dynamic Systems with Uncertainties. Special Issue of the International Journal of Applied Mathematics and Computer Science AMCS, Verified Methods: Applications in Medicine and Engineering 19(3), 425–439 (2009)

Rational Interpolation of Rigid-Body Motions

J.M. Selig

1 Interpolation in Groups

Let $g_1, g_2, \ldots, g_n$ be a sequence of elements of a Lie group, (knot points). Our problem is to find a smooth, parameterised curve in the group that passes through these elements at parameter values $t_1, t_2, \ldots, t_n$. There are many variations on this basic problem. For example we could take account of velocities. Perhaps we might only require the curve to be near the knot points.

There are many applications for this problem. In robotics, for instance, the Lie group will be $SE(3)$, the group of proper rigid-body transformations. Curves in this group are rigid-body motions. It is important to plan the motion of a robot's end-effector. We would like to choose a few knot positions, to avoid an obstacle say, and control the robot to move along an interpolated curve through these positions.

Other applications come from Computer Graphics. Again the relevant Lie group is $SE(3)$. In 3-D modelling, we often want to produce a 'fly-by'. Choose key positions for the camera, then interpolate a rigid motion between these. At present, skilled operator must use experience to choose good key positions to get acceptable results.

Another possible application is to the simulation of dynamics of rigid bodies. This is a key problem in computer games. Solving equations of motion is usually time consuming. So it might be advantageous to use expensive methods to solve for key positions and then interpolate between these positions. Alternatively, it might be possible to use interpolation methods to approximate the solution quickly and directly.

J.M. Selig
Faculty of Business, Computing and Information Management,
London South Bank University, U.K.

J. Lévine & P. Müllhaupt (Eds.): Adv. in Theory of Control, Signals, LNCIS 407, pp. 213–224.
springerlink.com © Springer-Verlag Berlin Heidelberg 2010

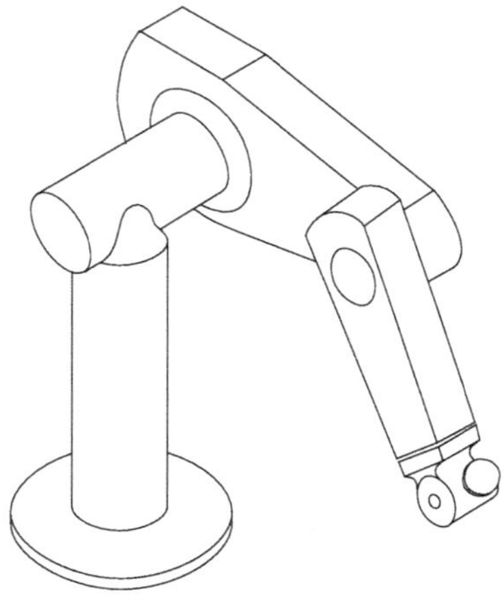

Fig. 1. A Typical Industrial Robot

With all these different applications it is understandable that the problem has received much attention in the literature. In fact far too much to review here, however see [2] for an extensive review.

1.1 A Simple Solution

In a vector space we could use well known techniques, for example Lagrange or Hermite interpolation and many others. In a group we can't add elements or multiply by scalars. So a possible strategy is,

1. Map knot positions to the Lie algebra.
2. Solve the interpolation problem in the Lie algebra, (a vector space.)
3. Map the interpolated curve back to the group.

The exponential and logarithm maps could be used to map between the Lie group and its Lie algebra, but this involves evaluating transcendental functions which is usually time consuming on a computer. An alternative would be to use Cayley maps and their inverses. Cayley maps, unlike the exponential map, depend on the representation of the group being used. So there are several Cayley maps to choose from. Cayley maps are rational maps meaning that the matrix entries in the result will be rational functions of the coordinates on the Lie algebra. This means that any interpolating curve will be a rational curve in the representation of the group being used. Rational functions are easy to evaluate on a computer and have other advantages from a computational point of view. To implement this strategy it is simplest to use dual quaternions to represent $SE(3)$, rather than matrices.

2 Dual Quaternions

Dual quaternions were invented by Clifford, and used by Study, Blaschke and others. In the latter half of the 20th century Mathematicians seems to forget about them, preferring matrix methods. However, they were always remembered and used in Kinematics. They give a very neat and succinct way to represent rigid-body transformations.

A general dual quaternion has the form,

$$q = h_0 + \varepsilon h_1, \tag{1}$$

where h_0, h_1 are standard quaternions $h = a_0 + a_1 i + a_2 j + a_3 k$. With $i^2 = j^2 = k^2 = -1$ and $ijk = -1$ as usual. The dual unit is ε and it commutes with i, j, k and squares to zero, $\varepsilon^2 = 0$. The algebra of dual quaternions is given by, $\mathbb{H} \otimes \mathbb{D}$ where $\mathbb{H}$ are the quaternions and $\mathbb{D}$ is the ring of dual numbers, generated by ε.

2.1 Quaternions and Rotations

It is well known that quaternions can be used to represent rotations about the origin,

$$r = \cos\frac{\theta}{2} + \sin\frac{\theta}{2}(l_x i + l_y j + l_z k), \tag{2}$$

where θ is the rotation angle and $\mathbf{l} = (l_x, l_y, l_z)$ is the unit vector along the rotation axis.

The action of these rotations on points given by conjugation,

$$x'i + y'j + z'k = r(xi + yj + zk)r^-, \tag{3}$$

where r^- is the quaternion conjugate of r. That is, if $h = a_0 + a_1 i + a_2 j + a_3 k$ then $h^- = a_0 - a_1 i - a_2 j - a_3 k$. Rotations are unit quaternions, $rr^- = 1$. But $+r$ and $-r$ give the same rotation, so the unit quaternions form a 2-to-1 cover of the rotation group.

2.2 Rigid-Body Transformations

In the dual quaternions a rigid transformation is represented by an element of the form,

$$g = r + \frac{1}{2}\varepsilon t r, \tag{4}$$

where r is a unit quaternion representing the rotational part of the transformation and $t = t_x i + t_y j + t_z k$ is a pure quaternion (that is a quaternion with no real part) representing the translational component of the transformation.

Here the action on points is given by,

$$1+\varepsilon(x'i+y'j+z'k)=(r+\frac{1}{2}\varepsilon tr)(1+\varepsilon(xi+yj+zk))(r^- +\frac{1}{2}\varepsilon r^- t). \quad (5)$$

Again it is clear that $+g$ and $-g$ give the same transformation.

Notice that these dual quaternions satisfy the equation $gg^- = 1$, where now the conjugate is given by,$(h_0+\varepsilon h_1)^- = h_0^- + \varepsilon h_1^-$. If we write,

$$g=(a_0+a_1i+a_2j+a_3k)+\varepsilon(c_0+c_1i+c_2j+c_3k), \quad (6)$$

then the real and dual parts of the equation give,

$$1 = a_0^2+a_1^2+a_2^2+a_3^2, \quad (7)$$

$$0 = a_0c_0+a_1c_1+a_2c_2+a_3c_3. \quad (8)$$

Now assume that $(a_0 : a_1 : a_2 : a_3 : c_0 : c_1 : c_2 : c_3)$ are homogeneous coordinates in a seven dimensional projective space, $\mathbb{P}^7$. This will identify $+g$ and $-g$. In a projective space only homogeneous equations are well defined, or at least the set of zeros for such an equation is well defined. The first quadric above (7), is not homogeneous so cannot be considered now. Only the second equation (8), is meaningful. Retaining only (8) it can be seen that the group elements lie on a six-dimensional quadric, called the Study quadric after its discoverer E. Study.

It can be shown that rigid transformations are in 1-to-1 correspondence with points in the Study quadric, with the exception of the 3-plane $a_0 = a_1 = a_2 = a_3 = 0$. That is the group $SE(3)$ of rigid-body transformations can be thought of as an open set in a six-dimensional quadric variety.

2.3 Lie Algebra

Elements of the Lie algebra to the group can also be represented as dual quaternions. Given a rigid-body motion $g(t)$ the corresponding Lie algebra element at parameter value t is given by,

$$\left(\frac{d}{dt}g\right)g^- = (\omega_x i+\omega_y j+\omega_z k)+\varepsilon(v_x i+v_y j+v_z k). \quad (9)$$

The angular velocity of the motion is given by the vector $\boldsymbol{\omega}$, with components $\boldsymbol{\omega} = (\omega_x, \omega_y, \omega_z)^T$. The other characteristic vector of the motion is a linear velocity $\mathbf{v}$ with components $\mathbf{v} = (v_x, v_y, v_z)^T$, physically this corresponds to the velocity of the origin.

2.4 Dual Quaternion and the 4 × 4 Representation

The homogeneous, or 4×4 representation of the group of rigid transformations uses matrices of the form,

$$\begin{pmatrix} R & \mathbf{t} \\ 0 & 1 \end{pmatrix}, \quad (10)$$

where R is the usual 3×3 rotation matrix and $\mathbf{t}$ is a translation vector.

Given a group element as a dual quaternion $g = (a_0 + \mathbf{a}) + \varepsilon(c_0 + \mathbf{c})$, that is with $\mathbf{a} = a_1 i + a_2 j + a_3 k$ and so forth, the corresponding 4×4 matrix has,

$$R = \frac{1}{\Delta^2}\left(\Delta^2 I_3 + 2a_0 A + 2A^2\right) \quad \text{and} \quad \mathbf{t} = \frac{2}{\Delta^2}\left(a_0\mathbf{c} - c_0\mathbf{a} + \mathbf{a} \times \mathbf{c}\right), \tag{11}$$

where $\Delta^2 = a_0^2 + a_1^2 + a_2^2 + a_3^2$ and A is the anti-symmetric matrix corresponding to $\mathbf{a}$, that is,

$$A = \begin{pmatrix} 0 & -a_3 & a_2 \\ a_3 & 0 & -a_1 \\ -a_2 & a_1 & 0 \end{pmatrix}. \tag{12}$$

It is also possible to recover the dual quaternion from a 4×4 matrix, (up to an overall sign). We have,

$$a_0 = \frac{1}{2}\sqrt{\mathrm{Tr}(R) + 1}, \tag{13}$$

and then

$$A = \frac{1}{2\sqrt{\mathrm{Tr}(R) + 1}}(R - R^T). \tag{14}$$

finally

$$(c_0 + \mathbf{c}) = \frac{1}{2}\mathbf{t}(a_0 + \mathbf{a}). \tag{15}$$

Notice, that vectors as pure quaternions and column vectors have not been distinguished,

$$\mathbf{t} = t_x i + t_y j + t_z k = \begin{pmatrix} t_x \\ t_y \\ t_z \end{pmatrix}. \tag{16}$$

More details on dual quaternions and their uses in robotics can be found in [4].

3 The 4 × 4 Cayley Map

In the 4×4 representation of $SE(3)$ the elements of the Lie algebra are given by matrices,

$$S = \begin{pmatrix} W & \mathbf{u} \\ 0 & 0 \end{pmatrix}, \tag{17}$$

where W is a 3×3 anti-symmetric matrix and $\mathbf{u}$ a vector. The Cayley map is defined to be,

$$\mathrm{Cay}_4(S) = (I_4 + S)(I_4 - S)^{-1}. \tag{18}$$

Since, by the Cayley-Hamilton theorem, the matrices S satisfies a degree 4 polynomial, the result of the map can be written as a cubic in S,

$$\mathrm{Cay}_4(S) = I_4 + 2S + \frac{2}{1+|w|^2}S^2 + \frac{2}{1+|w|^2}S^3, \tag{19}$$

where $|w|^2 = -(1/2)\,\mathrm{Tr}(S^2) = -(1/2)\,\mathrm{Tr}(W^2)$.

Writing the Lie algebra element as a dual quaternion $s = (w_1 i + w_2 j + w_3 k) + \varepsilon(u_1 i + u_2 j + u_3 k)$ this Cayley map can be written,

$$\mathrm{Cay}_4(s) = \frac{1}{2\sqrt{1+|w|^2}}\big((2+|w|^2) + 2s + s^2\big). \tag{20}$$

Notice that this is only quadratic in s.

The inverse Cayley map can also be written in terms of dual quaternions. Here g is a dual quaternion representing a group element, with rotation angle θ,

$$\mathrm{Cay}_4^{-1}(g) = \frac{1}{2\cos(\theta/2)}\big(g^3 - 4\cos\frac{\theta}{2}g^2 + (4\cos^2\frac{\theta}{2} + 3)g - 4\cos\frac{\theta}{2}\big). \tag{21}$$

More details on this map and other Cayley maps can be found in [5] and [6].

3.1 Cayley Map as a Rational Map

These maps can be thought of as birational transformations between the six-dimensional projective space $\mathbb{P}^6$ and the Study quadric in $\mathbb{P}^7$. To see this let us introduce a homogenising variable w_0, so the homogeneous coordinates for the $\mathbb{P}^6$ will be $(w_0 : w_1 : w_2 : w_3 : u_1 : u_2 : u_3)$ and the coordinates in the $\mathbb{P}^7$ will be $(a_0 : a_1 : a_2 : a_3 : c_0 : c_1 : c_2 : c_3)$ as above. Including the homogenising variable the Cayley map is then,

$$\mathrm{Cay}_4(s) = \frac{1}{2\sqrt{w_0^2+|w|^2}}\big((2w_0^2 + |w|^2) + 2w_0 s + s^2\big), \tag{22}$$

Explicitly in terms of coordinates, this is,

$$\begin{aligned}
a_0 &= w_0^2, & c_0 &= -(w_1u_1 + w_2u_2 + w_3u_3),\\
a_1 &= w_0w_1, & c_1 &= w_0u_1,\\
a_2 &= w_0w_2, & c_2 &= w_0u_2,\\
a_3 &= w_0w_3, & c_3 &= w_0u_3.
\end{aligned} \tag{23}$$

Note that the normalising factor, $1/2\sqrt{w_0^2+|w|^2}$, is irrelevant here since the coordinates are homogeneous. So the Cayley map is a quadratic transformation with exceptional set given by the 4-dimensional non-singular quadric,

$$w_0 = 0, \quad w_1u_1 + w_2u_2 + w_3u_3 = 0. \tag{24}$$

The inverse map, Cay_4^{-1} is given by a cubic polynomial,

$$\mathrm{Cay}_4^{-1}(g) = \frac{1}{2a_0(a_0^2+a_1^2+a_2^2+a_3^2)}\Big(g^3 - 4a_0g^2 + \\ \big(4a_0^2 + 3(a_0^2+a_1^2+a_2^2+a_3^2)\big)g - 4a_0(a_0^2+a_1^2+a_2^2+a_3^2)\Big). \quad (25)$$

Notice that here $\cos(\theta/2)$ has been replaced by a_0 and the factor $(a_0^2 + a_1^2 + a_2^2 + a_3^2)$ has been used to make the expression homogeneous. Recall from (7) above, that this factor can be chosen to be equal to 1 on physical group elements, it only vanishes on the ideal elements of the Study quadric where $a_0 = a_1 = a_2 = a_3 = 0$.

If we assign $w_0 = 2a_0(a_0^2 + a_1^2 + a_2^2 + a_3^2)$, the common denominator, then the other coordinates of $s = \mathrm{Cay}^{-1}(g)$ are given by expanding the polynomial in the dual quaternion g, see (6) above. The resulting expression can then be simplified using the relation (8), which defines the Study quadric. The result, after cancelling common factors, is,

$$\begin{aligned} w_0 &= a_0, \\ w_1 &= a_1, \quad u_1 = c_1, \\ w_2 &= a_2, \quad u_2 = c_2, \\ w_3 &= a_3, \quad u_3 = c_3. \end{aligned} \quad (26)$$

So this is in fact a linear projection, and is clearly the inverse to the Cayley map (23), above.

3.2 *Example — Two Positions*

Here we look at interpolation between two positions, the identity $g(0) = 1$ and $g(1) = \frac{1}{\sqrt{2}}(1+k) + \frac{1}{\sqrt{2}}\varepsilon(-\frac{3}{2} + i + \frac{3}{2}k)$. The interpolating curve is a line in the Lie algebra. As a rational map the Cayley map is quadratic so the line maps to a conic in the Study quadric, see figure 2. This looks similar to a finite screw motion, but the path of origin is a conic, not a helix.

4 Velocities

Using this approach it is possible to include information about velocities. The Cayley map can be used to pull-back velocities. Suppose that S is a function of time, then the Lie algebra element determined by the velocity of the motion produced by S is given by,

$$S_d = \left(\frac{d}{dt}\mathrm{Cay}_4(S)\right)\mathrm{Cay}_4(S)^{-1} = 2(I_4 - S)^{-1}\dot{S}(I_4 + S)^{-1}. \quad (27)$$

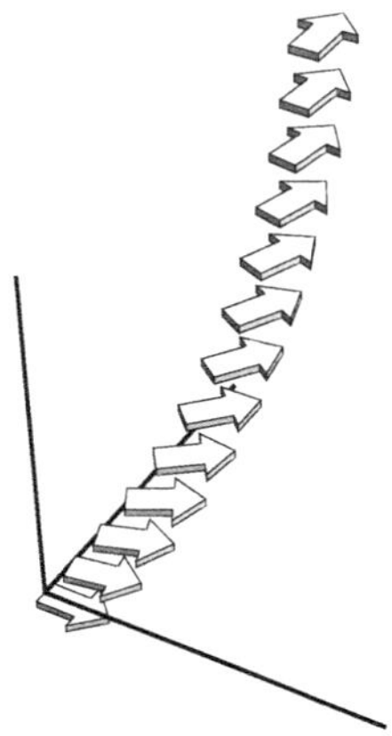

Fig. 2. Linear interpolation in the Lie algebra

The Lie algebra element S_d is usually known as the 'twist' of the motion. Rearranging this gives the result,

$$\dot{S} = \frac{1}{2}(I_4 - S)S_d(I_4 + S). \tag{28}$$

This is the tangent to the curve in the Lie algebra determined by S. Expanding this relation using the partitioned form of the matrices given in (17) above yields,

$$\dot{\mathbf{w}} = \boldsymbol{\omega} + \boldsymbol{\omega} \times \mathbf{w} + (\boldsymbol{\omega} \cdot \mathbf{w})\mathbf{w}, \tag{29}$$

$$\dot{\mathbf{u}} = \boldsymbol{\omega} \times \mathbf{u} + \mathbf{v} \times \mathbf{w} + (\boldsymbol{\omega} \cdot \mathbf{w})\mathbf{u} - \boldsymbol{\omega}(\mathbf{w} \cdot \mathbf{u}) + \mathbf{v}, \tag{30}$$

where $\boldsymbol{\omega}$ is the angular velocity vector and $\mathbf{v}$ the velocity of any point on the instantaneous screw axis. This does not seem to have a neat expression in terms of dual quaternions but this can now be used to interpolate with velocity constraints. The above expression can be used to determine the required time derivative in the Lie algebra given the desired angular and linear velocities of the motion.

4.1 Hermite Interpolation

In this example the previous example is re-examined but now we demand that the start and finish velocities are translational only and in the x and y direction respectively. The interpolating polynomial in the Lie algebra is now a cubic. This maps to a degree six curve in the Study quadric. See figure 3.

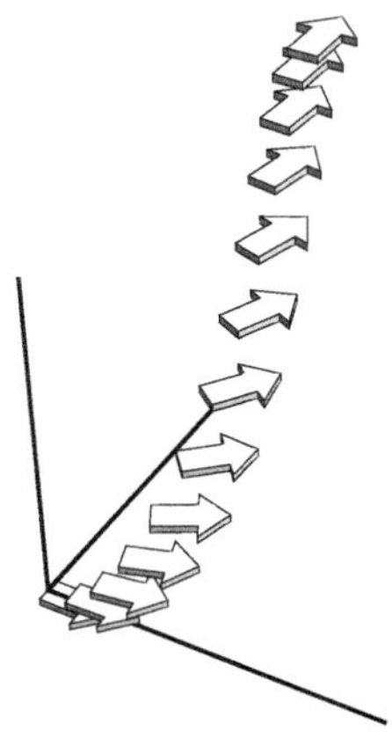

Fig. 3. Hermite interpolation in the Lie algebra

5 Bennett Motions

In 1903 G.T. Bennett discovered a mobile 4-bar mechanism, a crude diagram of this linkage is shown in figure 4. The motion of the coupler bar of this mechanism was found to be a conic curve in the Study quadric. Now any conic in the Study quadric is called a Bennett motion or a generalised Bennett motion, see [1].

Which curves in the Lie algebra will be mapped to Bennett motions in the Study quadric by the Cayley map? Clearly, since the Cayley map is quadratic, lines in $\mathbb{P}^6$ will be mapped to conics in the Study quadric. Ordinarily a conic in $\mathbb{P}^6$ would be mapped to a degree 4 rational curve in the Study Quadric. But if the conic meets the exceptional set at two points then the image will be a conic.

Recall from (24) above, that the exceptional set is a 4-dimensional quadric. This is essentially the Klein quadric of lines in $\mathbb{P}^3$. Hence elements of the exceptional set will be called 'lines at infinity'.

Now consider a conic in $\mathbb{P}^6$ given by,

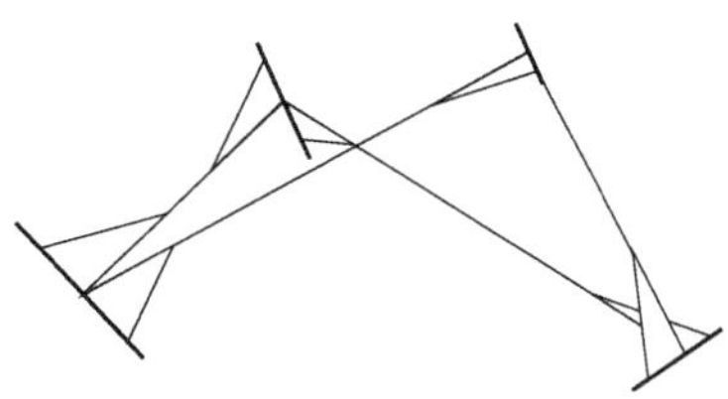

Fig. 4. The Bennett Mechanism

$$\begin{aligned} w_0 &= (t-\alpha)(t-\beta), \\ \begin{pmatrix} \mathbf{w} \\ \mathbf{u} \end{pmatrix} &= (t-\alpha)(t-\beta)\mathbf{z} + (t-\alpha)(t-\gamma)\mathbf{l}_1 + (t-\beta)(t-\delta)\mathbf{l}_2, \end{aligned}$$

where $\mathbf{z}$ is an arbitrary 6-vector and $\mathbf{l}_1$ and $\mathbf{l}_2$ are lines, that is they satisfy,

$$\mathbf{l}^T Q_0 \mathbf{l} = \mathbf{l}_i^T \begin{pmatrix} 0 & I_3 \\ I_3 & 0 \end{pmatrix} \mathbf{l}_i = 0, \qquad i = 1, 2. \tag{31}$$

This conic clearly meets the plane at infinity ($w_0 = 0$) in two lines at $t = \alpha$ and $t = \beta$. The result of the Cayley map on this conic will be,

$$\begin{aligned} a_0 &= (t-\alpha)^2(t-\beta)^2, \\ c_0 &= -(t-\alpha)^2(t-\beta)^2\mathbf{z}^T Q_0 \mathbf{z} - 2(t-\alpha)^2(t-\beta)(t-\gamma)\mathbf{z}^T Q_0 \mathbf{l}_1 \\ &\quad -2(t-\alpha)(t-\beta)^2(t-\delta)\mathbf{z}^T Q_0 \mathbf{l}_2 - 2(t-\alpha)(t-\beta)(t-\gamma)(t-\delta)\mathbf{l}_1^T Q_0 \mathbf{l}_2, \\ \begin{pmatrix} \mathbf{a} \\ \mathbf{c} \end{pmatrix} &= (t-\alpha)^2(t-\beta)^2\mathbf{z} + (t-\alpha)^2(t-\beta)(t-\gamma)\mathbf{l}_1 + (t-\alpha)(t-\beta)^2(t-\delta)\mathbf{l}_2. \end{aligned}$$

Cancelling the common factor $(t-\alpha)(t-\beta)$ give a conic in the Study quadric as expected,

$$\begin{aligned} a_0 &= (t-\alpha)(t-\beta), \\ c_0 &= -(t-\alpha)(t-\beta)\mathbf{z}^T Q_0 \mathbf{z} - 2(t-\alpha)(t-\gamma)\mathbf{z}^T Q_0 \mathbf{l}_1 \\ &\quad -2(t-\beta)(t-\delta)\mathbf{z}^T Q_0 \mathbf{l}_2 - 2(t-\gamma)(t-\delta)\mathbf{l}_1^T Q_0 \mathbf{l}_2, \\ \begin{pmatrix} \mathbf{a} \\ \mathbf{c} \end{pmatrix} &= (t-\alpha)(t-\beta)\mathbf{z} + (t-\alpha)(t-\gamma)\mathbf{l}_1 + (t-\beta)(t-\delta)\mathbf{l}_2. \end{aligned}$$

Finally here we look at an example interpolating three key points using a Bennett motion.

The strategy will be as follows, first we project the key points to $\mathbf{s}_0$, $\mathbf{s}_1$ and $\mathbf{s}_2$ in $\mathbb{P}^6$. Now three points determine a 2-plane in $\mathbb{P}^6$. This 2-plane intersects the hyperplane $w_0 = 0$ in a line, which in turn meets the Klein quadric in 2 points. Together with the original 3 points this gives 5 points in a 2-plane. So we can find a unique conic through these 5 points. Finally, the conic can be mapped back to the Study quadric with the Cayley map.

In figure 5 a Bennett motion is shown, the first and last points are the same as in the previous examples and the middle position is as shown on the left of the figure. In this case it is probably simpler to work in the Study quadric directly, since 3 points there will also determine a 2-plane and this 2-plane will intersect the Study quadric in a conic, through the 3 points.

Schröcker and Jüttler [3] have extended this to the construction of Bennett biarcs, pairs of Bennett motions through a common position with continuous derivative at the common point. Again these construction can be pulled-back to the Lie algebra but are simpler in the Study quadric itself.

Fig. 5. Interpolation using a Bennett Motion. The three key positions are shown on the left

6 Conclusions

The novelty of the technique presented here is that it treats the rotational and translation parts of the motion on an equal footing. In most of the previous solutions the rotations and translations are treated separately, and this leads to solutions which depend on the choice of origin in space. The methods given here do not suffer from this problem. Indeed the methods presented here satisfy both requirements proposed by Röschel, in [2]; that the interpolating motion be rational and independent of the coordinates chosen.

There are several other Cayley maps that could also have been used. The one used here appears to be the simplest, especially since the inverse of this Cayley map is just a linear projection. For the Cayley map based on the adjoint representation of the group, for example, the map is cubic and its inverse is quadratic. Using other Cayley maps will also give different solutions to the same problems.

Finally there are may other ways to specify a rigid-body motion. For example, it is often useful to specify a rigid-body motion as the solution to a variational problem as in dynamics or beam theory. It should be possible to approximate the solutions to such problems using rational motions as described here.

References

1. Bottema, O., Roth, B.: Theoretical Kinematics. Dover Publications, New York (1990)
2. Röschel, O.: Rational motion design — a survey. Computer-Aided Design 30(3), 169–178 (1998)
3. Schröcker, H.P., Jüttler, B.: Motion interpolation with Bennett biarcs. In: Kecskeméthy, A., Müller, A. (eds.) Proceedings of the 5th International Workshop on Computational Kinematics. Springer, Berlin (2009)

4. Selig, J.M.: Geometric Fundamentals of Robotics. Springer, New York (2005)
5. Selig, J.M.: Cayley Maps for SE(3). In: The International Fedaration of Theory of Machines and Mechanisms 12th World Congress, Besançon (2007)
6. Selig, J.M.: Exponential and Cayley maps for Dual Quaternions. Advances in Applied Clifford Algebras, published online (2010), doi:10.1007/s00006-010-0229-5

Contact Geometry and Its Application to Control

Peter J. Vassiliou

1 Introduction

The purpose of this note is to describe a recent generalisation of the well-known Goursat normal form and explore its possible role in control theory. For instance, we give a new, straightforward, general procedure for linearising nonlinear control systems, including time-varying, fully nonlinear systems and we illustrate the method by elementary pedagogical examples. We also exhibit an apparently non-flat control system[1] which can nevertheless be explicitly linearised and therefore posseses an infinite symmetry group.

The Goursat normal form is a local characterisation of the contact distribution on $J^k(\mathbb{R}, \mathbb{R})$ for all $k \geq 1$, which we denote $\mathcal{C}_1^{(k)}$. The original theorem is not due to Goursat who was its populariser [4]. It appears the theorem is originally due, in some form, to E. von Weber but the statement of it I give below essentially arises from a work of Cartan; see [9]. A good reference on this classical topic is [10].

The generalised Goursat normal form, presented in [11, 12] is a geometric characterisation of partial prolongations of the contact distribution $\mathcal{C}_q^{(1)}$ on the jet space $J^1(\mathbb{R}, \mathbb{R}^q)$ for all $q \geq 1$.[2] Given a Pfaffian system Ω or, equivalently, a vector field distribution $\mathcal{V}$ on manifold M, it solves the recognition problem up to a local diffeomorphism of M based solely on the derived type of $\mathcal{V}$. It is therefore analogous to the eponymous Goursat normal form. Additionally, in [12] it was shown how the generalised Goursat normal form gives rise to a procedure for explicitly identifying a distribution with some partial prolongation using the smallest number of integrations; this procedure is

Peter J. Vassiliou
Faculty of Information Sciences and Engineering
University of Canberra, Canberra A.C.T. 2601, Australia

[1] I do not yet have a proof that the system is not differentially flat.

[2] Recall that partial prolongations and Brunovsky normal forms are identical geometric objects.

J. Lévine & P. Müllhaupt (Eds.): Adv. in Theory of Control, Signals, LNCIS 407, pp. 225–237.
springerlink.com © Springer-Verlag Berlin Heidelberg 2010

summarised and illustrated below. Most of the calculations to be presented have been automated in the `Maple` package `DifferentialGeometry` and this makes the method particularly easy to use.

Note that the equivalences implied by the generalised Goursat normal form are not *a priori* static feedback transformations nor is the differential system $\mathcal{V}$ assumed to be a control system with predefined states and controls. In fact the theorem was not motivated by control theory in the first instance but rather purely as a result in exterior differential systems with applications in differential geometry and integrable systems theory. Indeed the main motivating examples arose from the theory of Darboux integrable partial differential equations [1]. Nevertheless, static feedback equivalences are not excluded and useful applications to problems of nonlinear control are possible, as will be demonstrated.

The problem of linearising nonlinear control systems has generated an important literature the review of which is well beyond the scope of this short paper. Nevertheless, in addition to the works previously cited and the citations therein, I would like to mention [3, 5, 6] which are close in aims and general techniques to the present work.

The derived flag. Suppose M is a smooth manifold and $\mathcal{V} \subset TM$ a smooth sub-bundle of its tangent bundle. The structure tensor is the map $\delta : \Lambda^2\mathcal{V} \to TM/\mathcal{V}$ defined by

$$\delta(X, Y) = [X, Y] \mod \mathcal{V}, \text{ for all } X, Y \in \mathcal{V}.$$

In more detail, suppose $X_1, \ldots, X_r$ is a basis for $\mathcal{V}$ and $\omega^1, \ldots, \omega^r$ is the dual basis for its dual $\mathcal{V}^*$. Suppose $Z_1, \ldots, Z_s$ is a basis for $TM/\mathcal{V}$ such that $[X_i, X_j] \equiv c^k_{ij} Z_k \mod \mathcal{V}$ for some functions c^k_{ij} on M. Then $\delta = c^k_{ij}\omega^i \wedge \omega^j \otimes Z_k$; that is, a section of $\Lambda^2\mathcal{V}^* \otimes TM/\mathcal{V}$. The structure tensor encodes important information about a sub-bundle, the most obvious of which is the extent to which it fails to be Frobenius integrable.

If δ has constant rank, we define the *first derived bundle* $\mathcal{V}^{(1)}$ as the inverse image of $\delta(\Lambda^2\mathcal{V})$ under the canonical projection $TM \to TM/\mathcal{V}$. Informally,

$$\mathcal{V}^{(1)} = \mathcal{V} + [\mathcal{V}, \mathcal{V}].$$

The derived bundles $\mathcal{V}^{(i)}$ are defined inductively:-

$$\mathcal{V}^{(i+1)} = \mathcal{V}^{(i)} + [\mathcal{V}^{(i)}, \mathcal{V}^{(i)}], \quad \mathcal{V}^{(0)} = \mathcal{V}, \quad i \geq 0,$$

assuming that at each iteration, this defines a vector bundle, in which case we shall say that $\mathcal{V}$ is *regular*. For regular $\mathcal{V}$, by dimension reasons, there will be a smallest k for which $\mathcal{V}^{(k+1)} = \mathcal{V}^{(k)}$. This k is called the *derived length* of $\mathcal{V}$ and the whole sequence of sub-bundles

$$\mathcal{V} \subset \mathcal{V}^{(1)} \subset \mathcal{V}^{(2)} \subset \cdots \subset \mathcal{V}^{(k)}$$

the *derived flag* of $\mathcal{V}$. We shall denote by $\mathcal{V}^{(\infty)}$ the smallest integrable sub-bundle containing $\mathcal{V}$.

Cauchy bundles. Let us define

$$\zeta : \mathcal{V} \to \mathrm{Hom}(\mathcal{V}, TM/\mathcal{V}) \text{ by } \zeta(X)(Y) = \delta(X, Y)$$

Even if $\mathcal{V}$ is regular, the homomorphism ζ need not have constant rank. If it does, let us write Char $\mathcal{V}$ for its kernel. The Jacobi identity shows that Char $\mathcal{V}$ is always integrable. It is called the *Cauchy bundle* or *characteristic bundle* of $\mathcal{V}$. If $\mathcal{V}$ is regular and each $\mathcal{V}^{(i)}$ has a Cauchy bundle then, we say that $\mathcal{V}$ is *totally regular.* Then by the *derived type* of $\mathcal{V}$ we shall mean the list $\{\mathcal{V}^{(i)}, \text{Char } \mathcal{V}^{(i)}\}$ of sub-bundles.

Theorem 1 (Goursat normal form). *Let $\mathcal{V} \subset TM$ be a smooth, totally regular, rank 2 sub-bundle over smooth manifold M such that*

a) $\mathcal{V}^{(\infty)} = TM$
b) $\dim \mathcal{V}^{(i+1)} = \dim \mathcal{V}^{(i)} + 1$, *while* $\mathcal{V}^{(i)} \neq TM$

Then there is a generic subset $\hat{M} \subseteq M$ such that in a neigbourhood of each point of $\hat{M}$ there are local coordinates $x, z_0, z_1, z_2, \ldots z_k$ such that $\mathcal{V}$ has local expression

$$\left\{\partial_x + \sum_{j=1}^{k} z_j \partial_{z_{j-1}},\ \partial_{z_k}\right\} \tag{1}$$

where $k = \dim M - 2$. That is, $\mathcal{V}$ is locally equivalent to $\mathcal{C}_1^{(k)}$ on $\hat{M}$.

A proof can be found in [10], pp157-159. For examples of application, see [13].

The Goursat normal form (Theorem 1) solves the problem of when a Pfafffian system or vector field distribution can be identified with the contact distribution (1) from information deduced algorithmically from the derived flag of the differential system. The generalised Goursat normal form (GGNF) does the same job in case distribution (1) is replaced by the *partial prolongations* of the contact distribution on jet space $J^1(\mathbb{R}, \mathbb{R}^q)$, with $q > 1$,

$$\mathcal{C}_q^{(1)} = \left\{\partial_x + z_1^1 \partial_{z^1} + z_1^2 \partial_{z^2} + \ldots + z_1^q \partial_{z^q}, \partial_{z_1^1}, \ldots, \partial_{z_1^q}\right\}. \tag{2}$$

An example of a partial prolongation of (2) is given by

$$\mathcal{C}\langle 0,1,1\rangle = \left\{\partial_x + z_1^2 \partial_{z^2} + z_2^2 \partial_{z_1^2} + z_1^3 \partial_{z^3} + z_2^3 \partial_{z_1^3} + z_3^3 \partial_{z_2^3}, \partial_{z_2^2}, \partial_{z_3^3}\right\} \tag{3}$$

in which there is one variable of order 2 and one of order 3 (so $q = 2$). The notation $\mathcal{C}\langle 0,1,1\rangle$ denotes one "dependent variable of order 2" (second element) and "one dependent variable of order 3" (third element). Note that in (3) the superscript 2 or 3 denotes the order of the variable.

2 Generalised Goursat Normal Form

In this section we describe the aforementioned theorem on partial prolongations. This leads to an optimal procedure for constructing equivalences to a partial prolongation. We begin with an introduction to the basic tools required.

2.1 The Singular Variety

For each $x \in M$, let

$$\mathcal{S}_x = \{v \in \mathcal{V}_x \backslash 0 \mid \zeta(v) \text{ has less than generic rank}\}$$

Then $\mathcal{S}_x$ is the zero set of homogeneous polynomials and so defines a subvariety of the projectivisation $\mathbb{P}\mathcal{V}_x$ of $\mathcal{V}_x$. We shall denote by $\mathrm{Sing}(\mathcal{V})$ the fibre bundle over M with fibre over $x \in M$ equal to $\mathcal{S}_x$ and we refer to it as the *singular variety* of $\mathcal{V}$. For $X \in \mathcal{V}$ the matrix of the homomorphism $\zeta(X)$ will be called the *polar matrix* of $[X] \in \mathbb{P}\mathcal{V}$. There is a map $\deg_{\mathcal{V}} : \mathbb{P}\mathcal{V} \to \mathbb{N}$ well defined by

$$\deg_{\mathcal{V}}([X]) = \text{rank } \zeta(X) \quad \text{for} \quad [X] \in \mathbb{P}\mathcal{V}.$$

We shall call $\deg_{\mathcal{V}}([X])$ the *degree* of $[X]$ (relative to $\mathcal{V}$). Function $\deg_{\mathcal{V}}([X])$ is a diffeomorphism invariant: $\deg_{\phi_*\mathcal{V}}([\phi_* X]) = \deg_{\mathcal{V}}([X])$. Hence the singular variety $\mathrm{Sing}(\mathcal{V})$ is also a diffeomorphism invariant.

The computation of the singular variety for any given sub-bundle $\mathcal{V} \subset TM$ is algorithmic. It involves only differentiation and commutative algebra operations. One computes the determinantal variety of the polar matrix for generic $[X]$; see [11, 12, 13] for examples.

The singular variety in positive degree. If $X \in \mathrm{Char}\,\mathcal{V}$ then $\deg_{\mathcal{V}}([X]) = 0$. For this reason we pass to the quotient $\widehat{\mathcal{V}} := \mathcal{V}/\mathrm{Char}\,\mathcal{V}$. We have structure tensor $\widehat{\delta} : \Lambda^2\widehat{\mathcal{V}} \to \widehat{TM}/\widehat{\mathcal{V}}$, well defined by

$$\widehat{\delta}(\widehat{X}, \widehat{Y}) = \pi([X, Y]) \mod \widehat{\mathcal{V}},$$

where $\widehat{TM} = TM/\mathrm{Char}\,\mathcal{V}$ and

$$\pi : TM \to \widehat{TM}$$

is the canonical projection. The notion of degree descends to this quotient giving a map

$$\deg_{\widehat{\mathcal{V}}} : \mathbb{P}\widehat{\mathcal{V}} \to \mathbb{N}$$

well defined by

$$\deg_{\widehat{\mathcal{V}}}([\widehat{X}]) = \text{rank } \widehat{\zeta}(\widehat{X}) \quad \text{for} \quad [\widehat{X}] \in \mathbb{P}\widehat{\mathcal{V}},$$

where $\widehat{\zeta}(\widehat{X})(\widehat{Y}) = \widehat{\delta}(\widehat{X}, \widehat{Y})$ for $\widehat{Y} \in \widehat{\mathcal{V}}$. Note that all definitions go over *mutatis mutandis* when the structure tensor δ is replaced by $\widehat{\delta}$. In particular, we have notions of polar matrix and singular variety, as before. However, if the singular variety of $\widehat{\mathcal{V}}$ is not empty, then each point of $\mathbb{P}\widehat{\mathcal{V}}$ has degree one or more.

The resolvent bundle. Suppose $\mathcal{V} \subset TM$ is totally regular of rank $c+q+1$, $q \geq 2, c \geq 0$, $\dim M = c + 2q + 1$. Suppose further that $\mathcal{V}$ satisfies

a) $\dim \operatorname{Char} \mathcal{V} = c$, $\mathcal{V}^{(1)} = TM$
b) $\widehat{\Sigma}_{|_x} := \operatorname{Sing}(\widehat{\mathcal{V}})_{|_x} = \mathbb{P}\widehat{\mathcal{B}}_{|_x} \approx \mathbb{RP}^{q-1}$, for each $x \in M$ and some rank q sub-bundle $\widehat{\mathcal{B}} \subset \widehat{\mathcal{V}}$. Then we call $(\mathcal{V}, \mathbb{P}\widehat{\mathcal{B}})$ (or $(\mathcal{V}, \widehat{\Sigma})$) a *Weber structure* of rank q on M.

Given a Weber structure $(\mathcal{V}, \mathbb{P}\widehat{\mathcal{B}})$, let $\mathcal{R}(\mathcal{V}) \subset \mathcal{V}$, denote the largest sub-bundle such that

$$\pi\big(\mathcal{R}(\mathcal{V})\big) = \widehat{\mathcal{B}}.$$

We call the rank $q + c$ bundle $\mathcal{R}(\mathcal{V})$ defined by (2.1) the *resolvent bundle* associated to the Weber structure $(\mathcal{V}, \widehat{\Sigma})$. The bundle $\widehat{\mathcal{B}}$ determined by the singular variety of $\widehat{\mathcal{V}}$ will be called the *singular sub-bundle* of the Weber structure. A Weber structure will be said to be *integrable* if its resolvent bundle is integrable.

An *integrable* Weber structure descends to the quotient of M by the leaves of $\operatorname{Char} \mathcal{V}$ to be the contact bundle on $J^1(\mathbb{R}, \mathbb{R}^q)$. For completeness, we record the following properties of the resolvent bundle of a Weber structure.

Proposition 1. [11]. *Let $(\mathcal{V}, \widehat{\Sigma})$ be a Weber structure on M and $\widehat{\mathcal{B}}$ its singular sub-bundle. If $q \geq 3$, then the following are equivalent*

a) *Its resolvent bundle $\mathcal{R}(\mathcal{V}) \subset \mathcal{V}$ is integrable*
b) *Each point of $\widehat{\Sigma} = Sing(\widehat{\mathcal{V}})$ has degree one*
c) *The structure tensor $\widehat{\delta}$ of $\widehat{\mathcal{V}}$ vanishes on $\widehat{\mathcal{B}}$: $\widehat{\delta}(\widehat{\mathcal{B}}, \widehat{\mathcal{B}}) = 0$.*

Proposition 2. [11]. *Let $(\mathcal{V}, \widehat{\Sigma})$ be an integrable Weber structure on M. Then its resolvent bundle $\mathcal{R}_{\widehat{\Sigma}}(\mathcal{V})$ is the unique, maximal, integrable sub-bundle of $\mathcal{V}$.*

Checking the integrability of the resolvent bundle is algorithmic. One computes the singular variety $\operatorname{Sing}(\widehat{\mathcal{V}}) = \mathbb{P}\widehat{\mathcal{B}}$. In turn, the singular bundle $\widehat{\mathcal{B}}$ algorithmically determines $\mathcal{R}_{\widehat{\Sigma}}(\mathcal{V})$.

For any totally regular sub-bundle $\mathcal{V} \subset TM$, we have the notion of its derived type. In section 2, we defined the *derived type* of a bundle as the list of all derived bundles together with their corresponding Cauchy bundles. We shall frequently abuse notation by using the term 'derived type of $\mathcal{V}$' for the ordered list of ordered pairs of the form

$$\mathfrak{d}(\mathcal{V}) = [[m_0, \chi^0], [m_1, \chi^1], \dots, [m_k, \chi^k]]$$

where $m_j = \dim \mathcal{V}^{(j)}$ and $\chi^j = \dim \mathrm{Char}\ \mathcal{V}^{(j)}$.

It is important to relate the type of a partial prolongation to its derived type. For this it is convenient to introduce the notions of *velocity* and *deceleration* of a sub-bundle.

Definition 1. Let $\mathcal{V} \subset TM$ be a totally regular sub-bundle. The velocity of $\mathcal{V}$ is the ordered list of k integers

$$\mathrm{vel}(\mathcal{V}) = \langle \Delta_1, \Delta_2, \dots, \Delta_k \rangle, \quad \text{where} \quad \Delta_j = m_j - m_{j-1}, \ 1 \le j \le k.$$

The deceleration of $\mathcal{V}$ is the ordered list of k integers

$$\mathrm{decel}(\mathcal{V}) = \langle -\Delta_2^2, -\Delta_3^2, \dots, -\Delta_k^2, \Delta_k \rangle, \quad \Delta_j^2 = \Delta_j - \Delta_{j-1}.$$

Note that total prolongations $\mathcal{C}_q^{(k)}$ have decelerations of the form $\langle 0, \dots, 0, q \rangle$, $q \ge 1$, where there are $k-1$ zeros before the final entry, q. The Goursat normal form is the case $q = 1$ in this family of decelerations. We denote the partial prolongation with deceleration σ expressed in standard jet coordinates by the symbol $\mathcal{C}(\sigma)$.

To recognise when a given sub-bundle has or has not the derived type of a partial prolongation we introduce one further canonically associated sub-bundle that plays a crucial role.

Definition 2. If $\mathcal{V} \subset TM$ is a totally regular sub-bundle of derived length k we let $\mathrm{Char}\ \mathcal{V}_{j-1}^{(j)}$ denote the intersections

$$\mathrm{Char}\ \mathcal{V}_{j-1}^{(j)} = \mathcal{V}^{(j-1)} \cap \mathrm{Char}\ \mathcal{V}^{(j)}, \ 1 \le j \le k-1.$$

Let

$$\chi_{j-1}^{j} = \dim \mathrm{Char}\ \mathcal{V}_{j-1}^{(j)}, \ \ 1 \le j \le k-1.$$

We shall call the the integers $\{\chi^0, \chi^j, \chi_{j-1}^j\}_{j=1}^{k-1}$ the type numbers of $\mathcal{V} \subset TM$ and the list

$$\mathfrak{d}_r(\mathcal{V}) = [[m_0, \chi^0], [m_1, \chi_0^1, \chi^1], [m_2, \chi_1^2, \chi^2], \dots, [m_{k-1}, \chi_{k-2}^{k-1}, \chi^{k-1}], [m_k, \chi^k]]$$

as the refined derived type of $\mathcal{V}$.

It is easy to see that in every partial prolongation sub-bundles $\mathrm{Char}\ \mathcal{V}_{j-1}^{(j)}$ are non-trivial and integrable, an invariant property of $\mathcal{V}$. Furthermore, there are simple relationships between the type numbers in any partial prolongation thereby providing further invariants for the local equivalence problem.

Proposition 3. [11]. *Let sub-bundle $\mathcal{V} \subset TM$ be totally regular, of derived length k, with velocity and $\langle \Delta_1, \Delta_2, \ldots, \Delta_k \rangle$ and $\langle -\Delta_2^2, \ldots, -\Delta_k^2, \Delta_k \rangle$, respectively. Then $\mathcal{V}$ has the derived type of a partial prolongation if and only if the type numbers of $\mathcal{V}$ satisfy*

$$\chi^j = 2m_j - m_{j+1} - 1,\ 0 \le j \le k-1,$$
$$\chi^i_{i-1} = m_{i-1} - 1,\ 1 \le i \le k-1.$$

If $\mathcal{V} \subset TM$ has refined derived type that satisfies the constraints of Proposition 3, then we say that it has the *derived type of a partial prolongation*.

Example 1. To illustrate all these notions we compute the refined derived type and all relevant bundles associated with the partial prolongation $\mathcal{V} = \mathcal{C}\langle 4,3,2 \rangle$ in standard (contact) coordinates

$$\Big\{X = \partial_x + \sum_{a_1=1}^{4} z_1^{a_1,1} \partial_{z^{a_1,1}} + \sum_{a_2=1}^{3}\sum_{l_2=0}^{1} z_{l_2+1}^{a_2,2}\partial_{z_{l_2}^{a_2,2}} + \sum_{a_3=1}^{2}\sum_{l_3=0}^{2} z_{l_3+1}^{a_3,3}\partial_{z_{l_3}^{a_3,3}},$$
$$\partial_{z_1^{a_1,1}}, \partial_{z_2^{a_2,2}}, \partial_{z_3^{a_3,3}}\Big\}$$

on $J^\sigma(\mathbb{R}, \mathbb{R}^9)$, $\sigma = \langle 4,3,2 \rangle$. The refined derived type is

$$[[10,0],[19,9,13],[24,18,21],[26,26]],$$

and the derived length is $k = 3$. The Cauchy bundles and intersections are

$$\text{Char } \mathcal{V}^{(1)} = \Big\{\partial_{z_1^{a_1,1}}, \partial_{z_2^{a_2,2}}, \partial_{z_3^{a_3,3}}, \partial_{z^{a_1,1}}\Big\},$$
$$\text{Char } \mathcal{V}_0^{(1)} = \Big\{\partial_{z_1^{a_1,1}}, \partial_{z_2^{a_2,2}}, \partial_{z_3^{a_3,3}}\Big\},$$

and

$$\text{Char } \mathcal{V}^{(2)} = \Big\{\partial_{z_1^{a_2,2}}, \partial_{z_2^{a_3,3}}, \partial_{z^{a_2,2}}, \partial_{z_1^{a_1,1}}, \partial_{z_2^{a_2,2}}, \partial_{z_3^{a_3,3}}, \partial_{z^{a_1,1}}\Big\},$$
$$\text{Char } \mathcal{V}_1^{(2)} = \Big\{\partial_{z_1^{a_2,2}}, \partial_{z_2^{a_3,3}}, \partial_{z_1^{a_1,1}}, \partial_{z_2^{a_2,2}}, \partial_{z_3^{a_3,3}}, \partial_{z^{a_1,1}}\Big\}.$$

The reader will find it easy to verify that the type numbers are in agreement with Proposition 3. The singular variety of $\widehat{\mathcal{V}}^{(2)} = \mathcal{V}^{(2)}/\text{Char } \mathcal{V}^{(2)}$ consists of lines $E = [\Xi]$, where $\Xi = e^1\pi(X) + e^2\pi(\partial_{z_1^{1,3}}) + e^3\pi(\partial_{z_1^{2,3}})$, whose degree is less than the generic degree which is 2. In practice, to compute the singular variety, one computes the *polar matrix* of E, the matrix of the vector bundle morphism $\widehat{\zeta}(\Xi)$. In this case it is given by

$$\begin{pmatrix} -e^2 & e^1 & 0 \\ -e^3 & 0 & e^1 \end{pmatrix},$$

whose rank is less than 2 if and only if $e^1 = 0$. We deduce that

$$\mathrm{Sing}(\widehat{\mathcal{V}}^{(2)})_{|\mathbf{z}} = \mathbb{P}\widehat{\mathcal{B}}_{|\mathbf{z}} = \mathbb{P}\left\{\pi(\partial_{z_1^{1,3}}), \pi(\partial_{z_1^{2,3}})\right\}_{|\mathbf{z}} \simeq \mathbb{RP}^1, \ \forall\ \mathbf{z} \in J^\sigma(\mathbb{R}, \mathbb{R}^9).$$

Consequently $\mathcal{V}^{(2)}$ is a Weber structure of rank 2 with resolvent bundle,

$$\mathcal{R}_{\widehat{\Sigma}_2}(\mathcal{V}^{(2)}) = \mathrm{Char}\ \mathcal{V}^{(2)} \oplus \left\{\partial_{z_1^{1,3}}, \partial_{z_1^{2,3}}\right\},$$

which is integrable. We note that $\mathcal{V}$ has $\chi^1 - \chi^1_0 = 4$ dependent variables of order 1; $\chi^2 - \chi^2_1 = 3$ dependent variables of order 2 and $\rho_3 := \Delta_3 = 2$ dependent variables of order 3. Finally, we observe that $\mathrm{decel}(\mathcal{V}) = \langle 4, 3, 2\rangle$.

Roughly speaking, the main result of [11] can be expressed "if $\mathcal{V}$ has the refined derived type of a partial prolongation and certain canonical bundles are (Frobenius) integrable then it *is* locally equivalent to the partial prolongation, uniquely prescribed by $\mathrm{decel}(\mathcal{V})$". More precisely, we are lead to make the following definition.

Definition 3. A totally regular sub-bundle $\mathcal{V} \subset TM$ of derived length k will be called a Goursat bundle with deceleration σ if

1. $\mathcal{V}$ has the derived type of a partial prolongation whose deceleration is $\sigma = \mathrm{decel}(\mathcal{V})$
2. Each intersection Char $\mathcal{V}^{(i)}_{i-1}$ is an integrable sub-bundle whose rank, assumed to be constant on M, agrees with the corresponding rank in $\mathcal{C}(\sigma)$
3. In case $\Delta_k > 1$, then $\mathcal{V}^{(k-1)}$ determines an integrable Weber structure of rank Δ_k on M.

Then the recognition problem for partial prolongations is solved by the generalised Goursat normal form.

Theorem 2 (Generalised Goursat Normal Form) [11]. *Let $\mathcal{V} \subset TM$ be a Goursat bundle over manifold M, of derived length $k > 1$ and deceleration $\sigma = \mathrm{decel}(\mathcal{V})$. Then there is an open, dense subset $\hat{M} \subseteq M$ such that the restriction of $\mathcal{V}$ to $\hat{M}$ is locally equivalent to $\mathcal{C}(\sigma)$. Conversely any partial prolongation of $\mathcal{C}^{(1)}_q$ is a Goursat bundle.*

The generalised Goursat normal form asserts that away from "singularities", the deceleration of any Goursat bundle is a complete local invariant. Hence partial prolongations are generically classified by their deceleration vector. For this reason the deceleration of a Goursat bundle $\mathcal{V}$ will sometimes be called its *signature*. If $\mathcal{V}$ is a Goursat bundle and non-negative integer ρ_j is the jth component of its signature, then $\mathcal{V}$ is locally diffeomorphic to a partial prolongation with ρ_j "dependent variables of order j". The theorem has a counterpart which provides an efficient procedure for constructing an equivalence to $\mathcal{C}(\sigma)$ where $\sigma = \mathrm{decel}(\mathcal{V})$ is the signature of $\mathcal{V}$.

Let $\mathcal{V} \subset TM$ be a Goursat bundle over M of derived length k. In fact there are two slightly different procedures depending upon whether $\Delta_k > 1$ or $\Delta_k = 1$; their proof of correctness is given in [12].

To describe them we introduce some notation. For all appropriate values of i and j, we shall denote the annihilators of $\mathcal{V}^{(i)}$, Char $\mathcal{V}^{(j)}$ and Char $\mathcal{V}^{(j)}_{j-1}$ by $\Omega^{(j)}$, $\Xi(\Omega)^{(j)}$ and $\Xi(\Omega)^{(j)}_{j-1}$, respectively. We begin with the case $\Delta_k = 1$.

Procedure Contact

(Case $\Delta_k = 1$)

INPUT: Goursat bundle $\mathcal{V} \subset TM$ of derived length k and signature $\sigma = \text{decel}(\mathcal{V}) = \langle \rho_1, \ldots, \rho_k \rangle$, $\rho_k = 1$.

a) Fix any invariant of Char $\mathcal{V}^{(k-1)}$ denoted x, and any section $\boldsymbol{Z}$ of $\mathcal{V}$ such that $\boldsymbol{Z}(x) = 1$.
b) Build distribution Π^k, defined inductively by

$$\Pi^{l+1} = [Z, \Pi^l], \quad \Pi^1 = \text{Char } \mathcal{V}^{(1)}_0, \qquad 1 \leq l \leq k-1.$$

c) Let $z^k = \varphi^{1,k}$ be an invariant of Π^k such that $dx \wedge d\varphi^{1,k} \neq 0$.
d) For each j, such that $\rho_j > 0$, compute the fundamental bundle $\Xi(\Omega)^{(j)}_{j-1}/\Xi(\Omega)^{(j)}$ of order j.
e) For each j, such that $\Xi(\Omega)^{(j)}_{j-1}/\Xi(\Omega)^{(j)}$ is non-trivial, compute the fundamental functions $\{\varphi^{l_j,j}\}_{l_j=1}^{\rho_j}$ of order j.
f) For each j, such that $\rho_j > 0$ let $z^{l_j,j} = \varphi^{l_j,j}$, $1 \leq l_j \leq \rho_j$.
g) For each j, such that $\rho_j > 0$ define functions

$$x, z_0^{l_j,j} := z^{l_j,j} = \varphi^{l_j,j}, \; z_{s_j+1}^{l_j,j} = \boldsymbol{Z} z_{s_j}^{l_j,j}, \; 0 \leq s_j \leq j-1, \; 1 \leq l_j \leq \rho_j.$$

OUTPUT: Contact coordinates for $\mathcal{V}$ identifying it with $\mathcal{C}(\sigma)$.

Case $\Delta_k > 1$. If the Goursat bundle $\mathcal{V}$ satisfies $\Delta_k > 1$ then steps a) and b) are replaced by the calculation of the resolvent bundle $\mathcal{R}(\mathcal{V}^{(k-1)})$ which is integrable and has $1 + \Delta_k$ invariants; these are fundamental functions of highest order, k. Any one of these can be taken to be the "independent" variable, x, in the canonical form. We then fix any section $\boldsymbol{Z}$ of $\mathcal{V}$ such that $\boldsymbol{Z}x = 1$ after which we proceed, as in the case $\Delta_k = 1$, to construct contact coordinates. Proofs of correctness of these procedures are given in [12].

3 Examples

In this section we illustrate the general theory with some pedagogical examples.

3.1 A Fully Nonlinear, Time Varying Control System

Consider the fully nonlinear, time varying control system on $M = \mathbb{R}^8$,

$$\frac{dx_1}{dt} = \exp x_2,\ \frac{dx_2}{dt} = x_1 x_3^2,\ \frac{dx_3}{dt} = v_1 v_2 + \left(\frac{x_4}{t}\right),$$
$$\frac{dx_4}{dt} = x_4\left(1 + \frac{1}{t}\right) + t^2(x_5 - x_1^3),\ \frac{dx_5}{dt} = \frac{1}{v_2}.$$

in fives states $x_1, \ldots, x_5$ and two control v_1, v_2. This is encoded by the distribution

$$\mathcal{V} = \Big\{\partial_t + \exp x_2 \partial_{x_1} + x_1 x_3^2 \partial_{x_2} + \left(v_1 v_2 + \left(\frac{x_4}{t}\right)\right)\partial_{x_3}$$
$$+ \left(x_4\left(1 + \frac{1}{t}\right) + t^2(x_5 - x_1^3)\right)\partial_{x_4} + \frac{1}{v_2}\partial_{x_5},\ \partial_{v_1},\ \partial_{v_2}\Big\}.$$

Computing the derived type we get

$$\dim \mathcal{V} = 3,\ \dim \mathcal{V}^{(1)} = 5,\ \dim \mathcal{V}^{(2)} = 7,\ \dim \mathcal{V}^{(3)} = 8$$

and

$$\text{Char}\ \mathcal{V}_0^{(1)} = \{\partial_{v_1},\ \partial_{v_2}\} = \text{Char}\ \mathcal{V}^{(1)}$$
$$\text{Char}\ \mathcal{V}_1^{(2)} = \text{Char}\ \mathcal{V}^{(1)} \oplus \{\partial_{x_3}, \partial_{x_5}\}$$
$$\text{Char}\ \mathcal{V}^{(2)} = \text{Char}\ \mathcal{V}_1^{(2)} \oplus \{\partial_{x_4}\}.$$

Hence the refined derived type is

$$\mathfrak{d}_r(\mathcal{V}) = \big[[3,0],[5,2,2],[7,4,5],[8,8]\big]$$

from which we deduce that the signature of $\mathcal{V}$ is $\langle 0,1,1\rangle$. The hypotheses of Theorem 2 are satisfied with $\text{decel}(\mathcal{V}) = \langle 0,1,1\rangle$. By the generalised Goursat normal form, Theorem 2, there is a local diffeomorphism that identifies it with the partial prolongation $\mathcal{C}\langle 0,1,1\rangle$. This settles the recognition problem for $\mathcal{V}$.

We now use procedure *Contact* to construct an equivalence. From the signature, $\langle 0,1,1\rangle$, we see that there is only one nontrivial fundamental bundle, namely

$$\Xi(\Omega)_1^{(2)}/\Xi(\Omega)^{(2)} = \{dx_4\}. \tag{4}$$

Here the derived length is $k = 3$ and since $\Delta_3 = 1$, we construct the bundle Π^3 inductively as in procedure *Contact*. We find that

$$\Pi^3 = \{\partial_{x_2},\ \partial_{x_3},\ \partial_{x_4},\ \partial_{x_5},\ \partial_{v_1},\ \partial_{v_2}\}$$

whose invariants are spanned by t, x_1. Continuing to follow *Contact*, we take t to be the "independent variable" and x_1 a fundamental function of (highest) order 3. Function x_4 spans the fundamental functions of order 2 by equation

(4). The contact coordinates for $\mathcal{V}$ are then obtained by differentiating the fundamental functions as follows: [*Contact*, step g)]

$$t,\ a = x_4,\ a_1 = \mathbf{Z}(a),\ a_2 = \mathbf{Z}(a_1),$$
$$b = x_1,\ b_1 = \mathbf{Z}(b_0),\ b_2 = \mathbf{Z}(b_1),\ b_3 = \mathbf{Z}(b_2).$$

Explicitly, we obtain the static feedback transformation

$$t = t,\ a = x_4,\ a_1 = \frac{1}{t}\big(x_4 + tx_4 + t^3(x_5 - x_1^3)\big),$$
$$a_2 = \frac{2v_2x_4 + 3t^2v_2x_5 - 3t^2v_2x_1^3 - 3t^3x_1^2v_2e^{x_2} + tv_2x_4 + t^3v_2x_5 - t^3v_2x_1^3 + t^3}{tv_2},$$
$$b = x_1,\ b_1 = e^{x_2},\ b_2 = x_1x_3^2\,e^{x_2}, b_2 = \frac{x_3e^{x_2}\big(tx_3e^{x_2} + tx_1^2x_3^3 + 2x_1(v_1v_2t + x_4)\big)}{t}.$$

This local equivalence to the canonical form (3) is easy to invert to obtain the required parametrisation of states and controls.

3.2 A Non-flat Control System

The previous example is easily seen to be differentially flat with flat outputs x_1 and x_4. Here we examine the nonlinear control system

$$\dot{x}_1 = e^{-x_5}(1 - u_1x_5),\ \dot{x}_2 = u_1e^{-x_5},\ \dot{x}_3 = e^{-2x_5}(1 - u_1x_5),\ \dot{x}_4 = u_1e^{-2x_5},\ \dot{x}_5 = u_2, \tag{5}$$

whose state space is a 5-dimensional (solvable) Lie group G with nonzero Lie algebra structure given by

$$[E_1, E_5] = -E_1, [E_2, E_5] = -E_1 - E_2, [E_3, E_5] = -2E_3, [E_4, E_5] = -E_3 - 2E_4.$$

These structure equations are realised by the Lie algebra of vector fields

$$\mathfrak{g} = \{e^{x_5}\partial_{x_1},\ e^{x_5}(x_5\partial_{x_1} + \partial_{x_2}),\ e^{2x_5}\partial_{x_3},\ e^{2x_5}(x_5\partial_{x_3} + \partial_{x_4}),\ \partial_{x_5}\} = \{E_1, E_2, \ldots, E_5\}.$$

The control system (5) determines the sub-bundle $\mathcal{K} \subset T(G \times \mathbb{R}^3)$ whose space of sections is spanned by

$$\mathcal{K} = \left\{\partial_t + E_1 + E_3 + u_1(E_2 + E_4) + u_2E_5,\ \partial_{u_1},\ \partial_{u_2}\right\}$$

which is invariant under the left-translations of G. A calculation shows that $\mathcal{K}$ has the same refined derived type as $\mathcal{V}$ of the previous example. An identical analysis leads to the fundamental functions that are required for its linearisation to be

$$\begin{aligned}
&\text{independent variable}: \quad \tau = -x_5\\
&\text{function of order } 2: \quad a = x_2 - x_4 e^{-x_5}\\
&\text{function of order } 3: \quad b = x_2 + 2x_1 - 2x_2x_5 - (x_3 + x_4 - x_4x_5)e^{-x_5} - te^{x_5}
\end{aligned} \tag{6}$$

where τ is the "independent variable" in the linearised system. Remark that since $\mathcal{K}$ is defined on a Lie group with known multiplication one can construct the fundamental functions (6) without carrying out any integration, using instead the Fels-Olver method of moving frames [2].

In contrast to the previous example, the transformation generated by (6) is not static feedback and, indeed, it can be checked that this system is not feedback linearisable. It can also be checked that none of its partial prolongations are feedback linearisable. Additionally, it does not appear to be differentially flat (though I have not yet proved this)[3]. Nevertheless, the above application of the generalised Goursat normal form shows that the system is locally equivalent to the partial prolongation $\mathcal{C}\langle 0, 1, 1\rangle$! As such its symmetry group consists of the relevant infinite Lie pseudogroup of contact transformations. If the system is indeed not flat then it provides a counter-example to a conjecture expressed in the literature to the effect that "a system is flat if and only if it has an infinite symmetry group", [7]. We end this example by noting that while the equivalence is not static feedback, nevertheless the system can be linearised explicitly and such linearisations may conceivably play a role in control theory.

Finally, we mention that, except in *Example 1*, we have not provided an illustration of procedure *Contact* in the case of control systems satisfying $\Delta_k > 1$ in this paper. For this we draw readers attention to references [11, 12] and [13].

Acknowledgements. I am grateful to the Institute Interfacultaire Bernoulli, École Polytechnique Fédérale de Lausanne and organisers Professors Jean Levine and Philippe Müllhaupt for gracious hospitality in April, 2009 during the "Mathematical Tools in Control, Signals and Systems Workshop", where this paper was presented.

References

1. Anderson, I.M., Fels, M.E., Vassiliou, P.J.: Superposition formulas for exterior differential systems. Advances in Mathematics 221, 1910–1963 (2009)
2. Fels, M., Olver, P.J.: Moving coframes: I. A practical algorithm. Acta Appl. Math. 51, 161–213 (1998), Moving coframes: II. Regularization and theoretical foundations. Acta Appl. Math. 55, 127–208 (1999)
3. Gardner, R.B., Shadwick, W.F.: The GS algorithm for exact linearization to Brunovsky normal form. IEEE Trans. Automat. Control 37, 224–230 (1992)
4. Goursat, E.: Leçons sur le problème de Pfaff. Hermann, Paris (1923)

[3] Note that the control system passes the "ruled manifold test" [9].

5. Martin, P., Rouchon, P.: Feedback linearization and driftless systems. Math. Control Signals Systems 7, 235–254 (1994)
6. Martin, P., Rouchon, P.: Any (controllable) driftless system with 3 inputs and 5 states is flat. Systems & Control Letters 25, 167–173 (1995)
7. Respondek, W.: Symmetries and minimal flat outputs of nonlinear control systems. In: Kang, W., et al. (eds.) New Trends in Nonlinear Dynamics and Control, and their Applications. LNCIS, vol. 295, Springer, Berlin (2003)
8. Sanstry, S.: Nonlinear Systems: Analysis, Stability and Control. Interdisciplinary Applied Mathematics Series, vol. 10. Springer, Heidelberg (1999)
9. Sluis, W.M.: Absolute Equivalence and its Applications to Control Theory, PhD thesis, University of Waterloo (1992)
10. Stormark, O.: Lie's Structural Approach to PDE Systems. Encyclopedia of Mathematics and its Applications, vol. 80. Cambridge University Press, Cambridge (2000)
11. Vassiliou, P.: A constructive generalised Goursat normal form. Differential Geometry and its Applications 24, 332–350 (2006)
12. Vassiliou, P.: Efficient construction of contact coordinates for partial prolongations. Foundations of Computational Mathematics, pp. 269–308 (2006)
13. Vassiliou, P.: Contact geometry of curves, Symmetry, Integrability and Geometry: Methods and Applications (SIGMA) 5(098), 27 pages (2009) (open access electronic journal)

Part III
Chemical Processes and Life Sciences

Piecewise Affine Models of Regulatory Genetic Networks: Review and Probabilistic Interpretation

Madalena Chaves and Jean-Luc Gouzé

Abstract. A formalism based on piecewise-affine (PWA) differential equations has been shown to be well-suited to modelling genetic regulatory networks. In this paper, we first review some results concerning the qualitative study of these models: we partition the phase space into *domains* bounded by the threshold hyperplanes. Inside each domain, the system is affine. To define solutions on the surfaces of discontinuity, we use the approach of Filippov, which extends the vector field to a differential inclusion. We obtain a transition graph, describing qualitatively the possible transitions of solutions between domains. In a second part of the paper, we give a new probabilistic interpretation of these transitions, by computing the proportion of the volume of the domain that crosses to one of its adjacent domains. We apply this idea to the model of the bistable switch and to parameter estimation from experimental transition probabilities.

1 Introduction

The regulation of gene expression plays a fundamental role in the functioning of cells. New mathematical modelling and computational techniques will be essential to the understanding of these genetic regulatory networks (see [4] for a review and [1] for biological aspects). The principal modelling challenges come from incomplete knowledge of the networks, and the dearth of quantitative data for identifying kinetic parameters required for detailed mathematical models. Qualitative methods overcome both of these difficulties and are thus well-suited to the modelling and simulation of genetic networks.

Madalena Chaves
e-mail: mchaves@sophia.inria.fr

Jean-Luc Gouzé
COMORE, INRIA, 2004 Route des Lucioles, BP 93, 06902 Sophia Antipolis, France
e-mail: gouze@sophia.inria.fr

J. Lévine & P. Müllhaupt (Eds.): Adv. in Theory of Control, Signals, LNCIS 407, pp. 241–253.
springerlink.com © Springer-Verlag Berlin Heidelberg 2010

The **first part** of the paper is a paraphrase of results obtained by the authors and collaborators, that are mostly taken from [2] and [10], and the references therein; it recalls the basis of the modelling of genetic regulatory networks with PWA differential equations. From a mathematical point of view, what is interesting in these dynamical systems (possibly of large dimensions, until several thousands) is that a global qualitative analysis (assisted by a computer) is possible and gives nontrivial results. This is to compare with classical nonlinear ordinary differential equations where, for dimensions greater than three, nothing is possible except a local analysis around the equilibria, if the equilibria are computable. Moreover, in the PWA case, the analysis is itself qualitative, and does not depend too much on the exact values of the parameters of the model; instead, it depends only on inequalities between these parameters.

In a **second part**, which is the original part of the paper, we build on the qualitative transition graph given by the above analysis. This graph describes the possible transitions between regions of the trajectories. We give a probabilistic interpretation of the transitions: often, the biologist can only measure the fact that a gene is highly or weakly expressed at some time. In this case, although the precise numerical values of the variables are not available to the biologist, he will be able to have an estimation (frequency) of the probability of transition from one domain to another. We compute these probabilities of transitions between domains, and show that it can give some informations about the parameters of the model: for the classical model of the bistable switch, we are able to estimate the expression rates.

2 Piecewise-Affine Models of Genetic Regulatory Networks

Piecewise affine models of genetic networks are built with discontinuous (step) functions; such models are originally due to Glass and Kauffman [8]. The use of such step functions has been motivated by the experimental observation that the activity of certain genes change in a switch-like manner at a threshold concentration of a regulatory protein. It is best illustrated with an example: the schematic diagram in Figure 1 describes a simple genetic regulatory network. In this example, the genes a and b code for the proteins A and B, which in turn control the expression of the two genes a and b. Protein A inhibits gene a and activates gene b above certain threshold concentrations, which are assumed to be different. Similarly protein B inhibits gene b and activates gene a above different threshold concentrations. Such a two-gene network could be found as a module of a more complex genetic regulatory network from a real biological system.

The equations modeling the example network in Figure 1 can be written down as

$$\begin{cases} \dot{x}_a = \kappa_a s^+(x_b, \theta_b^1) s^-(x_a, \theta_a^2) - \gamma_a x_a \\ \dot{x}_b = \kappa_b s^+(x_a, \theta_a^1) s^-(x_b, \theta_b^2) - \gamma_b x_b \end{cases} \tag{1}$$

where $s^+(x_s, \theta_s)$ is equal to 0 when $x_s < \theta_s$ and equal to 1 when $x_s > \theta_s$ and $s^-(x_s, \theta_s) = 1 - s^+(x_s, \theta_s)$. In this model, gene a is expressed at a rate κ_a if the concentration x_b of protein B is above the threshold θ_b^1 and the concentration x_a of

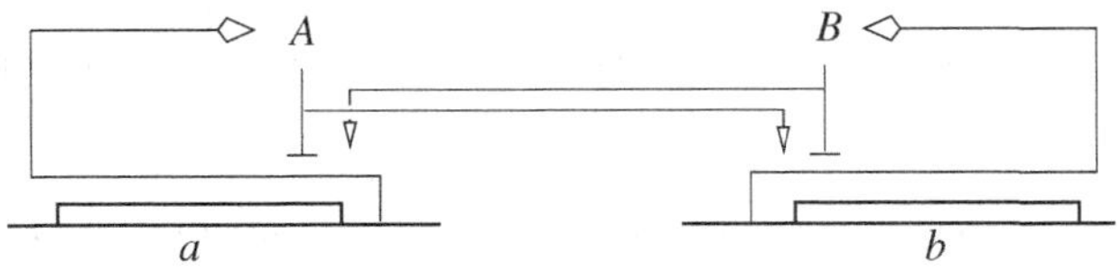

Fig. 1. Example of a genetic regulatory network of two genes (a and b), each coding for a regulatory protein (A and B)

protein A is below the threshold θ_a^2. Similarly, gene b is expressed at a rate κ_b if the concentration x_a of protein A is above the threshold θ_a^1 and the concentration x_b of the protein B is below the threshold θ_b^2. Degradation of both proteins is assumed to be proportional to their own concentrations, so that the expression of the genes a and b is modulated by the degradation terms $\gamma_a x_a$ and $\gamma_b x_b$ respectively. We suppose that $\theta_j^1 < \theta_j^2$ for $j = a,b$.

Such a model is readily generalized to models containing both expression and degradation terms for each gene:

$$\dot{x}_i = f_i(x) - \gamma_i x_i$$

where $f_i(x)$ represents the expression rate of gene i, depending on the whole state $x = (x_1, \cdots, x_n)^T$ and γ_i is the (relative) degradation rate. However, the expression rates of (1) have the additional property of being constant for values of x_a and x_b belonging to intervals that do not contain thresholds values θ_i^j. This can be rewritten by detailing $f_i(x)$ as follows:

$$f_i(x) = \sum_{l=1}^{L_i} \kappa_{il} b_{il}(x)$$

where $b_{il}(x)$ is a combination of step-functions $s^{\pm}(x_r, \theta_r^j)$ and $\kappa_{il} > 0$ is a rate parameter. The generalized form of (1) is a piecewise linear model

$$\dot{x} = f(x) - \gamma x \tag{2}$$

where the model is affine within hyper-rectangles of the state-space (γ is the diagonal matrix $(\gamma_1, \ldots, \gamma_n)$).

The dynamics of the piecewise-linear system (2) can be studied in the n-dimensional state-space $\Omega = \Omega_1 \times \Omega_2 \times \cdots \times \Omega_n$, where each Ω_i is defined by $\Omega_i = \{x_i \in I\!R_+ \mid 0 \leq x_i \leq max_i\}$ for some positive parameter $max_i > \max_x \left(\frac{f_i(x)}{\gamma_i}\right)$. A protein encoded by a gene will be involved in different interactions at different concentration thresholds, so for each variable x_i, we assume there are p_i ordered thresholds $\theta_i^1, \cdots, \theta_i^{p_i}$ (we also define $\theta_i^0 = 0$ and $\theta_i^{p_i+1} = max_i$). The $(n-1)$-dimensional hyperplanes defined by these thresholds partition Ω into hyper-rectangular regions we call *domains*. Specifically, a domain $D \subset \Omega$ is defined to be a set $D = D_1 \times \cdots \times D_n$, where D_i is one of the following:

$$
\begin{array}{ll}
D_i = \{x_i \in \Omega_i | 0 \le x_i < \theta_i^1\} & \\
D_i = \{x_i \in \Omega_i | \theta_i^j < x_i < \theta_i^{j+1}\} & \text{for } j \in \{1,\cdots,p_i - 1\} \\
D_i = \{x_i \in \Omega_i | \theta_i^{p_i} < x_i \le max_i\} & \\
D_i = \{x_i \in \Omega_i | x_i = \theta_i^j\} & \text{for } j \in \{1,\cdots,p_i\}
\end{array}
$$

Let $\mathscr{D}$ be the set of domains in Ω. A domain $D \in \mathscr{D}$ is called a regulatory domain if none of the variables x_i has a threshold value in D (it is the full hyperrectangle). In contrast, a domain $D \in \mathscr{D}$ is called a switching domain of order $k \le n$ if exactly k variables have threshold values in D [11]. The corresponding variables x_i are called switching variables in D. The two sets of domains are respectively denoted by $\mathscr{D}_r$ and $\mathscr{D}_s$.

2.1 Classical Solutions and Focal Points

For any regulatory domain D, the function $f(x)$ is constant for all $x \in D$, and it follows that the piecewise-affine system (2) can be written as an affine vector field

$$\dot{x} = f^D - \gamma x, \quad x \in D \tag{3}$$

where f^D is constant in D. Restricted to D, this is a classical linear ordinary differential equation. We assume that the parameters $\{\theta_i^j\}, \{\gamma_i\}, \{\kappa_{il}\}$ are all fixed. For any initial condition $x(t_0) \in D$, the unique solution is given by

$$x(t) = \phi(D) + e^{\gamma(t_0 - t)}(x(t_0) - \phi(D)), \tag{4}$$

where $\phi(D)$ satisfies the linear system $\gamma\phi(D) = f^D$. Clearly $x(t) \to \phi(D)$ monotonically until $x(t)$ reaches the boundary of the regulatory domain D.

Definition 1. Given a regulatory domain $D \in \mathscr{D}_r$, the point $\phi(D) = \gamma^{-1} f^D \in \Omega$ is called the *focal point* for the flow in D.

Generally we make the assumption that $\phi(D) \notin \text{supp}(D')$, for all $D' \subseteq \partial D$, for otherwise solutions can take infinite time to reach a focal point in the boundary of their domain ($\text{supp}(D')$ is the supporting hyperplane containing the domain D'). This is a special case of a more general assumption we make in Section 2.3. In the example network of Figure 1, it can easily be checked that for the regulatory domain D^{13} (see Figure 3(a)), the state equations reduce to

$$
\begin{aligned}
\dot{x}_a &= \kappa_a - \gamma_a x_a, \\
\dot{x}_b &= \kappa_b - \gamma_b x_b.
\end{aligned}
$$

Hence the focal point of D^{13} is $\phi(D^{13}) = (\kappa_a/\gamma_a, \kappa_b/\gamma_b)$, which lies outside D^{13}, in the domain D^{25} in fact, under some assumptions concerning the parameters. Thus solutions in D^{13} will flow towards $\phi(D^{13}) \in D^{25}$ until they leave the domain D^{13}. Different regulatory domains will usually have different focal points. In general, all solutions in a regulatory domain D flow towards the focal point $\phi(D)$ until they either reach it or leave the domain D. What happens when a solution leaves a regulatory domain D and enters a switching domain in the boundary of D? Since the step functions are not defined when a variable x_i takes some threshold value $\theta_i^{q_i}$, the vector field is undefined on the switching domains. We need to precise our definition of solutions.

2.2 Solutions in Switching Domains

In switching domains, the PWA system (2) is not defined, since in a switching domain of order $k \geq 1$, k variables assume a threshold value. If solutions do not simply go through a switching domain, it is necessary to give a definition of what a solution can be on that domain. Classically, this is done by using a construction originally proposed by Filippov [7] and recently applied to PWA systems of this form [9, 6].

The method consists of extending the system (2) to a differential inclusion,

$$\dot{x} \in H(x), \tag{5}$$

where H is a set-valued function (i.e. $H(x) \subseteq I\!R^n$). If D is a regulatory domain, then we define H simply as

$$H(x) = \{f^D - \gamma x\}, \tag{6}$$

for $x \in D$. If D is a switching domain, for $x \in D$, we define $H(x)$ as

$$H(x) = \overline{co}(\{f^{D'} - \gamma x \mid D' \in R(D)\}), \tag{7}$$

where $R(D) = \{D' \in \mathcal{D}_r | D \subseteq \partial D'\}$ is the set of all regulatory domains with D in their boundary, and $\overline{co}(X)$ is the closed convex hull of X. For switching domains, $H(x)$ is generally multi-valued so we define solutions of the differential inclusion as follows.

Definition 2. A solution of (5) on $[0,T]$ *in the sense of Filippov* is an absolutely continuous function (with respect to t) $\xi_t(x_0)$ such that $\xi_0(x_0) = x_0$ and $\dot{\xi}_t \in H(\xi_t)$, for almost all $t \in [0,T]$.

In order to more easily define these Filippov solutions, it is useful to define a concept analogous to the focal points defined for regulatory domains, extended to deal with switching domains.

Definition 3. Let $D \in \mathcal{D}_s$ be a switching domain of order k. Then its *focal set* $\Phi(D)$ is

$$\Phi(D) = supp(D) \cap \overline{co}(\{\phi(D') \mid D' \in R(D)\}). \tag{8}$$

Hence $\Phi(D)$ for $D \in \mathcal{D}_s$ is the convex hull of the focal points $\phi(D')$ of all the regulatory domains D' having D in their boundary, as defined above, intersected with the threshold hyperplane $supp(D)$ containing the switching domain D (Figure 2).

It is possible to show that

$$H(x) = \gamma(\Phi(D) - x) \tag{9}$$

which is a compact way of writing that $H(x) = \{y \in I\!R^n \mid \exists \phi \in \Phi(D)$ such that $y = \gamma(\phi - x)\}$. The Filippov vector field is defined by means of the focal set.

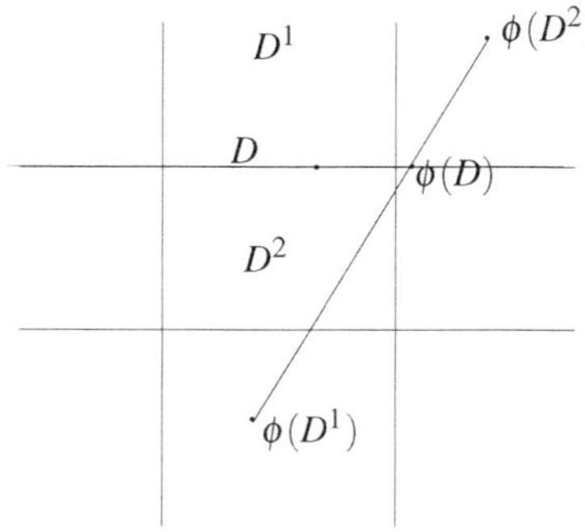

Fig. 2. Illustration of the definition of the focal set on a switching surface D according to the Filippov definition of solutions. The convex hull of the points $\phi(D^1)$ and $\phi(D^2)$ is simply the segment that links them, so that (8) implies that $\phi(D)$ is the intersection of this segment with supp(D). Starting from D^1, a typical trajectory will converge towards $\phi(D^1)$ and reach the surface D, then slide on D until the focal set $\phi(D)$.

If $\Phi(D) = \{\ \}$, with D a switching domain, solutions will simply cross D; otherwise, sliding mode is possible and convergence takes place "in the direction" of $\Phi(D)$. If $\Phi(D) \cap D = \{\ \}$, solutions eventually leave D. In the case where $\Phi(D) \cap D$ is not empty, it can be assimilated to an equilibrium set within D towards which all solutions will converge in the following sense (see [2]):

Lemma 1. *For every regulatory domain $D \in \mathscr{D}_r$, all solutions ξ_t of (2) in D monotonically converge towards the focal point $\Phi(D)$. For every switching domain $D \in \mathscr{D}_s$, the non-switching component $(\xi_t)_i$ of the solution ξ_t in D monotonically converges towards the closed interval*

$$\pi_i(\Phi(D)) = \{\phi_i \in \Omega_i \mid \phi \in \Phi(D)\},$$

the projection of $\Phi(D)$ onto Ω_i, if $(\xi_0)_i \notin \pi_i(\Phi(D))$. Every switching component $(\xi_t)_i$ of the solution ξ_t in D is a constant $(\xi_t)_i = \pi_i(\Phi(D)) = \theta_i^{q_i}$.

Basically, this means that convergence does not take place towards $\Phi(D)$, but towards the smallest hyper-rectangle that contains $\Phi(D)$. Indeed, if $\Phi(D)$ is neither empty, nor a singleton, and ξ_{t_0} belongs to $\Phi(D)$, the Filippov vector field at this point is defined as $H(\xi_{t_0}) = \gamma(\Phi(D) - \xi_{t_0})$ and there is no guarantee that no element of $H(\xi_{t_0})$ points outside of $\Phi(D)$ (we know however that a solution stays at ξ_{t_0}). Due to the structure of the differential equations, it is on the other hand certain that the transient solution does not leave the smallest hyper-rectangle $\Pi(D)$ containing $\Phi(D)$.

We then have the following corollary

Corollary 1. *All solutions ξ_t in D converge towards $\Pi(D)$, if $\xi_0 \notin \Pi(D)$. For all solutions ξ_t in D, $\Pi(D)$ is invariant.*

Corollary 2. *If $\Phi(D)$ is a point, all solutions ξ_t in D converge monotonically towards $\Phi(D)$.*

3 Stability and State Transition Graph

The stability analysis of the various equilibria is a direct consequence of the analysis in the previous section. It is easily seen that equilibria $\bar{x}_r$ in some $D \in \mathscr{D}_r$ are asymptotically stable. In a switching domain $D \in \mathscr{D}_s$, recall that solutions are defined by considering the differential inclusion $H(x)$. We say that a point $y \in \Omega$ is an equilibrium point for the differential inclusion if

$$0 \in H(y), \tag{10}$$

where H is computed using the Filippov construction in (7). In other words, there is a solution in the sense of Filippov, ξ_t, such that $\xi_t(y) = y$, $\forall t > 0$. We call such a point a *singular equilibrium point*. It is easily seen that, for y to be an equilibrium point inside D, it must belong to $\Phi(D)$. Also, since Assumption 1 below prevents $\Phi(D)$ from intersecting the border of D, we then have that $\Phi(D) \subset D$. Every element ϕ of $\Phi(D)$ is then an equilibrium when $\Phi(D) \subset D$ so that, for every $\phi \in \Phi(D)$, there exists a solution $\xi_t(\phi) = \phi$ for all t.

One of the results of [2] concerns the link between the configuration of the state transition graph and the stability of an equilibrium (there is a technical assumption, called Assumption 1, that the focal points are not located on the switching thresholds). This discrete, qualitative description of the dynamics of the PWA system that underlies the qualitative simulation of genetic regulatory networks was originally due to Glass. It indicates the passages between the different domains making up the phase space. A state transition graph is a directed graph whose vertices are the domains of the system and whose edges are the possible transitions between these domains (easily determined by examining the PWA model). The transition graph of system (1) is illustrated in Figure 3.

For a two-dimensional system, we show how this graph indicates the stability of singular equilibria:

Theorem 1. *Let the dimension of the PWA model be 2, and let D be a switching domain containing a singular equilibrium point $\phi(D)$. If for all regulatory domains $D' \in R(D)$ (that is, adjacent to D), there exists a transition from D' to D in the state transition graph, then $\phi(D)$ is asymptotically stable.*

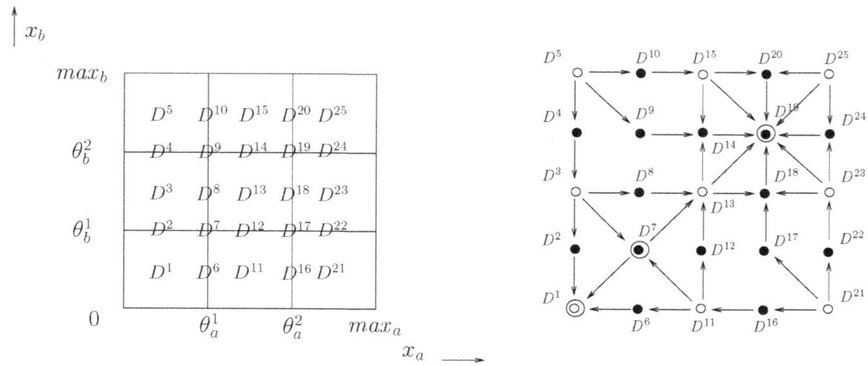

Fig. 3. Subdivision of the state-space in 25 domains and transition graph of system (1)

This result is purely qualitative: it only depends on some inequalities between the parameters (threshold and focal points), but their actual values are not needed. It can be directly applied to show that the singular equilibrium $(x_a, x_b) = (\theta_a^2, \theta_b^2)$, corresponding to D^{19} on Figure 3, is asymptotically stable because there are transitions to D^{19} from D^{13}, D^{15}, D^{23} and D^{25}, the regulatory domains adjacent to D^{19}.

A generalization, but in a weaker form, of this theorem to dimension n is also available.

Theorem 2. *Assume $\Omega \subset \mathbb{R}^n$. Let $D \in \mathscr{D}_s$ be a switching domain of order $p \geq 1$ containing a singular equilibrium set $\Phi(D)$ that satisfies Assumption 1. If for all $D' \in R(D)$, there is a transition from D' to D in the state transition graph, then $\Pi(D)$ is asymptotically stable.*

Corollary 3. *Under the conditions above, if, moreover, $\Phi(D)$ is a point, it is asymptotically stable.*

These results are helpful for the qualitative analysis of the genetic regulatory networks. Moreover, a software GNA was built to analyze genetic networks [5].

4 A Probabilistic Interpretation of the Transition Graph

In this part, we explore the new idea of associating a *probability* of transition to each of the edges in the transition graph. Since PWA systems are deterministic, one way to assign such a probability $D_0 \rightarrow D_1$ is to compute the volume of the region $C \subset D_0$ that switches to D_1.

The goal is to relate dynamical aspects determined by the system's parameters (here, synthesis and degradation rates) to *probabilities* of transition between two state space regions. The idea is to apply these probabilities to the estimation of (some) parameters. Focusing for the present paper on 2-dimensional systems, we will write an analytical expression for the probability of transition between two given regions in terms of the system's parameters.

Consider a piecewise affine system of dimension 2, verifying Assumption 1. Assume that there are r_i thresholds for each variable $i = 1,2$:

$$0 := \theta_i^0 < \theta_i^1 < \cdots < \theta_i^{r_i} < max_i := \theta_i^{r_i+1}, \tag{11}$$

where max_i is as defined in Section 2. Furthermore, assume that these thresholds are defined so that:

$$(\forall i = 1,2) \ \ (\forall k = 0,\ldots,r_i) \ \ \text{sign}(f_i(x) - \gamma_i x_i) = \text{const.}, \ \ \forall \ \theta_i^k < x_i < \theta_i^{k+1}. \tag{12}$$

This is a general condition, since *virtual* thresholds can be added (i.e., even if it is not an activation threshold from variable i to another variable). From now on, a regular domain will be called a *box* (to distinguish them from the switching domains). To label the regular domains, we will use the notation:

$$B_{k_1 k_2}: \quad k_i \in \{0,1,\ldots,r_i - 1\}, \quad \theta_i^{k_i} < x_i < \theta_i^{k_i+1}.$$

As an example, if $(r_1,r_2) = (1,3)$, B_{12} denotes the rectangle $x_1 \in (\theta_1^1, max_1)$, $x_2 \in (\theta_2^2, \theta_2^3)$.

Consider a trajectory that starts in box B_{ij}. The possible transitions from this box are given by the state transition graph (see Section 3).

By assumption (12), the Jacobian of the system is sign-invariant in each box B_{ij}. This implies that, according to the transition graph, any trajectory starting in a box B_{ij} can *switch to one of two neighbor boxes*: $B_{i+s_1,j}$ and $B_{i,j+s_2}$, where $s_k = \text{sign}(f_k(x) - \gamma_k x_k)$ $(k = 1,2)$ for $x \in B_{ij}$. Moreover, since solutions inside each box are uniquely defined, the initial condition in B_{ij} uniquely determines the next box to be visited. Let $\phi(t;x_0)$ denote the solution of system (2), for an initial condition x_0. Define

$$B_{ij}^1 = \{x_0 \in B_{ij} : \ \phi(t;x_0) \in B_{ij}, \forall t < T; \ \phi(t;x_0) \in B_{i+s_1,j}, T < t < T + \Delta T\}$$
$$B_{ij}^2 = \{x_0 \in B_{ij} : \ \phi(t;x_0) \in B_{ij}, \forall t < T; \ \phi(t;x_0) \in B_{i,j+s_2}, T < t < T + \Delta T\},$$

where T depends on x_0 and the various parameters $(\kappa_i, \gamma_i, \theta_i^k)$. B_{ij}^1 (resp., B_{ij}^2) is the set of initial conditions in B_{ij} generating trajectories for which the next visit is box $B_{i+s_1,j}$ (resp., $B_{i,j+s_2}$). We will say that the probability that a trajectory of the system switches from B_{ij} to $B_{i+s_1,j}$ is proportional to the volume of the region B_{ij}^1:

$$P_{ij \to i+s_1,j} = \frac{\text{Area}(B_{ij}^1)}{\text{Area}(B_{ij})}, \quad P_{ij \to i,j+s_2} = \frac{\text{Area}(B_{ij}^2)}{\text{Area}(B_{ij})}. \tag{13}$$

In Section 5 we will illustrate the computation of these probabilities as a function of the parameters κ_i and γ_i, for a simple example of the bistable switch.

To experimentally obtain measurements of the probabilities (13), one would need to perform N times the same experiment, with an initial state in B_{ij} (that is, initial concentrations of x in the region $[\theta_1^i, \theta_1^{i+1}] \times [\theta_2^j, \theta_2^{j+1}]$), and count the number of times N_1 (resp., N_2) that the system evolves to $B_{i+s_1,j}$ (resp., $B_{i,j+s_2}$). If $N_1 = N_2 = 0$, this means that the system remains in B_{ij} and $P_{ij \to i,j+s_2} = P_{ij \to i+s_1,j} = 0$. If $N_1 \neq 0$, then we expect $N = N_1 + N_2$ so that $P_{ij \to i+s_1,j} = N_1/N$ and $P_{ij \to i,j+s_2} = N_2/N = 1 - P_{ij \to i+s_1,j}$ (because we assume (12) which implies only two possible transitions).

These values can then be compared to the expressions in terms of the parameters, for estimation (see Section 5).

5 The Bistable Switch Example

Mathematical models of the bistable switch are characterized by the existence of two stable steady states (or two stable modes), representing two distinct outcomes of the biological system [1, 3]. We will study a general qualitative example of the bistable switch, $\dot{x} = \hat{\kappa}_1 s^-(y, \theta_2) - \gamma_1 x$ and $\dot{y} = \hat{\kappa}_2 s^-(x, \theta_1) - \gamma_2 y$, but considering that the two variables are normalized with respect to their respective thresholds (to reduce the number of free parameters) $x_1 = x/\theta_1$, $x_2 = y/\theta_2$, and $\kappa_i = \hat{\kappa}_i/\theta_i$ to obtain:

$$\begin{aligned} \dot{x}_1 &= \kappa_1 s^-(x_2, 1) - \gamma_1 x_1 \\ \dot{x}_2 &= \kappa_2 s^-(x_1, 1) - \gamma_2 x_2, \end{aligned} \tag{14}$$

with the assumption that (to guarantee existence of two steady states): $\frac{\kappa_i}{\gamma_i} > 1$, $i = 1, 2$. For each variable i, the thresholds (11) are: $\theta_i^0 = 0$; $\theta_i^1 = 1$; $\theta_i^2 = \frac{\kappa_i}{\gamma_i}$, so the state space for system (14) is partitioned into four boxes: B_{00}, B_{01}, B_{10} and B_{11}. It is not difficult to check that the system has two stable steady states, located in the regions B_{10} and B_{01}. Solutions starting in B_{00} or B_{11} will eventually cross to either B_{10} and B_{01} (depending on the exact initial condition). Moreover, we can compute the separatrix line $x_2 = \sigma_{00}(x_1)$ which divides region B_{00} into the two regions B_{00}^1 and B_{00}^2: solutions with initial conditions above (resp., below) the line σ_{00} will eventually converge to the steady state in B_{01} (resp., B_{10}). A similar separatrix line σ_{11} can be computed for the region B_{11}. These curves are given by:

$$\sigma_{00}(x) = \frac{\kappa_2}{\gamma_2} - \left(\frac{\kappa_2}{\gamma_2} - 1\right)\left(\frac{\frac{\kappa_1}{\gamma_1} - x}{\frac{\kappa_1}{\gamma_1} - 1}\right)^{\frac{\gamma_2}{\gamma_1}}, \qquad \sigma_{11}(x) = x^{\frac{\gamma_2}{\gamma_1}} \tag{15}$$

These separatrix lines are represented in Fig. 4, and correspond to the locus of the points that go through $(x_1, x_2) = (1, 1)$. To simplify the presentation, we will assume that:

$$\text{(A1) } \sigma_{00}(x=0) > 0 \Leftrightarrow \left(\frac{\frac{\kappa_1}{\gamma_1}}{\frac{\kappa_1}{\gamma_1} - 1}\right)^{\frac{\gamma_2}{\gamma_1}} < \frac{\frac{\kappa_2}{\gamma_2}}{\frac{\kappa_2}{\gamma_2} - 1};$$

$$\text{(A2) } \sigma_{11}\left(x = \tfrac{\kappa_1}{\gamma_1}\right) < \tfrac{\kappa_2}{\gamma_2} \Leftrightarrow \left(\frac{\kappa_1}{\gamma_1}\right)^{\frac{\gamma_2}{\gamma_1}} < \frac{\kappa_2}{\gamma_2};$$

where (A1) says that the line σ_{00} exits the box B_{00} through the axis $x_1 = 0$, and (A2) says that the line σ_{11} exits the box B_{11} through the axis $x_1 = \frac{\kappa_1}{\gamma_1}$. According to definition (13) we have:

$$P_{00\to10} = \int_0^1 \sigma_{00}(x_1)\,dx_1, \quad P_{00\to01} = 1 - P_{00\to10}, \tag{16}$$

Similarly, we can compute the probability of a transition from B_{11} to B_{10}. To obtain the correct probability, we need to subtract the area corresponding to the region B_{10} (which is part of the area below σ_{11}), and only then divide by the total area of B_{11}:

$$P_{11\to10} = \frac{1}{(\frac{\kappa_1}{\gamma_1}-1)(\frac{\kappa_2}{\gamma_2}-1)}\left\{\int_1^{\frac{\kappa_1}{\gamma_1}} \sigma_{11}(x_1)\,dx_1 - \left(\frac{\kappa_1}{\gamma_1}-1\right)\right\} \tag{17}$$

and $P_{11\to01} = 1 - P_{11\to10}$. For transitions from regions B_{10} or B_{01}, the theorical probability of transition to any other region is 0 so, in practice, we can expect very weak transition probabilities from these two regions, and one can say that $P_{01\to01} = 1$ and $P_{10\to10} = 1$. The expressions (16) and (17) can be written as:

$$P_{00\to10} = b + \frac{1}{c+1}(a-1)(b-1)\left(1-\left(\frac{a}{a-1}\right)^{c+1}\right) \tag{18}$$

$$P_{11\to10} = \frac{1}{(a-1)(b-1)}\left(\frac{1}{c+1}(a^{c+1}-1)-(a-1)\right), \tag{19}$$

in terms of the three parameters:

$$a = \frac{\kappa_1}{\gamma_1}, \quad b = \frac{\kappa_2}{\gamma_2}, \quad c = \frac{\gamma_2}{\gamma_1}.$$

Therefore, given measurements for the degradation rates and the probabilities of transition, it is possible to estimate the synthesis rates from (18) and (19). Let c be known, $P_{00\to10} = p_{00}$ and $P_{11\to10} = p_{11}$, then a is given by the solution of:

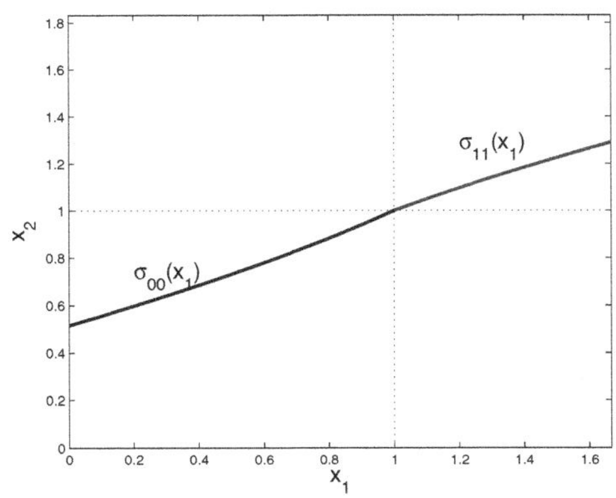

Fig. 4. Separatrix functions satisfying assumptions A1 and A2

$$\left(\frac{a^{c+1}-1}{c+1}-(a-1)\right)^{-1}(p_{00}-1)p_{11}=\frac{1}{a-1}+\frac{1}{c+1}\left(1-\left(\frac{a}{a-1}\right)^{c+1}\right) \quad (20)$$

and b is given by

$$b = 1+\frac{1}{(a-1)p_{11}}\left(\frac{1}{c+1}(a^{c+1}-1)-(a-1)\right) \quad (21)$$

in the domain of validity of the equalities (18)-(19) (assumptions A1 and A2 hold):

$$a^c < b < \frac{\left(\frac{a}{a-1}\right)^c}{\left(\frac{a}{a-1}\right)^c - 1}.$$

Note that the assumptions A1 and A2 can be dropped, but then the explicit expression for $P_{00\to10}$ and $P_{11\to10}$ must be modified according to the geometry of the separatrices, in particular the starting or ending points for the integrals will change. The general case can be easily written down, but for reasons of space and presentation we will not give it here. To give a numerical example, assume that $\gamma_1 = 0.9$, $\gamma_2 = 0.6$, $p_{00} = 0.9$, and $p_{11} = 0.25$, to obtain $c = 2/3$ and, from equations (20) and (21), respectively: $a \approx 1.48$ and $b \approx 1.61$, which are inside the region of validity ($b \in (1.29, 1.89)$). The estimated synthesis rates are thus: $\kappa_1 \approx 1.33$ and $\kappa_2 \approx 0.96$.

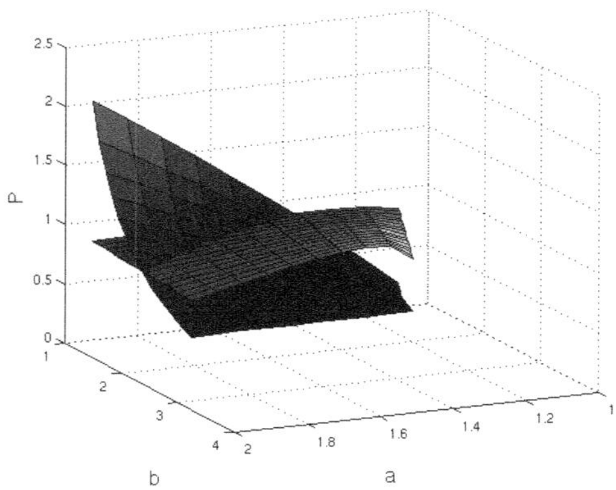

Fig. 5. Probabilities $P_{00\to10}$ (red surface) and $P_{11\to10}$ (blue surface), as a function of a and b, for $c = 0.5$

Finally, in Fig. 5 the probabilities are shown as functions of both a and b, for a fixed value of $c = 0.5$, in a domain where the functions (18)-(19) are valid. Observe that the probability $P_{00\to10}$ remains at a fairly constant high level, while $P_{11\to10}$ decreases significantly with b. This fact is interesting, because it shows that the dependence of the separatrix curve σ_{00} on b is in fact weak, and that increasing b leads essentially to increasing the area above the separatrix curve σ_{11} (see also Fig.4).

6 Conclusions

In this paper, after a first part dedicated to a review of results about PWA systems, we have given a probabilistic interpretation of the transitions in the second part. A method is suggested for parameter estimation, applicable to systems where the measurements are mostly qualitative. Assuming that the data consist of probabilities of transition between two different regions of the state space, and that (for instance) the degradation rates are known, one can estimate the synthesis rates. The method was described for 2-dimensional piecewise affine differential systems. Further work is needed for more complex systems.

Acknowledgments. This work is partly supported by ANR-Metagenoreg and INRIA-INSERM Colage.

References

1. Alon, U.: An introduction to systems biology: design principles of biological circuits. CRC Press, Boca Raton (2007)
2. Casey, R., de Jong, H., Gouzé, J.L.: Piecewise-linear models of genetic regulatory networks: equilibria and their stability. J. Math. Biol. 52, 27–56 (2006)
3. Chaves, M., Eissing, T., Allgöwer, F.: Bistable biological systems: a characterization through local compact input-to-state stability. IEEE Trans. Automat. Control 53, 87–100 (2008)
4. de Jong, H.: Modeling and simulation of genetic regulatory systems: A literature review. J. Comput. Biol. 9(1), 67–103 (2002)
5. de Jong, H., Geiselmann, J., Hernandez, C., Page, M.: Genetic Network Analyzer. Qualitative simulation of genetic regulatory networks 19(3), 336–344 (2003)
6. de Jong, H., Gouzé, J.-L., Hernandez, C., Page, M., Sari, T., Geiselmann, J.: Qualitative simulation of genetic regulatory networks using piecewise-linear models. Bull. Math. Biol. 66(2), 301–340 (2004)
7. Filippov, A.F.: Differential Equations with Discontinuous Righthand Sides. Kluwer Academic Publishers, Dordrecht (1988)
8. Glass, L., Kauffman, S.A.: The logical analysis of continuous non-linear biochemical control networks. J. Theor. Biol. 39(1), 103–129 (1973)
9. Gouzé, J.L., Sari, T.: A class of piecewise linear differential equations arising in biological models. Dyn. Syst. 17(4), 299–316 (2002)
10. Grognard, F., Gouzé, J.-L., de Jong, H.: Piecewise-linear models of genetic regulatory networks: theory and example. In: Queinnec, I., Tarbouriech, S., Garcia, G., Niculescu, S. (eds.) Biology and control theory: current challenges. Lecture Notes in Control and Information Sciences (LNCIS), vol. 357, pp. 137–159. Springer, Heidelberg (2007)
11. Mestl, T., Plahte, E., Omholt, S.W.: A mathematical framework for describing and analysing gene regulatory networks. J. Theor. Biol. 176(2), 291–300 (1995)

A Control Engineering Model for Resolving the TGF-β Paradox in Cancer

Seung-Wook Chung, Carlton R. Cooper, Mary C. Farach-Carson, and Babatunde A. Ogunnaike*

Abstract. Although TGF-β is widely known to appear to function paradoxically as a tumor suppressor in normal cells, and as a tumor promoter in cancer cells, the underlying mechanisms by which a *single* cytokine plays such a dual—and diametrically opposed—role are unknown. In particular, it remains a mystery why the level of TGF-β is unusually high in the primary cancer tissue and blood samples of cancer patients with the poorest prognosis, given that this cytokine is primarily a tumor suppressor. To provide a quantitative explanation of these paradoxical observations, we have developed, from a control theory perspective, a mechanistic model of TGF-β-driven regulation of cell homeostasis. Analysis of the overall system model yields quantitative insight into how the cell population is regulated, enabling us to propose a plausible explanation for the paradox: with the tumor suppressor role of TGF-β *unchanged* from normal to cancer cells, we demonstrate that the observed increased level of TGF-β is an *effect* of cancer cell characteristics (specifically, acquired TGF-β resistance), not the *cause*. We are thus able to explain precisely why the clinically observed correlation between elevated TGF-β levels and poor prognosis is in fact consistent with TGF-β's original (and unchanged) role as a tumor suppressor.

1 Introduction

Normal tissue homeostasis is maintained by a delicate and dynamic balance between the cellular processes of proliferation and death. In particular, too much growth and

Seung-Wook Chung · Babatunde A. Ogunnaike
University of Delaware, Department of Chemical Engineering, 150 Academy St. Newark, DE 19716
e-mail: ogunnaike@udel.edu

Carlton R. Cooper
University of Delaware, Department of Biological Sciences, Newark, DE 19716

Mary C. Farach-Carson
Rice University, Department of Biochemistry and Cell Biology, Houston, TX 77251

* Corresponding author.

J. Lévine & P. Müllhaupt (Eds.): Adv. in Theory of Control, Signals, LNCIS 407, pp. 255–268.
springerlink.com © Springer-Verlag Berlin Heidelberg 2010

too little death can lead to a severe condition that may ultimately result in cancer [1]. It is known that these cellular processes are affected by a variety of extracellular stimuli, each capable of inducing its own set of responses via specific intracellular signaling cascades. And among these extracellular signals, transforming growth factor β (TGF-β) has drawn much attention from cancer researchers because it plays a central role in regulating both cell proliferation and cell death [2, 3, 4].

TGF-β is known to be an important participant in a variety of physiological processes in both normal and malignant tissues [3], but considerable debate remains over its exact role during cancer progression. During the early stages of epithelial tumorigenesis, that TGF-β functions as a potent tumor suppressor primarily by inducing cell cycle arrest and programmed cell death (apoptosis) is unquestioned. However, the level of TGF-β is frequently elevated in many malignant tissues and in blood samples from cancer patients with the poorest prognosis. As such, the role of TGF-β in the late phases of tumor progression appears to become—somehow—one of tumor promotion, by appearing to support proliferation, by subverting immune-surveillance and also facilitating epithelial to mesenchymal transition (EMT), invasion, and angiogenesis. This has created the widely held perception that TGF-β is *simultaneously* a tumor suppressor under one condition and a tumor promoter under another. But how does a single stimulus produce multiple contradictory results? A clearer understanding of these apparently contradictory roles of TGF-β in cancer requires quantitative methods because TGF-β biology is simply too complex to be understood on the basis of qualitative descriptions.

While extensive physiological, biochemical, and clinical information is available on TGF-β, quantitative modeling of the TGF-β signaling system is still comparatively in its infancy. Furthermore, as briefly reviewed in [5], all published computational TGF-β models to date have focused only on the intracellular signal transduction pathway in a single cell. Such single-cell models, despite their significant contributions to our understanding of the dynamic behavior of TGF-β signaling, may not be sufficient for interpreting the contradictory clinical observations noted above. This is because the dynamic characteristics of *extracellular* molecules and signals depend strongly on the active interactions among cells and/or between cells and the surroundings; explaining such extracellular dynamics requires more that just studying *intracellular* events alone. Thus, a more realistic understanding of the role of TGF-β in cancer requires a more comprehensive examination of the TGF-β system encompassing the cells and their microenvironment.

To this end, we present in this study a macroscopic mechanistic model of TGF-β-driven regulation of cell homeostasis. The model deals, *not with a single cell*, but with the *cell population* as a systemic entity, and represents a control system characterization of how TGF-β achieves cell homeostasis via communication between the cells and their microenvironment. First, we identify the various functional components of the system, their respective input and output variables, and how they are connected to form the complete control system; each component is then modeled on the basis of available consensus information in the reported biological literature. Where the required information is unavailable, we state and employ reasonable assumptions, supporting our postulates adequately. An analysis of the resulting overall system model yields quantitative insight into how the biological processes of cell proliferation and death are regulated by TGF-β via the interactions between proliferating cells and their surroundings. The model also allows us to predict possible dynamic characteristics of the TGF-β-mediated control system in cancer tissues, from which we are able to present an alternative perspective of the TGF-β paradox in cancer.

2 Model Description

Of all the physiological processes that influence homeostasis in a cell population, none is as critical as the combined processes of cell proliferation and death. Maintaining the dynamic balance between proliferation and death is how cell population dynamics are regulated; and biological regulation is achieved in general by dedicated biological control systems. In the specific case of this study, we restrict our attention to the TGF-β-mediated system for achieving cell homeostasis, viewed as an automatic biological control system for rejecting "disturbances" that will otherwise cause a cell population to grow indefinitely and become cancerous. As with all control systems, engineered or biological, this control system will also consist of at least the following component subsystems: (i) *Sensor*: which receives information about the "controlled process" state and generates appropriate signals that are transmitted to the regulatory machinery; (ii) *Controller*: the regulatory machinery which receives "process state" signals and generates appropriate corrective action signals; (iii) *Actuator*: the "final control element" which implements the corrective action on the controlled process.

By representing each functional component with a block showing inputs and outputs as determined from mechanistic information available in the literature (and discussed in detail subsequently), the overall control system block diagram is shown in Fig. 1. What follows is a detailed discussion of model development for each component subsystem.

2.1 Controlled Process: Cell Proliferation and Death

In this study, the "controlled process" is the combined biological process of cell proliferation and death. The output of interest—the "controlled output"—is the total cell population count. Our study is restricted to cell proliferation and death *as regulated by TGF-β* via its ability to inhibit cell proliferation and induce apoptosis. As such, the "manipulated input" is the amount of active TGF-β to which the cell population is exposed. Because the specific pathology of interest is cancer, pro-proliferative signals (such as growth factors and hormones) constitute the "disturbance" of interest whose effects on proliferation are to be handled appropriately by the TGF-β-mediated control system, if normal cell growth and proliferation is to be kept under judicious restraint. The desired mathematical model therefore will represent the response of cell population to stimulation by growth factors on one hand and bioactive TGF-β on the other.

Cell population dynamics will be modeled under the following assumptions: (i) all cells of interest are capable of proliferating and do so at a uniform rate, p; (ii) all cells are homogeneously distributed so that each cell is readily and equally accessible to extracellular stimuli; (iii) cell death occurs at a uniform rate, d, for all cells; (iv) cell population dynamics are dominated by proliferation and death so that other cellular processes, including differentiation and migration, can be neglected; and (v) upon initial stimulation with growth signals, cells start to proliferate immediately, with no delay, and enter successive, synchronous, cell division rounds thereafter. The simplest model consistent with these assumptions is:

$$\frac{dX}{dt} = (p-d)X \tag{1}$$

where X is the total number of cells in the population. Observe that when $p = d$, the cell population is at steady state; when $p \neq d$, the population either grows or shrinks exponentially, depending on whether $p > d$ or vice-versa. The population dynamics are therefore clearly determined by the parameters p and d, which, in turn, are determined by the level of the extracellular cues that induce proliferation or death.

First, the rate of cell proliferation, p, is known to increase with the level of proliferation stimuli, but decreases with the level of anti-proliferation factors such as TGF-β which inhibits clonal expansion by arresting the cell cycle in the G1 phase [3, 6]. Therefore, we postulate the following functional relationship between the rate

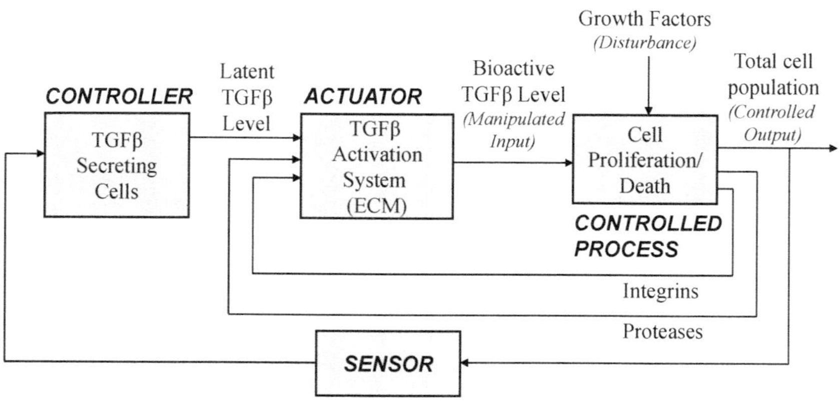

Fig. 1. Block diagrammatic representation of the TGF-β control of cell proliferation and death. Arrows indicate flow of information into and out of system blocks. Note the multiple feedback loops involved in the "actuator" subsystem for regulating the level of bioactive TGF-β.

of cell proliferation, p, and the concentrations of growth factors and TGF-β as two distinct terms, assuming no interaction effect between these distinct stimuli:

$$p(GF,\ TGF\beta) = \frac{p_a \cdot GF^r}{p_b{}^r + GF^r} - \frac{p_2 \cdot TGF\beta^m}{p_3{}^m + TGF\beta^m} \qquad p \geq 0 \tag{2}$$

Here, GF and $TGF\beta$ denote the concentrations of growth stimuli of any kind and of TGF-β, respectively; p_a is the maximum cell division rate; p_2 is the maximum anti-growth rate; p_b and p_3 are affinity constants; and r and m are Hill coefficients.

Next, d, the rate of cell death, is known to be influenced by the level of pro-apoptotic stimuli such as the TNF superfamily; it is also known that TGF-β promotes the death of unhealthy, damaged, and unnecessary cells by inducing apoptosis [3]. Thus, the rate of cell death should increase with increasing TGF-β level. Consequently, we represent the dependence of d on TGF-β level as follows:

$$d(TGF\beta) = d_1 + \frac{d_2 \cdot TGF\beta^n}{d_3{}^n + TGF\beta^n} \tag{3}$$

where, d_1 is the inherent rate of death due to endogenous pro-apoptotic factors; d_2 represents the maximum rate of TGF-β-induced apoptosis; d_3 is an affinity constant; and n is a Hill coefficient.

The overall model equation for the "controlled process" is therefore:

$$\frac{dX}{dt} = (p-d)X = \left(\frac{p_a \cdot GF^r}{p_b{}^r + GF^r} - \frac{p_2 \cdot TGF\beta^m}{p_3{}^m + TGF\beta^m} - d_1 - \frac{d_2 \cdot TGF\beta^n}{d_3{}^n + TGF\beta^n}\right)X \tag{4}$$

2.2 Sensor/Controller: TGF-β Production System

To elicit the well-established physiological response of healthy tissue to unusual changes in its cell population size, such changes will have to be detected by some sort of "sensor" system, which in turn will stimulate the required response from the "biological controller" responsible for maintaining cellular homeostasis. In this particular case, the response of interest is the production of TGF-β for the express purpose of restraining unusual growth. While the precise mechanisms by which this purpose is achieved is unknown, a growing body of knowledge is emerging to provide clues concerning the basic characteristics.

When cells undergoing unusual growth break the basement membrane, (i) they encounter the stroma, resulting in inflammation; (ii) in response, TGF-β is produced locally in latent form—known as large latent complexes (LLC)—in the stroma (as well as from other sources, including various immune cells such as macrophages, dendritic cells, T cells, B cells, etc. [7]); and finally, (iii) active TGF-β is made available by a subsequent multi-step process of activation (including secretion, interaction with ECM components, and proteolytic cleavage), with each step in the activation process under tight control [8]. Under normal circumstances, this action is sufficient to eliminate the errant cells, repair the damage, and promote normal healing.

Thus, even though not all of the mechanistic details of how the TGF-β producer cells "monitor and respond" to unusual changes in cell population are known, it is clear from what is known that such a TGF-β production system is the biological controller; and that it appears to be stimulated directly by changes in cell population. As such, we consider the total cell number, X, as the input to this combined sensor/controller, and the output is the level of inactive TGF-β complex. (Modeling how active TGF-β is produced from the latent form is discussed later.)

As an anti-growth cytokine, the level of TGF-β should *increase* with increasing cell population in order to inhibit abnormal cell growth. Conversely, a decrease in cell number should result in a commensurate reduction in the level of TGF-β. Therefore, as presented in many physiology textbooks (e.g., [9]), we employ the following sigmoidal response function for this biological controller:

$$LLC(X) = \frac{K}{1 + e^{[C_a \cdot (C_b - X)]}} \tag{5}$$

indicating how LLC, the inactive TGF-β complex concentration, changes as a function of X, the cell population. Here, K is the maximum level of latent TGF-β, C_a is a scaling parameter, and C_b is the sigmoid's "center parameter" at which the controller output is half of the maximum value, K. It can be shown that for this nonlinear controller, the effective controller gain, $\partial(LLC)/\partial X$, is maximum at $X = C_b$; also, it can be shown that C_b is the "implicit set-point" at which the controller appears to want to maintain X.

2.3 Actuator: TGF-β Activation System

As noted briefly in Section 2.2, in response to changes in cell population, TGF-β producer cells secrete an inactive form of the cytokine which is easily bound to and stored in extracellular matrix (ECM) proteins (e.g., fibrillin-1, perlecan, and fibronectin) via its latent TGF-β-binding protein (LTBP) component. In order to become bioactive (i.e., to be able to bind its cognate cell-surface receptors and effect intracellular signaling), TGF-β proteins sequestered within the ECM-bound LLC need to be released [8]. The dissociation of bioactive TGF-β from the LLC-ECM complex is mediated by two distinct mechanisms: enzymatic (or proteolytic) and mechanical. Enzymatic cleavage involves a variety of proteases including metalloproteinases (MMPs) and serine proteases (e.g., plasmin, thrombin, tryptases, etc.), and appears to be the most prominent of the two mechanisms [10]. With mechanical dissociation, TGF-β is released by cell traction forces generated via integrins[1] that bind to a LLC component known as latency-associated propetide (LAP) [11].

Our model of this TGF-β activation process is based on these two mechanisms, along with the following considerations: (i) TGF-β bioavailability depends most

[1] A large family of heterodimeric transmembrane proteins that function as adhesion receptors, promoting cell-cell adhesion or cell-matrix adhesion.

strongly on the final step of the activation process, the release of TGF-β from its latent complex; (ii) this protease-driven dissociation follows Michaelis-Menten kinetics; (iii) integrins that mediate cell traction force exhibit enzyme-like activity with Michaelis-Menten kinetics; and (iv) the released bioactive TGF-β can be irreversibly degraded by endogenous proteases.

This leads to the following "actuation dynamics" equation, representing the dynamics of activated TGF-β as a function of the concentration of the controller output, *LLC* concentration, and the levels of participating proteases and integrins.

$$\frac{dTGF\beta}{dt} = \frac{k_{cat1} \cdot P \cdot LLC}{K_{m1} + LLC} + \frac{k_{cat2} \cdot I \cdot LLC}{K_{m2} + LLC} - k_{deg} \cdot TGF\beta \tag{6}$$

Here, P and I denote the concentrations of TGF-β-activating proteases and integrins, respectively; k_{cat1} and k_{cat2} are turnover numbers; K_{m1} and K_{m2} are Michaelis-Menten constants; and k_{deg} is the rate of proteolytic degradation of bioactive TGF-β.

Finally, as part of the cell population regulation process, proliferating cells themselves produce proteases that promote the activation of TGF-β. In addition, proliferating cells also produce cell-surface integrins that potentially can associate directly with, and hence promote the activation of, latent TGF-β. The actuation process therefore involves feedback loops from the controlled process itself through the proteases and integrins from proliferating cells (see Fig 1). Consequently, in the absence of further mechanistic information, the simplest way to represent the concentrations of active proteases, $P(t)$, and of integrins, $I(t)$, is with the following equations:

$$P(t) = k_P \cdot X(t) + PRT_0 \tag{7}$$

$$I(t) = k_I \cdot X(t) \tag{8}$$

where X is the total cell number (as in Eq (1)); k_p is a proportional constant for protease synthesis/activation from proliferating cells; PRT_0 is the constitutive production level of proteases in the tissue; and k_I is a proportional constant for the number of integrins from a cell that can potentially bind to latent TGF-β.

2.4 Overall System Model and Parameters

These individual component models may now be connected as indicated in Fig. 1, resulting in the overall control system model which can now be used to simulate the closed-loop characteristics of the TGF-β-mediated regulation of cell population. The specific model parameter values selected for the simulation studies are listed in Table 1. The values indicated for the controller parameters (K, C_a, C_b) were chosen to obtain reasonable control system performance; K is subsequently subjected to parametric analysis in the next section. Although space limitation prevents a detailed discussion of how the other parameters were determined, we note that most were estimated from experimental observations reported in the literature (p_a, p_b, and r from [12]; d_1 and d_2 from [13]; k_{deg} from [14]; k_{cat1}, K_{m1}, k_{cat2}, and K_{m2}

Table 1. Model parameter values

Parameter	Value	Parameter	Value
p_a	0.0531 h^{-1}	C_a	4E-05
p_b	1.3451 nM	C_b	1.155E+05
r	1.4191	k_p	1E-05 nM/cell
p_2	0.0531 h^{-1}	PRT_0	2 nM
p_3	2 nM	k_{cat1}	34.83 h^{-1}
m	1	K_{m1}	8.5 μM
d_1	0.0355 h^{-1}	k_I	5E-06 nM/cell
d_2	0.0142 h^{-1}	k_{cat2}	8.1 h^{-1}
d_3	4.2 nM	K_{m2}	4.25 μM
n	1	k_{deg}	0.1155 h^{-1}
K	20 nM		

from [15]); the remaining few were chosen on the basis of biologically reasonable considerations.

A local sensitivity analysis was carried out based on the normalized sensitivity coefficient defined as:

$$NSC_i(t) = \left. \frac{\theta_i}{X} \frac{\partial X(t, \theta)}{\partial \theta_i} \right|_{\theta^*} \tag{9}$$

where θ and i are the vector of model parameters and parameter index, respectively. When time-averaged over the duration of the simulation, $T = 300$ hrs, the resulting time-averaged sensitivity defined as

$$< NSC_i >= \frac{1}{T} \int_0^T \left. \frac{\theta_i}{X} \frac{\partial X(t, \theta)}{\partial \theta_i} \right|_{\theta^*} dt \tag{10}$$

is shown in Fig. 2. As expected, the parameters that the system response is most sensitive to are related to the rates of cell growth and death (p_a, d_1), the responsiveness of the cells to TGF-β's cytostatic effects (p_2), and the controller parameters (C_b, and K).

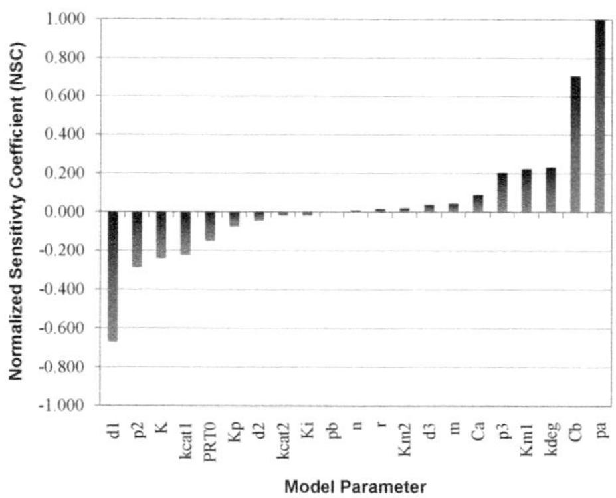

Fig. 2. Model parameter sensitivities

3 Results and Discussion

3.1 TGF-β-Driven Regulation of Normal Cell Growth/Death

The complete control system model may now be used to simulate the dynamic regulation of the cell proliferation/death process under various conditions.

Nominal Conditions. First, under nominal conditions indicated in Table 1, and from an initial condition of 1.0×10^5 cells in the population, the response of the overall system to a sustained step input of growth factor implemented at time $t = 0$ is shown in Fig. 3. Upon stimulation with growth factor, the cell number increases initially; but, as a result of an effective TGF-β-mediated controller and actuator, the cell population returns to a new steady state value not too far from the initial value (Fig. 3A). The dynamics of the amount of latent TGF-β complex (LLC) (controller output) and of bio-active TGF-β (actuator output) required to achieve this regulation are shown in Figs. 3B and C respectively. The net result is an increase in the bioactive TGF-β level to counterbalance the effect of the sustained growth factor stimulus. Observe that under these nominal conditions, the overall system is stable, and the control system regulates the cell population effectively.

Effect of controller parameter K. The performance of any control system depends on the values chosen for the controller parameters. Even though this particular biological controller is nonlinear and possesses three parameters, we choose to investigate the effect of the parameter K on the control system performance. This parameter can be shown to be related directly to the maximum possible controller gain, and is therefore most reminiscent of the "proportional gain" value in classical feedback control. In particular, we compare the nominal controller performance to the controller performance when a "high K" value (corresponding to twice the nominal value) is used, and also to the performance obtained when a "low K" value value (corresponding to half the nominal value) is used. The results are shown in Fig. 4. Compared to nominal performance, the figure shows that a higher K value results in a lower steady-state cell population, as a direct consequence of higher

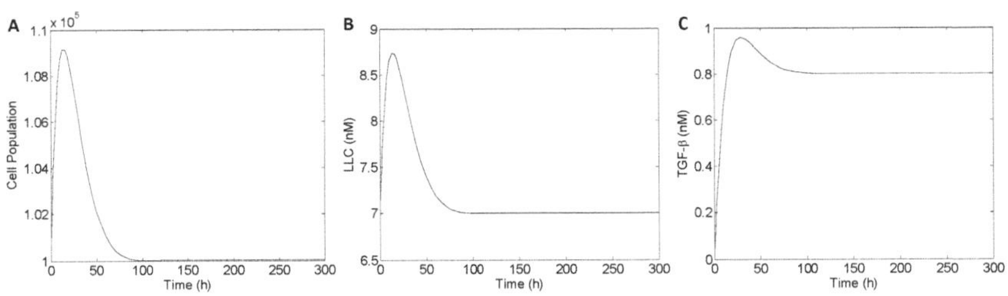

Fig. 3. Closed loop responses of the total cell number (A), the level of large latent TGF-β complex (B), and the bioactive TGF-β level (C), to a step input of growth factors (100 nM) implemented at time $t = 0$

production of latent TGF-β; a lower K value allows cells to grow more (because of the consequent lower overall production of latent TGF-β), resulting in a higher steady state value for the cell population.

3.2 The Dynamics of Cancer Cell Population

Compared to signalling in normal cells, TGF-β signaling in cancer cells is significantly different, primarily as a result of alterations in several components of the TGF-β signaling pathway that occur during cancer progression [16]. In particular, it is known that cancer cells, because of mutations, deletions, and downregulation, etc., have significantly fewer functional TGF-β receptors, thereby rendering cancer cells generally less responsive to TGF-β [17].

To investigate the effect of such reduced TGF-β responsiveness on the overall dynamics of TGF-β-mediated regulation in a cancer cell population, we start by observing that the cell population's responsiveness to TGF-β is represented in the system model by the parameters p_2 and d_2, respectively, the effect of TGF-β on the proliferation rate, p, and the death rate, d. Thus, we investigate the dynamic behavior of the TGF-β-mediated control system under cancerous conditions by comparing the performance under nominal conditions with the performance when the "responsiveness" parameters, p_2 and d_2, are both reduced simultaneously to 50%, 33.33%, and 25% of their respective nominal values (corresponding to a 2-, 3-, and 4- fold reduction in responsiveness).

The simulation results shown in Fig. 5 display several important features. First, Fig. 5A indicates that as cells become less sensitive to TGF-β, TGF-β naturally becomes less effective as a regulator of growth. As such, in response to growth factor stimulation, the total number of cells in the population increases, reaching progressively higher steady state values as the cells become progressively less responsive. Beyond a particular point (illustrated in this case by 25% of nominal responsiveness), the cells would become so unresponsive—and hence sufficiently resistant to the anti-growth effects of TGF-β—to the point where the cytokine is no longer effective in suppressing unwanted proliferation. Inevitably, the cell population will therefore grow without limit.

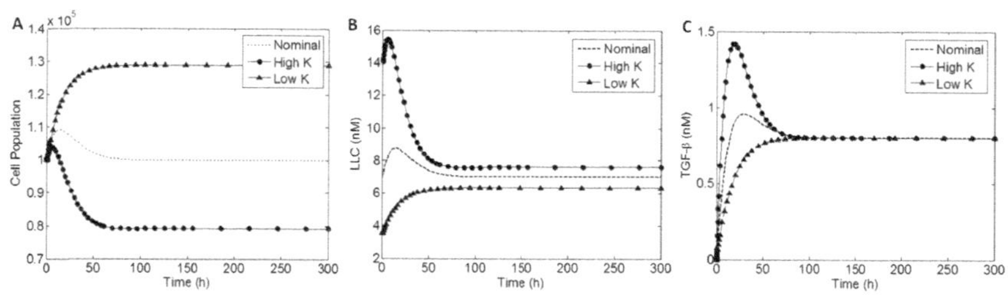

Fig. 4. Effect of changes in K: on the cell population (A); the level of inactive TGF-β, LLC (B); and the level of bioactive TGF-β (C), for "high K," twice the nominal value (circles), and "low K," half the nominal value (triangles)

These simulation results are supported by experimental observations. For example, as reported in [18], compared to normal prostatic cells, cells from pre-malignant prostatic tissue (i.e., benign prostatic hyperplasia) have fewer TGF-β receptors and tend to proliferate more rapidly. The signature enlargement of such glands corresponds to the higher steady state cell population values indicated by the simulation for cells with reduced (but still stabilizable) sensitivity to TGF-β. On the other hand, malignant prostate cancer cells (e.g., the LNCaP cell line) with their much higher proliferative potential, have even fewer TGF-β receptors [19], making them much more recalcitrant to TGF-β's tumor suppressor effects. This situation corresponds to the uncontrollable growth indicated in the simulation when the responsiveness is reduced to 25%.

Next, the results in the other plots (Figs. 5B, C and D) show another important feature of this control system: how the ineffectiveness of TGF-β in regulating the cell population has a compounding destabilizing effect that is typical of open loop unstable systems under ineffective feedback control. Observe that when the cell responsiveness is reduced to 25% of the nominal value, in response to growth factor stimulation, the cell population increases to a higher than normal level as a consequence of the lack of sensitivity to the anti-growth effect of TGF-β; the increasing cell population in turn causes the TGF-β producer cells to secrete an increasing amount of the latent TGF-β complex in an effort to regulate the rapid growth (Fig. 5B). Furthermore, the proliferating tumor cells themselves produce a growing amount of proteases which then participate in the enzymatic activation of TGF-β, in addition to promoting the synthesis and activation of such enzymes in

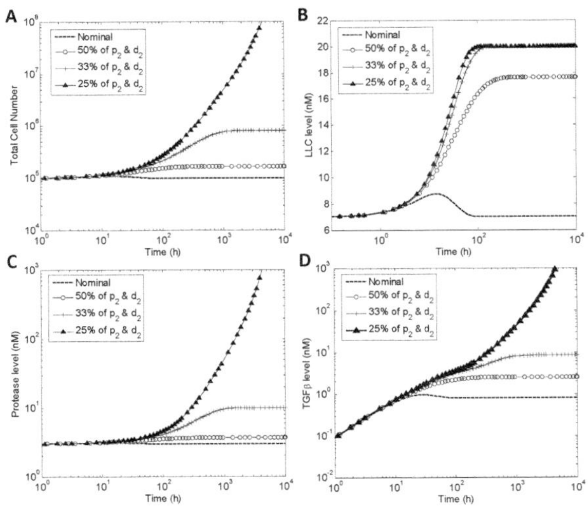

Fig. 5. Control system performance under abnormal conditions: total number of cells (A); the level of LLC (B); the level of proteases (C); and the level of bioactive TGF-β (D), in response to a step input of growth factor (100 nM). The values of parameter p_2 and d_2 are simultaneously reduced 2-fold (50% reduction) (open circles), 3-fold (33.33% reduction)(crosses), and 4-fold (25% reduction) (filled triangles).

the microenvironment (Fig. 5C). The increased level of active proteases, in turn, results in rapid degradation of the extracellular matrix (ECM) components, ultimately leading to increased and expedited release of bioactive TGF-β from the ECM. The increased number of cancer cells also provide more integrins that can bind to latent TGF-β, thereby facilitating TGF-β activation by cell traction force. The net effect is shown in Fig. 5D where the amount of bioactive TGF-β is seen to increase exponentially in an effort to suppress runaway growth in a population of cells that have become resistant to TGF-β.

This final point is crucial to our understanding of the TGF-β paradox. It now appears evident from these simulations that the observed increased level of TGF-β is a *consequence* of the acquired TGF-β resistance exhibited by the cancer cells, not the *cause*; as such, the correlation between the increased level of TGF-β and poor prognosis, while real, *should not have been interpreted as implying that the former caused the latter*. The implications are that as pre-malignant cells lose their responsiveness to TGF-β along the spectrum of tumor progression, a still-intact control system must secrete more of this cytokine in a futile attempt to achieve the level of tumor suppression attainable with normal, responsive cells. (On a single cell level, this is consistent with our previous mathematical model of intracellular TGF-β signaling [5] which showed, among other things, that the amount of TGF-β needed to produce a saturated Smad-mediated response in a cancer cell is far higher than that in healthy cells, implying that to elicit nuclear Smad-mediated (growth-inhibitory) activity, cancer cells require more TGF-β than normal.) Our control system model therefore indicates that there is no paradox: TGF-β remains a tumor suppressor; and observing increased levels of TGF-β in poor prognosis patients is entirely consistent with TGF-β's role as a tumor suppressor attempting to regulate the growth of aberrant cells that have lost their sensitivity to the cytokine.

4 Conclusions

We have studied the role of TGF-β in normal and cancerous cells using a control engineering model of TGF-β-mediated regulation of cell population dynamics. In particular, the results of our study indicate that the *correlation* between increased levels of TGF-β and poor prognosis may have been inadvertently misconstrued as *causality*, creating an apparent paradox. Our results indicate that the clinically observed increased TGF-β level in cancerous tissues is *not* an indication that the tumor suppressor role of TGF-β has changed fundamentally. Rather, the control system perspective supports the hypothesis that the role of TGF-β as a tumor suppressor is unchanged, and further stipulates that the level of TGF-β should in fact *increase* in an attempt to elicit normal responses from a tumor that is becoming increasingly resistant to this cytokine. Thus, the clinical observations are actually consistent with a TGF-β whose role as a tumor suppressor remains unchanged.

A key next step is to validate this hypothesis experimentally, *in-vitro*, using the following approach. Several cancer cell lines along the spectrum of cancer progression from normal to highly malignant, whose functional TGF-β receptor levels are well-characterized, will be stimulated with identical amounts of growth factors and allowed to begin proliferating. Measured amounts of TGF-β will then be added to each growing population progressively until growth is arrested. The amount of TGF-β required to suppress growth completely will then be noted for each cell line. If the hypothesis is true, it is expected that higher amounts of TGF-β will be required to suppress growth completely for the more malignant cell-lines.

If this control system hypothesis is confirmed, the consequences for how TGF-β ligand and TGF-β receptors are used as therapeutic agents could be significant. Specifically, it will mean that the current approach of targeting TGF-β ligand therapeutically may have to be abandoned in favor of re-sensitizing the cells to the tumor suppressive effect of the TGF-β, similar to treatment for diabetes mediated by prolonged insulin-resistance.

Acknowledgement. This work was supported by the Department of Defense grant PC050554, and by the Institute for Multiscale Modeling of Biological Interactions (IMMBI) (funded by the Department of Energy). Additional work was supported by National Institutes of Health/National Cancer Institute P01 CA098912, the University of Delaware Research Foundation, and the National Institutes of Health INBRE P20RR016472.

References

1. Hanahan, D., Weinberg, R.A.: The hallmarks of cancer. Cell 100, 57–70 (2000)
2. Massague, J.: TGF-β in cancer. Cell 134, 215–230 (2008)
3. Pardali, K., Moustakas, A.: Actions of TGF-β as tumor suppressor and pro-metastatic factor in human cancer. BBA-Rev Cancer 1775, 21–62 (2007)
4. Massague, J., Gomis, R.R.: The logic of TGF-β signaling. Febs Lett. 580, 2811–2820 (2006)
5. Chung, S.-W., Miles, F.L., Sikes, R.A., Cooper, C.R., Farach-Carson, M.C., Ogunnaike, B.A.: Quantitative modeling and analysis of the transforming growth factor-β signaling pathway. Biophys J. 96, 1733–1750 (2009)
6. Massague, J.: G1 cell-cycle control and cancer. Nature 432, 298–306 (2004)
7. Li, M.O., Wan, Y.Y., Sanjabi, S., Robertson, A.K.L., Flavell, R.A.: Transforming growth factor-β regulation of immune responses. Annu. Rev. Immunol. 24, 99–146 (2006)
8. ten Dijke, P., Arthur, H.M.: Extracellular control of TGF-β signalling in vascular development and disease. Nat. Rev. Mol. Cell Bio. 8, 857–869 (2007)
9. Guyton, A.C., Hall, J.E.: Textbook of medical physiology, 11th edn. Elsevier Saunders, Philadelphia (2006)
10. Jenkins, G.: The role of proteases in transforming growth factor-β activation. Int. J. Biochem. Cell Biol. 40, 1068–1078 (2008)
11. Wipff, P., Hinz, B.: Integrins and the activation of latent transforming growth factor-β1 - An intimate relationship. Eur. J. Cell Biol. 87, 601–615 (2008)
12. Deenick, E.K., Gett, A.V., Hodgkin, P.D.: Stochastic model of T cell proliferation: a calculus revealing IL-2 regulation of precursor frequencies, cell cycle time, and survival. J. Immunol. 170, 4963–4972 (2003)
13. Yoo, J., Ghiassi, M., Jirmanova, L., Balliet, A.G., Hoffman, B., Fornace, A.J., Liebermann, D.A., Bottinger, E.P., Roberts, A.B.: Transforming growth factor-β-induced apoptosis is mediated by Smad-dependent expression of GADD45b through p38 activation. J. Biol. Chem. 278, 43001–43007 (2003)

14. Hermonat, P.L., Li, D., Yang, B., Mehta, J.L.: Mechanism of action and delivery possibilities for TGF-β1 in the treatment of myocardial ischemia. Cardiovasc Res. 74, 235–243 (2007)
15. Aimes, R.T., Quigley, J.P.: Matrix metalloproteinase-2 is an interstitial collagenase. Inhibitor-free enzyme catalyzes the cleavage of collagen fibrils and soluble native type I collagen generating the specific 3/4 and 1/4 length fragments. J. Biol. Chem. 270, 5872–5876 (1995)
16. Levy, L., Hill, C.S.: Alterations in components of the TGF-β superfamily signaling pathways in human cancer. Cytokine Growth F R 17, 41–58 (2006)
17. Gerdes, M.J., Larsen, M., McBride, L., Dang, T.D., Lu, B., Rowley, D.R.: Localization of transforming growth factor-β1 and type II receptor in developing normal human prostate and carcinoma tissues. J. Histochem. Cytochem. 46, 379–388 (1998)
18. Claus, S., Wrenger, M., Senge, T., Schulze, H.: Immunohistochemical determination of age-related proliferation rates in normal and benign hyperplastic human prostates. Urol Res. 21, 305–308 (1993)
19. Shariat, S., Menesses-Diaz, A., Kim, I., Muramoto, M., Wheeler, T., Slawin, K.: Tissue expression of transforming growth factor-β1 and its receptors: Correlation with pathologic features and biochemical progression in patients undergoing radical prostatectomy. Urology 63, 1191–1197 (2004)

A Mathematical Model of Air-Flow Induced Regional Over-Distention during Mechanical Ventilation: Comparing Pressure-Controlled and Volume-Controlled Modes

P.S. Crooke*, A.M. Kaynar, and J.R. Hotchkiss

Abstract. In this paper we study a five compartment lung model to examine the effects of heterogeneity (*i.e.*, different portions of the lungs have different impedance characteristics) on physiologic outcomes using two common modes of mechanical ventilation: pressure-controlled (PCV) and volume-controlled (VCV). In particular, we attempt to answer the question: If heterogeneity exists in the lungs, then does one mode produce lower peak alveolar pressures, given a desired overall tidal volume? A third type of mechanical ventilation, decelerating flow ventilation (DFV), is also considered and it is shown that an optimal initial flow (a multiple of the desired minute ventilation) exists that will minimize peak compartmental pressures.

Keywords: model, mechanical ventilation, mutli-compartment, pressure-controlled ventilation, volume-controlled ventilation, decelerating flow ventilation.

AMS Classification: 92C30, 92C50.

1 Introduction

Mechanical ventilation can–by itself–damage the lungs. This contention is supported by many experiments in animals (see [1]). Elegant experimental work indicates that airspace overdistention (*i.e.*, volutrauma), rather that

P.S. Crooke
Department of Mathematics, Vanderbilt University, Nashville, TN 37240
e-mail: `philip.s.crooke@vanderbilt.edu`

A.M. Kaynar · J.R. Hotchkiss
Department of Critical Care Medicine University of Pittsburgh, Pittsburgh, PA 15261
e-mail: `kaynarm@upmc.edu`, `hotchkissjr@upmc.edu`

* Corresponding author.

J. Lévine & P. Müllhaupt (Eds.): Adv. in Theory of Control, Signals, LNCIS 407, pp. 269–282.
springerlink.com © Springer-Verlag Berlin Heidelberg 2010

elevated distending pressures (*i.e.*, barotrauma) *per se*, causes the injury [2]. Clinical studies are also consistent with the hypothesis that airspace overdistention during mechanical ventilation is injurious ([3, 4]). Restricting the tidal volume (V_T) during mechanical ventilation decreases mortality in Acute Respiratory Distress Syndrome (ARDS), as well as the risk of acute lung injury in other settings ([5]). Animal studies [6] have demonstrated the ventilatory-induced lung injury (VILI) is regionally heterogeneous and correlates with cyclical airway collapse and recruitment.

Disease/injured lungs are mechanically heterogeneous, both in the setting of ARDS and in the context of Chronic Obstructive Pulmonary Disease (COPD). Interaction between the applied pattern of ventilation and mechanical heterogeneity would be expected to cause differences in the peak strains (*i.e.*, peak airspace volumes) among different regions of the lungs. In the clinical setting, we can only control global strain during volume controlled ventilation (VCV) by changing tidal volume and/or positive end expiratory pressure (PEEP). During pressure controlled ventilation (PCV), we can only control global peak stress via modification of set inspiratory pressures and PEEP. Moreover, the outcome of our interventions can only be measured as global strain during VCV or global stress during PCV. Apart from global measures of stress and strain during mechanical ventilation, we cannot infer regional stresses or strains in the clinical setting using currently available methods. This may lead to unrecognized overdistention and subsequent injury in mechanically ventilated patients.

During VCV, one of the commonly referred guides is to restrict peak pressures along with tidal volumes, thus lessening the risk of lung injury ([7, 8]). However, analysis of neither experimental nor clinical data could not identify a *break point* in peak pressures below which restricting the tidal volume no longer decreased adverse consequences([9, 10]).

In this *in silco* study, we used a mathematical model to evaluate the hypothesis that VCV increases the incidence of regional overdistention as compared to PCV. This topic has recently been investigated from a clinical perspective ([11–15]). The model allows one to explore the magnitude of potential heterogeneity in peak regional strains during volume controlled as compared to pressure controlled ventilation. We also investigated the sensitivity with which common clinical measures would be expected to detect elevated regional strains. The *in silico* approach permits more detailed examination of a greater number of impedance configurations than would be practical in the experimental or laboratory settings.

2 The Mathematical Model

Mathematical models ([16, 17]) have been developed to predict clinical outcomes such as tidal volumes, mean alveolar pressures, end-expiratory pressures, given the physiologic parameters of the patient and the ventilator

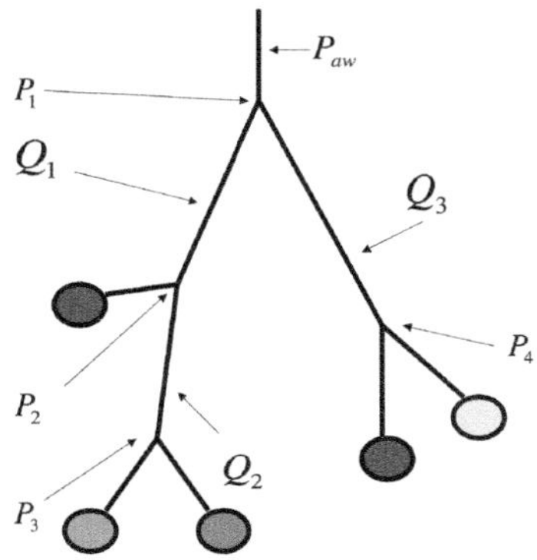

Fig. 1. Model lung configuration

settings. Some models ([19]) have incorporated two heterogeneous compartments (*i.e.*, two compartments with different compliances and resistances). To approximate the notion of heterogeneity of the lungs, our model is composed of five compartments as depicted in Figure 1. Each compartment has separate impedance characteristics. We studied the compartmental volumes over one cycle of mechanically controlled breathing which is composed of two segments: inspiration and expiration. The time of one cycle is denoted by t_{tot}. Inspiration occurs during the time-interval, $0 \leq t \leq t_i$ and expiration during $t_i \leq t \leq t_{tot}$. For PCV, we control the airway pressure (P_{aw}) and assume that during inspiration, $P_{aw} \equiv P_{set}$, $0 \leq t \leq t_i$. For VCV, the airway flow (Q_{aw}) is controlled during inspiration. The assumptions for expiration are identical in either mode of ventilation, namely, $P_{aw} \equiv P_{PEEP}$, $t_i \leq t \leq t_{tot}$.

The mathematical model for the five compartments as shown in Figure 1 is constructed using pressure and flow balances at junction points and along segments that are depicted in Figure 2. To illustrate the construction of the model, we consider the balances in the pressure controlled case when $P_{aw}(t) \equiv P_{set}$. In the compartments (A,B,C,D,E), the compliance of each compartment is denoted by C_j, the resistances by R_j, and end-expiratory pressures by P_{ex_j}. The pressure in each compartment is the sum of resistive (R_jQ_j), elastic (V_j/C_j) and residual (P_{ex_j}) where V_j is the instantaneous compartment volume and $Q_j = dV_j/dt$ is the flow in or out of the compartment. The resistances along the three connection segments are denoted by R_{LTR}, R_{LBR}, and R_{RMR}. Using this information, the following 13 equations involving the unknown compartment volumes (V_j) and compartment flows (dV_j/dt) are used to develop the system of differential equations for the compartmental volumes:

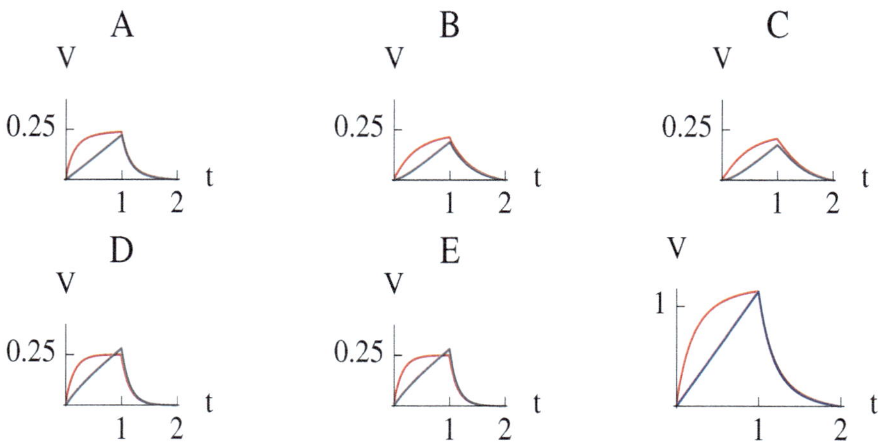

Fig. 2. Dynamic volume for each compartment for two modes of ventilation: pressure-controlled (PCV-red) and volume-controlled (VCV-blue). Here the compartmental compliances are 0.01 $L/cm\ H_2O$ and there is heterogeneity in the compartmental resistances: $R_A = R_B = 10$, $R_C = 15$, $R_D = R_E = 5\ cm\ H_2O/(L/sec)$ The overall volumes for both modes of ventilation are shown in the last panel

$$
\begin{aligned}
P_1 &= P_{set} \\
Q_{aw} &= Q_1 + Q_3 \\
P_2 &= P_1 - R_{LTR} Q_1 \\
Q_1 &= Q_A + Q_2 \\
P_2 &= R_A Q_A + (1/C_A) V_A + P_{ex_A} \\
Q_2 &= Q_B + Q_C \\
P_3 &= P_2 - R_{LBR} Q_2 \\
P_3 &= R_B Q_B + (1/C_B) V_B + P_{ex_B} \\
P_3 &= R_C Q_C + (1/C_C) V_C + P_{ex_C} \\
Q_3 &= Q_D + Q_E \\
P_4 &= P_1 - R_{RMR} Q_3 \\
P_4 &= R_D Q_D + (1/C_D) V_D + P_{ex_D} \\
P_4 &= R_E Q_E + (1/C_E) V_E + P_{ex_E}
\end{aligned} \tag{1}
$$

From the system of equations in (1) which governs the dynamics during inspiration, we can solve for the compartmental flows to obtain a system of differential equations:

$$
\begin{aligned}
dV_A^{(i)}/dt &= \alpha_A^{(i)} V_A^{(i)} + \alpha_B^{(i)} V_B^{(i)} + \alpha_C^{(i)} V_C^{(i)} + \alpha_D^{(i)} V_D^{(i)} + \alpha_E^{(i)} V_E^{(i)} + \tilde{P}_A^{(i)} \\
dV_B^{(i)}/dt &= \beta_A^{(i)} V_A^{(i)} + \beta_B^{(i)} V_B^{(i)} + \beta_C^{(i)} V_C^{(i)} + \beta_D^{(i)} V_D^{(i)} + \beta_E^{(i)} V_E^{(i)} + \tilde{P}_B^{(i)} \\
dV_C^{(i)}/dt &= \gamma_A^{(i)} V_A^{(i)} + \gamma_B^{(i)} V_B^{(i)} + \gamma_C^{(i)} V_C^{(i)} + \gamma_D^{(i)} V_D^{(i)} + \gamma_E^{(i)} V_E^{(i)} + \tilde{P}_C^{(i)} \\
dV_D^{(i)}/dt &= \delta_A^{(i)} V_A^{(i)} + \delta_B^{(i)} V_B^{(i)} + \delta_C^{(i)} V_C^{(i)} + \delta_D^{(i)} V_D^{(i)} + \delta_E^{(i)} V_E^{(i)} + \tilde{P}_D^{(i)} \\
dV_E^{(i)}/dt &= \epsilon_A^{(i)} V_A^{(i)} + \epsilon_B^{(i)} V_B^{(i)} + \epsilon_C^{(i)} V_C^{(i)} + \epsilon_D^{(i)} V_D^{(i)} + \epsilon_E^{(i)} V_E^{(i)} + \tilde{P}_E^{(i)}
\end{aligned} \tag{2}
$$

where $\tilde{P}_j^{(i)}$ depends on the end expiratory pressure (P_{ex_j}) and the airway pressure (P_{set}). The coefficients ($\alpha_j^{(i)}$, $\beta_j^{(i)}$, $\gamma_j^{(i)}$, $\delta_j^{(i)}$, and $\epsilon_j^{(i)}$) of the compartmental volume terms depend on the physiologic parameters (compliance and resistance of each compartment). The superscript on these constants, (i), denote the values of these constants during *inspiration*. At the start of inspiration we assume the compartmental volume is measured from its residual volume (P_{ex_j}/C_j) and hence, we assume $V_j^{(i)}(0) = 0$, $j = A, B, C, D, E$, which provide the initial conditions for the system of differential equations. We note that at this point in the model development, the end expiratory pressures are unknown. They will be determined after the expiratory part of the model is solved. We denote the solutions of (2) as $V_j^{(i)}(t)$, $0 \leq t \leq t_i$, $j = A, B, C, D, E$.

Using the same pressure and flow balance equations as in (1), except that we have $P_{aw} = P_{peep}$, $t_i \leq t \leq t_{tot}$, the corresponding system of differential equations can be obtained for the volumes in each compartment during expiration. In particular, we find:

$$\begin{aligned}
dV_A^{(e)}/dt &= \alpha_A^{(e)}V_A^{(e)} + \alpha_B^{(e)}V_B^{(e)} + \alpha_C^{(e)}V_C^{(e} + \alpha_D^{(e)}V_D^{(e)} + \alpha_E^{(e)}V_E^{(e)} + \tilde{P}_A^{(e)} \\
dV_B^{(e)}/dt &= \beta_A^{(e)}V_A^{(e)} + \beta_B^{(e)}V_B^{(e)} + \beta_C^{(e)}V_C^{(e)} + \beta_D^{(e)}V_D^{(e)} + \beta_E^{(e)}V_E^{(e)} + \tilde{P}_B^{(e)} \\
dV_C^{(e)}/dt &= \gamma_A^{(e)}V_A^{(e)} + \gamma_B^{(e)}V_B^{(e)} + \gamma_C^{(e)}V_C^{(e)} + \gamma_D^{(e)}V_D^{(e)} + \gamma_E^{(e)}V_E^{(e)} + \tilde{P}_C^{(e)} \\
dV_D^{(e)}/dt &= \delta_A^{(e)}V_A^{(e)} + \delta_B^{(e)}V_B^{(e)} + \delta_C^{(e)}V_C^{(e)} + \delta_D^{(e)}V_D^{(e)} + \delta_E^{(e)}V_E^{(e)} + \tilde{P}_D^{(e)} \\
dV_E^{(e)}/dt &= \epsilon_A^{(e)}V_A^{(e)} + \epsilon_B^{(e)}V_B^{(e)} + \epsilon_C^{(e)}V_C^{(e)} + \epsilon_D^{(e)}V_D^{(e)} + \epsilon_E^{(e)}V_e^{(e)} + \tilde{P}_E^{(e)}
\end{aligned} \tag{3}$$

where $\tilde{P}_j^{(e)}$ depends on P_{ex_j} and the applied PEEP (P_{PEEP}). The initial conditions for (3) are $V_j^{(e)}(t_i) = V_j^{(i)}(t_i)$ and the resulting solutions of this initial-value problem are denoted as $V_j^{(e)}(t)$, $t_i \leq t \leq t_{tot}$.

Having the solutions of (2) and (3), the end-expiratory pressures for each compartment can be determined by requiring that $V_j^{(e)}(t_{tot}) = 0$ for $j = A, B, C, D, E$. This is accomplished by solving a linear system of algebraic equations for $P_{ex_A}, P_{ex_B}, P_{ex_C}, P_{ex_D}, P_{ex_E}$. Once this is done, we have the solutions for the model equations. In particular,

$$V_j(t) = \begin{cases} V_j^{(i)}(t) & \text{if } 0 \leq t \leq t_i \\ V_j^{(e)}(t) & \text{if } t_i \leq t \leq t_{tot} \end{cases} \tag{4}$$

for $j = A, B, C, D, E$.

A similar exercise can be performed in the case of volume-controlled ventilation. In this case the flow at the airway is controlled by the ventilator *i.e.*, $Q_{aw} = Q_{set}$ where Q_{set} is a fixed flow that is related to the desired tidal volume, V_T. Since there is little difference in the derivation, we do not include the details here.

3 Model Simulations

In this section we perform simulations of the model developed in the previous section. In particular, we investigate the effect of compartmental resistances (R_A, R_B, R_C, R_D, R_E), compartmental compliances (C_A, C_B, C_C, C_D, C_E), and the resistances of connecting airways (R_{LTR}, R_{LBR}, R_{RMR}) on the end-expiratory and peak compartmental pressures using the two modes of ventilation (PCV and VCV) with both types of ventilation delivering the *same* tidal volume over one breath. We attempt to show that for certain combinations of parameters in each class (compartmental resistances, compartmental compliances, and connecting airway resistances), volume-controlled ventilation (with $Q_{aw}(t) = Q_{set}$) produces larger peak compartmental pressures and larger variations between highest and lowest peak compartmental pressures.

The mathematical model developed in the previous section allows one to compute tidal volumes ($V_j^{(i)}(t_i) = V_{T_j}$), end-expiratory pressures (P_{ex_j}) and peak-pressures ($P_{pk_j} = V_j^{(i)}/C_j + P_{ex_j}$) for each compartment and the overall system for each mode of ventilation. In the first set of simulations, we have chosen a setup where we vary **compartmental resistances**, but we keep the compartment compliances fixed along with the resistances of the connecting airway. Each compartment has the same compliance ($C_A = C_B = C_C = C_D = C_E = 0.01\ L/cm\ H_2O$) so that the overall compliance is $C = 0.05\ L/cm\ H_2O$, and the resistances of the individual connecting airways are set as $R_{LTR} = 4\ cm\ H_2O/L/sec$, $R_{LBR} = 9\ cm\ H_2O/(L/sec)$ and $R_{RMR} = 5\ cm\ H_2O/(L/sec)$. The length of a breath was set at 2 sec with a duty cycle of 0.5 so that $t_i = 1\ sec$. In the pressure-controlled mode, $P_{set} = 30\ cm\ H_2O$ with a peep of 5 $cm\ H_2O$. In the volume-controlled mode, the flow was chosen to produce the same tidal volume: $V_T = 1.15314\ L$. Simulations were performed using fixed compliances and connecting airway resistances while letting the compartmental resistances vary from $5 - 15\ cm\ H_2O/L$ in increments of 5. This calculation gives 243 combinations for which we have compartmental volumes and pressures. In Figures 2-3, we illustrate one of the simulations. The resistances for the individual compartments were assigned the values: $R_A = 10$, $R_B = 10$, $R_C = 15$, $R_D = 5$ and $R_E = 5$ (all with the units $cm\ H_2O/(L/sec)$). Figure 2 shows the dynamic volumes for inspiration and expiration over one breath for the individual compartments. It is interesting to note that the volume profiles for each compartment can be quite different in appearance than the overall volume profile. In Figure 3, we compare peak pressures and end-expiratory pressures for each compartment. For PCV ($P_{aw} \equiv 30\ cm\ H_2O$), the maximum compartmental peak pressure was 29.968 and the minimum compartmental peak pressure was 27.623. For VCV with the same tidal volume, the maximum compartmental peak pressure was 34.616 and the minimum 27.432. Hence, for this relatively small range of compartmental resistances, there is a

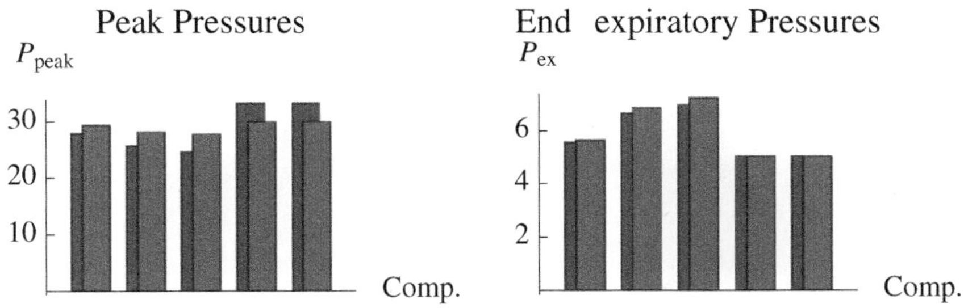

Fig. 3. Peak pressures and end-expiratory pressure for each compartment with each mode of ventilation (red-PCV, blue-VCV) for Figure 2

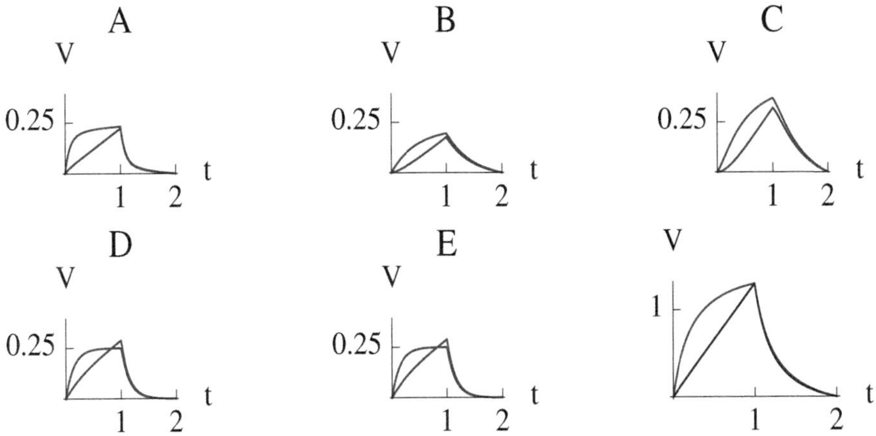

Fig. 4. Dynamic volume for each compartment for two modes of ventilation: pressure-controlled (CP-red) and volume-controlled (CF-blue). Here the compartmental resistances are 5 $cm\ H_2O$ and there is heterogeneity in the compartmental compliances: $C_A = C_B = C_D = C_E = 0.01$, $C_C = 0.02\ L/cm\ H_2O$.

difference of 4.648 $cm\ H_2O$ between the two modes of ventilation. For a clinician, having a peak pressures over 30 $cm\ H_2O$ in one of the compartments is significant.

A second set of simulations was performed where the compartmental resistances were fixed at $5/cm/H_2O/(L/sec)$ while letting the **compartmental compliances** vary between $0.01 - 0.03\ L/cm\ H_2O$. Simulations for one of these combinations are shown in Figures 4-5. In this case, the compartmental compliances were set at $C_A = C_B = C_C = C_D = C_E = 0.01\ L/cm\ H_2O$ and $C_C = 0.02\ L/cm\ H_2O$. The peak compartmental pressure for PCV was 29.97 $cm\ H_2O$ and for VCV it was 33.94 $cm\ H_2O$. As one can see from Figure 5, the end-expiratory pressures for both modes of ventilation were

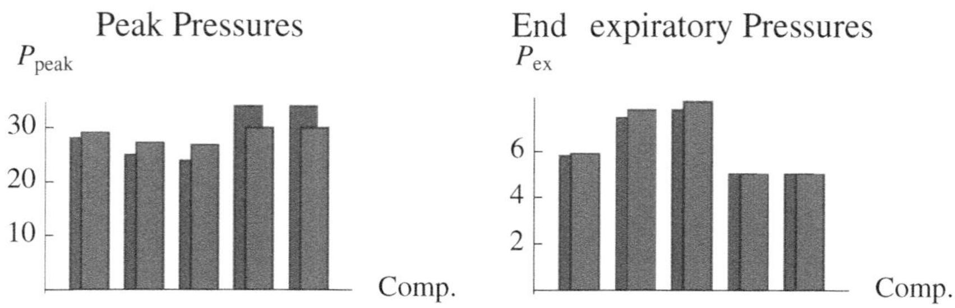

Fig. 5. Peak pressures and end-expiratory pressure for each compartment with each mode of ventilation for Figure 4

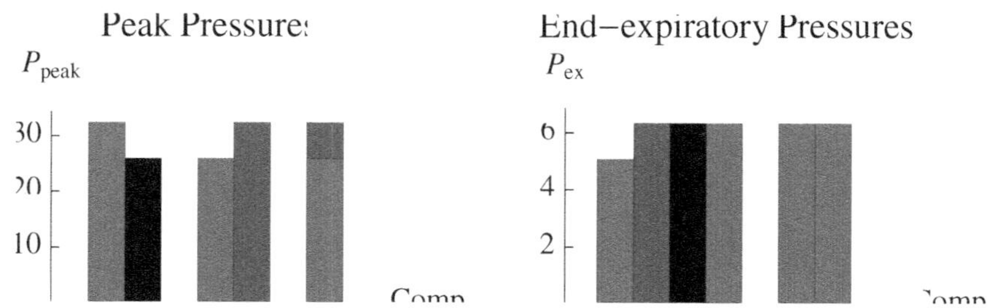

Fig. 6. Peak pressures and end-expiratory pressure for each compartment with each mode of ventilation for varying connecting airways resistances

very close and hence, the difference in peak pressures were due to dynamic effects.

A third set of simulations had the compartmental compliances set at 0.02 $L/cm\ H_2O$ and the compartmental resistances set at $5/cm/H_2O$. The **resistances for the connecting airways**, R_{LTR}, R_{LBR} and R_{RMR}, were then varied between $5-15$ $cm\ H_2O/(L/sec)$. For $R_{LTR}=5$, $R_{LBR}=10$, and $R_{RMR}=15$, the ratio between the maximum and minimum peak pressures was 1.04 for PCV and 1.45 for VCV. The peak pressures and end-expiratory pressures are shown in Figure 6.

In the simulations above, the PEEP was set at 5 $cm\ H_2O$. When no PEEP was used in the simulations, the differences between the smallest and largest compartmental peak pressures in VCV increased. For example in the third set of simulations, the ratio of the largest-to-smallest ratio is 1.45 in the case when $P_{PEEP}=5$ and 1.58 when $P_{PEEP}=0$. Hence, it appears that PEEP tends to mollify the heterogeneity effects. Comparisons in the P_{peak} and P_{ex} for the case of PEEP and no-PEEP are shown in Figure 7. When $P_{PEEP}=$ 10 $cm\ H_2O$, the ratio between the largest and smallest peak compartmental pressure was 1.34 for VCV. In Table 1, we summarize the effect of PEEP on the the maximum and minimum compartmental pressures and their ratio.

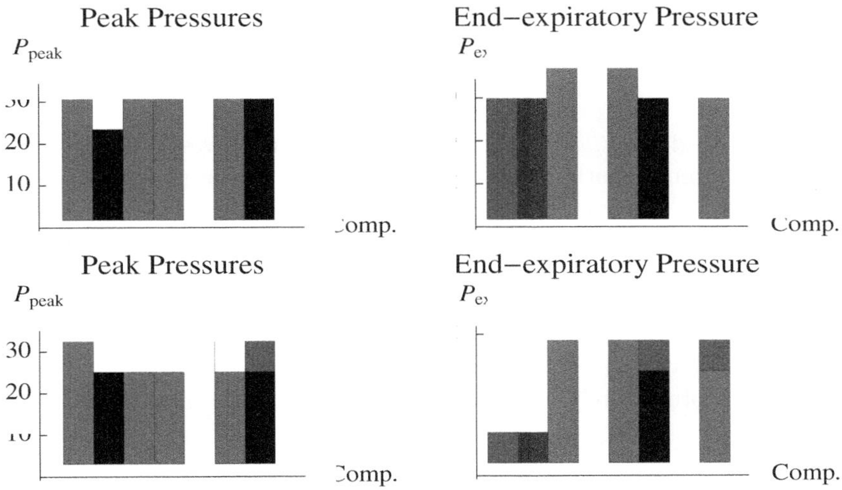

Fig. 7. Comparison of peak compartmental and end-expiratory pressures with (top) and without (bottom) PEEP

Table 1. The effect of PEEP on max. and min. peak compartmental pressures in VCV case

PEEP	Max-Min Ratio	Max P_{peak}	Min P_{peak}
0	1.58	34.36	21.25
2.5	1.52	34.00	22.43
5.0	1.45	33.64	23.12
7.5	1.39	33.27	23.81
10.0	1.34	32.91	24.50
12.5	1.29	32.55	25.18
15.0	1.24	32.18	25.87

Besides VCV and PCV, decelerating flow ventilation is often employed in the ICU. In this mode of ventilation, a controlled flow is given at the airway opening during inspiration. In particular, if $Q_{aw}(t)$ denotes the airway opening flow, then $Q_{aw}(t) = \alpha - \beta t$ where α and β are positive constants. We use our mathematical model to compare decelerating flow ventilation (DFV) against VCV (with $Q_{aw}(t) = Q_{set}$) in the five compartment system. Keeping the same inspiratory time and requiring $Q_{aw}(t_i) = 0$, we choose $\alpha = V_T/t_i$ and $\beta = V_T/t_i^2$. Using different combinations of compartmental compliances, compartmental resistances, and connecting airway resistances we studied the peak compartmental pressure and the ratio between the highest and lowest peak compartmental pressures. From these studies (by varying the impedance parameters of the compartments over clinically relevant ranges), it appears that DFV counteracts the effects of variation in the impedance characteristics.

In DFV, the clinician has the choice of the initial flow. The choice of this initial flow is often a multiple of the desired minute ventilation which we term, *minute ventilation factor* ($\dot{V}_E$ *factor*). We investigated the effects of the the minute ventilation factor on variation of the peak compartmental pressures. To illustrate this effect, we set the compartmental resistances set at 5 $cm\ H_2O/(L/sec)$, $R_{LTR} = R_{LBR} = 5\ cm\ H_2O(/L/sec)$, and $R_{RMR} = 10\ cm\ H_2O/L/sec$ and varied the minute ventilation factor. These computations are summarized in Figure 8. It appears that the largest compartmental peak pressure is minimized when $\dot{V}_E\ factor = 4$. This is in agreement with the ventilator management standard that the initial flow in DFV should be approximately 5 times the desired minute ventilation.

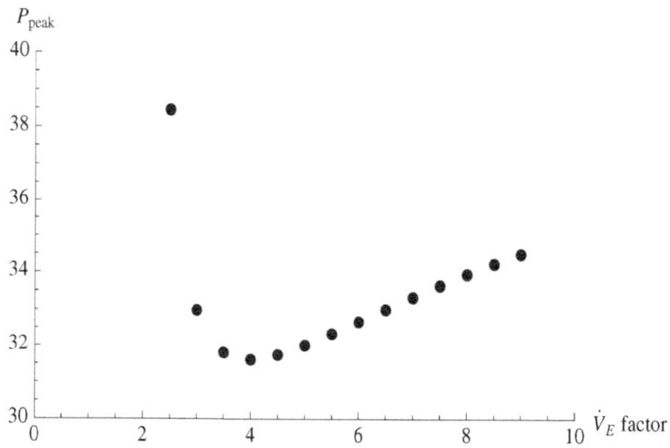

Fig. 8. Peak compartmental pressure as a function of minute ventilation factor

It is relatively easy to add compartments to the model and to change the topology of the network (*e.g.*, a three branch network). However, the present setup for the model illustrates that differences exists between the two basic modes of ventilation.

4 Discussion

Our results demonstrate that, at a fixed tidal volume and in the presence of compartmental heterogeneity, the following observations can be made from the model:

- There can be substantial heterogeneity in peak compartmental pressures on the order of several centimeters of water during both pressure control (PCV) and volume controlled ventilation (VCV);

- The peak compartmental pressure never exceeded the set airway opening pressure during pressure control ventilation (30 $cm\ H_2O$). In contrast, during VCV at the same tidal volume, peak compartmental pressures could be as high as 35 $cm\ H_2O$;
- The degree of heterogeneity in peak compartmental pressures was lower with PCV than with VCV;
- The addition of positive end expiratory pressure (PEEP) attenuates the heterogeneity of peak strain during VCV;
- There is sequential filling of compartments in volume VCV, as some compartments fill progressively more rapidly during inspiration, whereas during PCV flow into all compartments falls during inflation.

The heterogeneity of peak compartmental pressures, and presumably strains, would not be evident under current clinical monitoring approaches (primarily the measurement of peak pressure). Accordingly, clinical protocols or trials predicated on limiting tidal volume and/or monitoring peak pressures would not detect regional overdistention. Importantly, the magnitude of the differences is clinically relevant: in VCV, the peak compartmental pressures could be well above those targeted in the ARDSNet trial, despite a peak pressure that was of acceptable magnitude. These findings could, in part, explain the inability of Brower *et al.* [5] to identify a safe peak pressure (or a breakpoint in mortality at a specific level of peak pressure) during volume cycled ventilation. Because peak pressure is a global measure obtained after an end inspiratory pause, regional overdistention can be present even at a low peak pressure. Accordingly, particularly during VCV, ventilation at acceptable peak pressures could be accompanied by regional overdistention of a potentially injurious magnitude, triggering focal lung injury, increased regional permeability, and cytokine release. It is intriguing to speculate that such processes could promote more diffuse lung injury either via release of proinflammatory mediators, or by reducing the compliance of the most severely overstressed lung region and consequently increasing the stress on other, more remote, regions. Lung injury could thus progress considerably before being reflected in a clinically detectable change in peak pressure or compliance.

In contrast to the situation during VCV, the peak distending pressures during PCV never exceed the pressure at the airway opening. Accordingly, if P_{aw} is limited to less than 30 $cm\ H_2O$, no compartment will experience a peak stress greater than this value. As PCV has been shown to produce the same tidal volumes at lower set airway pressures than VCV ([13]), this suggests that it may have less potential for regional overdistention. Moreover, if impedance characteristics are changing (for example, due to mucous plugging, progressive edema formation, bronchospasm, or pleural processes), PCV ensures that no compartment will be even transiently exposed to unacceptable distending pressures, as long as adequate inspiratory times are ensured. These benefits must be weighed against the potential for greater shearing forces due to more rapid compartmental filling.

In this model, applied PEEP reduced heterogeneity of peak compartmental pressures during VCV, most likely as a result of modulating expiratory dynamics. The effect was less marked in PCV, where compartmental pressure gradients play a larger role in setting total initial inspiratory flow. This suggests that the application of PEEP, even in a linear system, might have salutary effects on the distribution of peak stresses and the accompanying strains. Notably, studies demonstrating no benefit of PEEP in ARDS have titrated this parameter based on oxygenation and FIO_2; our results suggest that there might be a mechanics-related benefit even in the absence of recruitment/derecruitment. Our model did not incorporate impedance characteristics (*e.g.*, C_j is a function of V_j) that change during the respiratory cycle (such as recruitment/derecruitment of lung regions). It seems likely that the lack of flow limitation during PCV (promoting more rapid inspiratory filling) would aid in recruitment; however, this is at present an unproven hypothesis. Moreover, although Chellboina *et al.* [18] have elegantly demonstrated that linear multicompartment models are dynamically stable, such stability might not obtain in multicompartment models having nonlinear compartmental impedance characteristics. The extent to which the benefits of PCV could be obtained by applying a linearly decelerating flow pattern is also not certain. Our preliminary work in this area suggests that decelerating flow VCV attenuates but does not eliminate compartmental heterogeneity, and PCV still results in less regional overdistention. Nonetheless, the advantages of PCV in the setting of unstable impedance characteristics remain.

5 Conclusions

Our results indicate that there can be substantial heterogeneity of peak compartmental pressures during VCV, and that this heterogeneity can be of a clinically relevant magnitude. PCV is accompanied by less regional heterogeneity and lower maximal compartmental distending pressures. Even with modest airway flows, there can be substantial regional overpressure during VCV. More detailed analyses, encompassing biological validation, techniques for detecting overdistention from airway opening flow/pressure tracings ([20–22]), and other flow patterns are warranted.

References

[1] Belperio, J.A., et al.: Critical role for CXCR2 and CXCR2 ligands during the pathogenesis of ventilator-induced lung injury. J. Clin. Invest. 110(11), 1703–1716 (2002)

[2] Hernandez, L.A., Peevy, K.J., et al.: Chest wall restriction limits high airway pressure-induced lung injury in young rabbits. J. Appl. Physiol. 66(5), 2364–2368 (1989)

[3] Amato, M.B.P., Barbas, C.S.V., et al.: Effect of a protective-ventilation strategy on mortality in the Acute Respiratory Distress Syndrome. N. E. J. Med. 338, 347–354 (1998)
[4] Brower, R.G., Lanken, R.N., et al.: Higher ver sus lower positive end-expiratory pressures in patients with the Acute Respiratory Distress Syndrome. N. E. J. Med. 351, 327–336 (2004)
[5] Brower, R.G.: Mechanical ventilation in acute lung injury and ARDS. Crit. Care Clinics 18, 1–13 (2002)
[6] Sinclair, S.E., Chi, E., Line, H., Altemeier, W.A.: Positive end-expiratory pressure alters the severity and spatial heterogeneity or ventilator-induced lung injury: An argument for cyclical airway collapse. J. Crit. Care (in press)
[7] Moran, J.L., Bersten, A.D., Solomon, P.J.: Meta-analysis of controlled trials of ventilator therapy in acute lung injury and acute respiratory distress syndrome: an alternative perspective. Intensive Care Med. 31, 227–235 (2005)
[8] Keszler, M.: Volume-targeted ventilation. Early Hum. Dev. 82, 811–818 (2006)
[9] Hager, D.N., Krishnan, J.A., et al.: ARDS Clinical Trials Network. Tidal volume reduction in patients with acute lung injury when plateau pressures are not high. Am. J. Respir. Crit. Care Med. 172(10), 1241–1245 (2005)
[10] Jardin, R., Vieillard-Baron, A.: Is there a safe plateu pressure in ARDS? The right heart only knows. Intensive Care Med. 33, 444–447 (2007)
[11] Boussarsar, M., Thierry, G., et al.: Relationship between ventilatory settings and barotraumas in the acute respiratory distress syndrome. Inten. Care Med. 28, 406–413 (2002)
[12] Hinz, J., Moerer, O., et al.: Regional pulmonary pressure volume curves in mechanically ventilated patients with acute respiratory failure measured by electrical impedance tomography. Acta Anaesthesial. Scand. 50, 331–339 (2006)
[13] Unzueta, M.C., Casas, J.I., Moral, M.V.: Pressure-controlled versus volume-controlled ventilation during one-lung ventilation for thoracic surgery. Anesth. Analg. 104, 1029–1033 (2007)
[14] Hinz, J., Gehoff, A., et al.: Regional filling characteristics of the lungs in mechanically ventilated patients with acute lung injury. European J. Anaesth. 24, 414–424 (2007)
[15] Satoru, I., Lutchen, K.R., Suki, B.: Effects of heterogeneities on the partitioning of airway and tissue properties in normal mice. J. Appl. Physiol. 102, 859–869 (2007)
[16] Marini, J.J., Crooke, P.S., Truwit, J.D.: Determinants and limits of pressure-preset ventilation: a mathematical model of pressure control. J. Appl. Physiol. 67, 1081–1092 (1989)
[17] Crooke, P.S., Kongkul, K., et al.: Mathematical models for pressure controlled ventilation of oleic acid-injured pigs. Math. Med. Biol. 22, 99–112 (2005)
[18] Chellaboina, V., Haddadit, W.M., et al.: Limit cycle stability analysis of a multi-compartment model for a pressure-limited respiratory and lung mechanics system. In: Proceedings of the 2007 American Control Conference, New York City, July 11-13, pp. 2024–2029 (2007)
[19] Crooke, P.S., Head, J.D., Marini, J.J.: A general two-compartment model for mechanical ventilation. Math. Comp. Mod. 24, 1–18 (1996)

[20] Ranieri, V.M., et al.: Pressure-time curve predicts minimally injurious ventilatory strategy in an isolated rat lung model. Anesthesia 93 (2000)
[21] Crooke, P.S., Marini, J.J., Hotchkiss, J.R.: A new look at the stress index for lung injury. J. Biol. Systems 13, 261–272 (2005)
[22] Wolf, G.K., Grychtol, B., et al.: Regional lung volume changes in children with acute respiratory distress syndrome during a derecruitment maneuver. Pediatric Grit. Care 35, 1972–1978 (2007)

Positive Feedbacks Contribute to the Robustness of the Cell Cycle with Respect to Molecular Noise

Didier Gonze and Marc Hafner

Abstract. Most cellular oscillators rely on interlocked positive and negative regulatory feedback loops. While a negative circuit is necessary and sufficient to have limit-cycle oscillations, the role of positive feedbacks is not clear. Here we investigate the possible role of positive feedbacks in the robustness of the oscillations in presence of molecular noise. We performed stochastic simulations of a minimal 3-variable model of the cell cycle. We compare the robustness of the oscillations in the 3-variable model and in a modified model which incorporates a positive feedback loop through an auto-catalytic activation. We find that the model with a positive feedback loop is more robust to molecular noise than the model without the positive feedback loop. This increase of robustness is parameter-independent and can be explained by the attractivity properties of the limit-cycle.

1 Introduction

Biological rhythms occur at various levels of the physiological organisation [16]. They are often generated at the cellular level through complex interactions among genes, proteins, and metabolites [32]. Most cellular

Didier Gonze
Laboratoire de Bioinformatique des Génomes et des Réseaux, CP 263, Faculté des Sciences, Université Libre de Bruxelles, Bvd du Triomphe, B-1050 Bruxelles, Belgium

Marc Hafner
Laboratory of Nonlinear Systems, School of Computer and Communication Sciences, Ecole Polytechnique Fédérale de Lausanne (EPFL), 1015 Lausanne, Switzerland
and
Wagner Lab, Department of Biochemistry, University of Zurich, 8057 Zurich, Switzerland
and
Swiss Institute of Bioinformatics, 1015 Lausanne, Switzerland

J. Lévine & P. Müllhaupt (Eds.): Adv. in Theory of Control, Signals, LNCIS 407, pp. 283–295.
springerlink.com © Springer-Verlag Berlin Heidelberg 2010

oscillators rely on interlocked positive and negative regulatory feedback loops [17, 18, 41]. Examples include the so-called Calcium-Induced Calcium Release mechanism responsible for the periodic calcium spiking [31], the p53/Mdm2 oscillator which induces oscillations of p53 in response to stress [6], the Delta/Notch oscillator involved in somitogenesis [34], the CDK/cyclin network controlling the cell cycle [7, 33], and the circadian clock controlling the daily rhythms of the organism [3, 5, 12, 37].

Mathematical modeling of biological oscillators has shown that a single delayed negative feedback loop is sufficient to generate self-sustained oscillations [13, 15, 22, 32]. An experimental demonstration of this prediction was recently brought by synthetic biology [8]: a minimal synthetic oscillator involving genetically engineered gene-promoter constructions was implemented in a bacterium and, in agreement with a theoretical model, exhibits oscillations in gene expression. Thus we may inquire into the role and advantage of additional positive feedback loops observed in most natural cellular oscillators.

Using several prototypical models, Tsai *et al* (2008) [41] performed a series of simulations showing that positive feedbacks lead to a greater tunability of the frequency and to an increase of the domain of conditions (region of the parameter space) which lead to limit-cycle oscillations. Hasty *et al* [25] proposed a theoretical model based on interlocked positive and negative feedback loops and showed that such a design, when coupled to another genetic oscillator, is capable of entrainment and of amplified oscillations. Recently, guided by the predictions of computational models, Stricker *et al* (2008) [40] designed and constructed an artificial oscillator based on interlinked positive and negative feedback loops. This study confirmed that the positive feedback loop enhances the tunability of the system's frequency and increases the robustness of the oscillations over a larger number of conditions.

In the present work, our aim is to check if the positive feedback loop may also lead to a higher robustness of the oscillations with respect to molecular noise. Several works already showed that oscillators based on positive and negative regulatory elements make oscillations more resistant to fluctuations [1, 42], but no comparative study showed how the addition of a positive feedback to an oscillator affects its robustness. We consider here two minimal models for the cell cycle. The first one is only based on a negative feedback. The second one has the same architecture, but incorporates an additional positive feedback. We performed stochastic simulations using the Gillespie algorithm [11] and we quantify the robustness of the oscillations using the auto-correlation function [20] and the distribution of the periods. We show that the positive feedback loop increases the robustness of the oscillations independently of the parameter values, and we provide a possible explanation for this observation.

2 Model

We consider here a minimal model proposed by Goldbeter (1991) for the frog embryonic cell cycle [13]. The model is schematized in Fig. 1 (left panel). The oscillator involves the activation of a cyclin-dependent kinase (CDK1) by Cyclin B, and the CDK1-induced degradation of Cyclin B by an ubiquitin ligase, which is part of the ubiquitin-mediated proteolysis system. Once activated, CDK1 triggers the entry into mitosis.

In an extension of the model, Goldbeter (1993) included an additional positive feedback loop, mediated by the CDC25 phosphatase [14]. In this work, the positive feedback was modeled with an additional variable, standing for the active fraction of CDC25. The latter is activated by CDK1 and, once active, CDC25 activates CDK1. Here we simplify this model by considering a direct feedback of CDK1 on itself (Fig. 1, right panel). This can be seen as an auto-catalytic process.

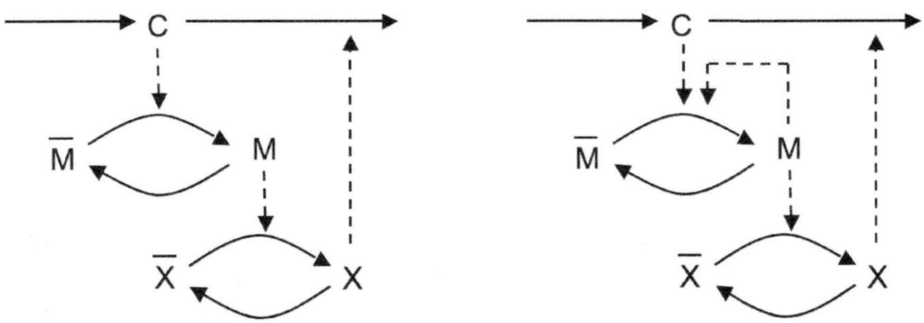

Fig. 1. Schemas of the two models. Left: 3-variable model [13]. Right: 3-variable model including a positive feedback loop (auto-catalysis) (adapted from [14]). Variables C, M, and X denote the Cyclin B, the active form of CDK1 kinase, and the active cyclin protease, respectively. The variables indicated with a bar refer to their inactive form. Solid arrows denote biochemical reactions, while dashed arrows indicate positive regulations.

The time evolution of the three variables is governed by the following system of kinetic equations (see refs. [13, 14] for a detailed description of the equations and the parameters):

$$\frac{dC}{dt} = v_i - v_d X \frac{C}{K_d + C} - k_d C \tag{1}$$

$$\frac{dM}{dt} = v_{m1}(a + bM) \frac{C}{K_c + C} \frac{M_{tot} - M}{K_1 + M_{tot} - M} - v_{m2} \frac{M}{K_2 + M} \tag{2}$$

$$\frac{dX}{dt} = v_{m3} M \frac{X_{tot} - X}{K_3 + X_{tot} - X} - v_{m4} \frac{X}{K_4 + X} \tag{3}$$

Table 1. Stochastic version of the cell cycle model. The variables indicated with a bar refer to their inactive form. Note also that M_{tot} and X_{tot} are here the total number of molecules of M and X. M_{tot} and X_{tot} are obtained by multiplying the concentration as given in the deterministic version by Ω.

No	Reaction	Propensity
1	$\rightarrow C$	$w_1 = v_i \Omega$
2	$C \rightarrow$	$w_2 = v_d X \frac{C}{K_d \Omega + C} + k_d C$
3	$\bar{M} \rightarrow M$	$w_3 = v_{m1}(a\Omega + bM) \frac{C}{K_c \Omega + C} \frac{M_{tot} - M}{K_1 \Omega + M_{tot} - M}$
4	$M \rightarrow \bar{M}$	$w_4 = v_2 \Omega \frac{M}{K_2 \Omega + M}$
5	$\bar{X} \rightarrow X$	$w_5 = v_{m3} M \frac{X_{tot} - X}{K_3 \Omega + X_{tot} - X}$
6	$X \rightarrow \bar{X}$	$w_6 = v_4 \Omega \frac{X}{K_4 \Omega + X}$

In these equations, the variables denote the concentration of Cyclin B (variable C), of active CDK1 kinase (M), and of active cyclin protease (X). Note that in the original version, M and X were the fraction of active CDK and protease but, in order to facilitate the conversion to the stochastic version of the model, we write here all the variables in terms of concentration and consider that the total amount of M and X are M_{tot} and X_{tot}. The term $a + bM$ has been introduced in this version. Parameter a, kept equal to 1 in all the following simulations, controls the negative feedback. The positive feedback is effective when $b > 0$. In the following we will compare the case where $b = 0$ (no positive feedback) and $b = 1$ (effective positive feedback). It is interesting to underly that, in this version of the model, the positive feedback can thus be added continuously through a progressive increase of one parameter (b).

To take into account the fluctuations arising from the limited number of molecules, we need to resort to stochastic simulations. We use here the Gillespie algorithm to simulate a stochastic version of the model as given in Table 1. This Table lists the six reaction steps that define the model as well as their corresponding propensities. These propensities are directly related to the kinetic rates and depend on the number of molecules present in the system, controlled by the system size Ω. Note that we use here directly the Michaelis-Menten functions to compute the propensities. An alternative would be to decompose these kinetics into a set of elementary reaction steps,

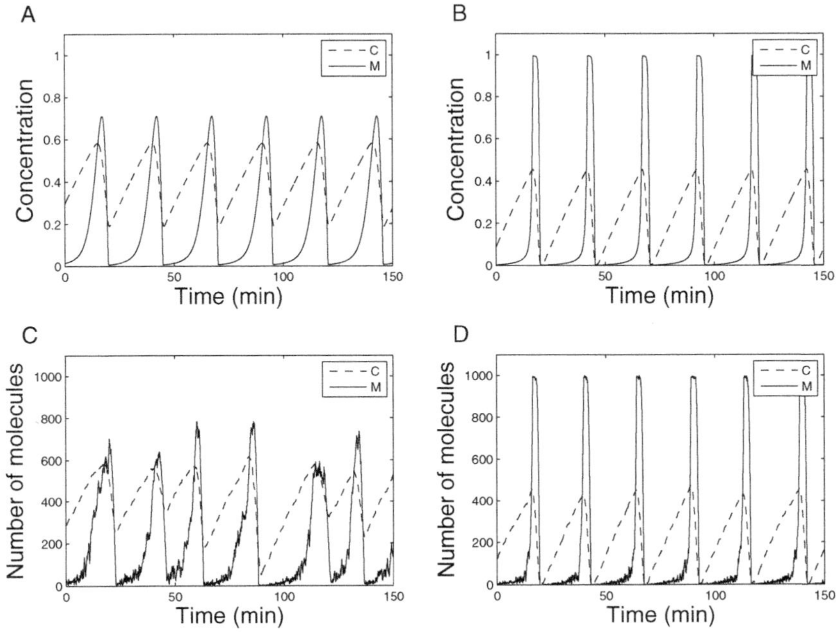

Fig. 2. Deterministic vs stochastic oscillations. (A,C) Model without auto-catalysis ($b = 0$). (B,D) Model with auto-catalysis ($b = 1$). (A,B) Deterministic oscillations. (C,D) Stochastic oscillations for $\Omega = 1000$. Parameter values: $v_i = 0.025$ nM/min, $v_d = 0.25$ nM/min, $K_d = 0.02$ nM, $k_d = 0.01$ min^{-1}, $v_{m1} = 3.0$ min^{-1}, $v_{m2} = 1.5$ min^{-1}, $v_{m3} = 1.0$ min^{-1}, $v_{m4} = 0.5$ min^{-1}, $K_1 = K_2 = K_3 = K_4 = 0.005$ nM, $K_c = 0.5$ nM, $M_{tot} = X_{tot} = 1$ nM, a=1nM. In panels A and B, the concentration is in nM.

but such a decomposition is not straightforward [36] and would lead to a large number of variables and reaction steps, resulting in a level of details unnecessarily high for such a simplified model. Furthermore, theoretical studies have shown that quasi-steady state approximations remain valid in the stochastic case [35].

3 Results

Deterministic simulations of the cascade-based model described above confirmed that oscillations can arise solely as a result of the negative feedback ($b = 0$) [13] (Fig. 2A). The period of the oscillations is around 30 min, which roughly corresponds to the duration of the mitotic cycle in frog embryos, and the shape of the oscillations matches those observed experimentally. Adding a positive loop ($b = 1$) preserves the oscillations but slightly changes their shape (Fig. 2B). In particular, variables M and X reach a small plateau, near their maximum value M_{tot} and X_{tot}.

Typical stochastic time series obtained by simulating our models with the Gillespie algorithm are shown in figures 2C (for $b = 0$) and 2D (for $b = 1$). For these simulations, we set $\Omega = 1000$, which leads to a number of molecules of few hundreds, a value in agreement with the estimation of the actual number of cell cycle molecules in a cell [27]. In presence of noise, the oscillations still persist but their amplitude and period show some variability. We can already notice that the model with auto-catalysis appears more robust than the model without auto-catalysis.

To understand this increase of robustness in the model with auto-catalysis, it is insightful to examine the dynamics in the phase space. The deterministic and stochastic limit cycles associated to the oscillations shown in figure 2 are given in figure 3 (see the thick close curves in panels A and B). To get a deeper understanding of the dynamics, it is useful to analyze the dynamics (in particular the steady states) of the reduced system obtained when the slow variable C is maintained constant, as described by Tsai et al [41]. In panels A and B, the thin line corresponds to the steady state of M as a function of C. These curves have been obtained by bifurcation analysis of the reduced model defined by eqs. (2) and (3) whith C taken as a parameter. The main difference between the two models is the appearance of a S curve in the reduced model with auto-catalysis. This S curve is associated with bistability. When the dynamics of C is considered (i.e. when the evolution of C is governed by eq. (1)), all variables oscillate, and the 3-ODE system converges to a limit cycle (thick curve on panels A and B). The trajectory follows the upper and lower parts of the S curve and periodically switches from the steady states of the corresponding reduced model (panel B). Two time scales thus appear: a slow motion when the system moves along the upper and lower branches of steady states and a rapid jump from one steady state to the other.

The stochastic trajectories corresponding to figures 3A and B are shown in panels C and D. For the model without auto-catalysis the trajectories are more spread than for the model with auto-catalysis, reflecting the higher robustness of the latter. The dual time scale generated by the positive feedback loop defines regions in the phase space where trajectories are strongly attracted, thereby reducing the spreading of the trajectories of the stochastic system.

To quantify the effect of noise, we computed the auto-correlation function [20] and the period distribution (Fig. 4). Since the entry into mitosis is controlled by the active CDK1, we computed these two measures using variable M. The periods (or, rather, the peak-to-peak intervals) were determined as the time interval separating two successive upward crossings of the mean level of variable M, an arbitrary value which can be seen as the threshold above which mitosis is triggered. We then use the half-life of the decorrelation and the standard deviation of the periods as quantifiers of the

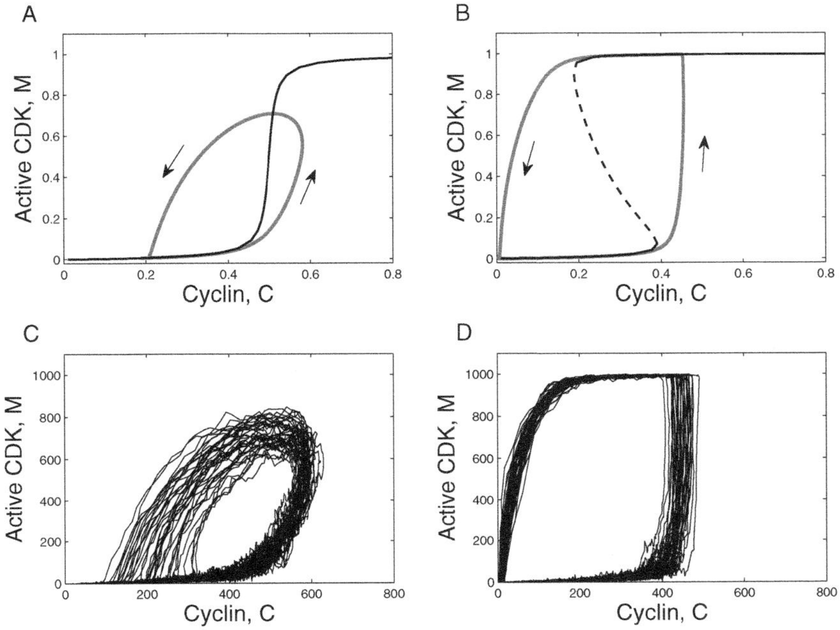

Fig. 3. Deterministic vs stochastic limit cycles. (A,C) Model without auto-catalysis ($b = 0$). (B,D) Model with auto-catalysis ($b = 1$). (A,B) The thin line corresponds to the steady state of M when C is taken as a parameter. The thick grey curve is the deterministic limit cycle of the 3-variable model. The arrows denote the direction along the limit cycle. (C,D) Stochastic trajectories obtained for $\Omega = 1000$. Parameter values are as in Fig. 2. In panel A and B, variables are concentrations (in nM), while in panels C and D, variables are numbers of molecules.

robustness [1, 19, 20]. Comparing the auto-correlation function and the period distribution, it is now obvious that the model with auto-catalysis is more robust than the model without auto-catalysis. Indeed, for the model without auto-catalysis ($b = 0$), the auto-correlation decreases more rapidly and the variability of the period is greater, reflecting a higher sensitivity to noise. Note that the two measures used here rather focus on the robustness of the period of the oscillations. We could have quantified the variation of the amplitude of the oscillations, but from a biological point of view we can hypothesize that mitosis is triggered when a threshold in the concentration of CDK1 is reached and that small variations of the amplitude would not affect the dynamics of cell cycle.

So far we have compared the two models for one parameter set only. However the dynamical properties of the oscillations (amplitude, period, etc) may depend on parameter values. To check if our observations are general, i.e. parameter-independent, we generated for each model about 100 parameter

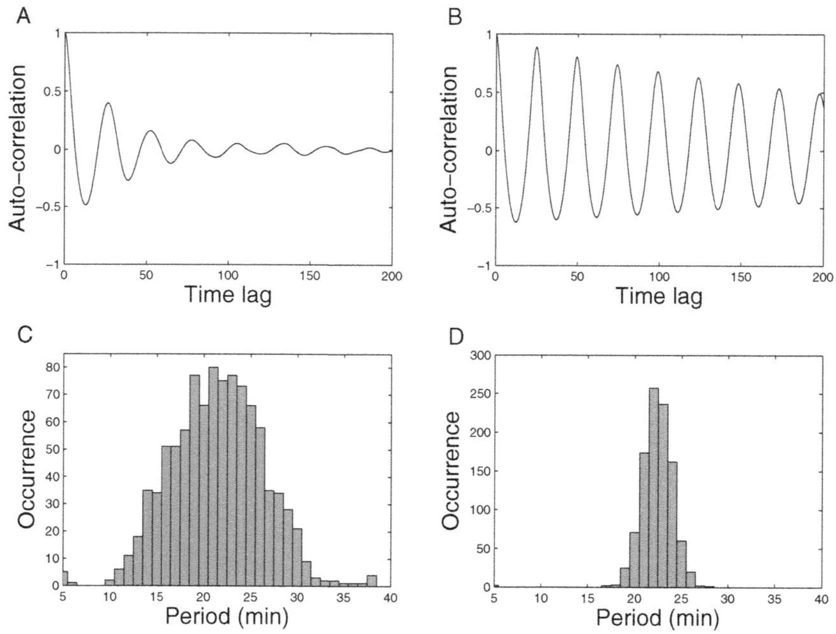

Fig. 4. Quantification of the robustness of the stochastic oscillations (obtained for $\Omega = 1000$). (A,C) Model without auto-catalysis ($b = 0$). (B,D) Model with auto-catalysis ($b = 1$). (A,B) Auto-correlation function. (C,D) Period distribution. These results have been obtained for the time series of M over a time interval of 10000 min. Parameter values are as in Fig. 2.

sets which give limit-cycle oscillations with a period within the range $[30, 40]$ min and a minimal amplitude for M of 0.6. The sets were found using a two-step sampling procedure that yields points uniformly distributed in the volume of the parameter space where these properties are fulfilled [23]. In order to avoid extreme, irrealistic values of some parameters, the sampling is restricted to a region of four orders of magnitude along each parameter, centered on the original parameter set defined in the legend of figure 2. This results in a uniform sampling of the possible parameter sets for which the model matches the predefined criteria. The oscillations were found to be sensitive to parameter a, which controls the negative feedback. In the sampling process we kept $a = 1$ (to maintain the negative feedback) and b was set either to 0(no positive feedback) or to 1 (effective positive feedback).

For both models, we performed stochastic simulations for each parameter set and systematically calculate the half-life of the decorrelation and the standard deviation of the period distribution. The distributions of these two quantifiers are given in Fig. 5. A Wilcoxon rank sum test returned $p-values$

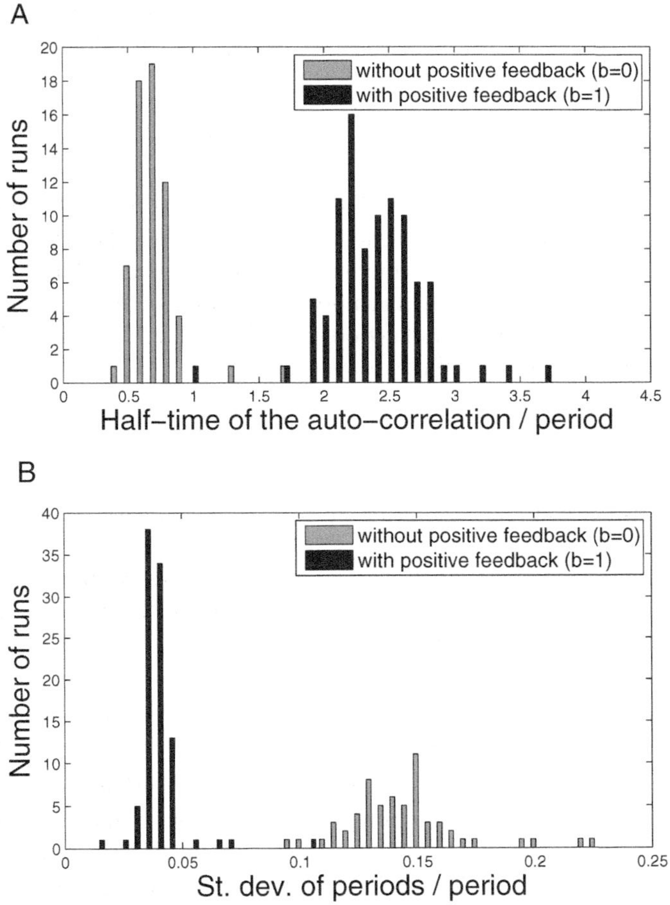

Fig. 5. Robustness of the stochastic oscillations for various parameter sets. (A) Distribution of the auto-correlation half-time and (B) distribution of the periods for the model without ($b = 0$) and with ($b = 1$) the positive feedback loop. For each model about 100 parameter sets have been generated as described in the text. One run has been performed for each parameter set and for each run the time series analysis has been done for variable M over a time interval of 10000 min.

of 2.01E-26 for the auto-correlation and 1.9E-26 for the period distribution, ensuring that we have two distinct distributions. These data thus confirm that the model with auto-catalysis is more robust than the model without auto-catalysis, regardless of parameter values.

4 Discussion

Understanding the design principles of biological oscillations is of general interest in molecular biology [32]. Many cellular oscillators rely on interlinked positive and negative feedback loops. Since a single delayed negative feedback loop has already the potential to generate self-sustained oscillations, the question of the role of additional, positive, feedback loops is open. Besides frequency tunability [41] and oscillations amplification [25], another possible role is illustrated here: positive feedback loops may increase the robustness of oscillators with respect to molecular noise. This role is here highlighted on a minimal three-variable model proposed several years ago for the embryonic mitotic cycle [13]. This model can be seen as a prototypical cascade model and is therefore useful to investigate questions about design. More elaborated models of the cell cycle [10, 26] exist. Stochastic simulations of such a detailed model have recently been carried out by Kar *et al* (2009) [27]. These authors assessed the relative level of intrinsic and extrinsic noise, but they do not address specifically how the oscillator design affects its robustness to molecular noise.

In the future, it would be interesting to perform similar analyses to other simple networks, such as the three-variable Goodwin oscillator, which represents a minimal genetic oscillator, and to extend this work to more detailed models for the cell cycle [10, 26] as well as for circadian clocks [2, 30], which incorporate positive and negative feedback loops.

Robustness to noise is related to regulatory networks topology [1, 4, 28, 38]. Stochastic simulations of minimal models of circadian clocks already put forward several robustness factors that contribute to the robustness of oscillations with respect to molecular noise [9, 19, 21]: cooperativity of gene repression, rate of binding-unbinding of the repressor protein to the gene promoter, forcing by a light-dark cycle, etc. The present study suggests that positive feedback loops, also occurring in circadian clocks [3, 5, 12, 37], may also play a role in the robustness of the oscillations with respect to molecular noise.

Positive feedbacks are typically associated with bistability and hysteresis. They induce several time scales which affect speed and attractivity of the limit cycle and, as already noticed in other works [20, 29, 42], the spreading of stochastic trajectories along the deterministic cycle is correlated with its attractivity properties. This may explain the increase of robustness observed in models based on interlocked positive and negative feedback loops.

Finally, these observations may have implications in understanding evolution of regulatory networks. Indeed, since sensitivity to noise may guide network topology, robustness to noise may be taken as a constraint to design regulatory networks and be added to other constraints such as period and amplitude of oscillations [39, 43]. Interestingly, our model has the property to be able to include or not the positive feedback loop (and to modify its strength) upon tuning of a single parameter and thus opens the possibility to use robustness to noise as a criterium to study evolution of biological network architectures. As shown in [24], it is possible to evolve in the parameter space continuously while maintaining some macroscopic properties. Our results thus suggest that robustness to noise has contributed to the emergence of additional positive feedback loops and that it may therefore serve as a selection criterion to simulate the evolution of biological oscillators in silico.

Acknowledgments. This work was supported by grant #3.4636.04 from the Fonds de la Recherche Scientifique Médicale (F.R.S.M., Belgium), by the European Union through the Network of Excellence BioSim (Contract No. LSHBCT- 2004-005137), by the Belgian Program on Interuniversity Attraction Poles, initiated by the Belgian Federal Science Policy Office, project P6/25 (BioMaGNet) and by SystemsX.ch through the grant for IPhD Project "Quantifying Robustness of Biochemical Modules to Parametric and Structural Perturbations".

References

[1] Barkai, N., Leibler, S.: Circadian clocks limited by noise. Nature 403, 267–268 (2000)
[2] Becker-Weimann, S., Wolf, J., Herzel, H., Kramer, A.: Modeling feedback loops of the Mammalian circadian oscillator. Biophys J. 87, 3023–3034 (2004)
[3] Brunner, M., Kaldi, K.: Interlocked feedback loops of the circadian clock of Neurospora crassa. Mol. Microbiol. 68, 255–262 (2008)
[4] Calatay, T., Turcotte, M., Elowitz, M.B., Garcia-Ojalvo, J., Suel, G.M.: Architecture-dependent noise discriminates functionally analogous differentiation circuits. Cell 139, 512–522 (2009)
[5] Cheng, P., Yang, Y., Liu, Y.: Interlocked feedback loops contribute to the robustness of the Neurospora circadian clock. Proc. Natl. Acad. Sci. USA 98, 7408–7413 (2001)
[6] Ciliberto, A., Novak, B., Tyson, J.J.: Steady states and oscillations in the p53/Mdm2 network. Cell Cycle 4, 488–493 (2005)
[7] Cross, F.R.: Two redundant oscillatory mechanisms in the yeast cell cycle. Dev. Cell 4, 741–752 (2003)
[8] Elowitz, M.B., Leibler, S.: A synthetic oscillatory network of transcriptional regulators. Nature 403, 335–338 (2000)
[9] Forger, D.B., Peskin, C.S.: Stochastic simulation of the mammalian circadian clock. Proc. Natl. Acad. Sci. USA 102, 321–324 (2005)
[10] Gérard, C., Goldbeter, A.: Temporal self-organization of the cyclin/Cdk network driving the mammalian cell cycle. Proc. Natl. Acad. Sci. USA 106, 21643–21648 (2010)
[11] Gillespie, D.T.: Exact Stochastic Simulation of Coupled Chemical Reactions. J. Phys. Chem. 81, 2340–2361 (1977)
[12] Glossop, N.R., Lyons, L.C., Hardin, P.E.: Interlocked feedback loops within the Drosophila circadian oscillator. Science 286, 766–768 (1999)
[13] Goldbeter, A.: A minimal cascade model for the mitotic oscillator involving cyclin and cdc2 kinase. Proc. Natl. Acad. Sci. USA 88, 9107–9111 (1991)

[14] Goldbeter, A.: Modeling the mitotic oscillator driving the cell division cycle. Comments on Theor. Biol. 3, 75–107 (1993)
[15] Goldbeter, A.: A model for circadian oscillations in the Drosophila period protein (PER). Proc. Biol. Sci. 261, 319–324 (1995)
[16] Goldbeter, A.: Biochemical Oscillations and Cellular Rhythms: The molecular bases of periodic and chaotic behaviour. Cambridge University Press, Cambridge (1996)
[17] Goldbeter, A., Gonze, D., Houart, G., Leloup, J.C., Halloy, J., Dupont, G.: From simple to complex oscillatory behavior in metabolic and genetic control networks. Chaos 11, 247–260 (2001)
[18] Goldbeter, A.: Computational approaches to cellular rhythms. Nature 420, 238–245 (2002)
[19] Gonze, D., Halloy, J., Goldbeter, A.: Robustness of circadian rhythms with respect to molecular noise. Proc. Natl. Acad. Sci. USA 99, 673–678 (2002)
[20] Gonze, D., Halloy, J., Gaspard, P.: Biochemical clocks and molecular noise: Theoretical study of robustness factors. J. Chem. Phys. 116, 10997–11101 (2002)
[21] Gonze, D., Goldbeter, A.: Circadian rhythms and molecular noise. Chaos 16, 026110 (2006)
[22] Goodwin, B.C.: Oscillatory behavior in enzymatic control processes. Adv. Enzyme Regul. 3, 425–438 (1965)
[23] Hafner, M., Koeppl, H., Hasler, M., Wagner, A.: 'Glocal' robustness analysis and model discrimination for circadian oscillators. PLoS Comput. Biol. 5, e1000534 (2009)
[24] Hafner, M., Koeppl, H., Wagner, A.: Evolution of Feedback Loops in Oscillatory Systems. In: Third International Conference on Foundations of Systems Biology in Engineering, pp. 157–160 (2009), http://arxiv.org/abs/1003.1231
[25] Hasty, J., Dolnik, M., Rottschaefer, V., Collins, J.J.: Synthetic gene network for entraining and amplifying cellular oscillations. Phys. Rev. Lett. 88, 148101 (2002)
[26] Kapuy, O., He, E., Lopez-Aviles, S., Uhlmann, F., Tyson, J.J., Novak, B.: System-level feedbacks control cell cycle progression. FEBS Lett. 583, 3992–3998 (2009)
[27] Kar, S., Baumann, W.T., Paul, M.R., Tyson, J.J.: Exploring the roles of noise in the eukaryotic cell cycle. Proc. Natl. Acad. Sci. USA 106, 6471–6476 (2009)
[28] Kollmann, M., Lovdok, L., Bartholomé, K., Timmer, J., Sourjik, V.: Design principles of a bacterial signalling network. Nature 438, 504–507 (2005)
[29] Kummer, U., Krajnc, B., Pahle, J., Green, A.K., Dixon, C.J., Marhl, M.: Transition from stochastic to deterministic behavior in calcium oscillations. Biophys J. 89, 1603–1611 (2005)
[30] Leloup, J.C., Goldbeter, A.: Toward a detailed computational model for the mammalian circadian clock. Proc. Natl. Acad. Sci. USA 100, 7051–7056 (2003)
[31] Meyer, T., Stryer, L.: Molecular model for receptor-stimulated calcium spiking. Proc. Natl. Acad. Sci. USA 85, 5051–5055 (1988)
[32] Novak, B., Tyson, J.J.: Design principles of biochemical oscillators. Nat. Rev. Mol. Cell. Biol. 9, 981–991 (2008)

[33] Pomerening, J.R: Positive-feedback loops in cell cycle progression. FEBS Lett. (in press)
[34] Pourquié, O.: The vertebrate segmentation clock. J. Anat. 199, 169–175 (2001)
[35] Rao, C., Arkin, A.: Stochastic chemical kinetics and the quasi steady-state assumption: application to the Gillespie algorithm. J. Chem. Phys. 118, 4999–5010 (2003)
[36] Sabouri-Ghomi, M., Ciliberto, A., Kar, S., Novak, B., Tyson, J.J.: Antagonism and bistability in protein interaction networks. J. Theor. Biol. 250, 209–218 (2002)
[37] Shearman, L.P., Sriram, S., Weaver, D.R., Maywood, E.S., Chaves, I., Zheng, B., Kume, K., Lee, C.C., van der Horst, G.T., Hastings, M.H., Reppert, S.: Interacting molecular loops in the mammalian circadian clock. Science 288, 1013–1019 (2002)
[38] Smolen, P., Baxter, D.A., Byrne, J.H.: Interlinked dual-time feedback loops can enhance robustness to stochasticity and persistence of memory. Phys. Rev. E 79, 31902 (2009)
[39] Stelling, J., Gilles, E.D., Doyle, F.J.: Robustness properties of circadian clock architectures. Proc. Natl. Acad. Sci. USA 101, 13210–13215 (2004)
[40] Stricker, J., Cookson, S., Bennett, M.R., Mather, W.H., Tsimring, L.S., Hasty, J.: A fast, robust and tunable synthetic gene oscillator. Nature 456, 516–519 (2008)
[41] Tsai, T.Y., Choi, Y.S., Ma, W., Pomerening, J.R., Tang, C., Ferrell, J.E.: Robust, tunable biological oscillations from interlinked positive and negative feedback loops. Science 321, 126–129 (2008)
[42] Vilar, J.M., Kueh, H.Y., Barkai, N., Leibler, S.: Mechanisms of noise-resistance in genetic oscillators. Proc. Natl. Acad. Sci. USA 99, 5988–5992 (2002)
[43] Wagner, A.: Circuit topology and the evolution of robustness in two-gene circadian oscillators. Proc. Natl. Acad. Sci. USA 102, 11775–11780 (2005)

Guaranteed and Randomized Methods for Stability Analysis of Uncertain Metabolic Networks

Heinz Koeppl, Stefano Andreozzi, and Ralf Steuer

Abstract. A persistent problem hampering our understanding of the dynamics of large-scale metabolic networks is the lack of experimentally determined kinetic parameters that are necessary to build computational models of biochemical processes. To overcome some of the limitations imposed by absent or incomplete kinetic data, structural kinetic modeling (SKM) was proposed recently as an intermediate approach between stoichiometric analysis and a full kinetic description. SKM extends stationary flux-balance analysis (FBA) by a local stability analysis utilizing an appropriate parametrization of the Jacobian matrix. To characterize the Jacobian, we utilize results from robust control theory to determine subintervals of the Jacobian' entries that correspond to asymptotically stable metabolic states. Furthermore, we propose an efficient sampling scheme in combination with methods from computational geometry to sketch the stability region. A glycolytic pathway model comprising 12 uncertain parameters is used to assess the feasibility of the method.

1 Modeling Metabolic Networks

Cellular metabolism, defined as the orchestrated biochemical interconversion of small molecules by dedicated proteins, is an important aspect of cellular physiology and of outstanding interest for many biotechnological and medical applications. In the past decades, great strides have been made to elucidate and compile the list of the biochemical reaction taking place in living cells and almost comprehensive

Heinz Koeppl · Stefano Andreozzi
Laboratory of Nonlinear Systems, School of Communication and Computer Sciences, Ecole Polytechnique Federale de Lausanne (EPFL), 1015 Lausanne, Switzerland
e-mail: {heinz.koeppl,stefano.andreozzi}@epfl.ch

Ralf Steuer
Institute for Theoretical Biology, Humboldt University of Berlin, Invalidenstrasse 43, D-10115 Berlin, Germany and Manchester Interdisciplinary Biocentre, The University of Manchester, Manchester M1 7DN, UK
e-mail: ralf.steuer@manchester.ac.uk

J. Lévine & P. Müllhaupt (Eds.): Adv. in Theory of Control, Signals, LNCIS 407, pp. 297–307.
springerlink.com © Springer-Verlag Berlin Heidelberg 2010

stoichiometric models for several model organisms such as the *S. cerevisiae* or *E. coli*, are now available [10]. However, to obtain a true understanding of cellular function and organization a mere list of parts is not enough. In this respect, the construction of mathematical models is an indispensable tool to study – and eventually understand – how the parts of a metabolic network interact and function as an integrated whole.

Current approaches to metabolic modeling are characterized by a dichotomy of large-scale constraint-based stoichiometric models on the one hand, and detailed kinetic models of small subsystems on the other hand. The advantage of topological and constraint-based methods is that they only require stoichiometric information, making them applicable to large, up to 'genome-scale', systems. Most prominently, flux-balance analysis (FBA) makes use of the mass conservation constraints to identify possible flux distributions that fulfill a given objective function, such as maximal ATP production or maximal biomass generation. Although one of the most successful approaches to date, the downside of FBA is that it cannot provide any information about the dynamical properties of the metabolic system. In contrast, the description of dynamics requires the construction of a detailed kinetic model of the network, usually in terms of ordinary differential equations. However, the construction of such explicit kinetic models of metabolism is based on detailed quantitative information on kinetic parameters and rate equations, information that is only rarely available in practice.

To overcome some of the difficulties imposed by the lack of information on kinetic parameters, there has been increasing interest in heuristic and semi-quantitative methods to describe the dynamics of large-scale metabolic networks in the face of uncertain kinetic data [23, 22, 20]. Specifically, structural kinetic modeling (SKM) proposes to augment the constraint-based analysis by a local stability analysis utilizing an appropriate parametrization of the Jacobian matrix [21]. The approach is based on the observation that in many cases a detailed kinetic model is not necessary. Rather, a large number of dynamical properties, such as control coefficients, the stability of states, transitions to oscillatory regions, among various others, are readily available using only a linear approximation of the system. SKM therefore seeks to derive stringent bounds on the entries of the Jacobian matrix, based on available phenotypic data and biophysical constraints, to enable a computational analysis in the absence of further kinetic information. We emphasize that SKM is a data-driven approach, taking another perspective than classical nonlinear dynamics. More specifically, SKM starts out with a given, *experimentally measured* steady state and asks for the underlying parameter region supporting this particular state.

In this work, we discuss an extension of SKM utilizing methods from robust stability theory [4, 1] that allows to determine subintervals of the Jacobian entries of a SKM model corresponding to stable metabolic states. To this end, we believe that the proposed reasoning about entire sets of models is an adequate semi-quantitative approach [18] to analyze biochemical models in general.

The paper is organized as follows. Section 2 provides an introduction to the SKM framework. The applied guaranteed methods from robust control as well as a novel random sampling scheme are discussed in Section 3. An application of the sampling method is given in Section 4, while Section 5 draws conclusions.

2 Structural Kinetic Modeling

SKM draws upon the fact that, even in the absence of detailed kinetic information, questions with respect to stability of the metabolic operating point can be addressed. To this end, we consider a metabolic network whose time-dependent behavior is described by an ordinary differential equation of the form

$$\frac{d\mathbf{S}}{dt} \equiv \dot{\mathbf{S}} = \mathbf{N}\mathbf{v}(\mathbf{S},\mathbf{k}) \,, \tag{1}$$

with $\mathbf{S} \in \mathbb{R}_+^N$ denoting the vector of concentration of all involved species, $\mathbf{N} \in \mathbb{Z}^{N\times L}$ the stoichiometric matrix, $\mathbf{v} : \mathbb{R}_+^N \times \mathbb{R}_+^M \to \mathbb{R}_+^L$ the parametric rate laws and $\mathbf{k} \in \mathbb{R}_+^M$ a vector comprising all kinetic parameters. We assume that the network has at least one non-zero steady state at concentration $\mathbf{S}^0$, which does not necessarily have to be stable. In this case, we can equivalently write

$$\frac{\dot{S}_i}{S_i^0} = \sum_{j=1}^{R} N_{ij} \frac{v_j(\mathbf{S}^0)}{S_i^0} \frac{v_j(\mathbf{S})}{v_j(\mathbf{S}^0)} . \tag{2}$$

Introducing concentrations that are normalized by the steady-state concentration $x_i = \frac{S_i}{S_i^0}$ one obtains

$$\dot{\mathbf{x}} = \boldsymbol{\Lambda}\boldsymbol{\mu}(\mathbf{x}), \tag{3}$$

with the constant matrix $\Lambda_{ij} \equiv N_{ij}\frac{v_j(\mathbf{S}^0)}{S_i^0}$ and the vector of normalized fluxes $\mu_j(\mathbf{x}) \equiv \frac{v_j(\mathbf{S})}{v_j(\mathbf{S}^0)}$. A linearization of the system at the steady state $\mathbf{x} = \mathbf{1}$ yields with $\boldsymbol{\Lambda}\boldsymbol{\mu}(\mathbf{1}) = \mathbf{0}$ a linear model with states z_i

$$\dot{z}_i = \sum_{j=1}^{L}\sum_{k=1}^{N} \Lambda_{ij} \left.\frac{\partial \mu_j(\mathbf{z})}{\partial z_k}\right|_{\mathbf{z}=\mathbf{1}} (z_k - 1). \tag{4}$$

Introducing the matrix $\Theta_{kj} \equiv \left.\frac{\partial \mu_j(\mathbf{z})}{\partial z_k}\right|_{\mathbf{z}=\mathbf{1}}$ we obtain

$$\dot{\mathbf{z}} = \boldsymbol{\Lambda}\boldsymbol{\Theta}(\mathbf{z} - \mathbf{1}). \tag{5}$$

The stability of the nonlinear system specified by (3) at $\mathbf{x} = \mathbf{1}$ is thus determined by the eigenvalues of the matrix $\boldsymbol{\Lambda}\boldsymbol{\Theta}$, which is equivalent to the (scaled) Jacobian matrix. Our further analysis rests upon a detailed interpretation of the matrices $\boldsymbol{\Lambda}$ and $\boldsymbol{\Theta}$. In particular, the matrix $\boldsymbol{\Lambda}$ is entirely specified by stoichiometric information, along with knowledge of a stationary metabolic state, characterized by a set of stationary concentrations $\mathbf{S}^0$ and fluxes $\mathbf{v}^0 = \mathbf{v}(\mathbf{S}^0)$. The latter satisfy the steady-state constraint $\mathbf{N}\mathbf{v}^0 = \mathbf{0}$. We note that large-scale measurements and the characterization of metabolic systems in terms of concentrations (*metabolomics*) and fluxes (*fluxomics*) are now almost standard techniques in the analysis of cellular metabolism [14, 24, 17], making the matrix $\boldsymbol{\Lambda}$ – at least in principle – accessible to direct experimentation.

The interpretation of the elements of $\boldsymbol{\Theta}$ is slightly more intricate. Every entry of the matrix $\boldsymbol{\Theta}$ specifies the derivatives of the normalized rate law with respect to the scaled concentrations, and can be interpreted as the (dimensionless) relative saturation level of one particular reaction with respect to one particular substrate concentration. Importantly, for most typical rate laws the elements of $\boldsymbol{\Theta}$ are confined to well-defined intervals that are independent of the respective metabolic state or mathematical details of the rate equation. We note that the elements of $\boldsymbol{\Theta}$ are analogous to logarithmic derivatives and are closely related to the scaled elasticity coefficients in Metabolic Control Analysis [12].

3 Stability of Uncertain Linear Systems

We are now in the position to apply the ideas of robustness analysis for linear systems to the Jacobian matrix $\mathbf{J} \equiv \boldsymbol{\Lambda}\boldsymbol{\Theta}$ of our linearized metabolic network. Allowing uncertainty in the kinetic rate law corresponds here to an uncertainty about the saturation matrix $\boldsymbol{\Theta}$. Thus we define the set of Jacobians as

$$\mathbf{J}([\boldsymbol{\Theta}]) = \left\{ \mathbf{J} \,|\, \mathbf{J} = \boldsymbol{\Lambda}\boldsymbol{\Theta}, \boldsymbol{\Theta} \in [\boldsymbol{\Theta}] \in \mathbb{IR}^{L\times N} \right\}, \tag{6}$$

where is $\mathbb{IR}$ is the set of all real intervals. Thus an element $[\boldsymbol{\Theta}] \in \mathbb{IR}^{N\times L}$ is an interval matrix

$$[\boldsymbol{\Theta}] \equiv \left\{ \boldsymbol{\Theta} \,|\, \Theta_{ij} \in [\underline{\Theta}_{ij}, \bar{\Theta}_{ij}], \underline{\Theta}_{ij} \leq \bar{\Theta}_{ij}, \forall (i,j) \right\},$$

the bounds of which are determined by biophysical constraints. In practice not every entry of $\boldsymbol{\Theta}$ is uncertain and one seeks a representation of the Jacobian as a function solely of the uncertain vector $\boldsymbol{\theta} \in \mathbb{R}^M$

$$\mathbf{J}(\boldsymbol{\theta}) = \mathbf{J}_0 + \sum_{i=1}^{M} \theta_i \mathbf{J}_i = \mathbf{J}_0 + \boldsymbol{\Lambda} \sum_{i=1}^{M} \theta_i \mathbf{T}_i \tag{7}$$

with template matrices $\mathbf{T}_i \in \{0,1\}^{N\times L}$. We do not exclude the case that one uncertain parameter controls multiple entries of $\boldsymbol{\Theta}$. Alternatively, the parametric Jacobian may be expressed as a convex matrix polytope with

$$\mathbf{J}(\boldsymbol{\theta}) \in \mathrm{co}\{\tilde{\mathbf{J}}_1, \ldots, \tilde{\mathbf{J}}_K\} \equiv \left\{ \mathbf{J} \,\middle|\, \mathbf{J} = \sum_{i=1}^{K} \alpha_i \tilde{\mathbf{J}}_i, \sum_{i=1}^{K} \alpha_i = 1, \alpha_i \geq 0,\ i \in \{1, \ldots, K\} \right\},$$

with $\mathrm{co}\{\cdot\}$, the convex hull. The image of the saturation hyper-rectangle $[\boldsymbol{\Theta}]$ under $\boldsymbol{\Lambda}$ is, in general, not a rectangle in the space of Jacobians and vertex points of $[\boldsymbol{\Theta}]$ can be mapped to the interior of the Jacobian polytope. Thus, we have $K \leq 2^M$ assuming that $L \geq N$, which is normally the case for reaction networks.

3.1 Guaranteed Methods

In the following, robust control methods are discussed that we consider particularly suitable for the SKM framework. They determine saturation subintervals, where stability of every member is guaranteed. The application of those methods to a model of the glycolytic pathway within the SKM framework is presented in [16].

Given a single Jacobian $\mathbf{J} \in \mathbb{R}^{N \times N}$ of a linearized dynamics, stability can be determined by checking the Hurwitz property, i.e., whether all roots of the characteristic polynomial $p(\lambda) = \det(\mathbf{J} - \lambda \mathbf{I})$ have negative real parts. For the case of parametric Jacobians $\mathbf{J}(\boldsymbol{\theta})$ the following theorem due to Kharitonov [15] can be utilized.

Theorem 1. *Every polynomial*

$$p(\lambda, \mathbf{c}) = c_0 + c_1\lambda + \cdots + c_{n-1}\lambda^{n-1} + c_n\lambda^n \tag{8}$$

of degree n which is an instance of the polynomial set $p(\lambda, [\mathbf{c}]) = \{p(\lambda, \mathbf{c}) \mid \mathbf{c} \in [\mathbf{c}]\}$ *and* $c_n > 0$ *is a Hurwitz polynomial, if and only if the associated following four Kharitonov polynomials*

$$\begin{aligned}
p^{+-}(\lambda, \mathbf{c}) &= \bar{c}_0 + \underline{c}_1\lambda + \underline{c}_2\lambda^2 + \bar{c}_3\lambda^3 + \bar{c}_4\lambda^4 + \underline{c}_5\lambda^5 + \cdots \\
p^{++}(\lambda, \mathbf{c}) &= \bar{c}_0 + \bar{c}_1\lambda + \underline{c}_2\lambda^2 + \underline{c}_3\lambda^3 + \bar{c}_4\lambda^4 + \bar{c}_5\lambda^5 + \cdots \\
p^{-+}(\lambda, \mathbf{c}) &= \underline{c}_0 + \bar{c}_1\lambda + \bar{c}_2\lambda^2 + \underline{c}_3\lambda^3 + \underline{c}_4\lambda^4 + \bar{c}_5\lambda^5 + \cdots \\
p^{--}(\lambda, \mathbf{c}) &= \underline{c}_0 + \underline{c}_1\lambda + \bar{c}_2\lambda^2 + \bar{c}_3\lambda^3 + \underline{c}_4\lambda^4 + \underline{c}_5\lambda^5 + \cdots
\end{aligned} \tag{9}$$

are Hurwitz polynomials.

The theorem gives a necessary and sufficient condition for stability. However, the necessity is lost if the coefficients $\mathbf{c}$ are not independent as it is the case for the characteristic polynomial $p(\lambda, \boldsymbol{\theta}) = \det(\mathbf{J}(\boldsymbol{\theta}) - \lambda \mathbf{I})$. Thus, the theorem just provides a sufficient condition, and gives conservative results in general. In practice, one can obtain the overbounding coefficient intervals $[\mathbf{c}]$ by computing the characteristic polynomial with $\boldsymbol{\theta} \in [\boldsymbol{\theta}]$ using interval arithmetic [13, 16].

Quadratic stability of a polytopic linear system with $\mathbf{J}([\boldsymbol{\theta}])$ is defined that for each member $\mathbf{J}(\boldsymbol{\theta}) \in \mathrm{co}\{\tilde{\mathbf{J}}_1, \ldots, \tilde{\mathbf{J}}_K\}$ one can find one common quadratic Lyapunov function. With that, quadratic stability is stronger than testing the Hurwitz stability of each member. Thus, for an uncertain system that is quadratically stable all members are Hurwitz stable, but a system that is not quadratically stable can still be stable for all members. Quadratic stability thus provides just another means to obtain conservative stability bounds. However, quadratic stability, by itself, can be determined without conservatism with a finite number of tests.

Theorem 2. *A linear polytopic system is quadratically stable if and only if all its vertex systems are stable.*

It remains to find a common Lyapunov function for all vertex systems. This can be done by solving the following K linear matrix inequalities simultaneously

$$\tilde{\mathbf{J}}_i^T \mathbf{P} + \mathbf{P}\tilde{\mathbf{J}}_i \prec \mathbf{0}, \tag{10}$$

for $i \in \{1, \ldots, K\}$ and $\mathbf{P} \succ 0$, the common positive-definite Lyapunov matrix. The proof of the theorem is based on the observation that any positive linear combination of negative definite terms is again negative definite

$$\tilde{\mathbf{J}}^T(\boldsymbol{\alpha})\mathbf{P}+\mathbf{P}\tilde{\mathbf{J}}(\boldsymbol{\alpha}) = \sum_{i=1}^{K} \alpha_i(\tilde{\mathbf{J}}_i^T\mathbf{P}+\mathbf{P}\tilde{\mathbf{J}}_i) \prec \mathbf{0} \quad \text{with} \quad \sum_{i=1}^{K} \alpha_i = 1 \quad \text{and} \quad \alpha_i \geq 0,$$

$i \in \{1,\dots,K\}$ and $\boldsymbol{\alpha} \equiv (\alpha_1,\dots,\alpha_K)$.

In contrast to quadratic stability, *affine quadratic stability* searches for a quadratic parameter-dependent Lyapunov function, where the parameter dependency is assumed to be affine

$$\mathbf{P}(\boldsymbol{\theta}) = \mathbf{P}_0 + \sum_{j=1}^{M} \theta_j \mathbf{P}_j. \tag{11}$$

Writing it in terms of polytopes we seek a Lyapunov matrix such that

$$\tilde{\mathbf{J}}(\boldsymbol{\alpha})^T\tilde{\mathbf{P}}(\boldsymbol{\alpha})+\tilde{\mathbf{P}}(\boldsymbol{\alpha})\tilde{\mathbf{J}}(\boldsymbol{\alpha}) \prec \mathbf{0} \tag{12}$$

and $\tilde{\mathbf{P}}(\boldsymbol{\alpha}) \succ \mathbf{0}$ for any convex combination $\boldsymbol{\alpha}$. We used the corresponding polytopic representation of the affine set (11)

$$\tilde{\mathbf{P}}(\boldsymbol{\alpha}) = \sum_{i=1}^{K} \alpha_i \tilde{\mathbf{P}}_i \quad \text{where} \quad \sum_{k=1}^{K} \alpha_k = 1 \quad \text{and} \quad \alpha_k \geq 0, \tag{13}$$

with the vertex matrices $\tilde{\mathbf{P}}_i$. Affine quadratic stability leads to *bilinear matrix inequalities* that are difficult to solve numerically. However, forcing another constraint on the Lyapunov function, namely multi-convexity [2, 11] one arrives at vertex conditions similar to the one of quadratic stability

$$\begin{aligned} \tilde{\mathbf{J}}_i^T\tilde{\mathbf{P}}_i+\tilde{\mathbf{P}}_i\tilde{\mathbf{J}}_i &\prec \mathbf{0} \\ \tilde{\mathbf{P}}_i &\succ \mathbf{0} \\ \mathbf{J}_j^T\mathbf{P}_j+\mathbf{P}_j\mathbf{J}_j &\prec \mathbf{0} \end{aligned} \tag{14}$$

for all $i \in \{1,\dots,K\}$ and $j \in \{1,\dots M\}$, where we used the affine representation of (7). The incorporation of multi-convexity (third inequality) introduces conservatism but yields a set of linear matrix inequalities that can now be solved efficiently using semi-definite programming.

3.2 Efficient Random Sampling

A downside of guaranteed methods of robust control is that they, in general, provide binary answers regarding stability. For instance the semidefinite program underlying quadratic stability qualifies the proposed parameter interval $[\boldsymbol{\theta}]$ as feasible or not. Thus, these methods do not lend themselves to locate the stable region or to determine the most constraining parameter dimensions. Quadratic stability can be extended to return a scalar variable, for which a proposal interval need to scaled uniformly around an expansion in order to meet quadratic stability [6]. However, also this requires *a priori* information about the proper expansion point and side-length ratios of the hyper-rectangle. Moreover, determination of the maximum-volume hyper-rectangle that can be inscribed into a closed surface, itself requires the solution of a nonlinear program. Multi-dimensional bisection methods is used in [16] to expand hyper-rectangles based on the binary decisions returned by the guaranteed methods of Section 3.1. However, such an approach does not scale well with the parameter dimension and also does not guarantee to converge to the maximal-volume rectangle, on top of the conservatism of those guaranteed methods.

The procedure outlined in the following aims to find a non-guaranteed hyper-rectangle through advanced random sampling of the stability region. Sketching the stability region in this way, also allows one to identify parameter combinations that are most constraining in terms of stability. This can be achieved through a minor component analysis (MCA) [9]. Besides its relevance in its own rights, the obtained rectangle can then be proposed to a guaranteed method. Even if a downscaling of the rectangle is necessary due to conservatism of the guaranteed method or due to the overapproximation of the stability region by the sampling method, the expansion center and the side-lengths ratios are likely to be representative.

Sampling. We randomly sample one-dimensional information through the following theorem that provides sufficient and necessary conditions in case of one-dimensional uncertainty [5].

Theorem 3. *Consider the affine uncertain system* $\mathbf{J}(\omega) = \mathbf{J}_0 + \omega\mathbf{J}_1$, *with* $\mathbf{J}_0$ *Hurwitz stable and* $\omega \in [\underline{\omega}, \bar{\omega}] \in \mathbb{IR}$. *The matrix* $\mathbf{J}(\omega)$ *is robustly stable if and only if* $\omega \in [\underline{\omega}^*, \bar{\omega}^*]$ *with*

$$\underline{\omega}^* = \frac{1}{\lambda^-_{\min}[-(\mathbf{J}_0 \oplus \mathbf{J}_0)^{-1}(\mathbf{J}_1 \oplus \mathbf{J}_1)]}$$

$$\bar{\omega}^* = \frac{1}{\lambda^+_{\max}[-(\mathbf{J}_0 \oplus \mathbf{J}_0)^{-1}(\mathbf{J}_1 \oplus \mathbf{J}_1)]}$$

with the Kronecker sum $\mathbf{J}_0 \oplus \mathbf{J}_0 \equiv \mathbf{J}_0 \otimes \mathbf{I}_N + \mathbf{I}_N \otimes \mathbf{J}_0$ *and with* $\lambda^-_{\min}(\cdot)$ *and* $\lambda^+_{\max}(\cdot)$ *the minimum and maximum of the strictly negative and strictly positive set of eigenvalues of a matrix.*

With a nominal parameter set that corresponds to a stable Jacobian $\mathbf{J}_0$ the theorem provides a means to sample the stability region around this expansion point without any conservatism. We do this by shooting *Bialas rays* in random directions $\boldsymbol{\theta}$ from this expansion point. In vector notation this reads

$$\mathrm{vec}(\mathbf{J}(\omega)) = (\mathbf{I} \otimes \boldsymbol{\Lambda})\,\mathrm{vec}(\boldsymbol{\Theta}) = \mathrm{vec}(\mathbf{J}_0) + \omega(\mathbf{I} \otimes \boldsymbol{\Lambda})\mathbf{R}\boldsymbol{\theta}, \tag{15}$$

with the appropriate rearrangement matrix $\mathbf{R} \in \{0,1\}^{LN \times M}$. The probability distribution over ray directions should be chosen such, that the intersection points between rays and stability boundary are distributed uniformly. Choosing random directions from an expansion point within a surface that result in a uniform distribution at that surface is a known problem; see for instance the problem of uniformly sampling the surface of a n-sphere [19]. However, in the absence of information on the surface to be sampled, we propose to resort to a sequential Monte-Carlo algorithm that generates a uniform distribution at a bounding rectangle $[\boldsymbol{\theta}] \subseteq [\boldsymbol{\theta}]_0$ that is updated during sampling. We refer to $[\boldsymbol{\theta}]_0$ as the outer interval determined by biophysical bounds.

Dimensionality Reduction. Several expansion centers are chosen according to a tree structure with predetermined depth and degree. Parameter combinations that are constrained in terms of stability are revealed by computing an eigendecomposition of the covariance matrix of the sampled line set (see for instance Fig. 2 in Section 4). The eigendirections corresponding to small eigenvalues indicate constrained parameter combinations (minor components) [9]. Inner products between eigendirections and basis vectors of the parameter coordinates allow to identify single parameters, that are maximally aligned with these constrained directions. This opens up the possibility for model reduction, where interval stability is investigated only for the most constrained parameters.

Rectangle Inscription. In order to be able to inscribe a hyper-rectangle into a sampled closed surface, the samples need to be connected to give closed surface. The most natural choice is to construct the convex hull, i.e. the smallest convex set containing the sampled points. The convex hull is a convex polytope – or bounded polyhedron and thus has besides its vertex representation also a representation as a set of half-spaces (see Minkowski-Weyl theorem). We define a polyhedron $\mathscr{P}$ as

$$\mathscr{P} \equiv \left\{ \boldsymbol{\theta} \in \mathbb{R}^M \,\middle|\, \mathbf{A}\boldsymbol{\theta} \leq \mathbf{b}, \mathbf{A} \in \mathbb{R}^{Q\times M}, \mathbf{b} \in \mathbb{R}^Q \right\},$$

with Q the number of half-spaces. The problem of inscribing the maximal-volume rectangle into $\mathscr{P}$ is convex and can thus be solved efficiently on polynomial time [7]. Denoting the interval of the inscribed box as $[\boldsymbol{\theta}] \in \mathbb{IR}^M$ we can write the convex program as

$$\begin{aligned} &\max_{[\boldsymbol{\theta}]} \log\det \mathbf{W}([\boldsymbol{\theta}]) \\ &\text{subject to} \\ &[\boldsymbol{\theta}] \subseteq \mathscr{P}, \end{aligned} \tag{16}$$

with the diagonal matrix $\mathbf{W}(\cdot)$ denoting the interval width $\mathbf{W}([\boldsymbol{\theta}]) = \text{diag}\{\sup([\boldsymbol{\theta}]) - \inf([\boldsymbol{\theta}])\}$. Instead of having $2^M Q$ linear inequalities due to the 2^M vertices of $[\boldsymbol{\theta}]$ and the M halfspaces, the constraint $[\boldsymbol{\theta}] \subseteq \mathscr{P}$ can be expressed more efficiently with $2MQ$ inequalities [7]. The plausibility of the obtained optimal $[\boldsymbol{\theta}]$ rest upon the assumption that the intersection of stability region and biophysical bounding box $[\boldsymbol{\theta}]_0$ can well be encoded through a convex polytope.

Utilizing the exact convex hull, i.e. the tightest convex enclosure, introduces scalability issues in high dimensions. The worst case complexity of an optimal convex hull algorithm was shown to be $\mathscr{O}(n^{\lfloor M/2 \rfloor})$ for $M \geq 4$, where n is the number of sample points. However, the worst-case is rarely encountered and the actual complexity depends on the number of necessary inequalities Q, the order of which can vary from $\mathscr{O}(1)$ to $\mathscr{O}(n^{\lfloor M/2 \rfloor})$. Taking Q into account, a polynomial algorithm in n, M and Q was shown to exist for the non-degenerate case [3].

4 Application

We apply the proposed sampling method of Section 3.2 to a medium-scale model of glycolysis, depicted in Fig. 1. Transformed to the SKM representation of Section 2, the model has 12 non-zero entries in the saturation matrix $\boldsymbol{\Theta}$. For all reactions, except one, we assume standard Michaelis-Menten kinetics giving rise to biophysical bounds $[\theta_i]_0 = [0,1]$ for $i \in \{2,\dots,12\}$. For the first reaction, the conversion of Glucose (Glc) into fructose-1,6-biphosphate (FBP) we implement the known inhibitory effect of ATP, resulting in $[\theta_1]_0 = [-3,1]$. We sketch the feasible region

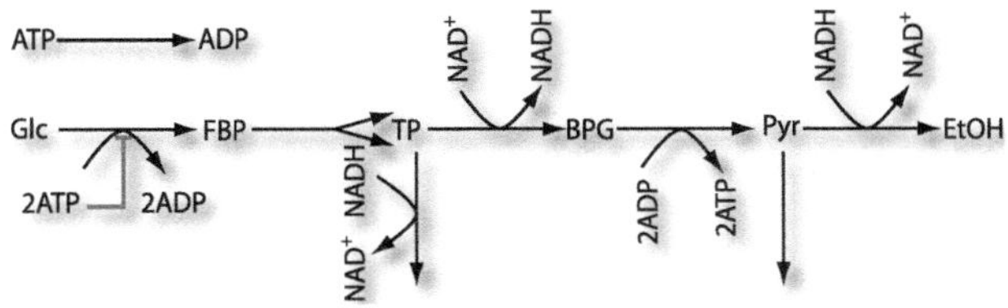

Fig. 1. Medium-scale model of the yeast glycolytic pathway comprising 8 reactions giving rise to 12 saturation parameters in the framework of structural kinetic modeling

characterized as the intersection of the stability domain with the biophysical bounding box using 10^4 Bialas rays with a flat tree configuration of depth one and degree 100. Figure 2(a) shows the eigendecomposition of the covariance matrix $\mathbf{C}$ in normalized coordinates, indicating one tightly constrained parameter combination. Inner product computation reveals that direction θ_1 is strongly aligned with the corresponding eigendirection.

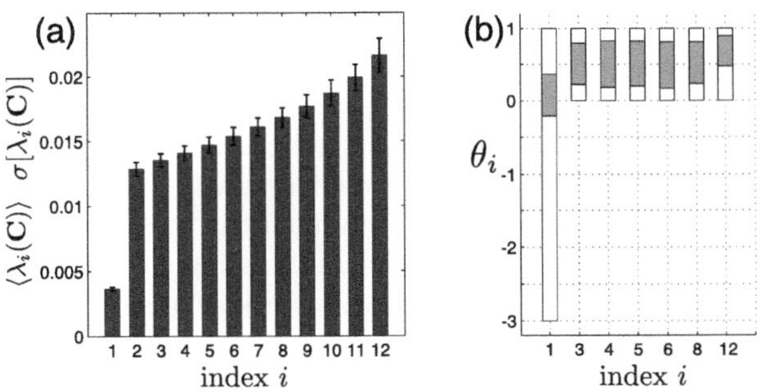

Fig. 2. (**a**) Mean and standard deviation of the spectrum for the covariance matrix $\mathbf{C}$ of the stable 12-dimensional parameter region – based on 100 runs each with 10^4 Bialas rays (tree of depths one and degree 100). (**b**) biophysical bounding boxes and obtained stable intervals (gray) for parameters retained after model reduction.

To illustrate the method we perform a model reduction retaining only the seven most constrained parameter dimensions as interval variables and adjusting the remaining ones to their nominal value, chosen to be the midpoint of $[\boldsymbol{\theta}]_0$. The obtained stability interval are depicted in Fig. 2(b). The convex hull and the obtained

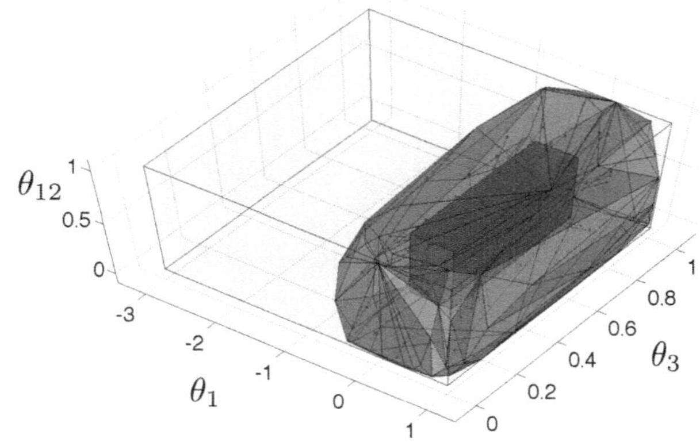

Fig. 3. Maximum-volume hyper-rectangle (green) inscribed in the convex hull (brown) and biophysical bounding box $[\boldsymbol{\theta}]_0$ (yellow); a few of the 10^4 rays (blue) used to sketch the stability region of the reduced 7-dimensional interval system; down-projection to coordinates that are the most aligned to the directions of the first three minor components

7-dimensional stability rectangle, down-projected onto the first three most constrained parameter dimension is shown in Fig. 3.

5 Conclusions

The scarcity of kinetic information for metabolic reactions rarely allows for the determination of detailed kinetic rate laws for a metabolic model. We combine the local stability analysis of structural kinetic modeling with interval methods to compute guaranteed and non-guaranteed stability intervals for the saturation levels of the involved reactions. We provide an efficient sampling algorithm to sketch high-dimensional stability regions and apply methods from statistics and computational geometry to obtain non-guaranteed stability intervals. The computed stability interval may serve as a proposal for the binary test of guaranteed methods from robust control. To alleviate scalability issues in the applied computational geometry methods, one may resort to randomized algorithms, for instance such as the randomized incremental construction of the convex hull [8].

Acknowledgement. HK acknowledges the support from the Swiss National Science Foundation, grant no. 200020-117975/1. SA was supported by the Laboratory of Nonlinear Systems, EPFL within the Summer@EPFL internship program. RS is supported by the grant FORSYS-Partner: Systems biology of cyanobacterial biofuel production, as well as by the rearch initiative SysMO: MOSES (Grant Reference: BBF0035281) and SulfoSys (Grant Reference: BBF0035361).

References

1. Ackermann, J., Bartlett, A., Kaesbauer, D., Sienel, W., Steinhauser, W.: Robust Control: Systems with Uncertain Physical Parameters. Springer, New York (2001)
2. Apkarian, P., Tuan, H.D.: Parametrized LMIs in control theory. SIAM J. Control Optim. 38(4), 1241–1264 (2000)
3. Avis, D., Bremner, D., Seidel, R.: How good are convex hull algorithms. Comput. Geom. Th. Appl. 7, 265–302 (1997)
4. Barmish, B.R.: New Tools for Robustness of Linear Systems. Macmillan Publishing Company, Basingstoke (1994)
5. Bialas, S.: A necessary and sufficient condition for the stability of convex combinations of stable polynomials or matrices. Bull. Pol. Acad. Sci. 33, 473–480 (1985)
6. Boyd, S., El Ghaoui, L., Feron, E., Balakrishnan, V.: Linear Matrix Inequalities in System and Control Theory. SIAM, Philadelphia (1994)
7. Boyd, S., Vandenberghe, L.: Convex Optimization. Cambridge University Press, Cambridge (2004)
8. Clarkson, K.L., Mehlhorn, K., Seidel, R.: Four results on randomized incremental construction. Comput. Geom. 3(4), 185–212 (1993)
9. Duda, R.O., Hart, P.E., Stork, D.G.: Pattern Classification. Wiley-Interscience Publication, Hoboken (2000)
10. Feist, A.M., Herrgard, M.J., Reed, J.L., Palsson, B.O.: Reconstruction of biochemical networks in microorganisms. Nat. Rev. Microbiol. 7(2), 129–143 (2009)
11. Gahinet, P., Apkarian, P., Chilali, M.: Affine parameter-dependent Lyapunov functions and real parametric uncertainty. IEEE Trans. Autom. Control 41(3), 436–442 (1996)
12. Heinrich, R., Schuster, S.: The regulation of cellular systems. Chapman Hall, New York (1996)
13. Jaulin, L., Kieffer, M., Didrit, O., Walter, E.: Applied Interval Analysis. Springer, London (2001)
14. Kell, D.B.: Metabolomics and systems biology: making sense of the soup. Curr. Opin. Microbiol. 7(3), 296–307 (2004)
15. Kharitonov, V.L.: Asymptotic stability of an equilibrium position of a family of systems of linear differential equations. Differential Equations 14, 1483–1485 (1979)
16. Koeppl, H., Hafner, M., Steuer, R.: Semi-quantitative stability analysis constrains saturation levels in metabolic networks. In: Proc. Int. Workshop on Comput. Syst. Biol., Aarhus, Denmark, pp. 91–94 (2009)
17. Kruger, N.J., Ratcliffe, R.G.: Insights into plant metabolic networks from steady-state metabolic flux analysis. Biochimie 91(6), 697–702 (2009)
18. Kuiper, B.J.: Qualitative Reasoning: Modeling and Simulation with Incomplete Knowledge. MIT Press, Cambridge (1994)
19. Marsaglia, G.: Choosing a point from the surface of a sphere. Ann. Math. Stat. 43, 645–646 (1985)
20. Schaber, J., Liebermeister, W., Klipp, E.: Nested uncertainties in biochemical models. IET Syst. Biol. 3(1), 1–9 (2009)
21. Steuer, R., Gross, T., Balsius, B.: Structural kinetic modeling of metabolic networks. Proc. Nat. Acad. Sci. U.S.A. 103(32), 11,868–11,873 (2006)
22. Steuer, R., Junker, B.H.: Computational models of metabolism: Stability and regulation in metabolic networks. Adv. Chem. Phys. 142 (2009)
23. Wang, L., Hatzimanikatis, V.: Metabolic engineering under uncertainty. I: framework development. Metab. Eng. 8(2), 133–141 (2006)
24. Zamboni, N., Sauer, U.: Novel biological insights through metabolomics and 13c-flux analysis. Curr. Opin. Microbiol. 12(5), 553–558 (2009)

Coexistence of Three Predators Competing for a Single Biotic Resource

Claude Lobry, Tewfik Sari, and Karim Yadi

Abstract. We construct a model of competition of three consumers for one single biotic resource ; simulations show that the three species coexist. Using singular perturbations theory we sketch a mathematical proof for this coexistence. The main mathematical tool used is an extension of the Pontryagin-Rodygin theorem on the "slow" motion of a "slow-fast" differential system when the "fast" motion possesses a stable limit cycle. The mathematical analysis is done within the framework of Non Standard Analysis.

1 Introduction

The question of coexistence of competing species for a single resource has a very long history that we shall not attempt to recall here. We just recall the two decisive papers by Armstrong and Mac Gehee [1, 6] where they pointed that *coexistence is not synonymous of coexistence at equilibrium*. These papers were the starting point of numerous papers showing complex behaviors of systems of competitors and evidence of coexistence on the basis of numerical simulations. Following this tradition we propose a model of coexistence of three species competing for one resource.

Claude Lobry
EPI-MERE, INRIA-Sophia Antipolis, FR
e-mail: claude.lobry@inria.fr

Tewfik Sari
EPI-MERE, INRIA-Sophia Antipolis and University of Haute Alsace, Mulhouse, FR
e-mail: Tewfik.Sari@sophia.inria.fr

Karim Yadi
Dynamical systems and Applications Laboratory, University Aboubekr Belkaïd, Tlemcen, DZ
e-mail: k.yadi@mail.univ-tlemcen.dz

J. Lévine & P. Müllhaupt (Eds.): Adv. in Theory of Control, Signals, LNCIS 407, pp. 309–321.
springerlink.com © Springer-Verlag Berlin Heidelberg 2010

The present paper has two parts. In the first part we construct our model and explain what is the rationale behind our construction ; each step is illustrated by simulations. In the second part we consider our model as a member of a more general "consumer-resource" model for which we explain how coexistence of species can be proved using singular perturbation analysis ; as an essential tool we use an extension of a theorem of Pontryagin and Rodygin.

Considered from the ecological point of view our model shows that oscillations in a "consumer-resource" relationship can open the door to coexistence with other species provided that the new introduced species do not perturb too much the oscillations. We do not know any example of an interaction between four species of the type of the model presented here but its existence is plausible. We shall explain it during the construction of a model. But we must acknowledge here that what we do is a kind of "virtual ecology" showing what is "theoretically possible" in a world of species respecting basic facts well established in concrete ecology. *It is not a description of the real world !*

From the mathematical point of view our paper can be considered as an application of singular perturbation methods to the mathematical proof of persistence for some specific system. Our contribution consists mainly in the analysis of the theorem of Pontryagin and Rodygin and its extension to a theorem which is more effective in some circumstances. Detailed proofs can be consulted at [14] and will be published elsewhere. The mathematical analysis is done within the framework of Non Standard Analysis, using the axiomatic of Nelson [7] and respecting the spirit of G. Reeb [9].

2 Construction of a Model

2.1 The Basic Oscillating Pair

We consider the system:

$$\begin{cases} \dfrac{ds}{dt} &= 3s(1-s) - \frac{s^2}{0.01+s^2}x_1 \\ \dfrac{dx_1}{dt} &= 0.1[\frac{s^2}{0.01+s^2} - 0.65]x_1 \end{cases} \tag{1}$$

Due to the presence of the factor 0.1, it can be considered as a "slow-fast" system of two differential equations of the following type:

$$\begin{cases} \dfrac{ds}{dt} &= \frac{1}{\varepsilon_1}[f(s) - g_1(s)x_1] \\ \dfrac{dx_1}{dt} &= \quad (g_1(s) - d_1)x_1 \end{cases}$$

where ε_1 is a "small" parameter. The real s represents the density of some biotic resource (prey) for a consumer which density is represented by x_1. This is a rather classical prey-predator model and it is well known that this kind of generalization of the Lotka-Volterra model can have sustained oscillations. The existence of oscillations in prey-predator interaction is clearly demonstrated in laboratory experiments (see [4]), if not in the real world. The choice of the function:

$$g_1(s) = \frac{s^2}{0.01 + s^2}$$

in place of the more classical Monod's function with s in place of s^2 was made in order that the nullcline $[f(s) - g_1(s)x_1] = 0$ has the "S-shape" shown on Fig. 1 which prevents the resource from extinction. This kind of assumption is sometimes called the "Allee" effect in ecological literature.

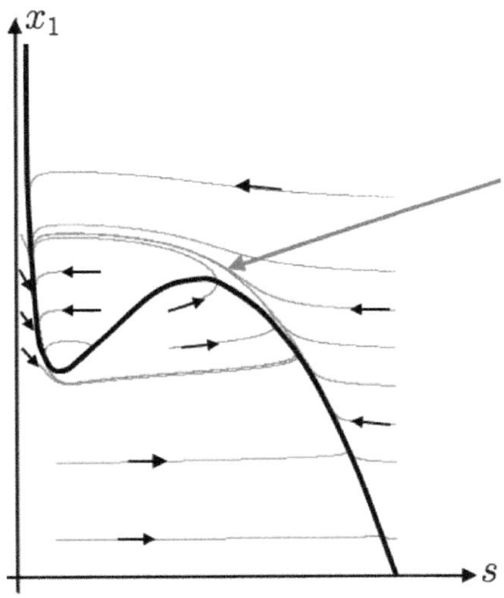

Fig. 1. On this picture one observes a simulation of few trajectories of the system (1). The black "S-shaped" curve is the nullcline $[f(s) - g_1(s)x_1] = 0$. The direction of the motion along trajectories is indicated by the black arrows and the limit cycle by the red arrow. By the way $s(t)$ oscillates between two values.

2.2 *Addition of a New Consumer "x_2"*

We want to add a new consumer and at the same time keep the oscillations of (s, x_1). It can be done by introducing a new species with a *very slow* dynamics compare to that of (s, x_1) like in the model:

$$\begin{cases} \dfrac{ds}{dt} &= \frac{1}{\varepsilon_1}[f(s) - g_1(s)x_1 - g_2(s)x_2] \\ \dfrac{dx_1}{dt} &= \quad (g_1(s) - d_1)x_1 \\ \dfrac{dx_2}{dt} &= \varepsilon_2\,(g_2(s) - d_2)x_2 \end{cases}$$

where ε_2 is small.

The existence of two consumers (which densities are represented by x_1 and x_2) of the same resource and having very different characteristic time seems to be common in nature. For instance small mammals and big mammals eating the same grass have a lifespan which may differ of an order of magnitude.

Denote by $\overline{x}_2$ some constant x_2 ; since $x_2(t)$ is quasi constant, for a while, the evolution of (s,x_1) is governed by:

$$\begin{cases} \dfrac{ds}{dt} &= \frac{1}{\varepsilon_1}[f(s) - g_1(s)x_1 - g_2(s)\overline{x}_2] \\ \dfrac{dx_1}{dt} &= \quad (g_1(s) - d_1)x_1 \end{cases}$$

In that system we see that when $\overline{x}_2$ is small the nullcline

$$f(s) - g_1(s)x_1 - g_2(s)\overline{x}_2 = 0 \tag{2}$$

is very close to the nullcline

$$f(s) - g_1(s)x_1 = 0 \tag{3}$$

and, thus, oscillations are preserved. The range of oscillations of s is slightly shortened as $\overline{x}_2$ increases. Now, if we look at the process from the point of view of x_2 during an oscillation of period T the growth is given by:

$$\int_t^{t+T} (g_2(s(\tau)) - d_2)x_2(\tau)d\tau$$

which varies monotonically according to the variation of amplitude of s. Thus the growth can be positive for small values of x_2 and negative for large ones ; in the middle there must be an equilibrium. This is the case for the model:

$$\begin{cases} \dfrac{ds}{dt} &= 3s(1-s) - \frac{s^2}{0.01+s^2}x_1 - g_2(s)x_2 \\ \dfrac{dx_1}{dt} &= 0.1[\frac{s^2}{0.01+s^2} - 0.65]x_1 \qquad with \quad \begin{cases} s < 0.5 \Rightarrow g_2(s) = 0 \\ s \geq 0.5 \Rightarrow g_2(s) = \frac{0.1(s-0.5)}{0.01+(s-0.5)} \end{cases} \\ \dfrac{dx_2}{dt} &= 0.01[g_2(s) - 0.025]x_2 \end{cases} \tag{4}$$

We choose $g_2 = 0$ on $[0, 0.5]$ in order to be sure that the nullcline (2) is exactly the same than the nullcline (3) and remains "S-shaped" ; thus oscillations are preserved. This artificial choice is convenient for simplicity but a model with g_2 being smoother would also work. Evidence of coexistence of x_1 and x_2 is given on Fig. 2. On this simulation three initial conditions where taken, keeping $s(0)$ and $x_1(0)$ constant and changing $x_2(0)$. On the picture we have superimposed the projections on the (s,x_1) plane (in green) and the (s,x_2) plane in red. The first initial condition for x_3 is a which is shown by the black arrow ; since the variation of x_2 is very slow the red projection looks like a point moving right to left and left to right very fast while moving up very slowly ; the thick red line corresponds to a dozen of oscillations. Starting from c we see that the point is slowly moving down. Starting from b we see that x_2 remains constant.

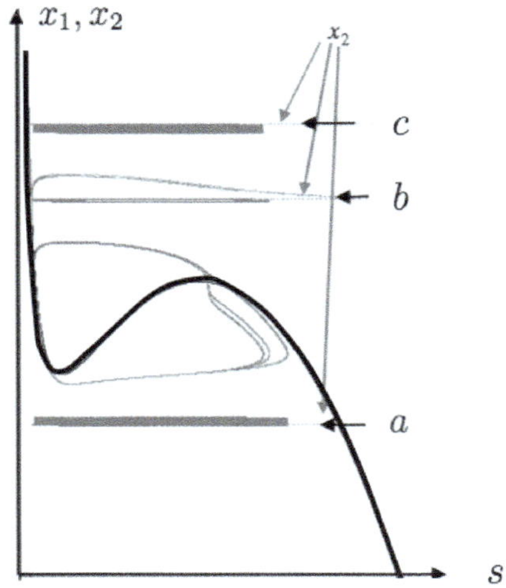

Fig. 2. Simulation of the system (4). The picture shows the two projections of the trajectories, in green on the plane (s,x_1) and in red on the plane (s,x_2). While the initial conditions where kept constant for s and x_1 they where changed for x_2. One sees that, starting from a small $x_2 = a$ then $x_2(t)$ is increasing and, conversely, starting from a big $x_2 = c$ then $x_2(t)$ is decreasing.

2.3 *Addition of a New Consumer* "x_3"

The idea is to have a new consumer with a g_3 growth rate such that the two graphs of g_2 and g_3 cross like on Fig. 3 in order that during an oscillation of (s,x_1) the species 2 and 3 take the advantage alternatively. This leads to the system:

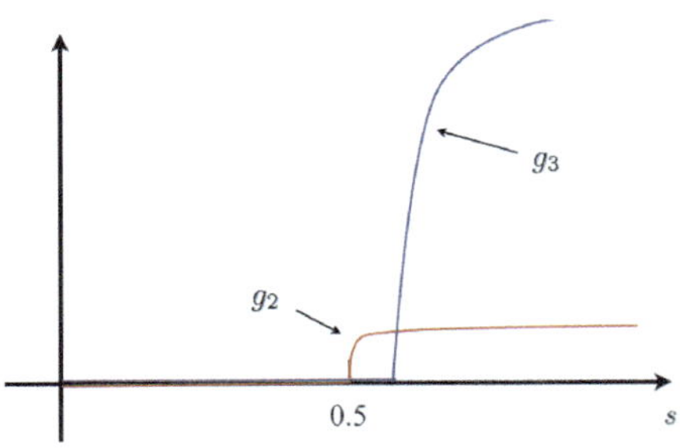

Fig. 3. The graphs of g_2 and g_3

$$\left\{\begin{aligned} \frac{ds}{dt} &= 3s(1-s) - \tfrac{s^2}{0.01+s^2}x_1 - g_3(s)x_3 \\ \frac{dx_1}{dt} &= 0.1[\tfrac{s^2}{0.01+s^2} - 0.65]x_1 \quad with \quad \left\{\begin{array}{l} s < 0.58 \Rightarrow g_3(s) = 0 \\ s \geq 0.58 \Rightarrow g_3(s) = \frac{2(s-0.58)}{0.01+(s-0.58)} \end{array}\right. \\ \frac{dx_3}{dt} &= 0.01[g_3(s) - 0.025]x_3 \end{aligned}\right. \tag{5}$$

On Fig.4 one sees that the behavior of the system (s,x_1,x_3) is similar to the behavior we observed for (s,x_1,x_2) ; the projection (in blue) on the (s,x_3) plane is similar to the projection (in red) observed for (s,x_2) in Fig.2.

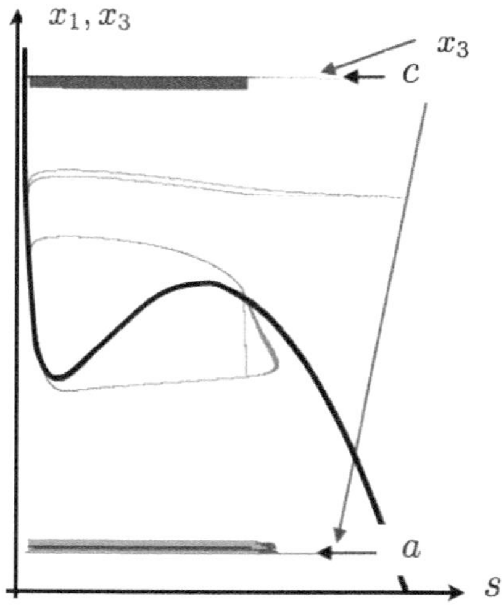

Fig. 4. Simulation of the system (5). The picture shows the two projections of the trajectories, in green on the plane (s,x_1) and in blue on the plane (s,x_3). The projection (in blue) on the (s,x_3) plane is similar to the projection (in red) observed for (s,x_2) in Fig.2. Starting from a small $x_3 = a$ then $x_3(t)$ is increasing and, conversely, starting from a big $x_3 = c$ then $x_3(t)$ is decreasing.
Between a and c there is some initial condition (not represented) for which x_3 do not increase nor decrease.

2.4 Coexistence of All the Three Consumers

Now we consider the complete system:

$$\begin{cases} \frac{ds}{dt} &= 3s(1-s) - \frac{s^2}{0.01+s^2}x_1 - g_2(s)x_2 - g_3(s)x_3 \\ \frac{dx_1}{dt} &= 0.1[\frac{s^2}{0.01+s^2} - 0.65]x_1 \\ \frac{dx_2}{dt} &= 0.01[g_2(s) - 0.025]x_2 \\ \frac{dx_3}{dt} &= 0.01[g_3(s) - 0.025]x_3 \end{cases} \quad (6)$$

Species x_2 and x_3 have a "slow motion" which can be approximated by computing suitable integrals along the basic cycle ; this determines a flow on (x_2, x_3) plane ; this flow is studied and proved to have a stable equilibrium which proves the persistence of both x_2 and x_3. On Fig.5 one sees the projection on the (x_2, x_3) plane of the full system (6) with the three competitors. The simulations from various initial conditions shows convergence to a point which actually corresponds to a periodic orbit in the full space (s, x_1, x_2, x_3). More details are given in the next section.

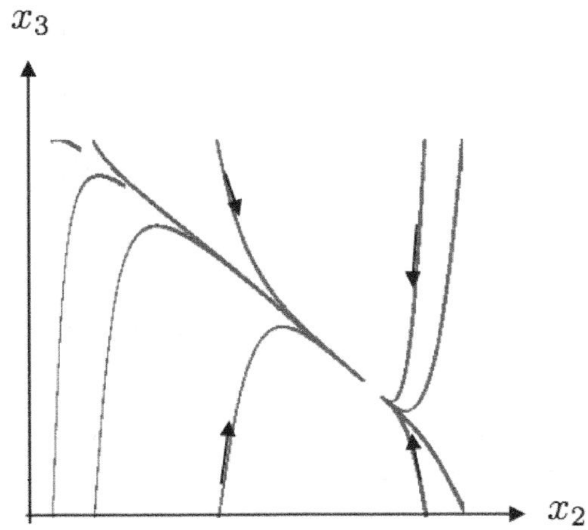

Fig. 5. On this picture we have represented the projection of simulations of the the complete system (6) on the plane (x_2, x_3). The variables (s_1, x_1) (not represented) present rapid oscillations while (x_2, x_3) evolves slowly. One sees that all the trajectories seem to converge to an equilibrium. Compare to the "theoretical" picture on Fig 8.

2.5 *Species 2 and 3 Alone*

Consider the system with s, x_2 and x_3 alone in the absence of the species represented by x_1. One easily checks that in this case, from the choice of g_2 and g_3, there is no oscillation and the stable equilibrium is the one for which species x_2 wins the competition.

3 Proof of the Persistence of the Three Competitors in a Model with Three Time Scales

In this section, we give the successive steps of a proof of the persistence in a model of the form

$$\left\{\begin{array}{ll}\frac{ds}{dt} &= \frac{1}{\varepsilon_1\varepsilon_2}(f(s) - g_1(s)x_1 - g_2(s)x_2 - g_3(s)x_3) \\ \frac{dx_1}{dt} &= \frac{1}{\varepsilon_1}(g_1(s) - d_1)x_1 \\ \frac{dx_2}{dt} &= (g_2(s) - d_2)x_2 \\ \frac{dx_3}{dt} &= (g_3(s) - d_3)x_3\end{array}\right. \tag{7}$$

where the occurring functions are differentiable, at least piecewise, ε_1 and ε_2 are **infinitesimal**[1]. The function f vanishes at 0 ; it is increasing then decreasing and vanishes at a value m. The functions g_i are zero at 0, increasing and bounded.

3.1 *Oscillations of s and x_1*

Let us consider the system

$$\left\{\begin{array}{ll}\frac{ds}{dt} &= \frac{1}{\varepsilon_1}(f(s) - g_1(s)x_1) \\ \frac{dx_1}{dt} &= (g_1(s) - d_1)x_1\end{array}\right. \tag{8}$$

- Suppose that the nullcline $ds/dt = 0$ is a curve φ that, when s increases from 0, decreases from $+\infty$ to a minimum value reached for $s = s^-$ then increases to a maximum value for $s = s^+$ to finally decrease and vanishes for $s = m$.
- Suppose that the value s^* such that $g_1(s^*) = d_1$ is between s^- and s^+.

Proposition 1. *For ε_1* ***infinitesimal****, the system (8) has a limit cycle* **close** *to the curve ABCD in Fig. 6.*

3.2 *The Pontryagin-Rodygin's Theorem*

Due to the lack of space, we shall try in these sections to avoid excessive mathematical formalism of the results and we refer to [12, 14, 13] for more details. To simplify, we suppose that all the occurring differential equations have the property of uniqueness of solutions. Let us consider the *slow and fast system*

$$\left\{\begin{array}{lll}\begin{bmatrix}\frac{ds}{dt} \\ \frac{dx_1}{dt}\end{bmatrix} &= \frac{1}{\varepsilon} & \begin{bmatrix}F(s,x_1,x) \\ G(s,x_1,x)\end{bmatrix} \\ \frac{dx}{dt} &= & H(s,x_1,x)\end{array}\right. \tag{9}$$

[1] We use N.S.A. terminology. Following the spirit of ([5]) when a word (like "infinitesimal") is written in bold character its meaning is the one used in the formal language of Nelson I.S.T. but the reader not familiar with this framework can use the intuitive meaning.

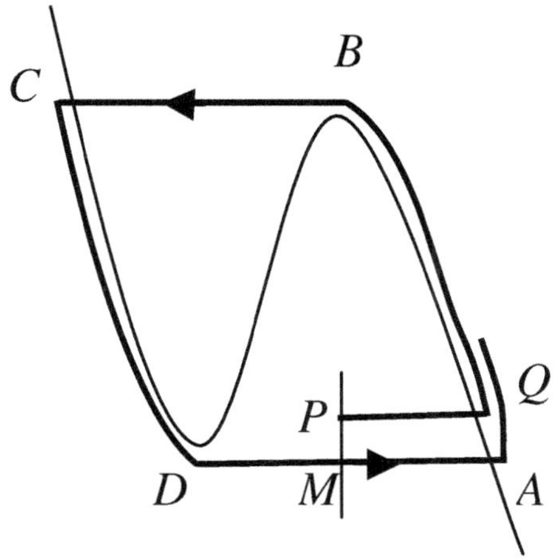

Fig. 6. Starting from the point P the solution is **quasi horizontal** and goes **fast** to the nullcline $ds/dt = 0$; then the solution stays **near** the nullcline and goes up until it reaches the maximum at B; then the solution is **quasi horizontal** and goes **fast** to the nullcline $ds/dt = 0$ at point C ; then the solution stays **near** the nullcline and goes down until it reaches the minimum at D ; then the solution goes **fast** to the right to the nullcline $ds/dt = 0$ and reaches it at A.

where the scalars s and x_1 are the *fast components*, and the vector x the *slow* one. The real number ε is positive and **infinitesimal**. The functions F, G and H are continuous. The following system, where x is considered as a parameter, is called the *fast equation*

$$\begin{bmatrix} \frac{ds}{dt} \\ \frac{dx_1}{dt} \end{bmatrix} = \frac{1}{\varepsilon} \begin{bmatrix} F(s,x_1,x) \\ G(s,x_1,x) \end{bmatrix} \tag{10}$$

Hence, the (s,x_1)-component of a solution of (9) varies very quickly according to (10) where x has been frozen at its initial value. When the fast equation (10) has stable limit cycles Γ_x for each x in a compact domain, Pontryagin-Rodygin's Theorem [8] gives the limiting behavior of the singularly perturbed problem (9): *Under suitable conditions, after a fast transition near the cycles described by the fast equation (10), the trajectories of (9) quickly roll up around the manifold generated by the cycles, with a slow evolution of the x-component according to the averaged system*

$$\frac{dx}{dt} = \frac{1}{P(x)} \int_0^{P(x)} H(s^*(\tau,x), x_1^*(\tau,x), x)\, d\tau \tag{11}$$

where $(s^*(\tau,x), x_1^*(\tau,x))$ *is a* $P(x)$*−periodic solution of the fast equation corresponding to the cycle* Γ_x. This result was originally obtained for at least C^2 vector fields, under the assumption that the cycles Γ_x are asymptotically stable in the linear approximation. However, the result obtained in [12] shows that Pontryagin-Rodygin description of solutions holds for C^0 vector fields under additional assumptions.

3.3 Extension of Pontryagin-Rodygin's Theorem

Note that System (7) has the form

$$\left\{ \begin{array}{lcl} \begin{bmatrix} \frac{ds}{dt} \\ \frac{dx_1}{dt} \end{bmatrix} & = \frac{1}{\varepsilon_2} & \begin{bmatrix} \frac{1}{\varepsilon_1} F(s,x_1,x) \\ G(s,x_1,x) \end{bmatrix} \\ \frac{dx}{dt} & = & H(s,x_1,x) \end{array} \right. \tag{12}$$

where $x = (x_2, x_3) \in \mathbb{R}^2$. The fast equation

$$\begin{bmatrix} \frac{ds}{dt} \\ \frac{dx_1}{dt} \end{bmatrix} = \frac{1}{\varepsilon_2} \begin{bmatrix} \frac{1}{\varepsilon_1} F(s,x_1,x) \\ G(s,x_1,x) \end{bmatrix} \tag{13}$$

admits a stable limit cycle for any value of x for **infinitesimal** values of ε_1. It is tempting to apply Pontryagin-Rodygin's Theorem to (12) but the main reason that makes this impossible is the fact that the fast equation is a **nonstandard** equation. It is itself a singularly perturbed equation. We can not avoid to take into account the three dynamics of the problem. In [14] Pontryagin-Rodygin's Theorem is extended to this kind of system the fast equation of which admits a *slow and fast limit cycle*. This new result has also the advantage to overcome a serious limitation of Pontryagin-Rodygin's Theorem: unlike the latter, it makes possible the localization of the cycles, the approximation of their periods and the calculation of the average along these cycles. *The functions F,G and H being continuous and the positive real numbers ε_1 and ε_2* ***infinitesimal****, suppose that there exists a compact domain K of $\mathbb{R}^2$ such that, for all $x \in K$ the nullclines $F = 0$ and $G = 0$ of (13) have the shape given in Figure 7. The (s,x_1)-plane is divided in four regions where the field has the indicated signs in the figure.* The limit cycle of (13) is **infinitesimally close** to the closed curve $(ABCD)$ in Fig. 6 formed by two "slow arcs" (AB) and (CD) and two "fast segments" (DA) and (BC). The two decreasing branches of the nullcline $F = 0$ are denoted $s = \psi_1(x_1,x)$ and $s = \psi_2(x_1,x)$. Let us define in the interior of K the *slow equation*

$$\frac{dx}{dt} = M(x), \tag{14}$$

where

$$\begin{aligned} M(x) &= \frac{1}{P(x)} \sum_{i=1}^{2} \int_{\xi_i(x)}^{\xi_{i+1}(x)} \frac{g(\psi_i(x_1,x),x_1,x)}{f_2(\psi_i(x_1,x),x_1,x)} dx_1, \\ P(x) &= \sum_{i=1}^{2} \int_{\xi_i(x)}^{\xi_{i+1}(x)} \frac{dx_1}{f_2(\psi_i(x_1,x),x_1,x)}, \text{ with } \xi_3(x) = \xi_1(x). \end{aligned} \tag{15}$$

Let $\gamma(t)$ be the trajectory of a solution of (12). Theorem 5.2.1 page 75 in [14] explains how $\gamma(t)$ behave in the same manner than in the classical Pontryagin-Rodygin

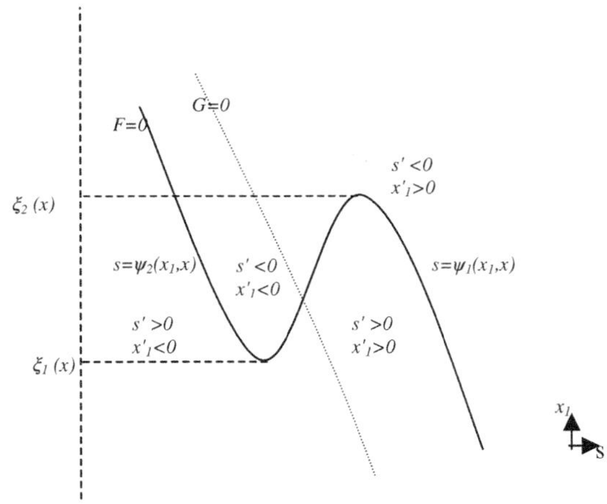

Fig. 7. Notations in equations (14)

theorem, the averaging on the cycles being now well approximated by the explicit formulas (15).

3.4 Application to the Model

Reconsider System (7).

- Assume that the functions g_2 and g_3 are zero until the respective thresholds s_2 and s_3 are reached such that $\min(s_2,s_3) \geq s^+$ and that g_2 and g_3 are increasing beyond.

This assumption allows us to assert that the subsystem

$$\begin{bmatrix} \frac{ds}{dt} \\ \frac{dx_1}{dt} \end{bmatrix} = \frac{1}{\varepsilon_2} \begin{bmatrix} \frac{1}{\varepsilon_1}(f(s) - g_1(s)x_1 - g_2(s)x_2 - g_3(s)x_3) \\ (g_1(s) - d_1)x_1 \end{bmatrix}$$

still admits, for every (x_2,x_3) and ε_1 small enough, a limit cycle Γ_{x_2,x_3} that differs from that of (8) for values of $s \geq \min(s_2,s_3)$. The more x_1 and x_2 are large, the more these cycles are distorted inwards in their right side. The minimum and maximum of the cycles remain unchanged. Here, the averaged equation (14,15) takes the form (see [14] for explicit formulas)

$$\begin{cases} dx_2/dt = x_2 M_2(x_2,x_3)/P(x_2,x_3), \\ dx_3/dt = x_3 M_3(x_2,x_3)/P(x_2,x_3). \end{cases} \tag{16}$$

A detailed study of equation (16) leads to the following conditions of persistence:

Theorem 1. *[14] Suppose that $s_3 > s_2 = s^+$, $\varphi(s_3) > \varphi(s^-)$ and that d_2 is below a certain constant well determined by the problem. Then, for $s_3 - s_2$ and d_3 small enough, there is persistence of the species x_2 and x_3 of (16).*

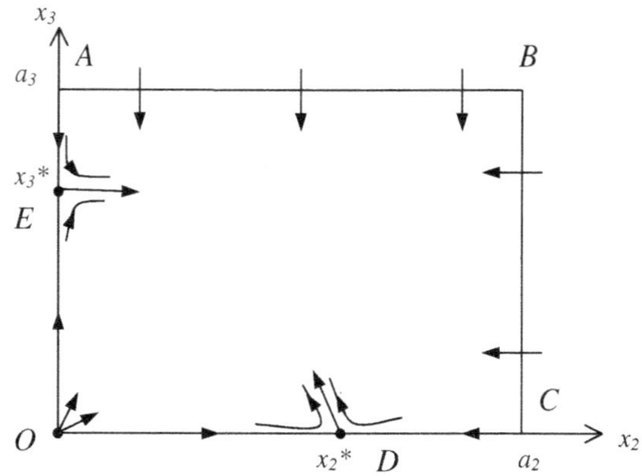

Fig. 8. Portrait of (x_2,x_3) obtained from the averaged system (16). Compare to the simulations presented on Fig.5

This result is reflected in Fig. 8 representing a positively invariant box *OABC* of (16) in which arrive all trajectories with positive initial conditions. The axes are invariant, the origin O is an unstable node and D and E are saddle points. A lemma due to Butler-McGehee [2] shows that the union of limit sets of positive half-trajectories is a compact subset Ω which does not meet the axes.

Theorem 2. *Under the assumptions of the preceding Theorem, there is persistence of the whole species of the model (7) for all positive initial conditions.*

This final result is obtained by using a nonstandard permanence lemma [10] which extends the approximation of the component $(x_2(t),x_3(t))$ of the solution of (7) by the solution of (16) to an **infinitely large** time interval $[0,\omega]$. We than prove[2] that $(x_2(t),x_3(t))$ remains **infinitely close to** Ω for all **infinitely large** values of t and all **infinitesimal** values of ε_1 and ε_2.

References

1. Armstrong, R.A., McGehee, R.: Coexistence of species competing for shared resources. Theoretical Pop. Biol. (9), 317–328 (1976)
2. Butler, G., Waltman, P.: Persistence in dynamical systems. J. Differential Equations (63), 255–263 (1986)
3. Callot, J.-L., Sari, T.: Stroboscopie et moyennisation dans les systèmes d'équations différentielles à solutions rapidement oscillantes. Mathematical Tools and Models for Control, Systems Analysis and Signal Processing, CNRS Paris (3), 345–353 (1983)
4. Jost, J.L., Drake, T.J.F., Frederickson, A.G., Tsuchiya, M.: Interaction of *Tetrahimen pyriformis, Escherichi coli, Azotobacter vinelandi* and Glucose in a Minimal Madium. J. of Bacteriology 113(2), 834–840 (1973)

[2] We say that the **standard** set $\{\Gamma_{x_2,x_3} \times \{(x_2,x_3)\} : (x_2,x_3) \in \Omega\}$ is *practically asymptotically stable* for (7) for all **infinitesimal** values of ε_1 and ε_2 (see [13] for more details).

5. Lobry, C., Sari, T.: Nonstandard analysis and representation of real world. International Journal on Control 80(3), 171–193 (2007)
6. McGehee, R., Armstrong, R.A.: Some mathematical problems concerning the ecological principle of competitive exclusion. Journal od Differential Equations (23), 30–52 (1977)
7. Nelson, E.: Internal Set Theory: a new approach to nonstandard analysis. Bull. Amer. Math. Soc. 83(6), 1165–1198 (1977)
8. Pontryagin, L.S., Rodygin, L.V.: Approximate solution of a system of ordinary differential equations involving a small parameter in the derivatives. Soviet. Math. Dokl. (1), 237–240 (1960)
9. Reeb, G.: La mathématique non standard vieille de soixante ans ? Troisième Colloque sur les Catégories, dédi é à C. Ehresmann, Amiens, 1980. Cahiers Topologie Géom. Différentielle 22(2), 149–154 (1981)
10. Robinson, A.: Nonstandard Analysis. American Elsevier, New York (1974)
11. Sari, T.: Averaging in Ordinary Differential Equations and Functional Differential Equations. In: van den Berg, I., Neves, V. (eds.) The Strength of Nonstandard Analysis, pp. 286–305. Springer, Wien (2007)
12. Sari, T., Yadi, K.: On Pontryagin–Rodygin's theorem for convergence of solutions of slow and fast systems. Electron. J. Diff. Eqns (139), 1–17 (2004)
13. Yadi, K.: Singular perturbations on infinite time interval. Revue Arima 9, 37–560 (2008)
14. Yadi, K.: Perturbations Singulières: Approximations, Stabilité Pratique et Applications à des Modèles de Compétition. Thèse de doctorat de l'Université de Haute-Alsace de Mulhouse (2008), `http://tel.archives-ouvertes.fr/tel-00411503/fr/`

Control Problems for One-Dimensional Fluids and Reactive Fluids with Moving Interfaces

Nicolas Petit

Abstract. The purpose of this paper is to expose several recent challenging control problems for mono-dimensional fluids or reactive fluids. These problems have in common the existence of a moving interface separating two spatial zones where the dynamics are rather different. All these problems are grounded on topics of engineering interest. The aim of the author is to expose the main control issues, possible solutions and to spur an interest for other future contributors. As will appear, mobile interfaces play key roles in various problems, and truly capture main phenomena at stake in the dynamics of the considered systems.

1 Introduction

The purpose of this paper is to expose several recent challenging control problems for mono-dimensional fluids or reactive fluids. These problems have in common the existence of a moving interface separating two spatial zones where the dynamics are rather different. All these problems are grounded on topics of engineering interest. The aim of the author is to expose the main control issues, possible solutions and to spur an interest for other future contributors. As will appear, mobile interfaces play key roles in various problems, and truly capture main phenomena at stake in the dynamics of the considered systems.

The paper contains a brief panorama. It is organized as follows. In Section 2, a Diesel oxidation catalyst for the automotive industry is considered. A boundary control problem is formulated for the outlet temperature control

Nicolas Petit
MINES ParisTech, Centre Automatique et Systèmes,
Unité Mathématiques et Systèmes
60, boulevard Saint-Michel, 75272 Paris, France
e-mail: `nicolas.petit@mines-paristech.fr`
`http://cas.ensmp.fr/~petit`

J. Lévine & P. Müllhaupt (Eds.): Adv. in Theory of Control, Signals, LNCIS 407, pp. 323–337.
springerlink.com © Springer-Verlag Berlin Heidelberg 2010

of this distributed reactive gaseous system. A model mobile interface separates the upstream reactive zone from the downstream transport zone. The location of this frontier is dependent on several variables, including measured disturbances and the control variable. In Section 3, a classic Stefan problem is presented. This system represents the melting of a solid phase into a liquid phase which is heated on its boundary. Heat propagates inside the system and generates a melting which occurs at a distance from the heat source which varies as the solid phase melts or grows. As will appear, both the temperature and the location of the moving interface can be controlled by the boundary actuation. In Section 4, some mixing models for stirring vessels are exposed. The proposed models use a mobile interface separating a distributed plug flow regime from a continuously stirred homogenous zone. The motion of the interface is generated by the variations of the blending speed which is a control variable. Finally, in Section 5, some recent developments on multiphasis slug flow are exposed. They appear in the petroleum industry. Slugs are large bubbles of gas separating pockets of liquid. They appear under certain flow conditions, and must be avoided as they have malicious effects. Models for them, involving a virtual choke which plays the role of a controlled interface, are discussed.

2 Diesel Oxidation Catalyst

This introductory example comes from the automotive engine control world. On most modern diesel vehicles, the increasing requirements regarding particulate matter emissions are satisfied using a particulate filter (DPF). This device is now widely spread among new vehicles. The filter, located in the vehicle exhaust line, stores particulate matter until it is burnt during an active regeneration process. This regeneration is achieved by raising the filter temperature (between 450 and 600 degrees) in the presence of oxygen in a diesel oxidation catalyst (DOC).

Historically, oxidation catalysts have been the first aftertreatment systems in the automotive industry. Catalysts used for diesel applications have appeared only recently because of the relative lower values of hydrocarbon reductants HC and CO emissions found in compression ignition engines compared against spark ignition engines. Because the HC and CO reactions are strongly exothermic, the DOC is also used to control the exhaust line temperature. In particular, it is used to generate the temperature required for the already mentioned DPF active regeneration. To increase the DPF inlet temperature, reductants are oxidized inside the DOC, which, in turn, increases its outlet temperature.

After treatment systems use monolith converters which are designed to maximize the mass transfer to the catalytic surface. To this end, the channels of the monolith are narrow and numerous (a typical order of magnitude is

400 cpsi). This geometric configuration (see Figure 1) also yields highly-efficient heat transfer between gas and solid. Hence, the solid phase (i.e. the monolith) acts as a spatially-distributed storage of energy and species. As can be experimentally observed, the induced propagation phenomenon leads to highly-delayed responses. Models for these devices are based on one-dimensional distributed parameter equations. These one-dimensional effects must be included in the modeling and further, they must be accounted for in the control strategies if performance is desired.

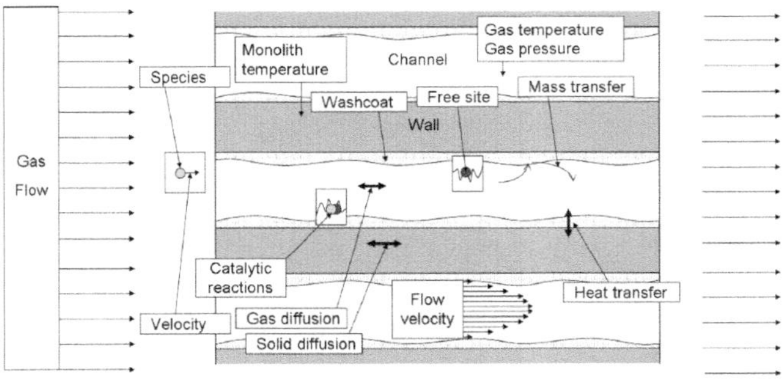

Fig. 1. Phenomena involved in the numerous channels of a Diesel Oxidation Catalyst. Reductant species in the exhaust gas are converted on the distributed catalyst surface.

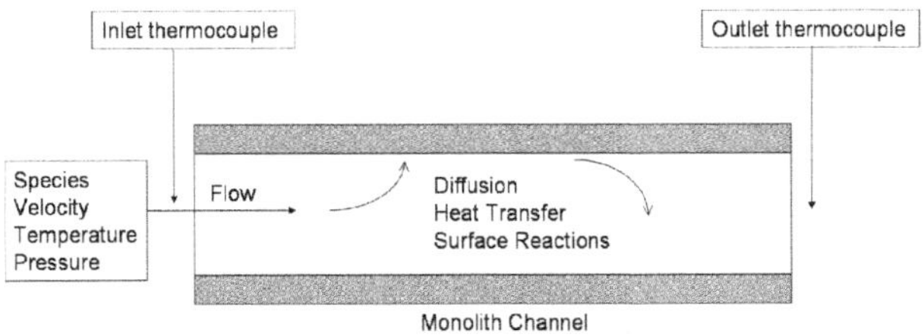

Fig. 2. Scheme of governing phenomena in a Diesel Oxidation Catalyst

Considering thermal effects, a simple model for the DOC consists of the following balance equations

$$\frac{\partial T}{\partial t} + v\frac{\partial T}{\partial z} = -k_1(T - T_s) \quad (1)$$

$$\frac{\partial T_s}{\partial z} = k_2(T - T_s) \quad (2)$$

which represent the dynamics of the temperature of the gas (T) and the temperature of the monolith (T_s). These equations are pictured in Figure 2. The control variable is the inlet temperature

$$T(0,t) = u(t)$$

Mathematically, this system of equations has a surprisingly long response time, which is consistent with experimental observations [16] discussed earlier.

In details, the input-output relationship can be easily calculated from the following transfer function (in the Laplace domain)

$$\hat{T}(z,s) = \hat{u}(s)\exp(-\frac{z}{v}s - \frac{k_1 z}{v} + \frac{m}{s+k_2}) \quad (3)$$

which gives, in the time-domain,

$$T(z,t) = H(t-\frac{z}{v})\exp(-\frac{k_1 z}{v}) \times \left(u(t-\frac{z}{v}) + \int_0^{t-\frac{z}{v}} \exp(-k_2\tau)\sqrt{\frac{m}{\tau}} I_1(2\sqrt{m\tau}) u(t-\frac{z}{v}-\tau)d\tau \right)$$

where H is the Heaviside function. The above formula, which involves a modified Bessel function, kindly fits experimental data, as can be observed in Figure 3.

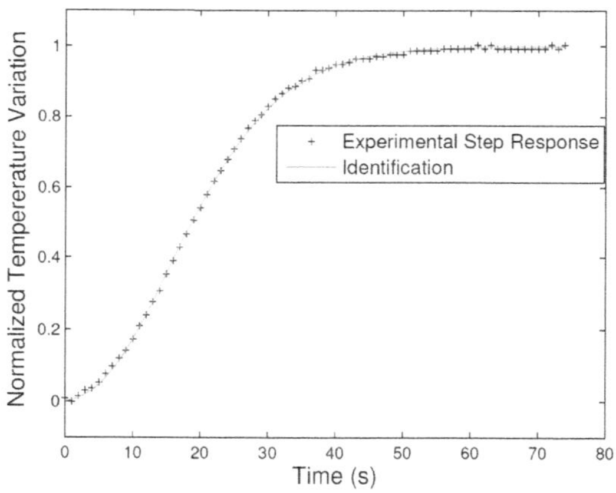

Fig. 3. Experimental data versus the DOC model

For (open-loop) control design, the transfer function (3) can be readily inverted. This gives

$$\hat{u}(s) = \exp(\frac{z}{v}s + \frac{k_1 z}{v} - \frac{m}{s+k_2})\hat{y}(s)$$

which, back in the time domain, yields an explicit formula using a Bessel function and a compact support convolution (see [16]). This formula provides a straightforward open-loop control law: given histories for the output temperature, one can simply determine the corresponding inlet temperature histories.

In practical applications, the true dynamics of the DOC systems is not simply a temperature-gas transport pass a solid monolith. In facts, the control variable is not the temperature, but, equivalently, the injected mass of fuel. These reductants are oxidized at the entry of the DOC system and, in turn, generate heat. One can model this heat generation using the mobile interface scheme of Figure 4. In fact, the DOC consists of two zones. An upstream reactive zone, and a temperature transport zone. The length of the (upstream) reactive zone directly depends on the amount of reductants under consideration, which is a control variable. It thus varies with the operating point. In turn, the complementary downstream transport zone also has a variable length.

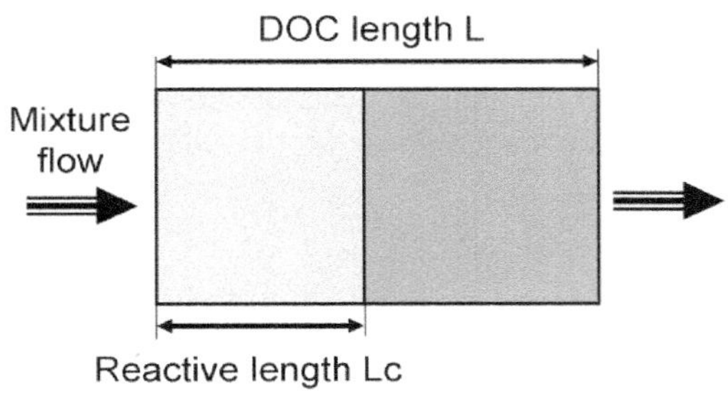

Fig. 4. DOC heat release model: reductants are spatially uniformly oxidized on the upstream part of the DOC. The system can be split into two zones separated by a moving interface: a reactive zone and a transport zone.

The location of the mobile interface can be identified quite accurately on experimental data. In practice, it is of great importance to account for the location of the interface in the derivation of control strategies. In particular, linear controllers reveal themselves to be efficient so long as they incorporate this variability in the computation of the gain scheduling and feed-forward actions [16, 17]. The reader can refer to [16, 18, 15] for practical vehicle applications relying on this model.

3 A Nonlinear Stefan Problem

In this second example, we study a heat diffusion equation with an endothermic reaction on a varying length. This can be seen as a crystal growth problem. Here, as in the previous example, the location of the mobile interface also depends on the control variable, but less directly, through the whole system dynamics.

In the papers [7, 8] it has been shown how to calculate open-loop trajectories for a nonlinear Stefan problem. It is a system governed by a nonlinear parabolic partial differential equation which has been vastly studied from a numerical analysis point of view, i.e. to compute solutions for future times knowing initial conditions and future control actions. We desire to solve the inverse problem, i.e. knowing the behavior of the free boundary *a priori* we seek a solution, here as a convergent series, to calculate the control and description trajectories between two stationary states.

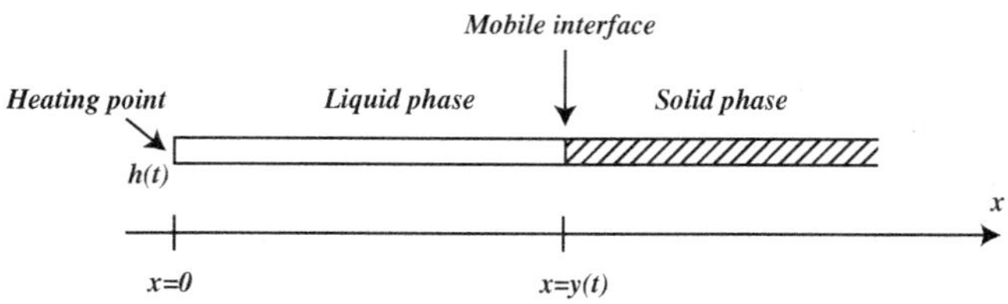

Fig. 5. Stefan problem with boundary control. Liquid phase with boundary control governed by a reaction-diffusion nonlinear partial differential equation in contact with a solid phase.

The classic Stefan problem considers a liquid phase column in contact at 0 degrees with an infinite phase solid, as shown in Figure 5. This problem is presented in detail in [1]. A list of problem reducing to this one can be found in [22] (including many processes formation and melting of crystals). Here, the Stefan problem is amended by adding a diffusion term and a nonlinear reaction term. This is a simplified model of reactant coolant fluid surrounded by solid phase.

Note $(x,t) \mapsto u(x,t)$ the temperature in the liquid phase, and $t \mapsto y(t)$ the varying location of the liquid/solid interface. The mappings $h(t)$ and $\psi(x)$ are the temperature on the fixed boundary ($x = 0$) and the initial condition, respectively ($t = 0$). The nonlinear Stefan problem consists in finding $u(x,t)$ and $y(t)$, for given $h(t)$ and $\psi(x)$ satisfying

$$\left.\begin{array}{ll} u_t = \ u_{xx} - \nu u_x - \rho u^2, & \forall (x,t) \in D_T \\ u(0,t) = h(t) \geq 0, & 0 < t \leq T \\ u(x,0) = \psi(x) \geq 0, & 0 \leq x \leq y(0) \\ u(y(t),t) = 0, \ u_x(y(t),t) = -\dot{y}(t), & 0 < t \leq T \end{array}\right\} \tag{4}$$

with

$$D_T \equiv \{(x,t) \ : \ 0 < x < y(t), \ 0 < t \leq T\}$$

where the boundaries are noted

$$B_T \equiv \{(0,t) : 0 < t \leq T\} \cup \{(x,0) : 0 \leq x \leq y(0)\} \cup \{(y(t),t) : 0 < t \leq T\}$$

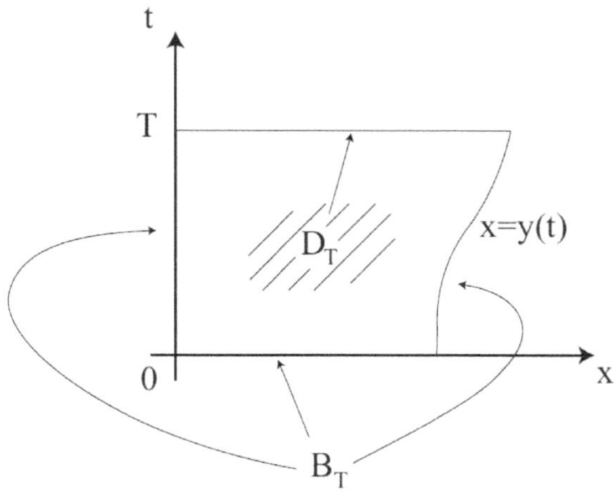

The boundary condition $u_x(y(t),t) = -\dot{y}(t)$ expresses that the heat flux at the interface is used for melting (or crystallization) of the solid phase. The parameters of conductivity and heat latent liquefaction are standard here, but without limitation, one may consider any factors in changes of variables x and t.

The inverse problem is to calculate the boundary control $h(t)$ allowing the transition between two stationary states. As noted in [11], it is a non-Cauchy problem characteristic with the Cauchy data. This nonlinear problem can be solved by the following method. One can seek solutions (4) under the form of the following series

$$u(x,t) = \sum_{n=0}^{\infty} \frac{a_n(t)}{n!}[x - y(t)]^n \tag{5}$$

where the coefficients $(a_n(t))$ satisfy the induction relations which are necessary and sufficient

$$a_n = \dot{a}_{n-2} - a_{n-1}\dot{y} + \nu a_{n-1} + \rho \sum_{k=0}^{n-2} \binom{n-2}{k} a_{n-2-k}\, a_k$$

for $n \geq 2$, with $a_0 = 0$ (from $u(y(t),t) = 0$) and $a_1 = -\dot{y}$ (from $-u_x(y(t),t) = \dot{y}(t)$).

By increments, one can show that the series (5) is *absolutely convergent* where there exists strictly positives parameters M, R, T such that

$$|y^{(l+1)}(t)| \leq M\frac{l!^{\alpha}}{R^l},\ \forall\, l = 0, 1, 2, ..., \forall t \in [0, T]$$

A lower limit to its radius of convergence can be easily determined. The main difficulties lie in the calculation of recurrence bounds on the successive derivatives of the coefficients $(a_n(t))$. This involves development of combinatorial derived cross terms from the nonlinear in u^2, for which one can use Chu-Vandermonde inequalities (see [21]). The lower bound on the radius of convergence is then calculated by analysis of roots of a polynomial of third degree. This lower bound can justify the use of this solution as series to solve the inverse problem of melting (or crystallization) of the solid phase by the control $h(t)$.

Suppose the liquid phase has an initial length L and that we wish to reach the length $L + \Delta L$ in finite time. It is a challenging problem because the actuator $h(t)$ is located at the opposite end fixed the liquid-solid interface which will move over time. The control must compensate the energy loss due to melting solid and that due to diffusion and reaction term. To solve this problem, simply use the function

$$y(\tau) = \begin{cases} L + \Delta L & \text{if } \tau \geq T, \\ L + \Delta L g(\tau/T) & \text{if } T > \tau > 0, \\ L & \text{if } \tau \leq 0, \end{cases}$$

where

$$g(\tau) = \frac{f(\tau)}{f(\tau) + f(1-\tau)}, \quad \tau \in [0, 1],$$

and

$$f(\tau) = \begin{cases} e^{-\frac{1}{\tau}} & \text{if } \tau > 0, \\ 0 & \text{if } \tau \leq 0. \end{cases}$$

This function defines a smooth transition between the lengths L and $L + \Delta L$. By choosing the parameter T depending on other physical parameters, one can guarantee that the radius of convergence of the series is larger than $L + \Delta L$ proving that that the series expansion, and therefore the solution to the inverse problem are valid.

This work follows [19] on reaction diffusion equation with fixed boundary. Besides convergence of this series for a very special class of Gevrey functions (as defined in [1]) used in an explicit assumption depending on physical parameters of the system, one can also prove a maximum principle stating that the maximum temperature is always achieved on the sides of the domain [8]. Asymptotic positivity property of the solution can also be established.

4 Mixing Models

We now pursue our panorama of distributed systems with mobile interfaces by considering mixing systems[1]. In this case, a rather unusual model can be

[1] The interested reader can refer to [20, 2] treating the related problem of blending systems.

proposed, where the motion of the moving interface separating a homogeneous zone and a distributed zone depends on the derivative of the input signal.

We expose ways to model mixing phenomena for Newtonian fluids under unsteady stirring conditions in agitated vessels using helical ribbon impellers. A model of torus reactor including a well-mixed zone and a transport zone is considered. The originality of the arrangement of ideal reactors developed in [5, 6] lies in the time-dependent location of the boundaries between the two zones. Interestingly, this concept is applied to model the positive influence of unsteady stirring conditions on homogenization process. It appears that this model allows the easy derivation of a control law, which is a great advantage when optimizing the dynamics of a mixing process. We now detail this model.

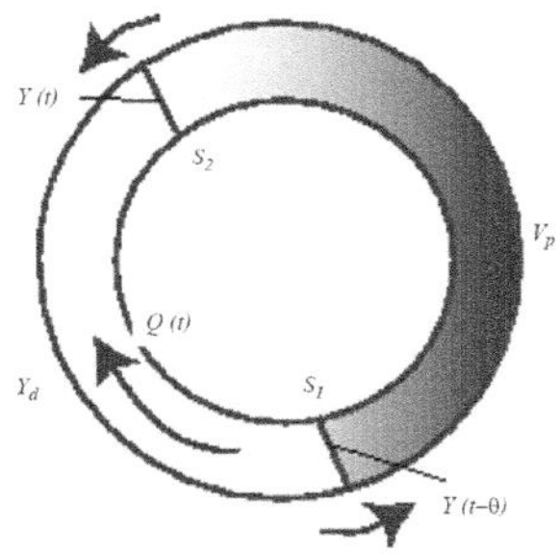

Fig. 6. Sketch of torus model proposed in [5, 6]

The mixing system is as follows. Consider a torus of fixed volume V divided into two ideal reactors (a constant stirred tank reactor of volume V_d and a plug flow zone of volume $V_p = V - V_d$) in which flows a Newtonian fluid with a uniform time-varying flow rate $\dot{Q}$ in a clockwise direction (see Figure 6). Further, y refers to the fluid concentration (kg/m3) which varies with time and space. It is assumed that the total material quantity of the component y in the reactor remains constant. The originality of the torus reactor arises from the time-dependent position of the interfaces (S1 and S2) which separate the two ideal flow zones. Indeed, it is assumed that S1 and S2 move alternately in a counter-clockwise direction to the flow rate fluctuations. Consequently, when the flow rate is non-steady, the volumes (V_d and V_p) of the two ideal reactors are time variant. In particular, it is assumed that S1 (respectively, S2) moves only when positive (respectively, negative) variations in the flow rate occur in the torus volume and is otherwise motionless. Note also, that when a variation of flow rate occurs, not only the volumes of the zones vary but their location within the torus evolves counter-clockwise. We assume that at each time t the flow rate $\dot{Q}(t)$ is proportional to the impeller rotational speed $N(t)$. For steady operations simulation results obtained with this model are close to

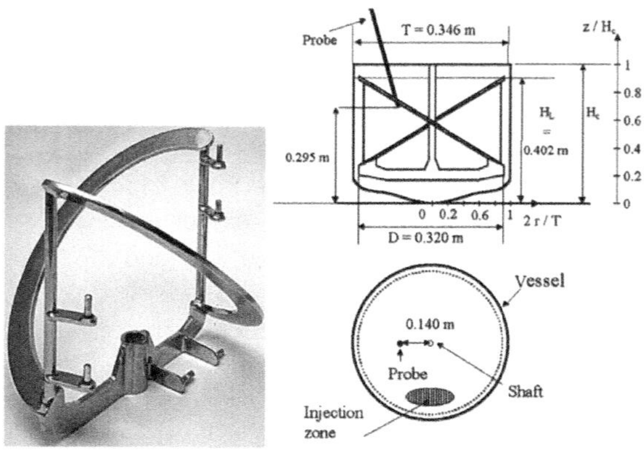

Fig. 7. A mixing vessel used for experimental validation of the proposed mixing model, from [6]

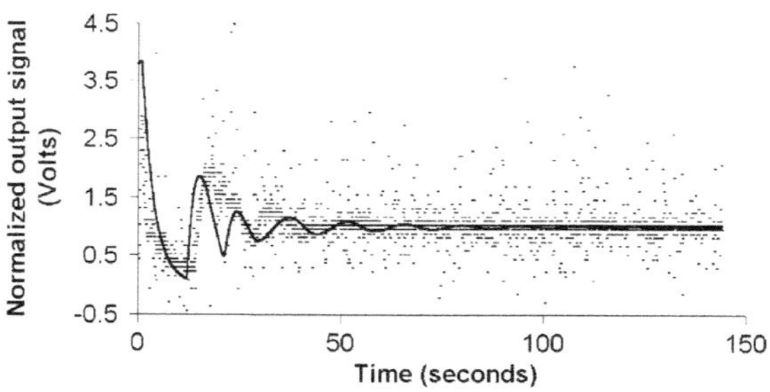

Fig. 8. The torus model reproduces well the experimentally observed mixing measured by a conductivity probe, from [6]

the reference results [14]. In the case of unsteady stirring, the model accounts for the experimental observation that an improvement in mixing occurs when a positive variation in the rotational speed is enforced. For example, in the case of a positive variation in impeller rotational speed, the volume of the stirred tank reactor increases while that of the plug flow decreases. As the whole volume of the torus loop is unchanged, an enhancement in mixing is expected.

Note $\dot{V}_d^+$ (respectively, $\dot{V}_d^-$) the variation of volume V_d due to the motion of S1 (resp., S2) in the torus, and let θ be the residence time of the particle leaving the plug flow zone at time t. Then, the whole system can be characterized by the following differential equations

$$
\begin{aligned}
&V = V_d(t) + V_p(t)\\
&\int_{t-\theta}^{t} \dot{Q}(\sigma)d\sigma = V - V_d(\dot{Q}(t-\theta)) - \int_{t-\theta}^{t} \dot{V}_d^{+}(\sigma)d\sigma\\
&V_d(\dot{Q}(t))\frac{d\,(y(t))}{dt} = (\dot{Q}(t) + \dot{V}_d^{+})(y(t-\theta) - y(t)),\\
&\dot{Q}(t) = \alpha N(t)
\end{aligned}
$$

with

$$
\begin{aligned}
\dot{V}_d^{+} &= k\frac{dN}{dt}, \text{ if } \frac{dN}{dt} > 0, \ \dot{V}_d^{+} = 0 \text{ otherwise},\\
\dot{V}_d^{-} &= -k\frac{dN}{dt}, \text{ if } \frac{dN}{dt} < 0, \ \dot{V}_d^{+} = 0 \text{ otherwise}.
\end{aligned}
$$

This model represents experimental data well. To check its validity, a mixing vessel pictured in Figure 7 was used. The agitated fluid is an aqueous solution of glucose. The rotational speed was controlled to reproduce increasing and decreasing ramps. A conductivity probe was used to obtain the circulation curves in the vessel. The rotational speed and the conductivity signal were recorded throughout the mixing process. The values of the rotational speed varied from 0.16 to 1.5 rev/s. Mixing and circulation times were determined from the response signal recorded after tracer injection. As is pictured in Figure 8, for the experimental conditions tested, the probe conductivity measurements are in close agreement with the expected behavior reproduced by the model.

5 Multiphasis Slug Flow

Finally, we wish to complete our catalogue of distributed systems with moving interfaces with a problem of multiphase flow. This problem is of great importance in the oil industry where long pipes (named risers, or flowlines) are used to transport large blends of gas, oil and water. The gas and the liquid phase do not mix, and, in the case when the dispersed bubbles gather, they form large bubbles, named "slugs" which induce malicious pressure variations which are highly detrimental for industrial facilities. In such cases, the interface is the boundary between liquid and gas phase. It is indirectly controlled by remote inputs.

In details, risers are long pipes connecting reservoirs to surface facilities for oil production. Severe slugging is a flow regime that arises mostly when entering tail production of an oil field. It is characterized by an unstable multiphase flow, where slugs of liquid accumulate before being pushed upwards by the gas. It is also characterized by oscillations of the pressure in the pipeline and oscillations of flow rates of gas and oil at the production end of the pipe. Although the phenomenon itself can be observed and sometimes reproduced on test rigs, its causes are not always known. Because the severe slugging flow regime can damage the installations (and most importantly reduce the oil production), various techniques have been investigated in view of suppressing it. The riser length typically ranges from a few hundred meters to several kilometers. To avoid instability, the most straightforward technique

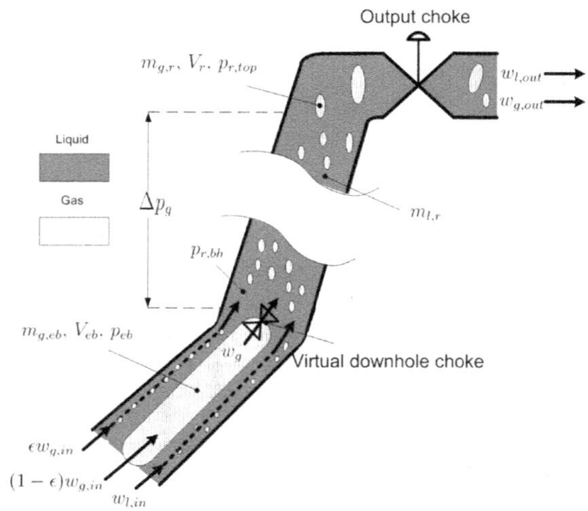

Fig. 9. A vertical riser carrying a multiphase flow, from [3]. An elongated bubble located at the bottom of the riser is subjected to a pressure buildup until it is released and generate a slugging flow. The interface between this elongated bubble and the rest of the riser is a virtually controlled interface.

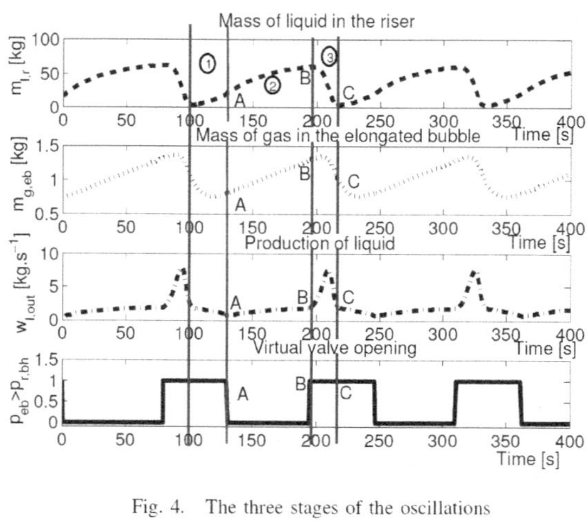

Fig. 10. The successive steps of the slugging cycle reproduced by the simple model

consists in choking down manually the pipes thanks to a choke located upstream the separator. Although this solution stabilizes the flow, it reduces significantly the oil production which, in turn, motivates the investigation of dynamic control of the valve. Indeed, it is possible, in closed-loop, to stabilize at higher flow rates.

Two classes of models can be found in the literature. The most accurate type is based on (usually nonlinear) partial differential equations representing two-phase (oil and gas) or three-phase (oil, water and gas) flows. These models are able to reproduce the slugging phenomenon in many cases, but fail to match the behavior of real-life wells in other cases, in particular when the instability comes from reservoir dynamics, for which there is little knowledge about. Unfortunately, it is not possible to derive control laws from these models because of their complexity. The second class of models is based on ordinary differential equations and represents a different trade-off between accuracy and complexity. A prime example is the model presented in [13], which, besides its numerous merits, does not sufficiently rely on physics to accurately reproduce the physical response of the system. Finally, the model is not general enough and is designed for a specific geometry. This is also the case for the model of gas-lift presented in [23, 26, 25].

Consider a vertical riser subjected to a constant input flow. The output flow of the riser is controlled by a choke. Unstable flow regime can occur, especially when the choke is largely opened, which, unfortunately, corresponds to a point of industrial interest. This kind of instability is also observed on related systems: oil wells with a gas reservoir [28, 12], risers with low-point [27, 10, 9]. Generally, switches of valves are reported to be at the birth of the oscillating phenomena: downhole choke plays a key role in the casing-heading in [23, 12, 24], while the geometric low-point acting as a valve is studied in [27]. In the riser considered here, no such valve exists or is even suggested by the geometry. Yet, one can model the riser using a virtual choke located at a well chosen point at the bottom of the riser. In this approach, the riser is modeled as a three-state set of ordinary differential equations. As is detailed in [3], one can tune the model analytically to fit most physical systems of interest. The equations reproduce the slugging flow regime as follows. The elongated bubble is subjected to a pressure buildup until its pressure get high enough so that the bubble is eventually released and travels through the rest of the vertical riser and produced a slugging flow regime. Then, the pressure buildup starts over again. The successive steps of this cycle are pictured in Figure 10. Further, this model suggest a control design that allows to stabilize the flow. One can refer to [3, 4] for an experimental study and a mathematical derivation of this control law.

6 Conclusion

In this paper, several distributed parameter systems with an internal mobile interface have been presented. In each case, the governing equations are relatively simple, and it appears that the introduction of a mobile interface is a key feature to make this model realistic. At various levels, the location or the nature of the mobile interface can be controlled by the input variables. In the case of the DOC system, it is directly dependent on the amount of reductants entering the system. In the Stefan problem, the liquid-solid interface moves as the heat flux travels through the whole system. In the mixing vessel, the interface moves according to sign of the time variations of the rotation speed of the blender. In the multiphase flow, the interface is virtually actuated by a the histories of the control variable. Interestlingly, all these models are simple enough to provide direct insight into the solution of control problems of engineering interest: thermal response of the DOC system, inverse control of crystal growth, optimization of blending dynamics, stabilization of slugging flows. The interested reader can refer to [16, 8, 6, 3] and the references therein for details.

Acknowledgements. The author wishes to thank several colleagues and associates who have greatly contributed to the works presented in this article: in alphabetical order, Y. Creff, O. Lepreux, F. di Meglio, L. Sinègre.

References

1. Cannon, J.R.: The one-dimensional heat equation. Encyclopedia of Mathematics and its applications, vol. 23. Addison-Wesley Publishing Company, Reading (1984)
2. Chèbre, M., Creff, Y., Petit, N.: Feedback control and optimization for the production of commercial fuels by blending. Journal of Process Control 20(4), 441–451 (2010)
3. Di Meglio, F., Kaasa, G.-O., Petit, N.: A first principle model for multiphase slugging flow in vertical risers. In: Proc. of the 48th IEEE Conf. on Decision and Control (2009)
4. Di Meglio, F., Kaasa, G.-O., Petit, N., Alstad, V.: Model-based control of slugging flow: an experimental case study. In: Proc. of the, American Control Conference (to appear 2010)
5. Dieulot, J.-Y., Petit, N., Rouchon, P., Delaplace, G.: An arrangement of ideal zones with shifting boundaries as a way to model mixing processes in unsteady stirring conditions in agitated vessels. Chemical Engineering Science 60(20), 5544–5554 (2005)
6. Dieulot, J.-Y., Petit, N., Rouchon, P., Delaplace, G.: A torus model containing a sliding well-mixed zone as a way to represent mixing process at unsteady stirring conditions in agitated vessels. Chemical Engineering Communications 192, 805–826 (2005)
7. Dunbar, W.B., Petit, N., Rouchon, P., Martin, P.: Boundary control for a nonlinear Stefan problem. In: Proc. of the 42nd IEEE Conf. on Decision and Control (2003)
8. Dunbar, W.B., Petit, N., Rouchon, P., Martin, P.: Motion planning for a nonlinear Stefan problem. ESAIM: Control, Optimisation and Calculus of Variations 9, 275–296 (2003)

9. Duret, E.: Dynamique et contrôle des écoulements polyphasiques. PhD thesis, École des Mines de Paris (2005)
10. Henriot, V., Duret, E., Heintz, E., Courbot, A.: Multiphase production control: Application to slug flow. Oil & Gas Science and Technology 57(1), 87–98 (2002)
11. Hill, C.D.: Parabolic equations in one space variable and the non-characteristic Cauchy problem. Comm. Pure Appl. Math. 20, 619–633 (1967)
12. Hu, B.: Characterizing gas-lift instablilities. PhD thesis, Norwegian University of Science and Technology (2004)
13. Kaasa, G.-O.: Attenuation of slugging in unstable oil by nonlinear control. In: Proc. of the 17th IFAC World Congress (2008)
14. Khang, S.J., Levenspiel, O.: New scale-up and design method for stirrer agitated batch mixing vessels. Chemical Engineering Science 31, 569–577 (1976)
15. Lepreux, O.: Model-based Temperature Control of a Diesel Oxidation Catalyst. PhD thesis, École des Mines de Paris (2009)
16. Lepreux, O., Creff, Y., Petit, N.: Motion planning for a Diesel oxidation catalyst outlet temperature. In: Proc. of the 2008 American Control Conference (2008)
17. Lepreux, O., Creff, Y., Petit, N.: Model-based control design of a diesel oxidation catalyst. In: ADCHEM 2009, International Symposium on Advanced Control of Chemical Processes (2009)
18. Lepreux, O., Creff, Y., Petit, N.: Warm-up strategy for a diesel oxidation catalyst. In: Proc. of European Control Conf. 2009 (2009)
19. Lynch, A.F., Rudolph, J.: Flatness-based boundary control of a nonlinear parabolic equation modelling a tubular reactor. In: Isidori, A., Lamnabhi-Lagarrigue, F., Respondek, W. (eds.) Lecture Notes in Control and Information Sciences 259: Nonlinear Control in the Year 2000, vol. 2, pp. 45–54. Springer, Heidelberg (2000)
20. Petit, N., Creff, Y., Rouchon, P.: Motion planning for two classes of nonlinear systems with delays depending on the control. In: Proc. of the 37th IEEE Conf. on Decision and Control, pp. 1007–1011 (1998)
21. Petkovsek, M., Wilf, H.S., Zeilberger, D.: A=B, Wellesley (1996)
22. Rubinstein, L.I.: The Stefan problem. Translations of mathematical monographs, vol. 27. AMS, Providence (1971)
23. Sinègre, L.: Dynamic study of unstable phenomena stepping in gaslift activated systems. PhD thesis, École des Mines de Paris (2006)
24. Sinègre, L., Petit, N., Lemétayer, P., Gervaud, P., Ménégatti, P.: Casing-heading phenomenon in gas-lifted well as a limit cycle of a 2d model with switches. In: Proc. of the 16th IFAC World Congress (2005)
25. Sinègre, L., Petit, N., Ménégatti, P.: Predicting instabilities in gas-lifted wells simulation. In: Proc. of the 2006 American Control Conference (2006)
26. Sinègre, L., Petit, N., Saint-Pierre, T.: Active control strategy for density-wave in gas-lifted wells. In: Proc. of the ADCHEM 2006, International Symposium on Advanced Control of Chemical Processes (2006)
27. Storkaas, E.: Control solutions to avoid slug flow in pipeline-riser systems. PhD thesis, Norwegian University of Science and Technology (2005)
28. Torre, A.J., Blais, R.N., Brill, J., Doty, D., Schmidt, Z.: Casing-heading in flowing wells. In: SPE Production Operations Symposium, no. SPE 13801 (1987)

A Port-Hamiltonian Formulation of Open Chemical Reaction Networks

Arjan van der Schaft and Bernhard Maschke

1 Introduction

This paper discusses the geometric formulation of the dynamics of chemical reaction networks within the port-Hamiltonian formalism [10, 9, 6]. The basic idea dates back to the innovative work of Oster, Perselson and Katchalsky [8, 7]. The main contribution concerns the formulation of a Dirac structure based on the stoichiometric matrix, which is underlying the port-Hamiltonian formulation. Interaction with the environment is modelled through the boundary metabolites and their boundary fluxes and affinities. This allows a compositional view on chemical reaction network dynamics.

2 Stoichiometry, 1-Complexes, and the Stoichiometry Dirac Structure

Consider a chemical reaction network involving N chemical species (metabolites), among which M chemical reactions take place. The basic structure underlying the dynamics of the concentrations $x_i, i = 1, \cdots, N$, of the metabolites is given by the balance laws

$$\dot{x} = Sv \tag{1}$$

Arjan van der Schaft
Johann Bernoulli Institute for Mathematics and Computer Science
University of Groningen, P.O. Box 407, 9700 AK Groningen, the Netherlands
e-mail: a.j.van.der.schaft@rug.nl

Bernhard Maschke
Lab. d'Automatique et de Genie des Procédés
Université Claude Bernard Lyon-1, F-69622 Villeurbanne, France
e-mail: maschke@lagep.univ-lyon1.fr

J. Lévine & P. Müllhaupt (Eds.): Adv. in Theory of Control, Signals, LNCIS 407, pp. 339–348.
springerlink.com
© Springer-Verlag Berlin Heidelberg 2010

where S is an $N \times M$ matrix, called the *stoichiometric matrix*. The elements of the vector $v \in \mathbb{R}^M$ are commonly called the (reaction) *fluxes*. The stoichiometric matrix S, which consists of (positive and negative) integer elements, describes the basic chemical structure of the reactions. For example, the single chemical reaction

$$2H_2 + O_2 \to 2H_2O$$

involving the species H_2, O_2, H_2O has the stoichiometric matrix

$$S = \begin{bmatrix} -2 \\ -1 \\ 2 \end{bmatrix}$$

Furthermore, for every set of chemical reactions one may define the so-called *elemental matrix* E, which captures the conservation of elements constituting the chemical species. In the above example the elemental matrix is given as

$$E = \begin{bmatrix} 0 & 2 & 1 \\ 2 & 0 & 2 \end{bmatrix}$$

where the first row corresponds to oxygen (O) conservation, and the second row to hydrogen (H) conservation. It follows that

$$ES = 0 \tag{2}$$

Chemical reaction networks do *not* immediately correspond to ordinary graphs. For example, if one associates to every chemical species a vertex of a graph then one cannot directly associate a chemical reaction to an edge, since a chemical reaction usually involves more than two species. Nevertheless, it *is* natural to associate to a chemical reaction network a 1-*complex*, which is a notion generalizing that of a graph[1]. Indeed, because of (2), a chemical reaction network with stoichiometric matrix S and elemental matrix matrix E defines the 1-complex

$$\Lambda_1 \overset{S}{\to} \Lambda_0 \overset{E}{\to} \Lambda_{-1} \tag{3}$$

Here Λ_1 is the vector space of all functions from the set of chemical species to $\mathbb{R}$, identified with $\mathbb{R}^N$, while Λ_0 is the vector space of all functions from the set of chemical reactions to $\mathbb{R}$, identified with $\mathbb{R}^M$. and finally Λ_{-1} is the vector space of all functions from the set of chemical elements to $\mathbb{R}$. (In the above example $\Lambda_1 = \mathbb{R}^1$, $\Lambda_0 = \mathbb{R}^3$, and $\Lambda_{-1} = \mathbb{R}^2$.)

Remark 1. A directed graph with N vertices, M edges, and incidence matrix B defines the 1-complex

[1] Other possibilities are to look at the chemical reaction network as a *species-reaction graph* [3], or as a *Petri-net* [4], with *transitions* corresponding to the reactions and *places* corresponding to species. Still another option is to use *hypergraphs*.

$$\Lambda_1 \xrightarrow{B} \Lambda_0 \xrightarrow{\mathbb{1}} \Lambda_{-1} = \mathbb{R}$$

with $\Lambda_1 = \mathbb{R}^M$, $\Lambda_0 = \mathbb{R}^N$, and $\mathbb{1}$ the vector consisting of all ones. Hence one can regard the stoichiometric matrix S to be the analogue of the incidence matrix of a graph.

In the example above $\operatorname{rank} E = \operatorname{corank} S$. However in general we only have $\operatorname{rank} E \leq \operatorname{corank} S$. In fact, if k is another N-dimensional row-vector satisfying $kS = 0$ then

$$\frac{d}{dt}(kx) = kSv = 0,$$

and kx is a conserved quantity (*conserved moiety*).

In many cases of interest, especially in biochemical reaction networks, chemical reaction networks are intrinsically *open*, in the sense that there is a continuous exchange with the environment (in particular, other reaction networks). This will be modelled by splitting the total vector of fluxes into a vector of *internal fluxes* v_i and a vector of *boundary* (or, exchange) *fluxes* v_b, corresponding to a splitting of the stoichiometric matrix S as

$$S = \begin{bmatrix} S_i & S_b \end{bmatrix}$$

whereby the dynamics takes the form

$$\dot{x} = S_i v_i + S_b v_b \tag{4}$$

Usually the boundary fluxes v_b are uptake (or demand) reactions for part of the metabolites, which we will call the *boundary metabolites.* Thus, boundary metabolites may also participate in other chemical reaction networks.

Example 1. Consider the reactions

$$\begin{aligned} &\to A \\ A + B &\to C + B \\ C &\to \end{aligned}$$

having stoichiometric matrix

$$S = \begin{bmatrix} S_i & S_b \end{bmatrix} = \left[\begin{array}{c|cc} -1 & 1 & 0 \\ 0 & 0 & 0 \\ 1 & 0 & -1 \end{array}\right]$$

with boundary metabolites A and C.

In the next section we will consider the (generalized) Hamiltonian formulation of chemical reaction networks. The generalized Hamiltonian formulation of dynamics involves two notions [10, 9, 6]. The first one concerns the constitutive relations, in particular those of energy storage and energy dissipation (resistive relations). The second one is an underlying geometric structure, which 'regulates' the power flow in the system. In classical mechanics this geometric structure is given by a symplectic form or a Poisson structure. In general, see e.g. [10, 9, 6] this geometric structure is given by a *Dirac structure*. For chemical reaction networks we will show how the Dirac structure is determined by the stoichiometric matrix.

Recall [2, 10, 9] that a subspace $\mathcal{D} \subset \mathcal{V} \times \mathcal{V}^*$ for some vector space $\mathcal{V}$ and its dual space $\mathcal{V}^*$ defines a (constant) Dirac structure if

$$\mathcal{D} = \mathcal{D}^{\perp} \tag{5}$$

where $^{\perp}$ denotes the orthogonal complement with respect to the indefinite inner product $\ll \cdot, \cdot \gg$ on $\mathcal{V} \times \mathcal{V}^*$ defined as

$$\ll (v_1, v_1^*), (v_2, v_2^*) \gg := < v_1^* \mid v_2 > + < v_2^* \mid v_1 >,$$

with $v_1, v_2 \in \mathcal{V}, v_1^*, v_2^* \in \mathcal{V}^*$, where $< \cdot \mid \cdot >$ denotes the duality product between $\mathcal{V}$ and $\mathcal{V}^*$.

In the finite-dimensional case an equivalent characterization of Dirac structures is given as follows [9], [6].

Proposition 1. *A subspace*

$$\mathcal{D} \subset \mathcal{V} \times \mathcal{V}^*$$

is a Dirac structure if and only if the following two conditions are satisfied:

$$\begin{aligned} &(i) < v^* \mid v >= 0, \qquad \textit{for all } (v, v^*) \in \mathcal{D} \\ &(ii) \dim \mathcal{D} = \dim \mathcal{V} \end{aligned} \tag{6}$$

The stoichiometric matrix $S = \begin{bmatrix} S_i & S_b \end{bmatrix}$ of a chemical reaction network with internal and boundary fluxes defines the following Dirac structure, called the *stoichiometry Dirac structure*. Recall that Λ_1 is the vector space of all functions from the set of chemical species to $\mathbb{R}$. The dual space of Λ_1, denoted by Λ^1, is going to define the space of *chemical potentials* μ. Recall furthermore that Λ_0 is the vector space of all functions from the set of chemical reactions to $\mathbb{R}$ (that is, Λ_0 is the vector space of fluxes). Corresponding to the splitting of the fluxes into internal and boundary fluxes we will write $\Lambda_0 = \Lambda_i \oplus \Lambda_b$, where $v_i \in \Lambda_i$ and $v_b \in \Lambda_b$. The dual spaces of Λ_i and Λ_b will be denoted by Λ^i, respectively Λ^b. They are going to define the space of thermodynamical *affinities*. Then the subspace

$$\begin{aligned} \mathcal{D} := \{ & (f, \mu, v_i, A_i, v_b, A_b) \in \Lambda_1 \times \Lambda^1 \times \Lambda_i \times \Lambda^i \times \Lambda_b \times \Lambda^b \mid \\ & -f = S_i v_i + S_b v_b, A_i = S_i^T \mu, A_b = S_b^T \mu \} \end{aligned} \tag{7}$$

defines a Dirac structure, as follows from the following general, easily proven, proposition [12]:

Proposition 2. *Let $A : \mathcal{V} \to \mathcal{W}$ be a linear map between the linear spaces $\mathcal{V}$ and $\mathcal{W}$ with adjoint mapping $A^* : \mathcal{W}^* \to \mathcal{V}^*$, that is*

$$< w^* \mid Av >=< A^*w^* \mid v > \tag{8}$$

for all $v \in \mathcal{V}, w^* \in \mathcal{W}^*$ (where as before $< \cdot \mid \cdot >$ denotes the duality product between the dual spaces $\mathcal{W}$ and $\mathcal{W}^*$, respectively $\mathcal{V}$ and $\mathcal{V}^*$). Then

$$\mathcal{D} := \{(v, w, v^*, w^*) \in \mathcal{V} \times \mathcal{W} \times \mathcal{V}^* \times \mathcal{W}^* \mid w = Av, v^* = -A^*w^*\} \tag{9}$$

is a Dirac structure.

3 The Port-Hamiltonian Formulation of Chemical Reaction Networks

The dynamics of the concentration vector x (or equivalently the vector n of mole numbers) is given once the internal fluxes v_i are specified as a function $r(x)$ of x, defining the *reaction rates.* The most basic possibility for specifying the reaction rates is *mass action kinetics.* For example, the reversible reaction

$$A + B \leftrightarrow C$$

is considered as a combination of the *forward reaction*

$$A + B \to C$$

with forward rate equation $r_f(a, b) = k_f ab$ (a, b denoting the concentrations of species A, B), and the *reverse reaction*

$$A + B \leftarrow C$$

with rate equation $r_r(c) = k_r c$, for certain constants k_f, k_r. The net reaction rate given by mass action kinetics is thus

$$v = r(a, b, c) = r_f(a, b) - r_r(c) = k_f ab - k_r c$$

More generally, the reversible reaction

$$mA + nB \leftrightarrow pC + qD$$

for positive integers m, n, p, q has according to mass action kinetics the net reaction rate

$$v = k_f a^m b^n - k_r c^p d^q$$

How can we write this into a port-Hamiltonian form ? Consider the chemical reaction part of Gibbs' law[2]

$$dG = \sum_{i=1}^{N} \mu_i(n) dn_i$$

with G the Gibbs free energy, μ_i the *chemical potential* of metabolite i, and n_i its mole number. In the ideal case the chemical potential μ_i is

$$\mu_i(x_i) = \mu_i^0 + RT \ln(\frac{n_i}{V}) = \mu_i^0 + RT \ln(x_i) \tag{10}$$

with μ_i^0 the reference potential, R the gas constant, T the temperature, V the volume, and $x_i = \frac{n_i}{V}$ the concentration. Equivalently, we have the inverse relation

$$x_i = \exp[\frac{(\mu_i - \mu_i^0)}{RT}] \tag{11}$$

In order to obtain a port-Hamiltonian description we would like to express the change in vector of concentrations $\dot{x}$ as a function of $\mu(x)$. Or better, we want to express the flux vector v as a function of the vector of the (thermodynamical) *affinities* A, defined as

$$A = S^T \mu \tag{12}$$

(For simplicity of exposition we first only consider internal fluxes $v_i = v$.) This will define the dynamics on $\mathbb{R}^M$, the space of reaction extents [8].

However, it is well-known, see e.g. [8, 7], that in general (far from thermodynamical equilibrium) it is *not* possible to express the flux vector v as a function of the affinities A. In particular, it is not possible to do this for mass action kinetics[3].

Nevertheless, see [7, 8, 1], the mass action reaction rate *can* be written as a function of the so-called *forward* and *reverse* affinities. Decompose the stoichiometric matrix S as the difference $S = S_r - S_f$ of the two matrices S_f, S_r with non-negative elements, where

$S_f =$ *forward stoichiometric matrix* corresponding to reactants

$S_r =$ *reverse stoichiometric matrix* corresponding to products

Example 2. The stoichiometric matrix

[2] For clarity of exposition we will only consider this part of Gibbs' law; see e.g. [5] for a treatment of the other parts.

[3] This is readily illustrated [7] by the simplest reaction $A \leftrightarrow B$ with reaction rate $r(a, b) = k_f a - k_r b$. When a and b are doubled, then so is the reaction rate $r(a, b)$. However, the thermodynamical affinity remains the same.

$$S = \begin{bmatrix} -1 & 0 & 0 \\ -1 & 0 & -1 \\ 1 & -1 & 0 \\ 0 & 2 & -3 \\ 0 & 0 & 2 \end{bmatrix}$$

corresponding to the reactions

$$A + B \leftrightarrow C, C \leftrightarrow 2D, 3D + B \leftrightarrow 2E$$

is decomposed into

$$S_f = \begin{bmatrix} 1\,0\,0 \\ 1\,0\,1 \\ 0\,1\,0 \\ 0\,0\,3 \\ 0\,0\,0 \end{bmatrix}, \quad S_r = \begin{bmatrix} 0\,0\,0 \\ 0\,0\,0 \\ 1\,0\,0 \\ 0\,2\,0 \\ 0\,0\,2 \end{bmatrix}$$

Then the mass action reaction rate of the j-th reaction is given by

$$v_j = k_f^j \prod_{i=1}^{N} x_i s_{ij}^f - k_r^j \prod_{i=1}^{N} x_i s_{ij}^r \tag{13}$$

where s_{ij}^f is the (i, j)-th element of the forward stoichiometric matrix S_f, s_{ij}^r is the (i, j)-th element of the reverse stoichiometric matrix S_r, and k_f^j, k_r^j are the forward/reverse reaction rate constants of the j-th reaction.

Define now the *forward* and *reverse affinities* A_f, A_r as

$$\begin{bmatrix} A_f \\ A_r \end{bmatrix} = \begin{bmatrix} S_f^T \\ S_r^T \end{bmatrix} \mu = \begin{bmatrix} S_f \; S_r \end{bmatrix}^T \mu, \tag{14}$$

and expand the dynamics as

$$\dot{x} = Sv = \begin{bmatrix} S_f \; S_r \end{bmatrix} \begin{bmatrix} v_f \\ v_r \end{bmatrix},$$

where the *forward* and *reverse fluxes* v_f, v_r satisfy

$$\begin{bmatrix} v_f \\ v_r \end{bmatrix} = \begin{bmatrix} -I \\ I \end{bmatrix} v, \tag{15}$$

with v the vector of fluxes. The relation dual to (15) is the following relation between the forward/reverse affinities A_f, A_r and the affinity A

$$A = \begin{bmatrix} -I \; I \end{bmatrix} \begin{bmatrix} A_f \\ A_r \end{bmatrix} = -A_f + A_r = S^T \mu \tag{16}$$

If we again assume ideal relations between concentrations x_i and chemical potentials μ_i given by (10,11) it follows that the mass action reaction rate (13) of the j-th reaction can be rewritten as, see [7, 8] for details,

$$v_j = \kappa^j (\exp[\frac{A_f^j}{RT}] - \exp[\frac{A_r^j}{RT}]) \tag{17}$$

where κ^j is a constant related to the forward and reverse rate constants k_f^j, k_r^j and the reference potentials [7], and with A_f^j, A_r^j denoting the j-th component of A_f, A_r. We conclude that although the mass action reaction rate cannot be written as a function of the thermodynamic affinity A it *can* be written as a function of the forward and reverse affinities A_f, A_r. Summarizing, we have expressed the vector of fluxes v as

$$v = v_r = -v_f = -J(A_f, A_r) \tag{18}$$

for the mapping J whose components are specified in (17)[4]. It can be shown [7, 8] that mass action kinetics is *passive*, and thus (18) corresponds to a kind of *resistive relation.*

By also taking into account the exchange or boundary fluxes v_b we may thus write the dynamics of the chemical reaction network, with reaction rates modeled by mass-action kinetics, as

$$\dot{x} = \begin{bmatrix} S_f & S_r \end{bmatrix} \begin{bmatrix} v_f \\ v_r \end{bmatrix} + S_b v_b = -SJ(A_f, A_r) + S_b v_b \tag{19}$$

This leads to the following *port-Hamiltonian description.* Note that the stacked matrix $\begin{bmatrix} S_f & S_r & S_b \end{bmatrix}$ defines a new 1-complex, as compared to the original 1-complex defined by the stoichiometric matrix $S = \begin{bmatrix} S_i & S_b \end{bmatrix}$. Correspondingly, the splitting into the forward and reverse affinities and fluxes leads to the *extended stoichiometry Dirac structure*

$$\begin{aligned} \mathcal{D}_e := \{ &(f, \mu, v_f, A_f, v_r, A_r, v_b, A_b) \in \Lambda_1 \times \Lambda^1 \times \Lambda_f \times \Lambda^f \times \Lambda_r \times \Lambda^r \times \Lambda_b \times \Lambda^b \mid \\ &-f = S_f v_f + S_r v_r + S_b v_b, A_f = S_f^T \mu, A_r = S_r^T \mu, A_b = S_b^T \mu \} \end{aligned} \tag{20}$$

Now all elements are in place for the port-Hamiltonian formulation of the chemical reaction network, following the basics of port-Hamiltonian theory [10, 9, 6]. Indeed, consider the extended stoichiometry Dirac structure (20), together with the constitutive relations for the energy storage

$$\mu(x) = \frac{\partial G}{\partial x}(x), \tag{21}$$

[4] As shown in [7], reduced-order reaction rates such as Michaelis-Menten reaction rates for enzymic reactions can also be written as a function of A_f and A_r.

interconnected to the variables in the definition of the modified stoichiometry Dirac structure via

$$f = -\dot{x}, \quad \mu = \mu(x) \tag{22}$$

Furthermore, consider the following resistive relation between the forward and reverse fluxes v_f, v_r and affinities A_f, A_r

$$\begin{aligned} v_f &= J(A_f, A_r) \\ v_r &= -J(A_f, A_r) \end{aligned} \tag{23}$$

where the components of J are given by (17). Substitution of the energy-storage constitutive relation (21,22) and resistive relation (23) into the extended stoichiometry Dirac structure leads to the following port-Hamiltonian system, with boundary variables being the boundary fluxes v_b and boundary affinities A_b given by

$$\begin{aligned} \dot{x} &= -SJ\left(S_f^T \tfrac{\partial G}{\partial x}(x), S_r^T \tfrac{\partial G}{\partial x}(x)\right) + S_b v_b \\ A_b &= S_b^T \tfrac{\partial G}{\partial x}(x) \end{aligned} \tag{24}$$

Note that for a boundary flux being the uptake reaction of some boundary metabolite the corresponding boundary affinity is simply the chemical potential of this metabolite.

As an immediate consequence of the port-Hamiltonian formulation we obtain the following energy balance

$$\frac{d}{dt}G(x) = -\mu^T(x)SJ(A_f, A_r) + A_b^T v_b = -(-A_f^T + A_r^T)J(A_f, A_r) + A_b^T v_b \tag{25}$$

where (see [8, 7]) $(-A_f^T + A_r^T)J(A_f, A_r) \geq 0$, thus showing *passivity.*

Note that $A_b^T v_b$ is the *power* provided to the chemical reaction network by the interaction with its environment (through the boundary metabolites and fluxes). This suggests, following basic ideas of port-based modeling and port-Hamiltonian theory, how to *interconnect* chemical reaction networks. Indeed, consider two reaction networks with boundary fluxes and affinities v_{b1} and A_{b1}, respectively v_{b2} and A_{b2} (all of equal dimension). Interconnection is achieved by setting

$$A_{b1} = A_{b2}, \quad v_{b1} + v_{b2} = 0 \tag{26}$$

4 Conclusions

In this paper we have aimed at merging the geometric approach to chemical reaction dynamics and the formulation of mass action kinetics as a resistive relation due the work of Oster, Perelson and Katchalsky, with a network approach based on port-Hamiltonian systems theory. This was achieved by defining a Dirac structure determined by the stoichiometry of the network.

This allows for a compositional view on complex chemical reaction network dynamics, and for the application of system-theoretic notions and tools to open chemical reaction networks, which is a subject of current research.

References

1. Couenne, F., Jallut, C., Maschke, B., Breedveld, P.C., Takayout, M.: Bond graph modelling for chemical reactors. Mathematical and Computer Modelling of Dynamical Systems 12(2), 159–174 (2006)
2. Courant, T.J.: Dirac manifolds. Trans. Amer. Math. Soc. 319, 631–661 (1990)
3. Craciun, G., Feinberg, M.: Multiple equilibria in complex chemical reaction networks: II. The species-reaction graph. SIAM J. Appl. Math. 66(4), 1321–1338 (2006)
4. De Leenheer, P., Sontag, E., Angeli, D.: A Petri net approach to the study of persistence in chemical reaction networks. Mathematical Biosciences 210, 598–618 (2007)
5. Eberard, D., Maschke, B., van der Schaft, A.J.: An extension of Hamiltonian systems to the thermodynamic space: towards a geometry of non-reversible thermodynamics. Reports on Mathematical Physics 60(2), 175–198 (2007)
6. Duindam, V., Macchelli, A., Stramigioli, S., Bruyninckx, H. (eds.): Modeling and Control of Complex Physical Systems; the Port-Hamiltonian Approach. Geoplex Consortium. Springer, New York (2009)
7. Oster, J.F., Perelson, A.S., Katchalsky, A.: Network dynamics: dynamic modeling of biophysical systems. Quarterly Reviews of Biophysics 6(1), 1–134 (1973)
8. Oster, J.F., Perelson, A.S.: Chemical reaction dynamics, Part I: Geometrical structure. Archive for Rational Mechanics and Analysis 55, 230–273 (1974)
9. van der Schaft, A.J.: L_2-Gain and Passivity Techniques in Nonlinear Control. Lect. Notes in Control and Information Sciences, vol. 218, p. 168. Springer, Berlin (1996)
10. van der Schaft, A.J., Maschke, B.M.: The Hamiltonian formulation of energy conserving physical systems with external ports. Archiv für Elektronik und Übertragungstechnik 49, 362–371 (1995)
11. van der Schaft, A.J., Maschke, B.M.: Conservation Laws and Lumped System Dynamics. In: Van den Hof, P.M.J., Scherer, C., Heuberger, P.S.C. (eds.) Model-Based Control; Bridging Rigorous Theory and Advanced Technology, pp. 31–48. Springer, Heidelberg (2009) ISBN 978-1-4419-0894-0
12. van der Schaft, A.J., Maschke, B.M.: Port-Hamiltonian dynamics on graphs (2010) (submitted for publication)

Bifurcations of Dynamical Systems, Logistic and Gompertz Growth Laws in Processes of Aggregation

Alex Shoshitaishvili and Andrei Raibekas

Abstract. From the systemic point of view protein aggregation is a compensatory mechanism allowing transition of a system (protein solution) from an initially stable equilibrium, which became unstable under a stress, to another stable equilibrium, which bifurcates from the initial one because of the stress. The simplest bifurcation of this type is Logistic bifurcation with a positive small parameter.

We realize this bifurcation as a model of protein aggregation through a large-dimensional Becker-Döring system with a one-dimensional Logistic attractor (BDL) containing two equilibria. BDL depends on the magnitude δ of stress as a small parameter. Kinetics on the attractor is transformed by the observable (which is a fewnomial, *i.e.*, a high-degree polynomial with a number of terms that is small relative to the degree) into the observed kinetics of the experiment. This model explains Gompertzian growth, unimodality of size distribution of aggregates, and relations between Rate, Plateau and time elapsed from onset to inflection moments. The explanation is based on the existence of a nonequilibrium partition function. It exists under the assumption of formation of aggregation-competent monomer as a precursor of the aggregation.

1 Protein Aggregation and Its Features

Protein aggregation is an important subject in medicine (neurological diseases) and pharmaceutics (degradation of large molecules in protein therapeutics). Numerous papers are devoted to this subject (see, for example, surveys [1, 2, 3, 4]). As a mathematical model for protein aggregation often the Becker-Döring system of differential equations [5] is considered. This system

Alex Shoshitaishvili · Andrei Raibekas
California State University Channel Islands

J. Lévine & P. Müllhaupt (Eds.): Adv. in Theory of Control, Signals, LNCIS 407, pp. 349–363.
springerlink.com © Springer-Verlag Berlin Heidelberg 2010

has been investigated in infinite-dimensional as well as in finite-dimensional settings ([6, 9, 10, 11, 12, 15, 16, 18] and many others). We treat the finite-dimensional Becker-Döring system from a different perspective of qualitative geometrical theory of dynamical systems and their bifurcations.

We also follow the principle which arises from singularity theory [13] and implies that any phenomenon of growth depending on one parameter (in our case the parameter is a small excess of protein concentration above the critical concentration) should be induced (*i.e.*, transformed by a change of parameters and by a change of the variables, depending on the parameters) from the corresponding versal deformation, which in our case is multidimensional Saddle-Node (SN) bifurcation with a positive parameter [14]. We will use the Logistic form of this bifurcation. In the 1D case, transition from SN bifurcation with a positive parameter $\dot{c} = \tau - c^2, \tau > 0$ to Logistic bifurcation with a positive parameter $\dot{\tilde{c}} = \tilde{c}\epsilon - \tilde{c}$ consists of the change of the parameters $\tau = (1/4)\epsilon^2$ and the change of the variables $\tilde{c} = c + (1/2)\epsilon$.

In this work multidimensional Logistic bifurcation appears as a large-dimensional Becker-Döring model with Logistic one-dimensional attractor (BDL). Mass of aggregated protein is an observable and is a function of states of BDL. The observable transforms the Logistic type of kinetics into experimentally observed kinetics (in our case into Gompertzian kinetics). To demonstrate this result we introduce new variables: coefficients of the Nonequilibrium Partition Function. These variables resolve a singularity in BDL.

We relate to each other the following features of aggregation which have been observed in experiments: Gompertzian growth [7, 8, 19]; linear dependences of $t_{i,o} = t_{inflection} - t_{onset}$, which is the duration of time between moments of onset and inflection on logarithm of the rate of growth (at the inflection point) or on the logarithm of the plateau (which is the maximum mass of aggregated protein) [7]; and finally, localization of the distribution of sizes of aggregates in the vicinity of some predominant size during the aggregation process [8].

In the frame of the suggested model the ratio of the slope and constant term in the $(t_{i,o}, log(Plateau))$-relationship is $-1/n$, where n is the predominant size of the aggregate or (in another scenario) is $-1/\nu$ where ν is the size of the nucleus.

2 A Phenomenological One-Dimensional Model

The following one-dimensional phenomenological model of aggregation will be justified in the subsequent sections as a reduction to a one-dimensional attractor of a large-dimensional Becker-Döring system of ordinary differential equations.

Let us introduce a small parameter δ which is the difference between the concentration of native monomer p and the critical aggregation concentration. δ and all other concentrations below are measured in mole fraction. Critical aggregation concentration (CAC) is the concentration of the protein above which the aggregation starts.

At the beginning of the experiment a sample of protein solution is heated quickly to a higher temperature; the critical aggregation concentration drops down and the difference between the concentration of native monomer p and the critical aggregation concentration instantly becomes a small positive number δ.

We will consider a case of aggregation in which the native monomer under thermal stress (or some other kind of stress) produces an active (or, in other terminology, aggregation-competent) monomer which in turn forms aggregates of different sizes (see, for example,[12, 21]).

Denote the mole-fraction of active monomer by c_1 and assume that c_1 changes in time according to differential equation

$$\dot{c}_1 = c_1(\epsilon - c_1)\gamma(c_1, \epsilon) \tag{1}$$

where $\epsilon = \epsilon(\delta)$ is a small constant which depends on $\delta, \epsilon(\delta) < \delta, \epsilon(0) = 0$, and $0 \leq c_1 \leq \epsilon$.

Here $\gamma(c_1, \epsilon)$ is a (sufficiently) smooth function, $\gamma = \gamma_0 + \gamma_1(c_1, \epsilon), \gamma_0 = constant > 0, \gamma_1(0, 0) = 0$. This differential equation is similar to a logistic differential equation, and can be transformed to it by a change of time. Therefore we will call it a logistic equation too.

According to the thermodynamics of aggregation (see, for example, [22]), at equilibrium $c_1 = \epsilon$ the relationships $c_m = Q_m c_1^m, (m > 1)$ have place, where c_m is the mole-fraction of the aggregates containing m monomers, and $Q_m, (m > 1)$, are positive constants (these constants are called coefficients of the partition function). This is not true for the kinetic part of the process. However, as we will show in other sections, it is almost true.

Assume that for some constant $\omega(\epsilon)$, depending on ϵ, where $\omega, 0 \leq \omega < 1$ and any integer $m > 1$, the mole fraction c_m of the aggregates of size m is

$$\begin{aligned} c_m &= q_m c_1^m, \omega \leq c_1 < \epsilon, \\ q_m = q_m(c_1, \epsilon) > 0, q_m &= q_{m,0}/\epsilon^{m-g_m} + q_{m,1}(c_1, \epsilon) \\ q_{m,0} &= constant > 0, q_{m,1} = O(c_1, \epsilon) \end{aligned}$$

Coefficients q_m are called coefficients of the non-equilibrium partition function. The presence of the powers of small parameter ϵ in the denominators for coefficients q_m of the partition function is an indication of the explosive character of aggregation in the aggregates of sizes favorable from a thermodynamical point of view. These powers could have different magnitudes, which are reflected in numbers g_m. We will consider two main cases: (1)

$g_m = m, m > 2$ (meaning that $q_{m,0}$ do not have ϵ in the denominators and c_m is of order ϵ^m) and (2) $g_m = 1, m > 2$ (meaning that c_m is of order ϵ).

Note that the mole fraction of the monomer which is consolidated in the aggregates of size m is $mc_m = mq_m c_1^m$.

Consider a scenario of aggregation with $g_m = 1, m > 1$. At the late stage of the process, when c_1 is sufficiently close to ϵ, the terms $\gamma_1(c_1, \epsilon)$ (see (1)), $c_1 q_{m,1}(c_1, \epsilon)$ and their derivatives become negligibly small. Also for this stage of the process, the population of the aggregates of sizes $m : n \geq m \geq n + k$ become abundant for some n, k and the population of aggregates of other sizes become scarce. Thus the mole-fraction of monomer $\sum_{m>1} mc_m$ which is consolidated in the aggregates (and which we observe in the experiment) can be for sufficiently small ϵ closely approximated by some polynomial F with nonnegative coefficients. So kinetics of the aggregation can be approximated by a one-dimensional Logistic system with observable F

$$\dot{c}_1 = c_1(\epsilon - c_1)\gamma_0 \tag{2}$$

$$F(c_1) = c_1^n \sum_{i=0}^{k} (n + i)(q_{n+i,0}/\epsilon^{n-1})c_1^i \tag{3}$$

$$\omega \leq c_1 \leq \epsilon \tag{4}$$

$$0 < \epsilon << 1, n >> 1, k << n, |\omega - e^{-e/n}\epsilon| << \epsilon \tag{5}$$

where n, k are positive integers, and the signs $>>, <<$ mean "much more" and "much less".

The exact estimations which are required for the above-mentioned $\gamma_1(c_1, \epsilon), q_{m,1}(c_1, \epsilon)$ and for their derivatives, as well as estimations showing the exact meaning for $<<$ and $>>$ in each of these cases (5) are omitted here for brevity.

In spite of the fact that we considered a late stage of the aggregation with respect to accumulation of the active monomer c_1, this stage is an early enough stage with respect to the observed process. It means that the whole onset-saturation period for the observable mass of aggregated protein is registered by the model. Namely the curve $F(c_1(t))$ will be observed as a Gompertzian curve on the time span from the onset moment of the Gompertzian curve to the saturation moment of this curve.

Statement 1. *The curve $F(c_1(t))$ considered on interval of $[t_{onset}, t_{saturation}]$ is uniformly close to a Gompertzian curve $G(t)$*
$|F(c_1(t)) - G(t)| = \epsilon(o(1/n) + o(\epsilon))$
where $G(t)$ is a solution of Gompertz equation
$\dot{G} = \gamma_0(G(ln(n) + ln(F(\epsilon)/\epsilon) - ln(G)))$

Remember that a Gompertzian curve is a solution of the Gompertz equation, which describes some type of limited growth (see, for example,[20])

$$\frac{d}{dt}g(t) = \beta g(t)(\alpha - ln(g(t))) \tag{6}$$

where α, β are constants.

The solution of Gompertz equation is

$$g(t) = a * e^{-e^{-\frac{t-x_0}{b}}} \tag{7}$$

where $a = e^{\alpha}, b = \frac{1}{\beta}, x_0 = b * ln(ln(a) - ln(g(0)))$. Constant a is equal to $g(\infty)$; constant x_0 is the moment when the inflection point of growth is achieved; constant b is the time elapsed from the onset point of the growth to its inflection point; and $x_0 - b$ is equal to the time (which is called lag) from the initial moment to the onset moment. The onset and saturation moments are the moments when the linear approximation of the Gompertzian curve by its tangential line taken at the inflection point intersects horizontal lines $y = o$ and $y = a$ respectively.

To prove the statement one should rescale time, c_1 and F:

$t = (1/\epsilon)\tau, c = (1/\epsilon)c_1, \tilde{F}(c, \epsilon) = F(\epsilon c)/(\epsilon \sum_{i=0}^{k}(n+i)q_{n+i,0})$,

consider mapping $\tilde{F} : s_1 \to s_1$ as a change of variable c_1 and then apply the following lemma.

For a constant $\Delta, 0 < \Delta < 1$ let us determine segment $s_\Delta = (c : \Delta \leq c \leq 1)$. Consider s_0 and a vector field L_v on s_0: $L_v c = c(1-c)\gamma_0$.

Consider function $P_{n,k} = c^n P_k$ where $P_k = (\sum_{i=0}^{k} p_n c^i)$. Here $\gamma_0 > 0, p_{n+i} > 0, i = 0, .., k$ are constants. Assume that $P_k(1) = 1$. Then function $P_{n,k}$ monotonically increases on s_1, and maps 0 to 0, 1 to 1. Hence it is a one-to-one mapping of the segment s_1 onto itself.

Lemma 1. *For $n \to \infty$ the sequence $R_{v,n,k} = (P_{n,k})_* L_v, n = 1, 2, ...$ of vector fields on $[0, 1]$ non-uniformly C^0-converges to the vector field $G_v = -uln(u)$ This convergence is uniform on any segment $s_\Delta, \Delta > 0$ and $max|R_{v,n} - G_v|$ on s_Δ is $O((1/n)ln(\Delta))$.*

(For mapping $f : s_0 \to s_1$, notation $f_* v(x)$ means a vector at a point $x \in s_0$ which is the image under the mapping f of a vector v at the point $f^{-1}(x)$.)

The lemma easily can be verified by application of Taylor approximation with respect to the exponent $y = 1/(n+i)$ at $y = 0$ for the functions $c^y = 1 + ln(c)y + 1/2ln(c)^2y^2 + ...$ and approximation $(1-c) = 1 - e^{ln(c)} = ln(c) + 1/2 * ln(c)^2 + ...$

2.1 $t_{i,o}$-Plateau Relation

From (3),(1),(6) one has that $a = F(\epsilon) = \kappa\epsilon^n, t_{i,o} = t_{inflection} - t_{onset} = b = 1/(\gamma_0\epsilon)$, where a is Plateau , *i.e.*, the saturation level of aggregated monomer observed. So κ is some large constant. Assume that magnitudes $a, 1/n$ are such that approximation $a^{1/n} = 1 + (1/n)ln(a)$ is accurate enough. Then

$t_{i,o} = (1/\gamma_0)\kappa^{1/n}/a^{1/n}$. Let us substitute in this equation an approximation $1/a^{1/n} = 1 - (1/n)ln(a)$. Comparing the result of the substitution with an empirical linear regression $lr_{io}(ln(a))$ which presents $t_{i,o}$ as a linear function of $ln(a)$, one has

Statement 2. *Ratio of the slope to the intercept of* $lr_{io}(ln(a))$ *is* $-1/n$.

3 Nonequilibrium Partition Function (NPF)

The system of Becker-Döring differential equations is a model for a process of coarsing (an aggregation). It represents the aggregation by a sequence of attachment and disattachment of a monomer to an aggregate by one monomer at a time

$$\frac{d}{dt}c_n = a_{n-1}c_1c_{n-1} - b_nc_n - a_nc_1c_n + b_{n+1}c_{n+1}, n \geq 2 \tag{8}$$

where c_n is the mole fraction of an aggregate comprised by n monomer particles, n is called an aggregation number, and a_n and b_n are rates of reactions of attaching and releasing of monomer particles to/from an aggregate and are determined by the thermodynamics of the aggregation.

Interactions of aggregates with differences of aggregation numbers greater than one are considered thermodynamically unfavorable and are forbidden (these interactions are considered in the more general Smoluchowski model of coursing).

Becker-Döring differential equations as well as their different limits have been investigated quite deeply mathematically (see [11] and references therein), and have been successfully applied for the precise description and prediction of the kinetics of many coarsing/coagulation/crystallization/aggregation phenomena [24, 17, 16, 3, 18]. However for some types of aggregation, attempts to describe and predict the kinetics of the aggregation in the frame of Becker-Döring differential equations have failed (see remark on page 143 [19]). These cases can be accurately treated through a phenomenological limited growth model, namely the Gompertz model.

We will connect Becker-Döring equations to a Gompertz equation by introducing a Logistic differential equation for c_1.

The equation for the derivative of c_1, which has to be a part of the BD model, traditionally is derived from additional requirements like the requirement of mass conservation. Below we introduce an equation for c_1 in a different way to meet a requirement of the existence of a nonequilibrium partition function.

One can see that at nonzero equilibrium $\bar{c}_1, .., \bar{c}_n$

$$\bar{c_n} = Q_n\bar{c}_1^n, n \geq 2 \tag{9}$$

where

$$Q_1 = 1, \frac{Q_{n+1}}{Q_n} = \frac{a_n}{b_{n+1}}$$

and thus

$$Q_n = \frac{a_1 a_2 ... a_{n-1}}{b_2 b_3 ... b_n} \tag{10}$$

Constants Q_n have thermodynamical meaning. They are called coefficients of the Partition Function (PF).

Suppose that $\dot{c}_1 = c_1 \phi(c_1, .., c_n, p, \epsilon)$ for some smooth function ϕ which depends on parameters ϵ and precursors p of the aggregation. Let us introduce new variables q_n such that

$$c_n(t) = q_n(t) c_1^n(t), n \geq 2, T_{start} \leq t \leq T_{finish} \tag{11}$$

Substituting (11) in (8) and dividing by c_1^n one has

$$\frac{d}{dt} c_1 = c_1 \phi(c_1, .., c_n, p, \epsilon) \tag{12}$$

$$q_1 = 1$$

$$\frac{d}{dt} q_n = -n q_n \phi(c_1, .., c_n, p, \epsilon) +$$

$$+ a_{n-1} q_{n-1} - b_n q_n - a_n c_1 q_n + b_{n+1} q_{n+1} c_1, n \geq 2 \tag{13}$$

We will show that under some conditions (12) can be approximated by Logistic $\dot{c_1} = c_1(\epsilon - c_1)\gamma_0$. In this case the system (12),(13) has two equilibria

$$c_1 = 0, q_1 = 1, q_n = \tilde{Q}_n = a_1..a_{n-1}/(b_2 + 2\epsilon)..(b_n + n\epsilon), n \geq 2 \tag{14}$$

$$c_1 = \epsilon, q_1 = 1, q_n = Q_n = a_1..a_{n-1}/b_2..b_n, n \geq 2 \tag{15}$$

The system (12),(13) will be called the Nonequilibrium Partition Function (NPF) System of differential equations.

4 1D Attractor in BD and NPF Systems

NPF system with Logistic-type equation for c_1 is justified for aggregation with aggregation-competent monomer-precursor.

To see this let us consider production of the active monomer c_1 (see section 2) from native monomer p as a precursor of aggregation and spending of c_1 on the aggregates:

The amount of mass $m_{tot} = p + c_1 + \sum_{n \geq 2} n c_n$ is conserved. So

$$\frac{d}{dt} c_n = J_{n-1} - J_n, n \geq 2, J_n = a_n c_n c_1 - b_{n+1} c_n$$

$$\frac{d}{dt} c_1 = \tilde{v}(p, c_1, \delta) - J_1 - \sum_{n \geq 1} J_n = f(c_1, .., c_N, \delta), \frac{d}{dt} p = -\tilde{v}(p, c_1, \delta) \tag{16}$$

for some $\tilde{v}(p, c_1, \delta)$.

Assume that: $a_n \geq 0, b_n \geq 0$ for all n, and that the system is finite, i.e., $a_n = 0, b_{n+1} = 0, n > N$ for some N.

Using the conserved quantity $p + \sum_1^N nc_n = m_{tot} = CAC + \delta$ where δ and CAC are defined as in section 2 above, one can exclude variable p from the system (16).

At the equilibrium $\bar{c}_1, .., \bar{c}_N$ one has $Obs(\bar{c}_2, ..., \bar{c}_N) = (\sum_{n>1}^N nQ_n\bar{c}_1^n) = a$ where a is the maximum of observed mass Obs of aggregated protein expressed in a mole fraction. The aggregation stops when the excess of the concentration p of native monomer above the critical concentration is zero. So $a + \bar{c}_1 = \delta$. It means that for a small enough δ, function $\epsilon = \bar{c}_1(\delta)$ is an analytical function of δ and vice-versa.

One obtains the following system which will be called a BD system (BDS)

$$\frac{d}{dt}c_n = a_{n-1}c_1c_{n-1} - b_nc_n - a_nc_1c_n + b_{n+1}c_{n+1}, n \geq 2 \tag{17}$$

$$\frac{d}{dt}c_1 = \tilde{f}(c_1, .., c_N, \epsilon) \tag{18}$$

where $\tilde{f}(c_1, .., c_N, \epsilon) = (\tilde{v}(p, c_1, \delta) - J_1 - \sum_{n \geq 1} J_n)_{p=p(c_1,..,c_N,\delta(\epsilon)),\delta=\delta(\epsilon)}$. We assume that $\tilde{v}$ and hence $\tilde{f}$ are sufficiently smooth.

Remark 1. *Sufficiently smooth means "having no fewer derivatives than are necessary for future calculations". In fact, for our purpose the necessary number of derivatives is 5.*

The following assumption guarantees that a NPF system exists and can be derived from (23,24) similar to (12).

Assumption 1. *If $c_1 = 0$ then $\dot{c_1} = 0$*
Thus $\dot{c_1} = \tilde{v}(p, c_1, \delta) = c_1v(p, c_1, \epsilon)$ for some function v.

Assumption 2. *For $\epsilon > 0$ the system (17,18) has two equilibria which become one of multiplicity 2 at $\delta = 0$*

Assumption 3. *Assume that $b_n \neq 0, n \leq N$.*

Because the assumption (2) has place, the following condition, which we formulate as a lemma, necessarily is true.

Lemma 2. *Eigenvalues of (17,18) at the point $(c_1 = c_n(n > 1) = \epsilon = 0)$ are split into group E_{neg} of negative eigenvalues $-b_n > 0, n > 1$ and group E_0 which consists of one zero eigenvalue and $(\partial_{c_1}\tilde{f})(c_1 = 0, .., c_N = 0, \epsilon = 0) = 0$.*

As it follows from Lemma 2 and Assumption (3) the following statement has place (see [14]).

Statement 3. *For sufficiently small ϵ the system BDS has in a neighborhood of the origin a 1D exponential attractor which is given by a sufficiently smooth system of equations*

$$c_n = f_n(c_1, \epsilon), n > 1 \tag{19}$$

and the restriction of the BDS on the attractor can be written in terms of coordinate c_1 as

$$\dot{c_1} = f(c_1, \epsilon) = \tilde{f}(c_1, f_2(c_1, \epsilon), .., f_N(c_1, \epsilon)) \tag{20}$$

From here on we consider only non-negative small parameters ϵ and are interested in the behavior of the BDS in the region $c_1 \geq 0, .., c_n \geq 0, 0 \geq c_1 \geq \epsilon$. This region is invariant under the BDS : any trajectory with an initial condition in the region always stays in the region.

The following assumptions guarantee that no more than two equilibria come together at $\epsilon = 0$ and that the final equilibrium is stable.

Assumption 4.

$$\partial^2_{c_1,\epsilon} f(0) > 0 \tag{21}$$

$$\partial^2_{c_1{}^2} f(0) < 0 \tag{22}$$

Under the assumptions the equation (20) can be written in the form (2). The system

$$\frac{d}{dt} c_n = a_{n-1} c_1 c_{n-1} - b_n c_n - a_n c_1 c_n + b_{n+1} c_{n+1}, n \geq 2 \tag{23}$$

$$\dot{c_1} = c_1(\epsilon - c_1)\gamma(c_1, \epsilon)$$

$$\gamma = \gamma_0 + \gamma_1(c_1, \epsilon), \gamma_0 = constant > 0, \gamma_1(0, 0) = 0 \tag{24}$$

where $\gamma(c_1, \epsilon)$ is a sufficiently smooth function, has the attractor which coincides with the attractor of the system BDS (17,18).

This system is called the BDL system.

Remark 2. *Because $c_m, m > 1$ is of order c_1^m (see below), Assumption 4 expresses the relationship between rates of production of active monomer from native monomer and creation of dimers from the active monomer and depend only on them.*

Consider the NPF system which is derived from (23,24) similar to (12). Eigenvalues of its linear part at the origin also meet Lemma 2. So it also has an exponential sufficiently smooth 1D attractor $q_n = q_n(c_1, \epsilon), n > 1, 0 \leq c_1 \leq \epsilon$ containing two equilibria $(c_1 = 0, \tilde{Q}), (c_1 = \epsilon, Q)$. It is clear that $q_n(c_1, \epsilon) = f_n(c_1, \epsilon)/c_1^n, n > 1$ and restriction of the NPF system on the attractor in terms of coordinate c_1 coincides with (20).

Remark 3. *A NPF system can be introduced also when production of active monomer is governed by any finite dimensional family of dynamical systems depending on multiple parameters α*

$$\dot{c}_1 = c_1 v_1(c_1, x, \alpha), \dot{x} = v_2(c_1, x),$$
$$v_1(0,0) = v_2(0,0) = 0, v_1(c_1(\alpha), x(\alpha)) = v_2(c_1(\alpha), x(\alpha)) = 0$$
$$\dot{p} = -(\sum_k^1 \dot{x}_i + \sum_{i>1} \dot{c}_i) \quad (25)$$

where $x = (x_1, ..., x_k)$ are prerequisites for active monomer and α are parameters of stresses.

4.1 NPF and Observable Restricted to the Attractor

The kinetics on the attractor of the NPF system is determined by

$$\begin{aligned} q_n &= m_n(c_1, \epsilon), n > 2 \\ \dot{c}_1 &= \epsilon c_1 - c_1^2 - c_1 O_2(c_1, \epsilon, m(c_1, \epsilon) c_1) = \\ &= (\epsilon c_1 - c_1^2)(\gamma_0 + \gamma_1(c_1, \epsilon)), \gamma(0,0) = 0 \\ & 0 < c_1 < \epsilon \end{aligned} \quad (26)$$

where m_n and γ are sufficiently smooth functions. The attractor contains both equilibria of the NPF system which are (see (14,15))

$$c_1 = \epsilon, q_n = 0, N \geq n > 2 \quad (27)$$
$$c_1 = 0, \tilde{q}_n = \tilde{Q}_n - Q_n, N \geq n > 2$$
$$Q_n = a_1..a_{n-1}/b_2..b_n, \tilde{Q}_n = a_1..a_{n-1}/(b_2 + 2\epsilon)..(b_n + n\epsilon) \quad (28)$$

One can approximate the observable $\sum_n nc^n$ by $\sum_n n(q_n + Q_n)c_1^n$ where q_n is on the attractor. Hence, the observable is close to $\sum nQ_n c_1^n$ because all variations of $q_n + Q_n$ with q_n on the attractor are between $\tilde{Q}_n$ and Q_n (see (27,28)) and can be made arbitrarily small together with first derivatives with respect to time by diminution parameter ϵ.

4.2 ^{c_1}Near-Saturated-^{c_n} Onset Stage of the Aggregation

The model of the aggregation described by formulas (17)-(23) and (26)-(28) is a complete model of the aggregation.

If the observable $\sum_n nc^n$ can be approximated by a polynomial $c_1^n P_k(c_1)$ (see section 2), then it will have a very small magnitude of order ϵ^n.

In order to obtain observables of greater magnitude we will modify this model and consider association and dissociation coefficients tending to infinity or to zero when ϵ tends to zero.

Consider the NPF system and rescale time, dissociation constants and variables

$$t = (1/\epsilon)\tau, C_1 = c_1/\epsilon, \tilde{b}_n = b_n/\epsilon, \tilde{q}_n = q_n \epsilon^{n-1}, n \geq 1 \quad (29)$$

After rescaling, the NPF system from the previous section will take form

$$\dot{C}_1 = C_1(1 - C_1)\gamma(C_1, \epsilon), \gamma(0) = \gamma_0 > 0,$$
$$\tilde{q}_1 = 1$$
$$\tilde{q}_n = -n\tilde{q}_n(1 - C_1)\gamma + a_{n-1}\tilde{q}_{n-1} - a_n\tilde{q}_n C_1 - \tilde{b}_n\tilde{q}_n + \tilde{b}_{n+1}\tilde{q}_{n+1} \qquad (30)$$

Let us assume that $b_n > 1(n \geq 2), 0 < a_1 << 1, a_i \approx 1(i \geq 2)$,
$a_{n+j} >> 1(1 \leq j \leq k), a_r << 1(r \geq n + k + 1)$.
Then for some Δ and $\Delta_l(l \geq 2)$ where
$0 < \Delta < 1,\ 0 < \Delta_l < \tilde{Q}_l = a_1 * .. * a_l/(\tilde{b}_2, *.. * \tilde{b}_l)(l \geq 2)$
there is 1D attractor $q_l = \phi(C_1)(l \geq 2)$ in a vicinity
$\Delta < C_1 \leq 1, \Delta_l < \tilde{q}_l \leq \tilde{Q}_l(l \geq 2)$.
Indeed the spectrum of the linearization of the system at the equilibrium $C_1 = 1, \tilde{Q}_l(l \geq 2)$ is split $\tilde{Q}_l(l \geq 2)$. Due to the choice of coefficients a_l one has $\tilde{Q}_l << 1, (2 \leq l < n)$,
$\tilde{Q}_l >> 1, (n + k \geq l \geq n), \tilde{Q}_l << 1, (l > n + k)$.

Thus for the C_1 late stage of the process (i.e.,$C_1 > \Delta$ or, in terms of $c_1, c_1 > \epsilon\Delta$) one has relationships $c_l \approx \tilde{Q}_l C_1^l = (\tilde{Q}_l/\epsilon^l)\epsilon^l c_1^l = Q_l c_1^l$ which also are relationships for the original BDL system. Because of the magnitudes of $\tilde{Q}_l$ we are now in the situation described by the phenomenological model in section 2.

To complete this construction one has to verify that the following two statements are true.

First, that Δ and n can be chosen such that $\Delta \leq e^{\frac{-1}{n}}$ in order for the c_1-late stage to include the c_n-onset-saturation (see section 2) segment of the aggregation. Because of $a_{n+j} >> 1, (1 \leq j \leq k)$, one needs to use the fact that the system (30) is close to linear (for the late stage): $a_{n+j}C_1, (1 \leq j \leq k)$ are almost constant because C_1 is almost 1, and, in fact, its eigenvalues are close to $-a_{n+j}C_1 - b_{n+j}$ which are as large as a_{n+j} are.

Second, that the system (23) still has a 1D attractor for all stages of the aggregation, *i.e.* for $0 \leq c_1 \leq 1$. To prove it for earlier stages of the aggregation one needs to use terms $n\tilde{q}_n(1 - C_1)\gamma$ which increase exponential convergence to the 1D attractor when C_1 is small. (Note that instead of dealing with this statement, one could introduce an assumption $\dot{c}_1 = v(p, c_1) = c_1(\epsilon - c_1)v_1(p, c_1, \epsilon)$ or assumptions similar to (21)(22) in terms of variable $c_{1,final} = \epsilon - c_1$ and consider only a final stage of the aggregation.)

Complete proofs of these statements will be provided elsewhere.

5 Nucleation and Richards Growth

In this section we keep Assumptions (1) and(2). We do not assume that size distribution of aggregates is concentrated around some n_0 and we do not suppose that Assumption (3) has place.

Assume that in (17,18) only the first ν dissociation constants b_n are positive and all others are zero. This means that above size ν the reaction of aggregation is kinetically irreversible. Formulas (10) for $n > \nu$ are not determined and c^n is not asymptotically equivalent to c_1^n for $n > \nu$. The aggregate of size ν is called a nucleus (see, for example, [18]). Under this condition the BD system (17),(18) with a small parameter can be presented as the following BDL system

$$\frac{d}{dt}c_1 = c_1((\epsilon - c_1) + O_2(c_1, \epsilon) + O(c_2, .., c_N)) \tag{31}$$

$$\frac{d}{dt}c_n = a_{n-1}c_1c_{n-1} - b_nc_n - a_nc_1c_n + b_{n+1}c_{n+1}, \nu - 1 \geq n \geq 2 \tag{32}$$

$$\frac{d}{dt}c_\nu = a_{\nu-1}c_1c_{\nu-1} - b_\nu c_\nu - a_\nu c_1 c_\nu + b_N c_N \tag{33}$$

$$\frac{d}{dt}c_n = a_{n-1}c_1c_{n-1} - a_nc_1c_n, N > n \geq \nu + 1 \tag{34}$$

$$\frac{d}{dt}c_N = a_{N-1}c_1c_{N-1} - b_Nc_N \tag{35}$$

The term $\pm b_N c_N$ with positive b_N is added to (35) and to (33) to avoid degeneracy of the system. Without this term the system will have only one equilibrium c=0 of infinite multiplicity. (Of course nondegeneracy of the system can be achieved by other deformations. They will require analysis similar to the following one.) The added terms mean that the aggregates of size N can break into nuclei (size ν) and active monomer.

The system has two equilibria (36) and (37,38)

$$c_n = 0, n \geq 1 \tag{36}$$

$$c_1 = \epsilon, c_n = Q_n\epsilon^n, Q_n = a_1..a_{n-1}/b_2..b_n(1 < n \leq \nu) \tag{37}$$

$$c_n = Q_n\epsilon^\nu, Q_n = a_1..a_\nu..a_n/b_2..b_\nu a_{\nu+1}..a_{n+1}(\nu < n \leq N - 1)$$

$$c_N = Q_N\epsilon^{\nu+1}, Q_N = a_1..a_{\nu-1}/b_2..b_\nu b_N \tag{38}$$

Consider the NPF system for the system (31)-(35). One has $c_n = q_n(t)c_1^\nu(t)$, $N > n \geq \nu, c_N = q_N c_1^{\nu+1}$ where q_n are solutions of the NPF system. There exists a neighborhood of final equilibrium (37,38) such that the onset-saturation segment of the observable process belongs to the neighborhood and the observable can be approximated by function

$$obs_\nu = \sum_{\nu}^{N-1} Q_n c_1^\nu \tag{39}$$

(under the assumption that $Q_m, m < \nu$ and Q_N are small).

For $(Q_n \sim 1, \nu \geq N)$ an estimate of the observable is $N^2\epsilon^\nu$ and for sufficiently large $N \sim (1/\epsilon_0)^{(\nu-1)/2}$ is of order ϵ. Here ϵ_0 is a value of small parameter and we assume that ϵ varies in a vicinity of ϵ_0.

Thus in this scenario the observable is $Kc_1^\nu(t)$ and $\dot{c}_1$ is close to logistic $c_1(\epsilon - c_1)$. So for $u = Kc_1^\nu$,
$\frac{d}{dt}(Kc_1^\nu) \approx K\nu c_1^\nu(\epsilon - c_1) = \nu u(\epsilon - (u/K)^{1/\nu})$,
which is Richards growth law ([23]). For large ν this law is close to Gompertzian law.

In the case when (35) is not the ending of the system and other equations are

$$\frac{d}{dt}c_N = J_{N-1} - J_N, .., c_n = J_{n-1} - J_n, .., c_{nf} = J_{nf-1} \tag{40}$$

one has $c_n = q_n c^{\nu+1+n-N}, nf \geq n \geq N$. Thus considering linear fit $t_{i,o} \approx d_1 ln(a/a_0) + d_0$ one has $d_1/d_0 \approx -1/\nu$.

Remark 4. *Further research in this direction will be concerned with bifurcations at zero of equilibria of the Becker-Döring system with a Logistic-like equation for c_1 when groups of b_i or a_i tend to zero (or infinity) driven by small parameters tending to zero. Different kinds of groups and asymptotics will be considered and their effect on asymptotics with respect to c_1 of the solutions c_n will be analyzed through Nonequilibrium Partition Function coefficients.*

Remark 5. *Consider equation*

$$\dot{\Delta} = -\Delta(\epsilon - \Delta) + ... \tag{41}$$

where $\Delta = (p - CAC), \epsilon = p_0 - CAC$ which is the Logistic part of the equation describing decrease of excess of monomer concentration for the reaction of aggregation. Note that we do not assume here that there is an aggregation-competent monomer precursor of the aggregation. Then Gompertzian curves of the mass of aggregate growth can be built through eigenfunctions of the corresponding differential operator $f(\Delta) \mapsto (\partial_\Delta f) * (-\Delta(\epsilon - \Delta))$. This provides a new scenario of logistic-based Gompertzian aggregation growth.

Acknowledgment. We would like to thank Jian Zhang-van Enk, Ed Maliski, Alexander Melyakhovetskiy and Janet Shoshitaishvili for useful discussion. We are thankful to Jean Lévine and Philippe Müllhaupt for the opportunity to talk on the subject at Bernoulli Center Workshop "Advances in the Theory of Control, Signals and Systems, with Physical Modeling".

References

[1] Morris, A.M., Watzky, M.A., Finke, R.G.: Protein Aggregation Kinetics, Mechanism, and Curve Fitting: A Review of the Literature. Biochimica et Biophysica Acta 1794, 375–397 (2009)

[2] Bernacki, J.P., Murphy, R.M.: Model Discrimination and Mechanistic Interpretation of Kinetic Data in Protein Aggregation Studies. Biophysical Journal 96, 2871–2887 (2009)
[3] Frieden, C.: Protein aggregation processes: In search of the mechanism. Protein Sci. 16(11), 2334–2344 (2007)
[4] Roberts, C.J.: Non-native protein aggregation: pathways, kinetics, and shelf-life prediction. In: Murphy, R.M., Tsai, A.M. (eds.) Misbehaving Proteins: Protein Misfolding, Aggregation, and Stability. Springer, New York (2006)
[5] Becker, R., Döring, W.: Kinetische Behandlung der Keimbildung in ubersattigten Dampfern. Ann. Phys. (Leipzig) 24, 719–752 (1935)
[6] Wegner, A., Engel, J.: Kinetics of the cooperative association of actin to actin filaments. Biophys. Chem. 3, 215–225 (1975)
[7] Raibekas, A.A.: Estimation of protein aggregation propensity with a melting point apparatus. Analytical Biochemistry 380, 331–332 (2008)
[8] Krishnan, S., Raibekas, A.A.: Multistep Aggregation Pathway of Human Interleukin-1 Receptor Antagonist: Kinetic, Structural, and Morphological. Biophysical Journal 96, 199–208 (2009)
[9] Ball, J.M., Carr, J., Penrose, O.: The Becker-Döring Cluster Equations: Basic Properties and Asymptotic Behaviour of Solutions. Commun. Math. Phys. 104, 657–692 (1986)
[10] Penrose, O.: The Becker-Döring equations for the kinetics of phase transitions, August 22 (2001)
[11] Niethammer, B.: A Vanishing Excess Density Limit of the Becker-D'oring Equations, `http://sfb611.iam.uni-bonn.de/uploads/159-komplett.pdf`
[12] Wattis, J.A.D., Coveney, P.V.: Renormalisation-theoretic analysis of non-equilibrium phase transitions I:The Becker-D'oring equations with power law rate coefficients, 1–18 (2001)
[13] Arnold, V.I.: Geometrical Methods in the Theory of Ordinary Differential Equations. In: Grundlehren der mathematischen Wissenschaften, vol. 250. Springer, New York (1983)
[14] Shoshitaishvili, A.N.: Bifurcations of topological type of a vector field near a singular point. Tr. Sem. Petrovskogo 1, 279–309 (1975); English translation in American Math. Soc. Translations 118(2) (1982)
[15] Wattis, J.A., King, J.R.: Asymptotic solutions of Becker-Döring equations. J. Phys. A: Math. Gen. 31, 7169–7189 (1998)
[16] Bishop, M.F., Ferrone, F.A.: Kinetics of Nucleation-controlled Polymerization, A Perturbation Treatment for Use with a Secondary Pathway. Biophysical Journal 46(5), 631–644 (1984)
[17] Cooper, J.A., Loren Buhle Jr., E., Walker, S.B., Tsong, T.Y., Pollard, T.D.: Kinetic Evidence for a Monomer Activation Step in Actin Polymerization. Biochemistry 22, 2193–2202 (1983)
[18] Powers, E.T., Powers, D.L.: The Kinetics of Nucleated Polymerizations at High Concentrations: Amyloid Fibril Formation Near and Above the Supercritical Concentration. Biophysical Journal 91, 122–132 (2006)
[19] Kuret, J., Congdon, E.E., Li, G., Yin, H., Yu, X., Zhong, Q.: Evaluating Triggers and Enhancers of Tau Fibrillization. Microscopy Research and Technique 67, 141–155 (2005)
[20] Winsor, C.P.: The Gompertz Curve as a Growth Curve. Proceedings of the National Academy of Sciences 18, 1 (1932)

[21] Wang, W.: Protein aggregation and its inhibition in biopharmaceutics. International Journal of Pharmaceutics 289, 1–30 (2005)
[22] Tanford, C.: Thermodynamics of Micelle Formation: Prediction of Micelle Size and Size Distribution. Proc. Nat. Acad. Sci. USA 71(5), 1811–1815 (1974)
[23] Richards, F.J.: A flexible growth function for empirical use. J. Exp. Bot. 10, 290–300 (1959)
[24] Wegner, A., Savko, P.: Fragmentation of Actin Filaments. Biochemistry 21, 1909–1913 (1982)

Global Uncertainty Analysis for a Model of TNF-Induced NF-κB Signalling

Steffen Waldherr, Jan Hasenauer, Malgorzata Doszczak,
Peter Scheurich, and Frank Allgöwer

Abstract. In this work, we study the problem of computing outer bounds for the region of steady states of biochemical reaction networks modelled by ordinary differential equations, with respect to parameters that are allowed to vary within a predefined region. An improved implementation of an algorithm which we presented earlier is developed in order to increase the computational efficiency. The gain in efficiency enables the analysis of medium scale biochemical network models. The applicability of the algorithm to such networks is illustrated by studying a newly developed model for a tumor necrosis factor signalling pathway. This pathway is of major importance for the inflammatory response in mammals and therefore of high biomedical interest. The proposed uncertainty analysis algorithm is applied to the model in order to understand how variations in the parameters and co-stimulation of different receptor types may affect the signalling response in this pathway.

1 Introduction

In an effort to obtain further understanding of intracellular processes, researchers are now constructing detailed computational models of biochemical signalling pathways. One major problem with these models is that there are large uncertainties concerning the values of reaction parameters and initial

Steffen Waldherr · Jan Hasenauer · Frank Allgöwer
Universität Stuttgart, Institute for Systems Theory and Automatic Control,
Pfaffenwaldring 9, 70569 Stuttgart, Germany

Malgorzata Doszczak · Peter Scheurich
Universität Stuttgart, Institute for Cell Biology and Immunology, Allmandring 31,
70569 Stuttgart, Germany

J. Lévine & P. Müllhaupt (Eds.): Adv. in Theory of Control, Signals, LNCIS 407, pp. 365–377.
springerlink.com © Springer-Verlag Berlin Heidelberg 2010

conditions. The reason for this type of uncertainty is that there is only a very limited amount of experimental data available to determine the parameters. This problem motivates the development of uncertainty analysis methods which allow for a quantitative evaluation of the effect that large parametric uncertainties have on the model predictions.

In this work, we consider specifically the effect of parametric uncertainty on the steady state concentration values in a biochemical reaction network. The goal is to compute certified bounds on steady state values being feasible for a given set of possible parameter values. Such a result effectively gives an upper bound on the steady state uncertainty that is generated by the considered parametric uncertainty. In contrast to local approaches, which for example are based on an extrapolation from a local sensitivity analysis, the global analysis gives stringent bounds on the steady state uncertainty [27].

The first part of the paper aims to refine and to speed up the uncertainty analysis algorithm presented in [30] in order to achieve increased computational efficiency. In the second part, we develop and analyse a medium scale model of a tumor necrosis factor (TNF) signalling pathway [29] with this algorithm, which is made possible by the increased efficiency compared to the previous implementation.

The paper is structured as follows. In Section 2, we first give a formal problem formulation for global uncertainty analysis of steady states, and second present an algorithm to compute an upper bound on the uncertainty based on previous results in [30]. In Section 3, we develop a model of TNF signal transduction, and analyse this model with the proposed uncertainty analysis algorithm. Details about the proposed model of TNF signal transduction are given in Appendix 4.

2 Global Uncertainty Analysis

2.1 Problem Formulation and Solution by Infeasibility Certificates

Let us consider a dynamical model of the network given by the differential equation

$$\dot{x} = F(x, p), \tag{1}$$

where $x \in \mathbb{R}^n$ is the state vector of the network, $p \in \mathbb{R}^m$ the parameter vector, and $F : \mathbb{R}^n \times \mathbb{R}^m \to \mathbb{R}^n$ a vector field describing the network's dynamics. We assume that F contains only polynomial and rational terms in x and p. This assumption holds for most models of biochemical signal transduction networks, for example those modelled with mass action or Michaelis-Menten kinetics.

The problem under consideration can be formulated as follows. Given a parametric uncertainty set $\mathcal{P} \subset \mathbb{R}^m$, define the set $\mathcal{X}_s^*$ of all feasible steady states of the system (1) as

$$\mathcal{X}_s^* = \{x \in \mathbb{R}^n \mid \exists p \in \mathcal{P} : F(x,p) = 0\} . \tag{2}$$

Due to the possible non-linearity of F, there are at present no general methods available to compute $\mathcal{X}_s^*$ efficiently and reliably. In order to get an upper bound on the uncertainty in the steady states, we therefore aim at computing an outer approximation $\mathcal{X}_s$ of $\mathcal{X}_s^*$, such that $\mathcal{X}_s \supseteq \mathcal{X}_s^*$. Then it can be guaranteed that all feasible steady states for any parameter from $\mathcal{P}$ are contained in $\mathcal{X}_s$. Hence, $\mathcal{X}_s$ constitutes an upper bound for the steady state uncertainty induced by the parametric uncertainty $p \in \mathcal{P}$.

We have previously developed a method to solve the formulated problem [30], based on the application of so called infeasibility certificates. There, the feasibility problem

$$\begin{array}{ll} \text{find} & x \in \widehat{\mathcal{X}},\ p \in \mathcal{P} \\ \text{subject to} & F(x,p) = 0, \end{array} \tag{3}$$

has been constructed, which is employed for the classification of a test set $\widehat{\mathcal{X}} \subset \mathbb{R}^n$. If (3) is infeasible, then the set $\widehat{\mathcal{X}}$ cannot contain any steady states for parameters $p \in \mathcal{P}$, and thus $\widehat{\mathcal{X}} \cap \mathcal{X}_s^* = \emptyset$ holds. To compute infeasibility certificates for (3) the Lagrange dual problem of (3) is derived. Exploiting the properties of Lagrangian duality [3], existence of particular solutions in the dual problem are certificates for infeasibility of the primal problem. Due to convexity of the dual problem, we get a computationally efficient test for non-existence of steady states in $\widehat{\mathcal{X}}$. The implementation of the infeasibility test is based on the freely available Matlab toolbox SeDuMi [28], which is a general solver for the type of optimisation problems encountered here, namely semidefinite programs. For details on the computation of infeasibility certificates, we refer to [30]. The extension from polynomial to rational terms is described in [15].

2.2 An Algorithm for Global Uncertainty Analysis

As discussed in [30], the infeasibility certificates for test regions $\widehat{\mathcal{X}}$ can be applied to compute an outer approximation $\mathcal{X}_s$ for the set $\mathcal{X}_s^*$ of all steady states. The algorithm requires an initial estimate in the form of a compact set $\mathcal{X}_0 \subset \mathbb{R}^n$, where all steady states have to be contained. Such an estimate can typically be derived from physical constraints and biochemical conservation relations, and may be a very coarse outer bound on the steady state set. In order to refine the set $\mathcal{X}_0$ towards a tighter outer approximation of $\mathcal{X}_s^*$, the uncertainty analysis algorithm iteratively constructs suitable test regions $\widehat{\mathcal{X}}$ and tries to obtain infeasibility certificates for these. If an infeasibility certificate can be computed, the test region $\widehat{\mathcal{X}}$ is subtracted from the steady state approximation set.

In this paper, we consider the specific problem of computing lower and upper bounds on the steady state values of each state variable for a parametric uncertainty given by the set $\mathcal{P}$. These bounds are denoted by $x_{min}, x_{max} \in \mathbb{R}^n$ and correspond to the approximation of the steady state set given by

$$\mathcal{X}_s(x_{min}, x_{max}) = \{x \in \mathbb{R}^n \mid x_{i,min} \leq x_i \leq x_{i,max},\ i = 1, \dots, n\}. \quad (4)$$

A bisection algorithm to compute lower and upper bounds is proposed in [30]. Let us denote the two versions of this algorithm for lower and upper bounds as $x_{i,min} = \text{computelowerbound}_i(\mathcal{X}_0, \mathcal{P})$ and $x_{i,max} = \text{computeupperbound}_i(\mathcal{X}_0, \mathcal{P})$, respectively. For the results presented in this paper, we use these algorithms iteratively in order to compute tighter upper and lower bounds on the steady state values. The resulting algorithm to compute an outer approximation $\mathcal{X}_s$ for the steady state set $\mathcal{X}_s^*$ is briefly described as follows.

Algorithm: $\mathcal{X}_s = \text{computebounds}(\mathcal{X}_0, \mathcal{P})$

1. Initialize $\mathcal{X}_s = \mathcal{X}_0$.
2. Repeat as long as there is an improvement in the bounds x_{min} and x_{max}:
   a. $x_{i,min} = \text{computelowerbound}_i(\mathcal{X}_s, \mathcal{P})$ for all $i = 1, \dots, n$
   b. $x_{i,max} = \text{computeupperbound}_i(\mathcal{X}_s, \mathcal{P})$ for all $i = 1, \dots, n$
   c. $\mathcal{X}_s = \{x \in \mathbb{R}^n \mid x_{i,min} \leq x_i \leq x_{i,max},\ i = 1, \dots, n\}$

3 Case Study: The TNF Signalling Model

In this section, we apply the previously described algorithm to a model of tumor necrosis factor (TNF) signal transduction. We thereby illustrate the analysis of uncertain signal transduction models with the proposed approach, and discuss the biological conclusions that can be drawn from such an analysis.

3.1 Overview of TNF Signalling and Model Development

The tumor necrosis factor (TNF) is a cytokine which coordinates the mammalian immune response. TNF activates several intracellular pathways, notably apoptotis via the caspase cascade and the NF-κB, JNK, and MAPK pathways [29]. A misregulation of TNF and the associated pathways is involved in various high-impact diseases, such as cancer or autoimmune diseases [14, 24, 10]. The interplay between the apoptotic and anti-apoptotic pathways activated by TNF also makes these networks worth studying from a more theoretical perspective.

TNF signalling is mediated by membrane receptors of the TNF receptor family, comprising over twenty different receptors, of which the two receptors TNF receptor 1 (TNFR1) and TNF receptor 2 (TNFR2) are the binding partners for the TNF ligand [13]. TNFR1 plays a major role in apoptosis induction by signal transduction to the caspase cascade [29, 7], but it also activates anti-apoptotic pathways. TNFR2 usually does not directly signal to the caspase cascade, but may have a strong influence on the results of TNFR1 signalling by crosstalk effects [12] and induction of TNF expression, leading to autocrine signalling. The anti-apoptotic effects of TNF signalling are mainly mediated via the transcription factor NF-κB, which is a known inducer of several anti-apoptotic proteins like the inhibitor-of-apoptosis proteins (IAP) [31, 26]. Physiologically, TNFR1 is mostly stimulated by soluble TNF, while TNFR2 is usually activated by juxtacrine signalling with a membrane bound form of the TNF ligand. Importantly, the analysis done in this paper is for an experimental setup where the two TNF receptor types can be stimulated selectively [4], which allows to study the contributions of the individual receptor types to the signalling outcome.

We construct a biochemical reaction network model to describe the response in nuclear NF-κB activity to the separate or combined stimulation of TNFR1 and TNFR2. The structure of the model has been derived from basic knowledge of relevant proteins which are involved in the signalling network, and from literature data on their interactions. For the NF-κB pathway downstream of the receptor complexes, we rely mainly on previous modelling efforts. The structure of this part of the model is adapted from [20]. Other sources are the models described in [17], [21], and [2]. For the receptor complex formation, the construction of mathematical models is not as advanced as for the NF-κB pathway. TNFR1 complex formation has been modelled previously [25], although focusing on different adaptor proteins than considered here. For the formation of the TNFR2 complex and its signalling, no previous mathematical models are known to the authors.

Upon ligand binding, TNF receptors start to recruit adaptor proteins to form the relevant signalling complexes. TNFR1 first recruits TRADD [22, 23, 8], but for simplicity, this step is not explicitly included in the model. Rather, TRADD is assumed to bind instantly, or to be already associated to TNFR1. In the next step, TNFR1 recruits the adaptor proteins RIP1 and TRAF2. From available biological data, it is not clear whether these adaptor proteins can only bind sequentially, and, if so, what the sequence is, or whether RIP1 and TRAF2 can independently bind to TNFR1 under *in vivo* conditions. In the model proposed here, TRAF2 is recruited to the receptor complex only after RIP1, as suggested in [11].

For TNFR2, fewer adaptor proteins seem to be relevant: the receptor directly recruits TRAF2, and the thus formed complex transmits the signal towards the NF-κB pathway [33, 5]. A relevant additional effect is the ubiquitination and subsequent proteasomal degradation of TRAF2 at the TNFR2 complex [19, 32]. Such observations have not been made for TRAF2 when recruited to the TNFR1 complex. The fact that both receptor complexes require TRAF2 for efficient signal transduction constitutes a crosstalk between the two receptor complexes, with potentially important effects on TNF signal transduction [5].

The crucial mediator between the TNF receptor complexes and the NF-κB pathway is the I-κB kinase complex (IKK) which is activated at the TNF receptor complexes by TRAF2. Active IKK then phosphorylates I-κBα, which is subsequently degraded, thus liberating NF-κB to move into the nucleus [17]. For the NF-κB, IKK and part of the gene expression modules, the species and reactions to be included in the model are adapted from a previous model [20].

The resulting overall model structure is illustrated in Figure 1, while all details are given in Appendix 4.

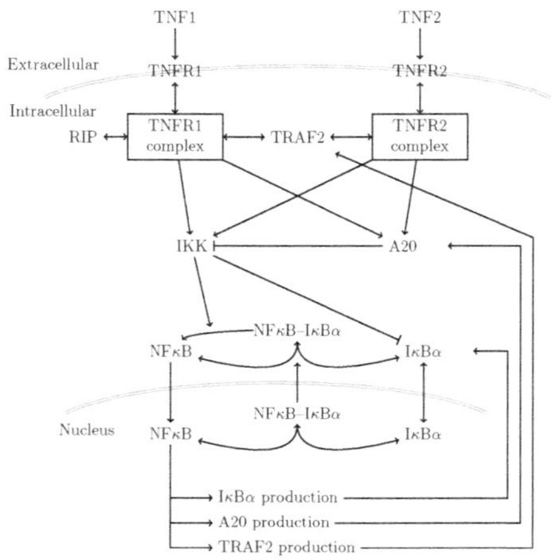

Fig. 1. Structure of the TNF signalling model

3.2 *Model Modularisation for Increased Computational Efficiency*

The methods developed in [30] and [15] to determine an outer approximation of the set of feasible steady states rely on the formulation of a single optimisation problem for the whole system. Unfortunately, for medium and large scale systems, this approach yields a huge optimisation problem and becomes computationally inefficient. In order to improve the computational efficiency of the uncertainty analysis algorithm presented in Section 2.2, we split the TNF network model in two interconnected blocks, as shown in Figure 2. The first block (TNF receptor module) contains the two TNF receptor species and the complexes they form with ligands and adaptor proteins, subject to reactions A1 – A9. The second block (NF-κB module) contains IKK, A20, and the downstream components of the NF-κB signalling pathway, including the transcripts of NF-κB inducible genes, subject to reactions B1 – D6.

The modularisation allows to solve significantly smaller uncertainty analysis problems, one for each block, where the uncertain parameters in the analysis are given by the intrinsic model parameters within each block and, in addition, the input variables coming from the other block. In order to obtain tight bounds on all state variables as well as the signals that are exchanged among the two blocks, we iterate the uncertainty analysis over the two blocks until no further improvement in the uncertainty bounds is made.

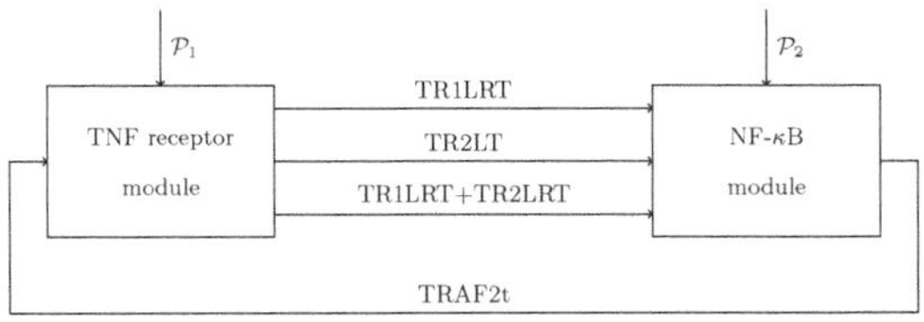

Fig. 2. Modularisation of the TNF network model

In the considered uncertainty cases, the computation time for the steady state bounds could be reduced by about 98 % by applying the modularisation approach. In the full model, each optimisation problem uses 2527 optimisation variables in SeDuMi, whereas with the modularised model, two optimisation problems with 510 and 1081 variables need to be solved. The significant reduction in computation time is due to both the reduction of the total number of optimisation variables and the non-linear dependency of the computation time for a single problem on the number of variables.

3.3 Results of the Global Uncertainty Analysis

In this section, we report the results of the global uncertainty analysis, as described in Section 2, to the TNF signal transduction model. The specific focus from the biological side will be on the mechanism of TRAF2 recruitment to the TNF receptors, and how this affects the outcome of the signalling process.

In a first step, we consider parametric uncertainty in the four parameters that determine the binding affinity of TRAF2 to the two TNF receptor types, namely the parameters $k_{A3,f}$, $k_{A3,b}$, $k_{A5,f}$, and $k_{A5,b}$ for the forward and backward reaction rates A3 and A5. For different values of this uncertainty, we compute upper and lower bounds for each of the state variables in the model. These bounds are illustrated in Figure 3. While the considered uncertainty affects the states in the receptor part significantly, it has hardly any effect on the activity of NF-κB in the nucleus. We thus see that the system has an insulation property [6], in the sense that even a large uncertainty in TRAF2 binding for the receptor part of the pathway does not have a significant influence on the core NF-κB part. While this can also be verified with classical Monte-Carlo sampling techniques, they are in this case computationally less efficient, and do not provide stringent upper bounds on the steady state uncertainty.

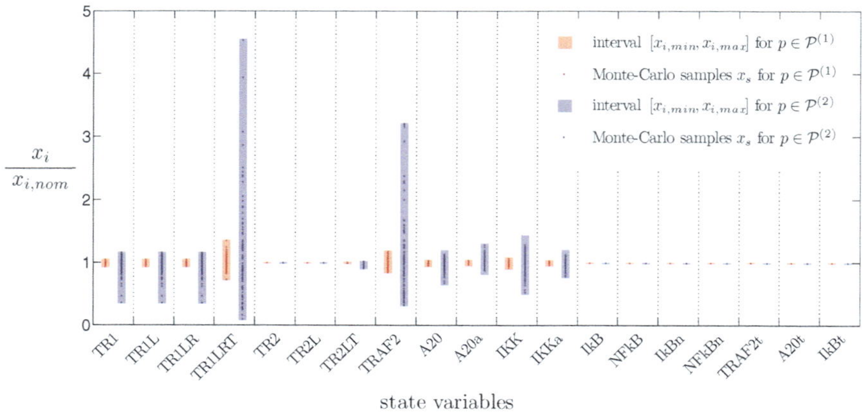

Fig. 3. Upper and lower bounds on individual state variables for uncertainty in the binding affinity of TRAF2 to the two TNF receptor types ($k_{A3,f}$, $k_{A3,b}$, $k_{A5,f}$, and $k_{A5,b}$). Proposed approach vs. Monte-Carlo sampling for a parametric uncertainty of 10% ($\mathcal{P}_1$), and a parametric uncertainty of factor 2 ($\mathcal{P}_2$).

In the second step, we study the effects of variations in the stimulus strength on the signalling outcome. Note that we have to consider two external stimuli: first the specific stimulation of TNFR1, denoted by TNF1, and second the specific stimulation of TNFR2, denoted by TNF2, as described in Section 3.1. As signalling output, we consider the activity of the fully activated TNFR1 complex. Since TNFR1 can, in contrast to TNFR2, directly activate the caspase cascade [29], this output is relevant for potential initiation of programmed cell death. Using the algorithm presented in Section 2.2, we compute lower and upper bounds on the steady state value of TR1LRT, the TNFR1 complex with RIP and TRAF2 associated. The resulting bounds for a range of variations in the two stimuli is depicted in Figure 4. Interestingly, we observe that the uncertainty in the output TR1LRT is much higher for variations in the TNF2 stimulus, compared to variations in the TNF1 stimulus. Biologically, this result can be interpreted in the way that TNFR2 signalling seems to modulate the activity of the TNFR1 signalling complex significantly. This interaction establishes strong crosstalk from the TNFR2 to the TNFR1 via the shared adaptor protein TRAF2, but not vice versa.

4 Discussion and Conclusions

The application of infeasibility certificates for global uncertainty analysis of steady states in biochemical networks has been proposed in [30]. Here, we refined the previously proposed algorithm in order to obtain both tighter approximations to the steady state set and increased computational efficiency,

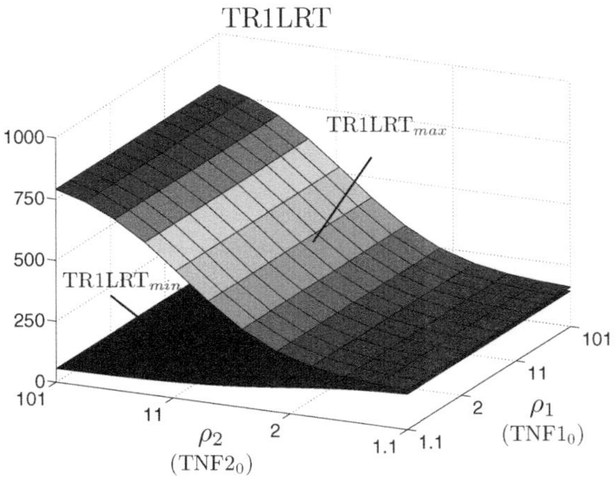

Fig. 4. Upper and lower bounds of the TR1LRT steady state value shown by surfaces. The ρ_1 and ρ_2 axis show the multiplicative uncertainty in the stimulus strength, i.e. the stimuli are $\text{TNF1}_0 = [\frac{1}{\rho_1}, \rho_1] \cdot 3.94 \cdot 10^5$ molecules and $\text{TNF2}_0 = [\frac{1}{\rho_2}, \rho_2] \cdot 3.94 \cdot 10^5$ molecules.

making the approach useful for medium scale biochemical networks on the order of tens of molecular species and reactions.

With these improvements in hand, we did steady state uncertainty analysis for a model of tumor necrosis factor signal transduction, consisting of 24 molecular species and 35 reactions, mostly mass action. With the proposed extensions, our uncertainty analysis algorithm could be applied successfully to this model. In terms of new biological insight, our analysis first indicated that the TNF network seems to have significant insulation from uncertainty in TRAF2 binding characteristics to NF-κB induced gene expression. As a second result, we observed that the model shows strong crosstalk from TNFR2 stimulation to the signalling activity of the TNFR1 complex via the adaptor protein TRAF2. The second result has implications for the process of programmed cell death, as it indicates how stimulation of TNFR2, which does not by itself directly activate the caspase cascade, still may have a profound effect on cell death by its influence on TNFR1. The crosstalk via TRAF2 as discussed here is one example, others have been described in literature, e.g. TNFR2-mediated stimulation of TNF production, which then acts via TNFR1, signaling via the molecule RIP, and others [9].

Further extensions of the described uncertainty analysis algorithm consider hybrid differential algebraic systems in process control, also with general non-polynomial terms [15]. Secondly, the infeasibility certificates on which the analysis is based can also be applied in feasible parameter set estimation from uncertain dynamic measurements [18, 16].

Acknowledgments. We acknowledge partial funding of the project by the Landesstiftung Baden-Württemberg in the Center for Systems Biology at the University of Stuttgart. S.W., J.H., P.S., and F.A. would like to thank the German Research Foundation (DFG) for financial support of the project within the Cluster of Excellence in Simulation Technology (EXC 310/1) at the University of Stuttgart.

A Description of the TNF Signalling Model

The following 24 molecular species are included in the TNF signalling model. Initial conditions are given in parenthesis, unless they are zero. The unit for the initial conditions is in molecules per cell. TNF1 – TNF ligand specific for receptor 1 ($[\text{TNF1}]_0 = 3.94 \cdot 10^5$); TR1 – TNF receptor type 1 ($[\text{TR1}]_0 = 10^3$); TR1L – complex of TR1 and TNF1; TR1LR – complex of TR1L and RIP; TR1LRT – complex of TR1LR and TRAF2; TNF2 – TNF ligand specific for receptor 2 ($[\text{TNF2}]_0 = 3.94 \cdot 10^5$); TR2 – TNF receptor type 2 ($[\text{TR2}]_0 = 10^4$); TR2L – complex of TR2 and TNF2; TR2LT – complex of TR2L and TRAF2; RIP – receptor interacting protein ($[\text{RIP}]_0 = 3.3 \cdot 10^5$); TRAF2 – TNF receptor associated protein 2; A20 – A20 ubiquitin ligase, inactive form; A20a – active form of A20; IKK – I-κB kinase, inactive form; IKKa – active form of IKK; IkB – inhibitor of NF-κB, cytosolic; NFkB – nuclear factor κB, cytosolic; NI – complex of NF-κB and I-κB, cytosolic ($[\text{NI}]_0 = 10^5$); IkBn – I-κB, nuclear; NFkBn – NF-κB, nuclear; NIn – complex of NF-κBn and I-κBn; IkBt – $i\text{-}\kappa B$ transcript; A20t – *a20* transcript; TRAF2t – *traf2* transcript.

The considered TNF signalling pathway is described by the following set of reactions. Most reactions are modelled according to the law of mass action [1], meaning that if A and B are reactands, then the reaction rate is constructed as $v = k[\text{A}][\text{B}]$, with reaction parameter k. Reactions with a forward and backward direction, for example $\text{A} + \text{B} \leftrightarrow \text{C}$, have a forward and backward reaction parameter, k_f and k_b, respectively, and the rate is constructed as $v = k_f[\text{A}][\text{B}] - k_b[\text{C}]$. The transcription reactions D1, D3, and D5 are not modelled by the law of mass action, but with Hill reaction kinetics. The corresponding reaction rate is given directly in the list of reactions. In all cases, the given reaction rate parameters are considered as nominal values for the uncertainty analysis in this study, and are taken from literature and own measurements. Physical units for parameters are in molecules per cell for concentrations and seconds for time.

#	Reaction	Parameters (forward/backward)
A1	$TR1 + TNF1 \leftrightarrow TR1L$	$9.13 \cdot 10^{-9}$ / $3.5 \cdot 10^{-4}$
A2	$TR1L + RIP \leftrightarrow TR1LR$	$1.04 \cdot 10^{-8}$ / $2 \cdot 10^{-7}$
A3	$TR1LR + TRAF2 \leftrightarrow TR1LRT$	$6.43 \cdot 10^{-7}$ / $5 \cdot 10^{-3}$
A4	$TR2 + TNF2 \leftrightarrow TR2L$	$1.25 \cdot 10^{-8}$ / $1.05 \cdot 10^{-2}$
A5	$TR2L + TRAF2 \leftrightarrow TR2LT$	$6.43 \cdot 10^{-7}$ / $5 \cdot 10^{-3}$
A6	$TRAF2 \rightarrow \emptyset$	$1.75 \cdot 10^{-5}$
A7	$TRAF2t \rightarrow TRAF2t + TRAF2$	$8.47 \cdot 10^{-1}$
A8	$TR1LRT \rightarrow TR1LR$	$1.75 \cdot 10^{-5}$
A9	$TR2LT \rightarrow TR2L$	$1.75 \cdot 10^{-3}$
B1	$IKK \leftrightarrow \emptyset$	$1.25 \cdot 10^{-4}$ / $1.2 \cdot 10^{1}$
B2	$IKK + TR1LRT \rightarrow IKKa + TR1LRT$	$1.14 \cdot 10^{-5}$
B3	$IKK + TR2LT \rightarrow IKKa + TR2LT$	$1.14 \cdot 10^{-5}$
B4	$IKKa \rightarrow \emptyset$	$1.25 \cdot 10^{-4}$
B5	$IKKa + A20a \rightarrow A20a$	$1 \cdot 10^{-6}$
B6	$A20t \rightarrow A20t + A20$	0.5
B7	$A20 \rightarrow \emptyset$	$5 \cdot 10^{-4}$
B8	$A20 + TR1LRT \rightarrow A20a + TR1LRT$	0.1
B9	$A20 + TR2LT \rightarrow A20a + TR2LT$	0.1
B10	$A20a \rightarrow \emptyset$	$5 \cdot 10^{-4}$
C1	$NFkB + IkB \rightarrow NI$	$1.04 \cdot 10^{-6}$
C2a	$IkB + IKKa \rightarrow IkB\_IKKa$	$4.15 \cdot 10^{-7}$
C2b	$IkB\_IKKa \rightarrow IKKa$	0.1
C3a	$NI + IKKa \rightarrow NI\_IKKa$	$2.08 \cdot 10^{-6}$
C3b	$NI\_IKKa \rightarrow NFkB + IKKa$	0.1
C4	$IkBt \rightarrow IkBt + IkB$	0.5
C5	$IkB \rightarrow \emptyset$	$1 \cdot 10^{-4}$
C6	$NI \rightarrow NFkB$	$2 \cdot 10^{-5}$
C7	$NFkB \rightarrow NFkBn$	$2.5 \cdot 10^{-3}$
C8	$IkB \leftrightarrow IkBn$	$2 \cdot 10^{-3}$ / $5 \cdot 10^{-3}$
C9	$NFkBn + IkBn \rightarrow NIn$	$4.15 \cdot 10^{-6}$
C10	$NIn \rightarrow NI$	0.01
D1	$NFkBn \rightarrow IkBt + NFkBn$	$v_{D1} = \frac{0.1[NFkBn]}{(5 \cdot 10^5)^2 + [NFkBn]^2} \frac{1}{s}$
D2	$IkBt \rightarrow \emptyset$	$4 \cdot 10^{-4}$
D3	$NFkBn \rightarrow A20t + NFkBn$	$v_{D3} = \frac{0.1[NFkBn]}{(5 \cdot 10^5)^2 + [NFkBn]^2} \frac{1}{s}$
D4	$A20t \rightarrow \emptyset$	$7.5 \cdot 10^{-4}$
D5	$NFkBn \rightarrow TRAF2t + NFkBn$	$v_{D5} = \frac{0.02[NFkBn]}{(5 \cdot 10^5)^2 + [NFkBn]^2} \frac{1}{s}$
D6	$TRAF2t \rightarrow \emptyset$	$4 \cdot 10^{-4}$

References

1. Aldridge, B.B., Burke, J.M., Lauffenburger, D.A., Sorger, P.K.: Physicochemical modelling of cell signalling pathways. Nat. Cell. Biol. 8(11), 1195–1203 (2006)

2. Ashall, L., Horton, C.A., Nelson, D.E., Paszek, P., Harper, C.V., Sillitoe, K., Ryan, S., Spiller, D.G., Unitt, J.F., Broomhead, D.S., Kell, D.B., Rand, D.A., Se, V., White, M.R.H.: Pulsatile stimulation determines timing and specificity of NF-kappaB-dependent transcription. Science 324(5924), 242–246 (2009)
3. Boyd, S., Vandenberghe, L.: Convex Optimization. Cambridge University Press, Cambridge (2004)
4. Bryde, S., Grunwald, I., Hammer, A., Krippner-Heidenreich, A., Schiestel, T., Brunner, H., Tovar, G., Pfizenmaier, K., Scheurich, P.: Tumor necrosis factor (TNF)-functionalized nanostructured particles for the stimulation of membrane TNF-specific cell responses. Bioconjug. Chem. 16, 1459–1467 (2005)
5. Bryde, S.: Characterisation of TNF receptor-2 mediated signal initiation and transduction. Ph.D. thesis, Universität Stuttgart (2004)
6. Del Vecchio, D., Ninfa, A.J., Sontag, E.D.: Modular cell biology: retroactivity and insulation. Molec. Syst. Biol. 4, 161 (2008)
7. Eissing, T., Waldherr, S., Allgöwer, F.: Modelling and analysis of cell death signalling. In: Queinnec, I., Tarbouriech, S., Garcia, G., Niculescu, S.I. (eds.) Biology and Control Theory: Current Challenges. LNCIS, vol. 357, pp. 161–180. Springer, Berlin (2007)
8. Ermolaeva, M.A., Michallet, M.C., Papadopoulou, N., Utermhlen, O., Kranidioti, K., Kollias, G., Tschopp, J., Pasparakis, M.: Function of TRADD in tumor necrosis factor receptor 1 signaling and in TRIF-dependent inflammatory responses. Nat. Immunol. 9(9), 1037–1046 (2008)
9. Faustman, D., Davis, M.: TNF receptor 2 pathway: drug target for autoimmune diseases. Nat. Rev. Drug Discov. 9(6), 482–493 (2010)
10. Feldmann, M., Maini, R.N.: TNF defined as a therapeutic target for rheumatoid arthritis and other autoimmune diseases. Nat. Med. 9(10), 1245–1250 (2003)
11. Festjens, N., Berghe, T.V., Cornelis, S., Vandenabeele, P.: RIP1, a kinase on the crossroads of a cell's decision to live or die. Cell Death Differ. 14(3), 400–410 (2007)
12. Fotin-Mleczek, M., Henkler, F., Samel, D., Reichwein, M., Hausser, A., Parmryd, I., Scheurich, P., Schmid, J.A., Wajant, H.: Apoptotic crosstalk of TNF receptors: TNF-R2-induces depletion of TRAF2 and IAP proteins and accelerates TNF-R1-dependent activation of caspase-8. J. Cell Sci. 115(Pt 13), 2757–2770 (2002)
13. Grell, M., Wajant, H., Zimmermann, G., Scheurich, P.: The type 1 receptor (CD120a) is the high-affinity receptor for soluble tumor necrosis factor. Proc. Natl. Acad. Sci. 95(2), 570–575 (1998)
14. Hanahan, D., Weinberg, R.A.: The hallmarks of cancer. Cell 100(1), 57–70 (2000)
15. Hasenauer, J., Rumschinski, P., Waldherr, S., Borchers, S., Allgöwer, F., Findeisen, R.: Guaranteed steady-state bounds for uncertain chemical processes. In: Proc. Intern. Symp. Adv. Contr. Chem. Proc. (ADCHEM), Istanbul, Turkey (2009)
16. Hasenauer, J., Waldherr, S., Wagner, K., Allgöwer, F.: Parameter identification, experimental design and model falsification for biological network models using semidefinite programming. IET Systems Biology 4(2), 119–130 (2010)
17. Hoffmann, A., Levchenko, A., Scott, M.L., Baltimore, D.: The IkappaB-NF-kappaB signaling module: temporal control and selective gene activation. Science 298(5596), 1241–1245 (2002)

18. Kuepfer, L., Sauer, U., Parrilo, P.: Efficient classification of complete parameter regions based on semidefinite programming. BMC Bioinform. 8(1), 12 (2007)
19. Li, X., Yang, Y., Ashwell, J.D.: TNF-RII and c-IAP1 mediate ubiquitination and degradation of TRAF2. Nature 416(6878), 345–347 (2002)
20. Lipniacki, T., Paszek, P., Brasier, A.R., Luxon, B., Kimmel, M.: Mathematical model of NF-κB regulatory module. J. Theor. Biol. 228(2), 195–215 (2004)
21. Lipniacki, T., Puszynski, K., Paszek, P., Brasier, A.R., Kimmel, M.: Single TNFalpha trimers mediating NF-kappaB activation: Stochastic robustness of NF-kappaB signaling. BMC Bioinform. 8, 376 (2007)
22. Micheau, O., Tschopp, J.: Induction of TNF receptor I-mediated apoptosis via two sequential signaling complexes. Cell 114(2), 181–190 (2003)
23. Pobezinskaya, Y.L., Kim, Y.S., Choksi, S., Morgan, M.J., Li, T., Liu, C., Liu, Z.: The function of TRADD in signaling through tumor necrosis factor receptor 1 and TRIF-dependent Toll-like receptors. Nat. Immunol. 9(9), 1047–1054 (2008)
24. Rae, C., Langa, S., Tucker, S.J., Macewan, D.J.: Elevated NF-κB responses and FLIP levels in leukemic but not normal lymphocytes: reduction by salicylate allows TNF-induced apoptosis. Proc. Natl. Acad. Sci. 104, 12790–12795 (2007)
25. Schliemann, M., Eissing, T., Scheurich, P., Bullinger, E.: Mathematical modelling of TNF-α induced apoptotic and anti-apoptotic signalling pathways in mammalian cells based on dynamic and quantitative experiments. In: Proc. of the 2nd Found. Syst. Biol. Engin. (FOSBE), Stuttgart, Germany, pp. 213–218 (2007)
26. Stehlik, C., de Martin, R., Binder, B.R., Lipp, J.: Cytokine induced expression of porcine inhibitor of apoptosis protein (iap) family member is regulated by NF-κB. Biochem. Biophys. Res. Commun. 243(3), 827–832 (1998)
27. Streif, S., Waldherr, S., Allgöwer, F., Findeisen, R.: Steady state sensitivity analysis of biochemical reaction networks: A brief review and new methods. In: Jayaraman, A., Hahn, J. (eds.) Systems Analysis of Biological Networks, pp. 129–148. Methods in Bioengineering. Artech House, Boston (2009)
28. Sturm, J.F.: Using SeDuMi 1.02, a Matlab toolbox for optimization over symmetric cones. Optim. Meth. Softw. 11(1), 625–653 (1999)
29. Wajant, H., Pfizenmaier, K., Scheurich, P.: Tumor necrosis factor signaling. Cell Death Differ. 10(1), 45–65 (2003)
30. Waldherr, S., Findeisen, R., Allgöwer, F.: Global sensitivity analysis of biochemical reaction networks via semidefinite programming. In: Proc. of the 17th IFAC World Congress, Seoul, Korea, pp. 9701–9706 (2008)
31. Wang, C.Y., Mayo, M.W., Korneluk, R.G., Goeddel, D.V., Baldwin, A.S.: NF-kappaB antiapoptosis: induction of TRAF1 and TRAF2 and c-IAP1 and c-IAP2 to suppress caspase-8 activation. Science 281(5383), 1680–1683 (1998)
32. Wu, C.J., Conze, D.B., Li, X., Ying, S.X., Hanover, J.A., Ashwell, J.D.: TNF-alpha induced c-IAP1/TRAF2 complex translocation to a Ubc6-containing compartment and TRAF2 ubiquitination. EMBO J. 24(10), 1886–1898 (2005)
33. Ye, H., Wu, H.: Thermodynamic characterization of the interaction between TRAF2 and tumor necrosis factor receptor peptides by isothermal titration calorimetry. Proc. Natl. Acad. Sci. 97(16), 8961–8966 (2000)

Author Index

Allgöwer, Frank 365
Andreozzi, Stefano 297
Antritter, Felix 127
Arimoto, Suguru 3, 17
Aschemann, Harald 31, 201

Basic, Duro 41
Bloch, Anthony M. 141
Butcher, Mark 49

Chang, Dong Eui 153
Chaves, Madalena 241
Chiasson, John 65
Chung, Seung-Wook 255
Cooper, Carlton R. 255
Crooke, P.S. 269

Doszczak, Malgorzata 365

Farach-Carson, Mary C. 255

Gonze, Didier 283
Gouzé, Jean-Luc 241

Hafner, Marc 283
Hasenauer, Jan 365
Hotchkiss, and J.R. 269

Jebai, Al Kassem 41
Jiang, Zhong-Ping 77

Karimi, Alireza 49
Kaynar, A.M. 269
Koeppl, Heinz 297
Krause, Matthias 89
Krstic, Miroslav 161

Lévine, Jean 127
Lobry, Claude 309

Malrait, François 41
Martin, Philippe 41
Maschke, Bernhard 339
Moreau, Xavier 99
Mounier, Hugues 179

Nihtilä, Markku 189

Ogunnaike, Babatunde A. 255
Oteafy, Ahmed 65
Oustaloup, Alain 99

Petit, Nicolas 323

Raibekas, Andrei 349
Rauh, Andreas 201
Rouchon, Pierre 41
Rudolph, Joachim 89, 179

Sari, Tewfik 309
Schaft, Arjan van der 339
Scheurich, Peter 365
Schindele, Dominik 31
Selig, J.M. 213
Shoshitaishvili, Alex 349
Steuer, Ralf 297

Vassiliou, Peter J. 225

Waldherr, Steffen 365
Woittennek, Frank 89, 179

Yadi, Karim 309

Zhu, Guchuan 113

Lecture Notes in Control and Information Sciences

Edited by M. Thoma, F. Allgöwer, M. Morari

Further volumes of this series can be found on our homepage:
springer.com

Vol. 407: Lévine, J., Müllhaupt, P.:
Advances in the Theory of Control, Signals and Systems with Physical Modeling
380 p. 2010 [978-3-642-16134-6]

Vol. 406: Bemporad, A., Heemels, M., Johansson, M.:
Networked Control Systems
appro. 371 p. 2010 [978-0-85729-032-8]

Vol. 405: Stefanovic, M., Safonov, M.G.:
Safe Adaptive Control
appro. 153 p. 2010 [978-1-84996-452-4]

Vol. 404: Giri, F.; Bai, E.-W. (Eds.):
Block-oriented Nonlinear System Identification
425 p. 2010 [978-1-84996-512-5]

Vol. 403: Tóth, R.;
Modeling and Identification of Linear Parameter-Varying Systems
319 p. 2010 [978-3-642-13811-9]

Vol. 402: del Re, L.; Allgöwer, F.; Glielmo, L.; Guardiola, C.; Kolmanovsky, I. (Eds.):
Automotive Model Predictive Control
284 p. 2010 [978-1-84996-070-0]

Vol. 401: Chesi, G.; Hashimoto, K. (Eds.):
Visual Servoing via Advanced Numerical Methods
393 p. 2010 [978-1-84996-088-5]

Vol. 400: Tomás-Rodríguez, M.; Banks, S.P.:
Linear, Time-varying Approximations to Nonlinear Dynamical Systems
298 p. 2010 [978-1-84996-100-4]

Vol. 399: Edwards, C.; Lombaerts, T.; Smaili, H. (Eds.):
Fault Tolerant Flight Control
appro. 350 p. 2010 [978-3-642-11689-6]

Vol. 398: Hara, S.; Ohta, Y.; Willems, J.C.; Hisaya, F. (Eds.):
Perspectives in Mathematical System Theory, Control, and Signal Processing
appro. 370 p. 2010 [978-3-540-93917-7]

Vol. 397: Yang, H.; Jiang, B.; Cocquempot, V.:
Fault Tolerant Control Design for Hybrid Systems
191 p. 2010 [978-3-642-10680-4]

Vol. 396: Kozlowski, K. (Ed.):
Robot Motion and Control 2009
475 p. 2009 [978-1-84882-984-8]

Vol. 395: Talebi, H.A.; Abdollahi, F.; Patel, R.V.; Khorasani, K.:
Neural Network-Based State Estimation of Nonlinear Systems
appro. 175 p. 2010 [978-1-4419-1437-8]

Vol. 394: Pipeleers, G.; Demeulenaere, B.; Swevers, J.:
Optimal Linear Controller Design for Periodic Inputs
177 p. 2009 [978-1-84882-974-9]

Vol. 393: Ghosh, B.K.; Martin, C.F.; Zhou, Y.:
Emergent Problems in Nonlinear Systems and Control
285 p. 2009 [978-3-642-03626-2]

Vol. 392: Bandyopadhyay, B.; Deepak, F.; Kim, K.-S.:
Sliding Mode Control Using Novel Sliding Surfaces
137 p. 2009 [978-3-642-03447-3]

Vol. 391: Khaki-Sedigh, A.; Moaveni, B.:
Control Configuration Selection for Multivariable Plants
232 p. 2009 [978-3-642-03192-2]

Vol. 390: Chesi, G.; Garulli, A.; Tesi, A.; Vicino, A.:
Homogeneous Polynomial Forms for Robustness Analysis of Uncertain Systems
197 p. 2009 [978-1-84882-780-6]

Vol. 389: Bru, R.; Romero-Vivó, S. (Eds.):
Positive Systems
398 p. 2009 [978-3-642-02893-9]

Vol. 388: Jacques Loiseau, J.; Michiels, W.; Niculescu, S-I.; Sipahi, R. (Eds.):
Topics in Time Delay Systems
418 p. 2009 [978-3-642-02896-0]

Vol. 387: Xia, Y.;
Fu, M.; Shi, P.:
Analysis and Synthesis of
Dynamical Systems with Time-Delays
283 p. 2009 [978-3-642-02695-9]

Vol. 386: Huang, D.;
Nguang, S.K.:
Robust Control for Uncertain
Networked Control Systems with
Random Delays
159 p. 2009 [978-1-84882-677-9]

Vol. 385: Jungers, R.:
The Joint Spectral Radius
144 p. 2009 [978-3-540-95979-3]

Vol. 384: Magni, L.; Raimondo, D.M.;
Allgöwer, F. (Eds.):
Nonlinear Model Predictive Control
572 p. 2009 [978-3-642-01093-4]

Vol. 383: Sobhani-Tehrani E.:
Khorasani K.;
Fault Diagnosis of Nonlinear Systems
Using a Hybrid Approach
360 p. 2009 [978-0-387-92906-4]

Vol. 382: Bartoszewicz A.;
Nowacka-Leverton A.:
Time-Varying Sliding Modes for Second
and Third Order Systems
192 p. 2009 [978-3-540-92216-2]

Vol. 381: Hirsch M.J.; Commander C.W.;
Pardalos P.M.; Murphey R. (Eds.):
Optimization and Cooperative
Control Strategies: Proceedings of the 8th
International Conference on Cooperative
Control and Optimization
459 p. 2009 [978-3-540-88062-2]

Vol. 380: Basin M.:
New Trends in Optimal Filtering and Control for
Polynomial and Time-Delay Systems
206 p. 2008 [978-3-540-70802-5]

Vol. 379: Mellodge P.; Kachroo P.:
Model Abstraction in Dynamical Systems:
Application to Mobile Robot Control
116 p. 2008 [978-3-540-70792-9]

Vol. 378: Femat R.; Solis-Perales G.:
Robust Synchronization of Chaotic Systems
Via Feedback
199 p. 2008 [978-3-540-69306-2]

Vol. 377: Patan K.:
Artificial Neural Networks for
the Modelling and Fault
Diagnosis of Technical Processes
206 p. 2008 [978-3-540-79871-2]

Vol. 376: Hasegawa Y.:
Approximate and Noisy Realization of
Discrete-Time Dynamical Systems
245 p. 2008 [978-3-540-79433-2]

Vol. 375: Bartolini G.; Fridman L.;
Pisano A.; Usai E. (Eds.):
Modern Sliding Mode Control Theory
465 p. 2008 [978-3-540-79015-0]

Vol. 374: Huang B.; Kadali R.:
Dynamic Modeling, Predictive Control
and Performance Monitoring
240 p. 2008 [978-1-84800-232-6]

Vol. 373: Wang Q.-G.; Ye Z.; Cai W.-J.;
Hang C.-C.:
PID Control for Multivariable Processes
264 p. 2008 [978-3-540-78481-4]

Vol. 372: Zhou J.; Wen C.:
Adaptive Backstepping Control of Uncertain
Systems
241 p. 2008 [978-3-540-77806-6]

Vol. 371: Blondel V.D.; Boyd S.P.;
Kimura H. (Eds.):
Recent Advances in Learning and Control
279 p. 2008 [978-1-84800-154-1]

Vol. 370: Lee S.; Suh I.H.;
Kim M.S. (Eds.):
Recent Progress in Robotics:
Viable Robotic Service to Human
410 p. 2008 [978-3-540-76728-2]

Vol. 369: Hirsch M.J.; Pardalos P.M.;
Murphey R.; Grundel D.:
Advances in Cooperative Control and
Optimization
423 p. 2007 [978-3-540-74354-5]

Vol. 368: Chee F.; Fernando T.
Closed-Loop Control of Blood Glucose
157 p. 2007 [978-3-540-74030-8]

Vol. 367: Turner M.C.; Bates D.G. (Eds.):
Mathematical Methods for Robust and
Nonlinear Control
444 p. 2007 [978-1-84800-024-7]

Vol. 366: Bullo F.; Fujimoto K. (Eds.):
Lagrangian and Hamiltonian Methods for
Nonlinear Control 2006
398 p. 2007 [978-3-540-73889-3]

Vol. 365: Bates D.; Hagström M. (Eds.):
Nonlinear Analysis and Synthesis
Techniques for Aircraft Control
360 p. 2007 [978-3-540-73718-6]